Climate Change

Kristen St. John · Lawrence Krissek
Editors

Climate Change

A Geoscience Perspective

 Springer

Editors
Kristen St. John
Department of Geology
and Environmental Science
James Madison University
Harrisonburg, VA, USA

Lawrence Krissek
School of Earth Sciences
The Ohio State University
Columbus, OH, USA

ISBN 978-3-031-82868-3   ISBN 978-3-031-82869-0 (eBook)
https://doi.org/10.1007/978-3-031-82869-0

© The Editor(s) (if applicable) and The Author(s) 2025. This book is an open access publication.

Open Access This book is licensed under the terms of the Creative Commons Attribution 4.0 International License (http://creativecommons.org/licenses/by/4.0/), which permits use, sharing, adaptation, distribution and reproduction in any medium or format, as long as you give appropriate credit to the original author(s) and the source, provide a link to the Creative Commons license and indicate if changes were made.
The images or other third party material in this book are included in the book's Creative Commons license, unless indicated otherwise in a credit line to the material. If material is not included in the book's Creative Commons license and your intended use is not permitted by statutory regulation or exceeds the permitted use, you will need to obtain permission directly from the copyright holder.
The use of general descriptive names, registered names, trademarks, service marks, etc. in this publication does not imply, even in the absence of a specific statement, that such names are exempt from the relevant protective laws and regulations and therefore free for general use.
The publisher, the authors and the editors are safe to assume that the advice and information in this book are believed to be true and accurate at the date of publication. Neither the publisher nor the authors or the editors give a warranty, expressed or implied, with respect to the material contained herein or for any errors or omissions that may have been made. The publisher remains neutral with regard to jurisdictional claims in published maps and institutional affiliations.

This Springer imprint is published by the registered company Springer Nature Switzerland AG
The registered company address is: Gewerbestrasse 11, 6330 Cham, Switzerland

If disposing of this product, please recycle the paper.

Acknowledgments

Climate Change—A Geoscience Perspective is truly a team effort. Co-editors, Kristen St. John and Lawrence Krissek, are tremendously grateful to the entire author team for their knowledge, expertise and perseverance. We are so pleased to have worked with each of you on this collaborative textbook. An additional thank you to author institutions that contributed to the cost of open access publishing: The Ice Core Salvage Fund of the Byrd Polar Climate and Research Center at The Ohio State University, Indiana University of Pennsylvania, James Madison University, Pennsylvania State University, Texas A&M University at Galveston, and the University of Dayton. Our gratitude also extends to publishers and other copyright holders who gave us permission and commonly waived fees to use figures and tables in this textbook.

A well-worn adage says that "we stand on the shoulders of giants," thereby recognizing the tremendous contributions of those who have come before us. This recognition is especially true for a field as diverse as climate science, and even more the case when the fields of environmental humanities, ethics, and climate geoengineering are included. The references cited in this book are similarly diverse, but are a very small subset of the total existing body of knowledge about climate science (writ large). As a result, we take this opportunity to gratefully acknowledge all who have contributed to our present understanding of Earth's past, present, and future climate.

Chapter "Modern Climate Change in the Context of the Last Two Millennia" benefited from the invaluable graphics assistance of Anna Golub, whose many talents suggest a bright future in paleoclimatology. Chapter "Modern Climate Change in the Context of the Last Two Millennia" also benefited from the many public online paleoclimate and instrumental databases; these include World Data Center for Paleoclimatology, NOAA Paleoclimatology Program, the Met Office Hadley Centre, NOAA National Centers for Environmental Information, National Snow and Ice Data Center, and Past Global Changes (PAGES) Program (PAGES 2K Consortium). Support for Tom Cronin to work on Chapter "Modern Climate Change in the Context of the Last Two Millennia" came from the US Geological Survey Climate Research and Development (Land Change Program). Any use of trade, firm, or product names is for descriptive purposes only and does not imply endorsement by the US Government.

Chapter "The Roles of Rock Formation and Weathering in Long-Term Climate Change" benefited from graphics assistance of Grace Berg—it was greatly appreciated.

Finally, a very special thank you to our families; your support and patience as we directed our attention to writing and editing was immeasurable.

Contents

Climate Change in Geoscience and Social Contexts 1
Kristen St. John, Dónal O'Mathúna, and Melissa Burt

The Earth's Climate System 49
Lawrence Krissek and Kristen St. John

Modern Climate Change in the Context of the Last Two Millennia 105
Thomas M. Cronin

Icy Secrets Preserved in Earth's Glaciers 139
Lonnie G. Thompson and Ellen Mosley-Thompson

The Sedimentary Record of Past Climate Change 183
Adriane R. Lam, R. Mark Leckie, Steven Hovan, and Dana Royer

Plate Tectonics and Long-Term Climate Change 239
Mary H. Schultz

The Roles of Rock Formation and Weathering in Long-Term Climate Change 279
Lee Kump and James Kasting

Abrupt Climate Change: The PETM 315
Debbie Thomas and Kristen St. John

Climate Cycles 351
Steven Clemens and Lawrence Krissek

Climate Models as Tools for Understanding Earth's Climate System ... 385
Shuang-Ye Wu

Large-Scale Climate Interventions: Carbon Dioxide Removal and Solar Radiation Management 435
Walker Raymond Lee, Douglas MacMartin, and Amanda Borth

Going Beyond the Science: Climate Ethics 473
Greg Hitzhusen and Jill Schneiderman

Editors and Contributors

About the Editors

Kristen St. John is from the Department of Geology and Environmental Science, James Madison University. Intent on being a political science major in college, Kristen's path changed after taking physical and historical geology as her science requirements. She's found that piecing together the stories that are preserved in climate archives of Earth's history is like working on a 10,000+ piece jigsaw puzzle. The challenges and rewards of figuring things out are much more fun—and solutions more attainable—when working with others. Synergistic research collaborations, and seeking ways to make geoscience more accessible and meaningful for students, have been guideposts for her career path. Her research focuses on reconstructing the history of land ice and sea ice by looking at clues in marine sediments. Kristen's degrees are from Furman University and The Ohio State University; she is Professor at James Madison University. Kristen is the primary author of Chap. 1, a primary author of Chap. 2, and the author of Box 1 in Chap. 8. e-mail: stjohnke@jmu.edu

Lawrence Krissek is from the School of Earth Sciences, The Ohio State University. Lawrence Krissek became intrigued with the oceans while completing the Boy Scouts' Oceanography merit badge and a middle school science fair project, even though he lived far from the coast. He has been fortunate to be able to pursue that interest throughout his career, combining his initial focus on scientific research with more-recent efforts to improve earth- and climate-science literacy across the K-16 spectrum. Lawrence's research interests focus on the information about past climates carried by land-derived sediments deposited in the ocean, especially the history of past ice sheets. Lawrence has spent his academic career in the School of Earth Sciences (and its predecessors) at Ohio State University. Larry is a primary author of Chaps. 2 and 9. e-mail: krissek.1@osu.edu

Contributors

Amanda Borth is from the Consortium for Science, Policy and Outcomes at Arizona State University. 16-year-old Amanda was a climate change denier. But in a determined attempt to avoid taking high school Physics, she enrolled in Advanced Placement Environmental Science: A choice that forever changed her career path. Since then, Amanda has dedicated her studies and work to climate change governance along with public participation and community engagement in science. While receiving her BA in International Studies, concentrating on Global Environmental Policy, at American University (AU), she worked as a Museum Educator at the Smithsonian National Air and Space Museum and held internships related to science policy. She then worked as a Project Coordinator for the Institute for Carbon Removal Law and Policy at AU before entering her PhD program in Communication at George Mason University (GMU). Specializing in science and climate change communication during her Ph.D., Amanda worked as a Graduate Research Assistant for GMU's Center for Climate Change Communication and focused her research on public engagement in carbon removal governance. Post-Ph.D., Amanda is now an Associate Research Professional at Arizona State University's Consortium for Science, Policy and Outcomes, where she focuses on data collection and analysis for Participatory Technology Assessments of climate intervention strategies. Amanda is the author of Box 1 in Chap. 11. e-mail: amanda.borth@asu.edu

Melissa Burt is from the Department of Atmospheric Science, College of Engineering, Colorado State University. Melissa's interest in weather started from her fear of tornadoes, which sparked a curiosity to learn more about the environment, weather, and climate. Melissa Burt is an Associate Professor in Atmospheric Science and Associate Dean for Diversity and Inclusion at Colorado State University. Her research spans the intersection of atmospheric science and social justice issues, climate change, and science communication. She also leads and facilitates diversity, equity, and inclusion efforts to strengthen a culture of inclusion and belonging. Melissa is the Vice President for the non-profit organization, the Earth Science Women's Network, and co-founder of Science Moms, a non-partisan group of climate scientists working to demystify climate science and solutions that preserve the planet. Melissa has a B.S. degree in Meteorology from Millersville University and a M.S. and Ph.D. in Atmospheric Science from Colorado State University. Melissa is the author of Box 2 in Chap. 1. e-mail: melissa.burt@colostate.edu

Steven Clemens is from the Department of Earth, Environmental and Planetary Sciences, Brown University. Steven Clemens grew up in Texas, Colorado, Florida, New Jersey, Alaska, and Washington State. Each place offered new and different opportunities for exploration of the natural environment; from urban-influenced to untouched and natural, from mountains to oceans. These led to an interest in the sciences and in particular, the science that integrates all others, Earth Science. Steve is a Professor of Research at Brown University, studying how and why monsoon rainfall in Asia changes through time. He has planned, participated in, and led ocean drilling expeditions to the Arabian Sea, the South China Sea, the East China Sea, the Sea of Japan, and the Bay of Bengal. Each expedition is followed by years of research in the lab, reconstructing monsoon wind and rainfall patterns by analyzing changes in the chemical, physical, and biological properties preserved in the sea floor sediments and publishing the results so that other scientists can evaluate and build upon them. Nearing retirement, Steve looks forward to spending even more time on and in the ocean. Steve is a primary author of Chap. 9. e-mail: steven_clemens@brown.edu

Thomas M. Cronin works at the United States Geological Survey, Reston, VA. Thomas Cronin [USGS emeritus research geologist] became interested in glacial geology and climate change growing up in Connecticut where evidence of the last great ice sheets is everywhere. He was lucky to go to Colgate University to play college basketball where he met excellent professors who encouraged him to pursue a geology major. After getting his BA from Colgate he again was lucky to pursue an MA and Ph.D. in Geology from Harvard University and then a National Research Council Post-doc at the USGS and Museum of Natural History, Washington DC. His research since at USGS, which coincided with a growing international awareness of the importance of geological records of climate change, has focused on paleoclimatology, sea-level change, biostratigraphy, geochemistry and ecosystems. He was an NSF-sponsored visiting researcher at Shizuoka University, Japan (1991), taught at the Urbino (Italy) Summer School for Paleoclimatology Faculty (2008–2016), was adjunct faculty at Georgetown University's Science Technology International Affairs Program, Walsh School of Foreign Service (2005–2021), and served in the White House Office of Science, Technology and Policy (OSTP) (1996–97). He has participated in sediment coring expeditions including four to the Arctic Ocean. His awards include the Brady Medal (TMS London), Duke of Montefeltro Medal (USSP Urbino), Fellow, American Association for Advancement of Science, Wilmot H Bradley lecture (Geological Society Washington), US Coast Guard Service Medal, AGU Citation for Excellence Reviewing, Bolin Climate Center Annual Lecturer (Stockholm), USGS Leadership, Meritorious Service, and Excellence Award. During his career, he worked extensively with high school, bachelors, masters and Ph.D. students leading to many fruitful careers in the sciences. Tom is the primary author of Chap. 3. e-mail: tcronin@usgs.gov

Greg Hitzhusen is from the School of Environment and Natural Resources, The Ohio State University. Growing up in Ohio during the heyday of acid rain, young Greg suddenly realized that flicking on his light switch was causing sulfuric acid to fall from the skies in the Adirondacks. And that just seemed *wrong*. He studied ecology at Cornell to learn about how to help solve problems like acid rain, and realized it was the *ethical* dimensions of environmental problems (the part about something being "wrong," and why) that really drew his interest. He worked for the National Wildlife Federation's Outdoor Ethics division before pursuing joint theology-environment graduate studies at Yale, and then returned to Cornell for Ph.D. research about environmental ethics in North American faith communities. After serving as Land Stewardship Specialist for the National Council of Churches

Eco-Justice Programs in Washington, DC, and becoming the founding executive director of Ohio Interfaith Power and Light, he joined the faculty in the School of Environment and Natural Resources at The Ohio State University, where he teaches about religion and environmental values, environmental writing, and the philosophies of environmental and natural resource sciences. He serves on the steering team of OSUs Center for Ethics and Human Values, and coordinates a sustainability capstone program where student research teams collaborate with community partners to solve local and regional sustainability challenges; his students have now published over 130 reports and 30 book chapters pushing the boundaries of sustainability science and values. Greg is the primary author of Chap. 12. e-mail: hitzhusen.3@osu.edu

Steven Hovan is from the Kopchick College of Natural Sciences and Mathematics, Indiana University of Pennsylvania. Steven Hovan became an oceanographer by chance. Growing up in the Motor City (Detroit, MI), he set off to college to become an automotive engineer. Or so he thought until an introductory oceanography class piqued his interest enough to join a 4-week research expedition in the Pacific Ocean. Since then, he has never looked back. He switched his major and earned a Ph.D. in Marine Geology and Geophysics from the University of Michigan and has enjoyed a career as an ocean scientist focused on teaching, research and leadership and currently serves as the Dean of Natural Sciences and Mathematics at Indiana University of Pennsylvania. His research interests surround the paleoclimatic record of terrigenous inputs to the deep sea, particularly those involving dust transport to map global winds patterns throughout time and understand how they relate to changes in the global climate system. In all aspects of his work, providing undergraduate students with the opportunity to become involved in genuine research activities is a primary goal. Steve is a primary author of Chap. 5. e-mail: hovan@iup.edu

James Kasting is from the Department of Geosciences and Meteorology, Pennsylvania State University. James Kasting is a retired colleague of Lee Kump's at Penn State. Like Lee, he was an early affiliate of the Earth System Science Center. He worked at NASA Ames for seven years before that. James' research interests are in planetary science and astrobiology. He has published numerous papers about ancient atmospheric photochemistry and climate, with an emphasis on the Archean Earth. He is known by astronomers for his work on habitable zones around stars. James is the author of Box 2 in Chap. 7. e-mail: jfk4@psu.edu

Lee Kump is from the College of Earth and Mineral Sciences, Pennsylvania State University. Lee Kump has always struggled to define his scientific discipline in conventional terms, but was attracted to Earth system science in its fledgling years and has maintained an interdisciplinary approach to the study of Earth ever since. He has spent his entire career at Penn State as a geosciences faculty member and affiliate of the Earth System Science Center (now Earth and Environmental Systems Institute), department head, and now dean of the College of Earth and Mineral Sciences. Through his research he and his students strive to unlock the mysteries of Earth's climate and biotic evolution, with a focus on the causes and consequences of the establishment of an oxygen-rich atmosphere, and climate and ocean change during past episodes of global warming and mass extinction. Lee is the primary author of Chap. 7. e-mail: lrk4@psu.edu

Adriane R. Lam is from the Department of Earth Sciences, Binghamton University. A person who hated high school and education—that was Adriane R. Lam before she attended Reynolds Community College, where she earned her AS and found her love of rocks and super old dead things. She completed her BS in geology at James Madison University, MS in invertebrate paleontology at Ohio University, and Ph.D. in paleoceanography and micropaleontology at University of Massachusetts Amherst before landing a postdoc at Binghamton University, where she is now an Assistant Professor in the Department of Earth Sciences. Adriane is the co-founder and co-President of Time Scavengers, a non-profit organization, which supports the next generation of Earth stewards. Adriane's research interests include science communication, reconstructing ocean surface currents across major climate events, inferring how biota respond to Earth systems changes, and biostratigraphy, using fossils to put time into the sedimentary record. When she's not in the lab, Adriane enjoys spending time with her friends, cats and dog; hiking, reading, and pretending she knows a lot about gardening. Adriane is a primary author of Chap. 5. e-mail: alam@binghamton.edu

R. Mark Leckie is from the Department of Earth, Geographic, and Climate Sciences, University of Massachusetts Amherst. Mark Leckie has been interested in rocks and fossils since he was knee-high to a grasshopper, but he wasn't aware that 'geology' was a major until applying to college. He became a micropaleontologist and stratigrapher with research interests in the Cretaceous Western Interior Seaway of North America, paleoceanography of the Cretaceous and Cenozoic deep sea, and Antarctic ocean-climate change. He has been active in scientific ocean drilling throughout his career. Mark earned a BS and MS at Northern Illinois University and a Ph.D. at the University of Colorado. He is a Professor at the University of Massachusetts Amherst where he teaches Oceanography, Earth History, Field Methods, and Paleoceanography. Mark is a primary author of Chap. 5. e-mail: leckie@umass.edu

Walker Raymond Lee is from the National Center for Atmospheric Research. Walker grew up in Minnesota, where he spent lots of time outdoors with the Boy Scouts of America and became passionate about the environment. As an engineering student at university, he studied fluid mechanics and heat transfer and worked on renewable hydropower research. He expected to continue on a similar path in graduate school, but during his first week of orientation at Cornell, he was fascinated by a presentation on climate intervention research, and he immediately changed tracks to join the group. In 2023, he completed his Ph.D. on climate model simulations of solar geoengineering, in which aerosols (tiny droplets or solid particles) are scattered into the middle atmosphere to reflect sunlight and mitigate the impacts of global warming. Walker is now a postdoctoral fellow in the Climate and Global Dynamics research group at the National Center for Atmospheric Research, where he continues to study climate intervention. Walker is a primary author of Chap. 11. e-mail: walkerl@ucar.edu

Douglas MacMartin is from the Sibley School of Mechanical and Aerospace Engineering, Cornell University. Douglas grew up in Ottawa, Canada, spending all of his summers outside, developing a love of nature. Not knowing what he wanted to do, he originally trained as an aerospace engineer, with a Bachelors in Engineering Science at the University of Toronto, a Ph.D. at MIT, followed by six years in industry at United Technologies. After a break that included hiking in New Zealand and Nepal, and diving in Australia and Thailand, he left industry for Caltech in 2000, determined to do something to help with climate change. A chance encounter in 2004 introduced him to solar geoengineering (or solar radiation modification). After initially thinking this was just a curious academic idea, he gradually realized both that it may actually be helpful to limit damages from climate change, and that his background as an engineer was essential for thinking about a problem that isn't just studying how the climate works, but rather how one might deliberately intervene in the climate system. He has been at Cornell University since 2015, focused on better understanding what the options are for solar geoengineering, what the effects would be if it were ever deployed, and the risks and uncertainties. Douglas is a primary author of Chap. 11. e-mail: dgm224@cornell.edu

Ellen Mosley-Thompson is from the Department of Geography, Atmospheric Sciences, and the Byrd Polar and Climate Research Center, The Ohio State University. Ellen Mosley-Thompson grew up in West Virginia and graduated from Nitro High. She has had a life-long love of science and received her B.S. in physics and math from Marshall University and her M.A. and Ph.D. in atmospheric science and paleoclimatology from The Ohio State University. Early in her career she participated in an ice core drilling project at South Pole Station, Antarctica where she was one of only two women. She was quickly "hooked" on the excitement of using the chemical and physical properties preserved in the ice to reconstruct Earth's climate history. She is a Distinguished University Professor, who, along with Lonnie Thompson and their research team, have acquired ice cores from both polar ice sheets and numerous high mountain glaciers to reconstruct Earth's complex climate history. Collectively, this global collection of paleoclimate histories confirms that Earth's climate has moved outside the range of natural variability experienced over at least the last 2000 years. Ellen has had the privilege of leading nine expeditions to Antarctica and six to Greenland to retrieve these valuable ice cores. Ellen teaches an undergraduate honors course on global climate and environmental change that focuses on understanding the forces driving these climate changes and facilitating collective and individual actions to change the direction in which Earth's climate system is currently heading. Ellen is a primary author of Chap. 4. e-mail: thompson.4@osu.edu

Dónal O'Mathúna is from the College of Nursing and Center for Bioethics, The Ohio State University. Dónal grew up in Ireland and developed an interest in geosciences because of the deeper appreciation it gave him of the mountains he loved to hike and the waters he kayaked. Course selection in college brought him to other natural products—the ones pharmacy uses to develop medicinal agents. This brought him into uncharted territories as he grappled with the ethical issues related to medicinal products and then to healthcare ethics more broadly. Things have come full circle as now he conducts research on the ethical issues at the intersection of human, animal and environmental factors in global health, aka One Health. Dónal is an Associate Professor at The Ohio State University College of Nursing and the Center for Bioethics. He has spoken and published widely in bioethics, especially disaster ethics, and has contributed to ethics initiatives with the World Health Organization, UNICEF and other international agencies. Dónal is the author of Box 1 in Chap. 1. e-mail: omathuna.6@osu.edu

Dana Royer is from the Department of Earth and Environmental Sciences, Wesleyan University. In college, Dana Royer was drawn to big-picture ideas like global biogeochemical cycling. This led to an interest in plants, both living and fossil, because plants play such an important role in these cycles. Dana uses fossil plants as tools to unlock information about the climate and ecology of ancient terrestrial ecosystems. He is particularly interested in ways to reconstruct CO_2 and temperature from information preserved in fossil leaves, and in turn how CO_2 and temperature are related to one another on geologic time scales. Dana is a Professor at Wesleyan University. Dana is the author of Box 2 in Chap. 5. e-mail: droyer@wesleyan.edu

Jill Schneiderman is from the Department of Earth Science, Vassar College. Concerned about the welfare of humans and other living beings on Earth, Jill Schneiderman applies her geological knowledge in order to raise awareness about systemic injustices in the arena of the natural and built environment. Although she has conducted research on topics such as the formation of mountain belts, heavy minerals in deltas, and terrestrial microplastics, she considers herself primarily to be a teacher of the next generation who will need deep knowledge of the Earth System in order to secure a livable planet. She teaches courses such as The Solid Earth, Environmental Justice in the Anthropocene, and Feminist Approaches to Science and Technology. She earned her S.B. at Yale University and her Ph.D. at Harvard University and is Professor of Earth Science in the Department of Earth Science and Geography at Vassar College in Poughkeepsie, New York. Jill is the author of Box 1 in Chap. 12. e-mail: schneiderman@vassar.edu

Mary H. Schultz is from the Department of Physical Sciences, San Jacinto College. After becoming captivated by plate tectonics as an undergraduate student at Bryn Mawr College, Mary Schultz went on to earn her doctoral degree in Geological Sciences at Arizona State University where she explored the relationship between climate and tectonics in the evolution of the central Himalayan Mountains. In graduate school, Mary became interested in how earth processes and the environment impact communities in the United States and around the world. This led her to Washington, DC, where she served as a congressional fellow, working in the United States Senate on policy issues ranging from climate change to the space industry. She then turned to one of her primary passions, teaching, and was a Visiting Assistant

Professor of Geology at James Madison University, a position she held for two years. Mary's fellowship experience and interest in translating complex scientific ideas to broader audiences led her to a position as a scientific research writer at Texas Children's Hospital in Houston. Mary remained in Houston and is now a Professor of Geology at San Jacinto College. Mary is the primary author of Chap. 6. e-mail: Mary.Schultz@sjcd.edu

Debbie Thomas is from the College of Marine Sciences and Maritime Studies, Texas A&M University. Debbie Thomas was born and raised in Cincinnati, Ohio but somehow decided at age 10 that she wanted to be an oceanographer when she grew up. After earning her bachelor's degree in Geological Sciences from Brown University, she earned her MS in Marine Sciences and her Ph.D. in Geological Sciences from the University of North Carolina—Chapel Hill. In 2004 she joined the faculty of the Department of Oceanography at Texas A&M University, fulfilling her childhood dream, and most recently is the founding dean of the College of Marine Sciences and Maritime Studies at Texas A&M University (located at the Galveston Campus). Debbie's disciplinary specialty of paleoceanography allowed her to apply her training in geological and marine science in a truly interdisciplinary manner. Debbie is the primary author of Chap. 8. e-mail: dthomas@tamug.edu

Lonnie G. Thompson is from the School of Earth Sciences and the Byrd Polar and Climate Research Center, The Ohio State University. Lonnie G. Thompson grew up in the mountains of West Virginia and attended Marshall University where he received his B.S. degree in Geology. He began his graduate work at The Ohio State University studying coal geology, although he became interested in polar studies and glaciology after participating in an expedition to Antarctica. His Ph.D. dissertation focused on the development and interpretation of the first records of wind-blown dust that extended back into the last ice age from ice cores from Greenland and West Antarctica. Since then he has dedicated his career to collecting and interpreting ice core records from the world's highest mountains in 16 countries in order to develop a global history of climate and environmental changes archived in glaciers. He has spent a total of over 4 years of his life at altitudes higher than 18,000 ft, and after surviving a heart transplant in 2012 he conducted research in 2015 on the 22,000 ft Guliya ice cap, Western Kunlun Mountains, a record elevation for a heart transplant recipient. He is an elected member of the National Academy of Sciences and a recipient of the United States Medal of Science. Lonnie is a Distinguished University Professor in the School of Earth Sciences and Senior Research Scientist in the Byrd and Polar Climate Research Center at The Ohio State University. Lonnie is a primary author of Chap. 4. e-mail: thompson.3@osu.edu

Shuang-Ye Wu is from the Department of Geology and Environmental Geosciences, University of Dayton. Shuang-Ye Wu often considers herself an accidental scientist. Working for the National Environmental Protection Agency in China after her Master's degree in linguistics, she became fascinated by the multi-faceted and interconnected nature of climate science. She went on to obtain her Master's and Doctoral degrees in geosciences from the University of Cambridge, focusing on the impact of climate change on the hydrological cycle both in the past and future. She has worked in the Department of Geology and Environmental Geosciences at the University of Dayton since 2004, and is currently a professor and the Chair of the Department. Her research uses both historical proxies and observations and climate models to examine the changing patterns of precipitation, floods, droughts under different climate drivers, and how such changes affect vegetation and land cover. Shuang-Ye is the primary author of Chap. 10. e-mail: swu001@udayton.edu

Climate Change in Geoscience and Social Contexts

Kristen St. John, Dónal O'Mathúna, and Melissa Burt

Guiding Questions: Why does climate change matter? How does examining climate change in geoscience and social contexts help our understanding?

1 Key Take-Away Points

- Like a pandemic, climate change is not just a scientific concern, and it concerns and affects the well-being of individuals and society. Also similar to a pandemic is the fact that climate change has disparate effects; everyone does not have same vulnerabilities and readiness to deal with the effects of climate change. Climate change is a threat multiplier of already existing societal challenges.
- The climate system is complex, with multiple drivers and feedbacks operating on different temporal and spatial scales.
- Paleoclimatology is the study of past climate. It provides valuable context to understand modern climate change and to ground-truth models of future climate change.

K. St. John (✉)
Department of Geology and Environmental Science, James Madison University, Harrisonburg, USA
e-mail: stjohnke@jmu.edu

D. O'Mathúna
College of Nursing & Center for Bioethics, The Ohio State University, Columbus, USA
e-mail: omathuna.6@osu.edu

M. Burt
Department of Atmospheric Science, College of Engineering, Colorado State University, Fort Collins, USA
e-mail: melissa.burt@colostate.edu

- The rate of human-driven global warming in recent decades is faster than even the onset rate of the most extreme naturally driven warming event of the last 65 million years.
- Even small changes in global average temperature can have large consequences on Earth systems and human systems. Regional variability in climate conditions means some locations will experience more extreme changes than other locations.
- Modern climate change involves interconnected Earth systems and human systems. Both mitigation and adaptation are necessary in order to avoid the worst impacts of climate change now and in the future.
- For scientific findings to benefit humanity they need to be communicated effectively, trusted, and understood by human systems decision-makers and the public-at-large. If there are fundamental disconnects between the scientific and the public understandings of climate change, mitigating actions to counter the negative effects of climate change on human systems can be delayed or impaired, and status quo behaviors may persist even though preparations to adapt are urgently needed.
- Geoscience is a broad discipline that describes the study of Earth systems (i.e., the ocean, atmosphere, solid Earth, ice, and life) and human interactions with the Earth. The geoscience discipline is also time-transcendent; it encompasses studying records of Earth's past, processes of the present, and uses the results to make evidence- and theory-based forecasts for the future. A geoscience perspective on climate change is therefore important to the study of climate change and has implications for predicting future conditions on Earth.
- Misconceptions about climate change are very common. These form in part due to the complexity of the climate system, and in part due to the fundamental difference between how scientists and non-scientists draw conclusions about the natural world.
- It is going to take more than accurate and reliable scientific evidence to address misconceptions about climate change; challenges such as groupthink and other cognitive biases, media biases affecting journalistic standards, the social media filter bubble, and deliberate disinformation must be overcome as well.

2 Introduction

2.1 Why a Pandemic Can Serve as a Wake-Up Call for the Challenges of Climate Change

During the early writing stages of this book about climate change, the world was facing a multi-year pandemic. The COVID-19 disease caused by a coronavirus outbreak took the world by surprise, though it was not a surprise to experts in the field who had been warning for years that a pandemic was highly likely given their knowledge of zoonotic diseases.[1] It was a global emergency, the effects of which extended

---

[1] Madhav et al. (2017).

beyond country borders. It affected how we live, where we could go, our social interactions, our physical health, our mental health, our economic stability, and even our political stability. Individual, community, and governmental decisions affected the spread of the virus. Misinformation (i.e., inaccurate information) and disinformation (i.e., *intentionally* inaccurate information) created confusion and some resistance to taking mitigating steps. Vulnerability to the disease, and the readiness to mitigate it, varied among individuals and across communities and countries, and therefore we saw disproportionate impacts demographically and geographically. While COVID vaccines were developed at record pace, their availability was uneven across the globe. And even in places where the vaccine was readily available, not everyone was willing to be vaccinated or to do what it takes to limit the spread of the disease.

An individual's or community's vulnerability to a disease and their readiness to mitigate it were important to understand the disparate impacts of the pandemic. The concepts of vulnerability and readiness are also important to understanding and predicting the impacts of climate change across the globe. In this book, we use these terms as defined by the university-based research group, Notre Dame Global Adaptation Initiative (ND-GAIN[2]), which focuses on helping countries counter the risks of climate change:

- **Vulnerability** is the propensity or predisposition of human societies to be negatively impacted by climate-related hazards (e.g., drought, sea level rise, food scarcity, excessive heat). Vulnerability depends on a range of factors, such as the physical exposure to climate-related hazards and the level of stress (or sensitivity) that a hazard would pose.
- **Readiness** is the ability to make effective use of investments for adaptation actions. Readiness depends on economic, government, and social factors.

Taken together, vulnerability and readiness can describe a community's resilience to climate change. Adopting ND-GAIN definition, **climate resilience** is the ability to prepare for, recover from, and adapt to climate change impacts.

Research on the coronavirus, and human behavior related to it, is an example of high-stakes science about a complex and nuanced problem, and an example of the intersection of science with human judgment and decisions. The coronavirus pandemic was also a wake-up call about global emergencies. As recognized by a number of editorial cartoons and op-ed pieces in major news outlets (e.g., The Economist[3]), while the world focused on fighting the coronavirus, a larger and more complex challenge exists in the wings (and has been there for some time). Recognizing climate change as an urgent and even more powerful foe that can't be ignored and for which we must prepare is the global community's next epic battle.

---

[2] For definitions go to https://gain.nd.edu/our-work/country-index/ and select "View Technical Document". Chen, C., Noble, I., Hellmann, J., Coffee, J., Murillo, M., & Chawla, N., (2015). University of Notre Dame Global Adaptive Index: Country Index Technical Report. University of Notre Dame, 46 p.

[3] For example, see the editorial cartoon by Kevin Kallaugher (KAL), in The Economist, April 23, 2020, https://www.economist.com/the-world-this-week/2020/04/23/kals-cartoon.

There are both contrasting characteristics and parallel traits when comparing climate change to a global pandemic. Unlike the pandemic, the 'foe' is obviously not a discrete package of genetic code in protein coats. Rather, climate change has a complex collection of natural and human causes that range in relative importance across space and time. And, unlike the pandemic, climate change cannot be prevented by a vaccine, or treated with a pill, although innovative technological solutions may help. Humans also have some ability to influence the driver of climate change (e.g., reduce greenhouse gas emissions) even while climate change is underway; an equivalent influence on the driver of the pandemic, once the virus emerged in the globally connected society, was unlikely.

Clear parallels between climate change and a pandemic include their global reach, yet disproportionate effects depending on different levels of risk and preparedness by individuals, communities and countries; the complexity of the science; and the need for society to both mitigate negative effects and adapt to 'new normals.' Importantly, successfully addressing the challenges of climate change will—like for the pandemic—depend on scientific research, effective governmental leadership, modified individual and community behaviors, heeding expert advice, and overcoming the effects of misinformation.

2.2 Why a Climate Change Textbook Begins with a Chapter About Social Science

One of the big takeaways of the above comparison between the coronavirus pandemic and climate change is that these high-stakes scientific topics have tendrils that extend well beyond their immediate scientific fields of study (e.g., biology and Earth science, respectively). These topics are not only scientific concerns—they are societal concerns. But advances in scientific studies on the coronavirus pandemic or on climate change have little impact on society unless the findings are shared beyond scientific institutions and are used to inform evidence-based decision-making at the individual, local, national, and global community levels. In the case of climate change, there is a need to understand how Earth systems and human systems interrelate (Fig. 1) in order to inform climate change mitigation and adaptation. In this textbook, we adopt the following definitions from the IPCC[4] for these two important terms:

- **Mitigation** is a human intervention to control factors that result in climate change. Much of its focus is on reducing the sources or enhancing the sinks of greenhouse gases.[5] It can also focus on controlling other substances, such as particulate matter (aerosols) that affect Earth's energy balance and thus climate.

---

[4] IPCC Data Distribution Centre glossary https://www.ipcc-data.org/guidelines/pages/glossary/glossary_lm.html.

[5] A greenhouse gas is an atmospheric gas that absorbs infrared (IR) energy emitted by the Earth, reradiating some of it back to the surface. This process adds heat to the troposphere and is referred to as the greenhouse effect. The primary greenhouse gases in Earth's atmosphere are water vapor, carbon dioxide, methane, and nitrous oxide.

Climate Change in Geoscience and Social Contexts

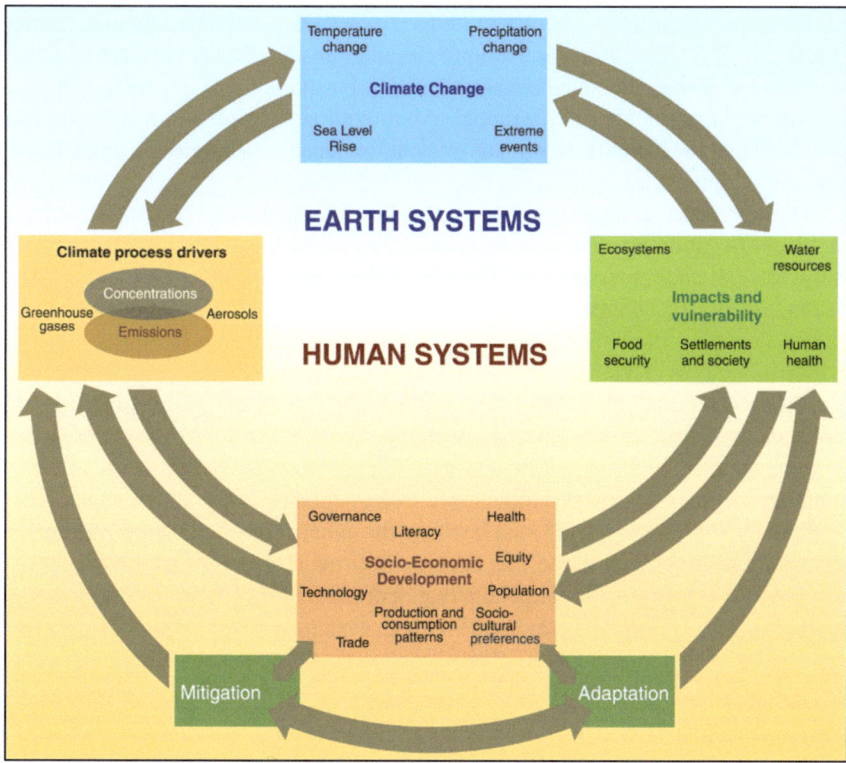

Fig. 1 Schematic framework representing anthropogenic drivers, impacts of and responses to climate change, and their linkages. Modern climate change involves many linkages between Earth systems and human systems. Socio-economic factors are connected to anthropogenic (human-caused) drivers, as well as the impacts and vulnerabilities to climate change. Addressing the challenges of climate change will therefore also involve the human systems via mitigation and adaptation. Figure from: IPCC, (2007)[6] Figure used with permission from the IPCC. This figure is excluded from Creative Commons license

- **Adaptation** is the process of adjustment to actual or expected climate and its effects. In human systems, adaptation seeks to moderate harm or exploit beneficial opportunities.

Mitigation and adaptation are not mutually exclusive strategies (i.e., it is not an 'either/or' situation). As the IPCC noted,[7] effective mitigation reduces the impact of climate change and thus reduces the adaptive challenges. Adaptation can take advantage of positive changes and reduce negative ones. Both mitigation and adaptation to the adverse effects of climate change (e.g., temperature rise, sea level rise) will require

---

[6] IPCC, 2007. *Climate Change 2007: Synthesis Report. Contribution of Working Groups I, II and III to the Fourth Assessment Report of the Intergovernmental Panel on Climate Change* (p. 26, Figure 1), IPCC. https://www.ipcc.ch/site/assets/uploads/2018/02/ar4%5Fsyr.pdf.

[7] Klein et al. (2007).

collaborative efforts among human systems (i.e., social, political, economic, technological, scientific, corporate) and among communities at all scales—local, regional, national, and global. Personal decisions matter, but policy changes by government and industry leaders are also essential. Also, both mitigation and adaptation take time, such that the benefits may not be evident for decades (a timeframe much longer than an election cycle).

Clearly, climate change is not to be put in a box just for scientists to address. For scientific findings to benefit humanity, they need to be communicated effectively, trusted, and understood by decision-makers and the public-at-large. If there are fundamental disconnects between the scientific and the public understandings of climate change, mitigating actions to counter the negative effects of climate change on human systems can be delayed or impaired, and status quo behaviors may persist even though preparations to adapt are urgently needed. So what do scientists want the public and policy-makers to know about climate change? How does the public understanding compare to the scientific understanding? What are the barriers to effective communication, trust, and scientifically sound understanding of climate change by the public and policy-makers? And how can those barriers be overcome? These are important questions that guide this chapter. You may have noticed that several of these questions are more about people (e.g., communication, intake of information, feelings, and behaviors) than about the climate system itself. We can consider these human-system factors as 'social science' topics. **Social science** is a broad umbrella term for scientific study of human society. Choosing to examine social science topics in the first chapter of a natural science (specifically, a geoscience) textbook is unusual, but it is purposeful—as is the fact that the concluding chapter of this textbook examines ethics, another social science topic. We have already introduced ethics in this chapter, though implicitly, when we referred to trust, misinformation, and disparities. See Box 1 for a more explicit introduction to ethics. We bookend the collection of geoscience chapters with social science context because social science factors affect our ability to understand and respond to the threat of climate change. We think this framing will give you important awareness about the intersectionality of Earth systems and human systems (Fig. 1) which will make learning about climate change more meaningful—and actionable—for you.

2.3 Why a Climate Change Textbook Uses a Geoscience Perspective

In addition to framing the study of climate change with a social science context, this textbook examines climate change through a geoscience lens. What is geoscience? **Geoscience** is a broad discipline that describes the study of Earth systems. The term is inclusive of geology, Earth science, physical geography, and much of oceanic and atmospheric science. The American Geosciences Institutes (AGI), a national network representing geoscientists, describe geoscience as the *study of Earth's ocean, atmosphere, rivers and lakes, ice sheets and glaciers, soils, its complex surface, rocky* *interior, and metallic core. This includes many aspects of how living things, including humans, interact with* the Earth.[8] There is a lot in common between this definition and the climate-related relationships between Earth systems and human systems that are shown in Fig. 1. Thus, while geoscience is a much broader than the study of climate change (e.g., it also studies hazards such as earthquakes and the formation and mining of resources such as mineral ores), geoscience is the primary disciplinary field that studies climate change.

Importantly, the geoscience discipline is also time-transcendent; it encompasses studying records of Earth's past, processes of the present, and uses the results to make evidence-based and theory-based predictions for the future. Many records of Earth's past have bearing on our understanding of the Earth systems' processes involved in climate change. Just as it is valuable to study past (historical) records of human systems (e.g., politics, economics) to make sense of how these systems operate, it is also valuable to study past records of the climate system to make sense of how the climate system operates, and what drives (or inhibits) change over different scales of time and place (e.g., local to global). **Paleoclimatology** is the study of climate from time periods prior to the widespread availability of instrumental records. It depends on the use of natural **archives**, such as sediment, rock, ice, coral, cave, and tree ring records that preserve evidence of past climatic conditions. Much of that evidence is indirect; it is chemical, physical, or biological data that serves as surrogates, or '**proxies**,' for past climatic conditions (e.g., CO_2 levels, temperatures, sea level, extent of sea and land-based ice). Therefore, paleoclimate archives, and the proxies they contain, provide scientific evidence that helps us evaluate the potential magnitudes and impacts of modern and future climate change. Paleoclimate research allows us to examine the causes, rates, and consequences of climate change on Earth systems. Paleoclimate records can be compared with instrumental records of historical and modern climate change, and these can be used to test the accuracy and validity of climate change models. Thus, a geoscience perspective means we have a longer lens of time to see beyond the human-time window, and situate what is happening now into a longer context. That longer perspective can be both humbling and sobering, as we see how dramatically climate has changed in the geologic past, but also how the rate of human-caused climate change in the present (and expected for the near future) is outpacing the rate of climate changes caused by natural drivers alone.

Let's look at an example. Figure 2 shows global average temperature trends for the past 65 million years, historical temperatures, modern temperatures, and future temperature projections for the next few centuries. The timeline (horizontal axis) is segmented into five parts, each with its own scale suitable to view the changes in temperature for that time range. The two left-most timeline segments span millions of years (Myr) before present. The next two timeline segments span hundreds of thousands of years (kiloyears or kyr) before present. The right-most timeline segment

---

[8] American Geosciences Institute (AGI; accessed 2021, June 7). *What is geoscience?* American Geosciences Institute. https://www.americangeosciences.org/critical-issues/faq/what-is-geoscience.

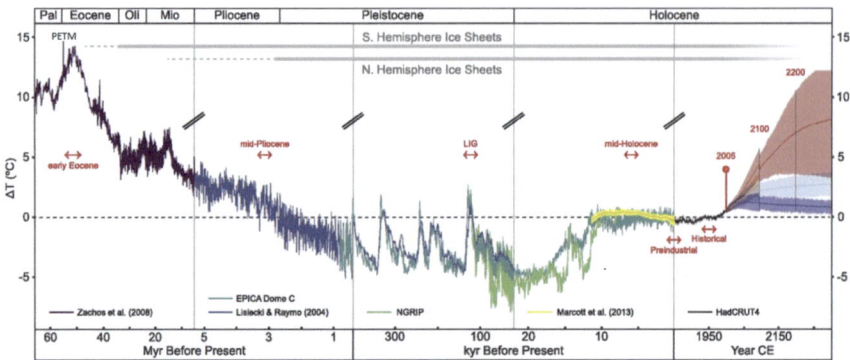

Fig. 2 Temperature trends for the past 65 Myr and projected to the year 2250 CE. Six potential geologic analogs for future climates are noted with red labeled arrows (e.g., early Eocene). Major patterns include a long-term cooling trend, periodic fluctuations driven by changes in the Earth's orbit, and recent and projected warming trends. Temperature anomalies (ΔT) are relative to 1961–1990 global means and are composited from proxy-based reconstructions, modern observations (instrumental measurements), and future temperature projections for four emissions pathways. The presence of ice sheets in the Southern and Northern Hemisphere are shown as gray horizontal lines. Geologic time epochs are noted along the top, with Paleocene abbreviated as Pal; Oligocene as Oli; and Miocene as Mio. The Paleocene-Eocene Thermal Maximum event is abbreviated as PETM. Figure modified (PETM label added) and used with permission from PNAS, Burke et al. 2018.[9] This figure is excluded from Creative Commons license

starts in the year 1900 Common Era (CE; equivalent to A.D. 1900) and ends in the year 2250 CE. The temperature data (vertical axis) are reported as **temperature anomalies** (ΔT) in degrees Celsius (°C). This means the ΔT values are changes (or **anomalies**) relative to some reference value; in this case, relative to the mean global temperature between the years 1961 and 1990. The perspective that the paleoclimate data provides in contextualizing modern and future climate change is profound. We can see that in the past the Earth was both much warmer and much colder than it is currently. We can see there were past times of gradual change and of rapid change. We can see there were times when temperature appears to have cycled back and forth in a complicated pattern. But importantly, we can also see that the potential pathways for future global average temperatures are taking us well outside of the recent historical, pre-industrial, and mid-Holocene norms. Rather, the last interglacial (LIG; 120 thousand years ago) period, the mid-Pliocene (3 million years ago), and the early Eocene (50 million years ago) are among the possible geologic analogs for the conditions we are expected to face in the future. Therefore, looking into the geologic past gives us windows to possible futures and context to understand why and how climate change occurs.

---

[9] Burke et al. (2018).

2.4 Why Misconceptions Are Stubborn Problems that Are Important to Address

There is an important tangential benefit to the longer-term time window that a geoscience perspective brings; it allows us to address several common misconceptions about climate change that are held by the public and policy-makers, such as the misconceptions that temperature and CO_2 levels have never been higher than today, or that modern global warming is just part of a natural cycle. Learning that these statements are **misconceptions** (i.e., that they are incorrect, or incomplete to a point of being fundamentally flawed) may surprise you. Correcting misconceptions is an essential aspect of education, and a primary goal of this textbook. However, education research shows that correcting misconceptions isn't easy. That is because misconceptions learned early on stick with us. A classic demonstration of the stubborn problem of misconceptions was presented in an award-winning short video documentary[10] by Annenberg Media and the Harvard-Smithsonian Center for Astrophysics that first aired in the late 1980s and is still relevant today. In the film, education researchers visit the graduation ceremony of a well-known U.S. university and ask the happy graduates a simple science question: *What causes the seasons?* This is a climate-related question that you (and they) likely learned about at least once during your elementary or secondary school years and again in college. As you will learn more about later in this textbook, the seasons in a region are caused by regular changes in the amount of incoming solar radiation striking that portion of the Earth. These changes occur as the Earth travels on its orbital path around the Sun, while maintaining a consistent axial tilt angle and tilt direction. Distance from the Sun is not the cause of the seasons. In fact, currently Earth is closest to the Sun in early January, and farthest from the Sun in July, which are the times of Northern Hemisphere winter and summer, respectively. Yet, the common *incorrect* idea that the planet is hotter in summer because the Earth is nearer to the Sun than in winter persisted in the graduates' responses. The results from this education research study demonstrated how our sense of beliefs (self-efficacy) could be at odds with scientific evidence even after years of education.

Widely held incorrect or incomplete beliefs can be harmless if believing in them has no consequence; however, holding on to misconceptions about scientific issues, such as climate change, can have serious consequences. It may result in confusion, denial, flawed reasoning in decision-making, and delayed action toward climate change mitigation and adaptation. Table 1 highlights some of the most common misconceptions about global warming which were primarily compiled by Skeptical Science,[11] a non-profit science education organization that focuses on correcting misconceptions about climate change. Notice that the left column lists the misconception and the middle column refutes the misconception by stating (in brief) what

---

[10] See "A Private Universe": https://www.learner.org/series/a-private-universe/ [and useful discussion of it https://www.scienceinschool.org/2010/issue17/privateuniverse].

[11] Skeptical Science (accessed 2021, September 16). *Global Warming & Climate Change Myths.* https://skepticalscience.com/argument.php.

the science says. This direct contrast of the false belief with correct scientific information is an example of a refutation-style approach to addressing the stubborn problem of misconceptions. Science education researchers Christine Tippett and John Cook[12] have demonstrated that refutation-style approaches help with critical thinking and can be effective at overcoming misconceptions. Each of the climate change misconceptions included in Table 1 is addressed in detail in one or more of chapters in this textbook.

3 Why Climate Change Matters

Climate change is one of the most widespread geoscience challenges facing society today. Like a pandemic, it is high-stakes science which requires a broad response by a scientifically informed populace to avoid harmful and costly outcomes. These are serious claims that have their roots in the relationships between Earth systems and human systems (Fig. 1), but let's put these ideas into context to better understand why climate change matters. Regular assessments by the Intergovernmental Panel on Climate Change (IPCC)[13] and the United States Global Change Research Program[14] (among others) help do that. These international and national scientific bodies are well respected in the scientific communities and have been tasked with synthesizing independent peer-reviewed scientific studies to inform governmental policy-makers, industry leaders, and the public about what science says about climate change, the implications and potential future risks, and options for mitigation and adaptation.

As an example of why climate change matters, let's consider temperature data from the recent past and projected for the near future, along with impacts that rising temperatures can have on things that matter to people. Global average surface temperature was 1.1 °C (2 °F) higher in 2011–2020 than in 1850–1900, with larger increases (1.6 °C) over land.[15] For the last 60 + years, every decade has been hotter than the last.[16] Note too that the corresponding *rate* of temperature rise (i.e., the amount of temperature rise divided by the time period of change, such as °C/year or °C/century) for this century thus far is faster than the average rate of temperature rise in the last century, and faster than the rate of temperature rise associated with the most extreme warming event of the entire Cenozoic Era. That extreme warming event occurred ~ 56 million years ago. It can be seen as a spike in Fig. 2 and is the focus of Chap. 8.

---

[12] Tippet (2010).

[13] Intergovernmental Panel on Climate Change (IPCC), https://www.ipcc.ch/.

[14] U.S. Global Change Research Program https://www.globalchange.gov/. In 2025, executive actions by the U.S. federal government may change the focus of the USGRP.

[15] Allan et al. (2021).

[16] Same as above.

Table 1 Examples of commonly held misconceptions about global warming, contrasted with what science says, and addressed in relevant textbook chapters

Common global warming misconceptions	What the science says	Where this common misconception is addressed in this textbook
Temperature and CO_2 levels have never been higher than today	For much of Earth's history the world was warmer and had higher CO_2 levels than today, but the *rate* of current warming is unprecedented in at least the last 65 million years	Chaps. 1, 5, 6, 8
Global warming today is just part of a natural cycle	No known natural forcing fits the fingerprints of observed warming. The observed pattern can only be explained with forcing by anthropogenic greenhouse gases	Chap. 9
Climate has changed before; therefore we can't be the cause of global warming now	Climate reacts to whatever forces it to change at the time; humans are now the dominant forcing	Chaps. 1, 2, 3, 4, 5, 6, 7, 8, 9, 10, 11, 12
Modern global warming is caused by the Sun	In the last 50 + years of global warming, solar forcing and climate have been going in opposite directions	Chaps. 3, 9, 11
Global warming isn't that bad	Negative impacts of global warming on vulnerable populations, water resources, agriculture, health & the environment far outweigh any positives	Chaps. 1, 3, 4, 8, 11, 12
Climate change will only harm plants and animals, future generations, other people, but not me	Climate change affects everyone, not just in the distant future but now too. Impact will depend on the vulnerability and readiness to deal with the effects of climate change	Chaps. 1, 12

(continued)

Table 1 (continued)

Common global warming misconceptions	What the science says	Where this common misconception is addressed in this textbook
There is no consensus about the cause of modern global warming	97 to > 99% of climate experts agree humans are causing modern global warming	Chaps. 1, 3, 11, 12
Climate models are unreliable	Climate models successfully reproduce temperatures since 1900 globally, by land, in the air and the ocean	Chaps. 3, 10, 11
Temperature records are unreliable	The warming trend is observed in rural and urban areas, measured by thermometers and satellites. Past temperatures can be reconstructed by multiple proxies and calibrated to modern instrumental records	Chaps. 3, 4, 5
Animals and plants can adapt	Global warming will cause mass extinctions of species that cannot adapt on short time scales. We can see the effect of climate change on life of the past from the climate proxies	Chaps. 4, 5, 6, 7
CO_2 lags temperature changes	CO_2 didn't initiate warming from past ice age cycles, but it did amplify the warming	Chaps. 7, 9
Ocean acidification isn't serious	Ocean acidification threatens entire marine food chains	Chaps. 5, 8
Sea level rise is exaggerated	A variety of different measurements find rising sea levels over the past century	Chaps. 3

From Skeptical Science,[17] Global Weirding[18] Lynas and others (2021)[19] and author observations of student-held misconceptions

---

[17] Skeptical Science. (Accessed 2021, September 16). *Global Warming & Climate Change Myths*, Skeptical Science. https://skepticalscience.com/argument.php.

[18] Hayhoe, K. (Accessed 2021, September 21). *Is Global Warming Causing All Of These Hurricanes?* https://www.youtube.com/watch?v=yfkS7LqCMDQ&t=158s In PBS Digital Series, Global Weirding with Katherine Hayhoe, PBS. https://www.youtube.com/channel/UCi6RkdaEqgRVKi3AzidF4ow.

[19] Lynas et al. (2021).

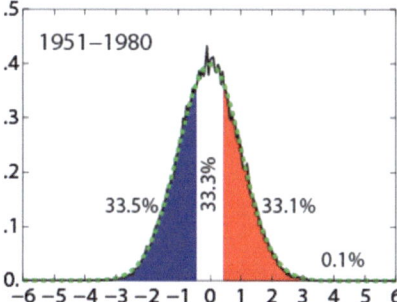
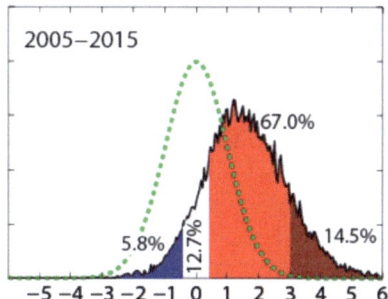

Fig. 3 Frequency of occurrence (vertical axis, with higher indicating more frequent) of Northern Hemisphere summertime land surface temperatures for two time periods: 1951–1980 (left, and in green dashed line on right) and 2005–2015 (in color on right) plotted by increments of the local standard deviation (horizontal axis). The time period 1951–1980 is used as a baseline to compare more recent temperature distributions. From Hansen and Sato (2016).[23] Figure used with permission from Jim Hansen. This figure is excluded from Creative Commons license

Because of the shift in average temperature that has already taken place in recent decades, we also see greater summertime extreme conditions. For example, what used to be considered extremely hot outliers half a century ago can now be found on land in about 14.5% of the Northern Hemisphere (Fig. 3).[20] In nearly every part of the world, heatwaves (defined as prolonged periods of excessive heat) have been increasing in intensity, frequency, and duration since the 1950s, while cold extremes have become less frequent and less extreme.[21] Human health, agriculture, and economies suffer, but these impacts are not distributed equally geographically or demographically. Poor and otherwise disadvantaged people and nations are more vulnerable to the adverse effects of heatwaves because the resources and options available to adapt to living and working conditions in extreme heat are often very limited.[22]

---

[20] From Hansen, J. & Sato, M. (2016). *Regional Climate Change and National Responsibilities.* http://www.columbia.edu/~jeh1/mailings/2016/20160301_Dice2.PopSci.pdf, which summarizes a paper of the same authors and title: Hansen, J., & Sato, M. (2016). Regional climate change and national responsibilities. *Environmental Research Letters 11* (3), 034009. https://doi.org/10.1088/1748-9326/11/3/034009 https://iopscience.iop.org/article/10.1088/1748-9326/11/3/034009.

[21] Same as above, and also from Perkins-Kirpatrick, S.E., & Lewis, S.C. (2020). Increasing trends in regional heatwaves, *Nature Communications 1*, 3357. https://www.nature.com/articles/s41467-020-16970-7.

[22] Perkins-Kirpatrick & Lewis (2020).

[23] From a figure highlighted on https://csas.earth.columbia.edu/our-work/climate-data, which is derived from Hansen and Sato (2016) Regional Climate Change and National Responsibilities, http://www.columbia.edu/~jeh1/mailings/2016/20160301_Dice2.PopSci.pdf which summarized a paper of the same title in Environmental Research Letters, v 11, no. 3, https://iopscience.iop.org/article/10.1088/1748-9326/11/3/034009.

Precipitation patterns have also changed in recent decades. A general pattern of greater extremes has emerged, with wet areas getting wetter, and dry areas getting drier. The changes observed in the contiguous United States are a good example of this shift. The western and southwestern U.S. are typically dry, as the prevailing westerly winds blowing from the Pacific Ocean eastward lose most of their moisture on the upwind side of the many mountain chains in the American West (e.g., the Cascades, Sierra Nevada, and Rockies), hence the prevalence of deserts and plains (grasslands rather than forests) on the leeward side of these mountains. Meanwhile, the eastern U.S. is typically wet, as the prevailing winds coming up from the Gulf of Mexico bring moisture inland and to the east and are unimpeded by large mountains. This natural west–east contrast in the U.S. has become exaggerated in recent decades: there has been a decrease in precipitation in the west and an increase in the east in the last 30 years compared to the twentieth-century average (Fig. 4 lower panel). Additionally, the west is also experiencing greater temperatures, so it is both increasingly drier and warmer (Fig. 4 upper panel). The changes set up other challenges. Less rainfall in the west sets up the potential for extreme drought, and dry vegetation in drought-stricken areas then becomes tinder for wildfires. The U.S. EPA[24] tracks wildfires as a climate change indicator, noting that wildfire season length and burned area have both increased since the 1980s. In contrast, flooding and other damage from hurricanes is increasing in the southeast, as warmer sea surface temperatures power larger and more intense (higher winds and higher precipitation) storms.[25] The ability to recover or rebuild from any of these climate-related events is not the same for everyone; just as heat waves disproportionately impact disadvantaged peoples, so do the precipitation-related climate phenomena.

As we can see just from these brief examples above and the essays in Boxes 1 and 2, increases in global average temperatures and changes in regional precipitation affect things that matter to society; these affect people and places that we care about or have empathy toward, whether they are close to home or across the globe.

---

[24] U.S. EPA (accessed 2023, May 22). Climate Change Indicators: Wildfires. U.S. EPA. https://www.epa.gov/climate-indicators/climate-change-indicators-wildfires.

[25] Kossin et al. (2020).

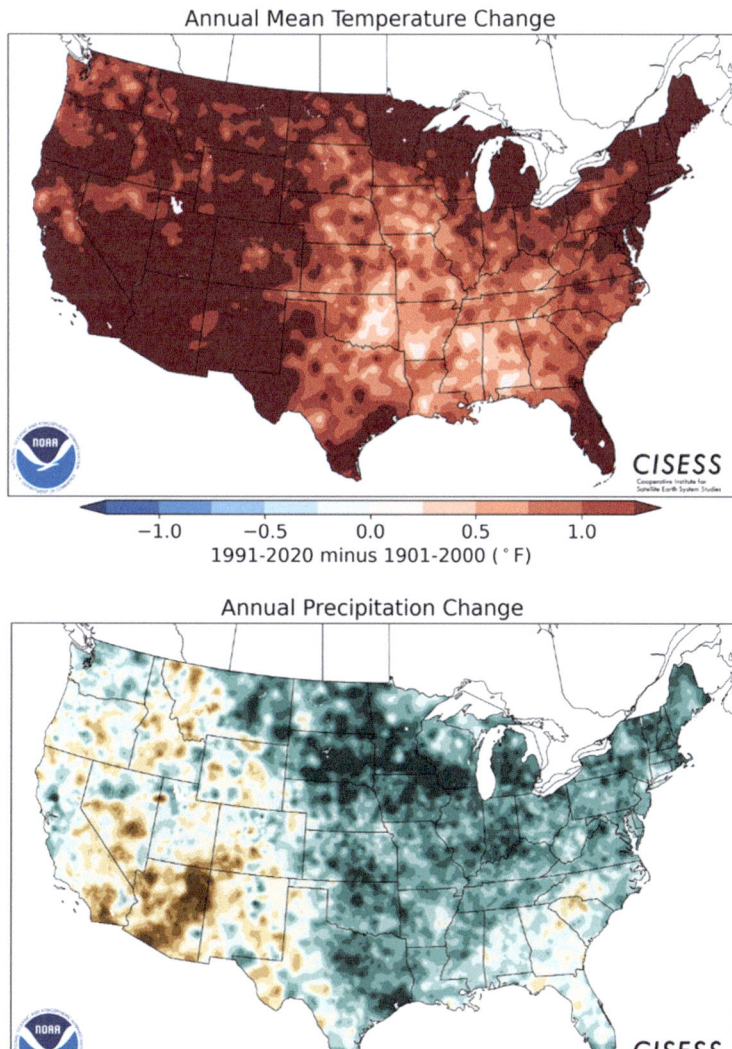

Fig. 4 Upper: Change (in degrees Fahrenheit) of average annual temperature for the 1991–2020 period relative to the twentieth century. Lower: Percent change in annual precipitation for 1991–2020 compared to the twentieth century (1901–2000) baseline. From NOAA[26]

---

[26] NOAA (2021). *NOAA Releases Updated Climate Normals.* NOAA National Weather Service. https://www.weather.gov/psr/19912020Normals, and NOAA (2021). *U.S. Climate Normals.* NOAA National Centers for Environmental Information. https://www.ncei.noaa.gov/products/land-based-station/us-climate-normals.

Box 1 Ethics for disasters and climate change

By Dónal O'Mathúna, College of Nursing & Center for Bioethics, The Ohio State University.

Ethics is the study of right and wrong, good and bad, what we should do and the sort of people we should be or become. Science tells us how the world is and how it works, and the social sciences describe the way people and communities behave. We enter the realm of ethics when we say that people should do certain things in response to climate change, or that society has ethical duties to other people or the Earth, or that people's rights place ethical responsibilities on others. The scientific information in this book helps us understand the way our world is and what is happening. You may find it interesting; it may satisfy your curiosity, but it should also lead you to think about ethics: what should you do in response to learning this information?

The ethics of climate change have global dimensions and long timeframes, which will be addressed in Chap. 12. Climate change also contributes to events with shorter timeframes and more localized impacts: disasters. These are events that lead to large-scale damage, destruction, disease and injuries, and death. They typically overwhelm a region's capacity to respond to the harms, losses, and suffering. This often leads to requests for external assistance (Fig. 5), but ethical challenges can arise when different cultures are suddenly thrown together after disasters.

Fig. 5 Responses to disasters raise challenging ethical issues. European Master Disaster Medicine (EMDM) disaster drill, Novara, Italy. Photo by Dónal O'Mathúna

Hurricanes, heat waves, floods, and other disasters have existed since long before climate change was in the news. Geoscience research shows that such disasters are increasing in frequency, intensity, and duration. Disasters touch a deep ethical sense within us that those with the resources should help those who are suffering. This empathy in the face of suffering is a prime motivating force for humanitarian aid. The devastation of disasters often leads to an out-pouring of help, much of it welcome. But, as the impact of local and international humanitarian aid has started to be evaluated, data shows that sometimes the help provided does little good and might even cause harm. Good intentions are not enough: humanitarian responses need to be evidence-based.

Disaster bioethics is a relatively new discipline seeking to understand and address the ethical issues that arise regarding disasters and humanitarian crises. Even before disasters hit, an ethical responsibility exists to plan and prepare. Communities and organizations should examine the disaster risks they face and put mitigation plans into place. These will vary with locale and point to the importance of regular risk assessment. For example, such planning could have shown that locating hospital back-up generators below flood levels would lead to the problems experienced during Hurricane Katrina and elsewhere. Needs assessments should be carried out so that risk reduction and response plans are in place and people know how to react if a disaster occurs.

When a disaster hits, the help provided should be that which actually benefits people. Case studies have documented medicines being sent to disasters where they were not needed, warm clothes being sent to hot climates, and other poorly informed responses. Key to avoiding such well-intentioned blunders is responders listening to local communities as they explain what they need and when. Sleeping mats and mosquito nets may be more important than what we think of from far away. Liaising with local authorities and organizations is essential for effective responses, rather than flying in as poorly informed saviors.

Some of the most challenging ethical issues in disasters involve decisions to allocate scarce resources ethically. The COVID-19 pandemic has shown that sometimes the public good must be prioritized over individual preferences. Such decisions are very difficult and should be made with great care and careful deliberation. They range from decisions about public shut-downs and lock-ups, what personal freedoms to limit, and who receives scarce resources, whether those be life-saving medical treatments, the little food and clean water available, or the limited funds to rebuild homes and lives.

Permeating all these decisions should be concern for respect, dignity and justice. Box 2 addresses this dimension of ethics more specifically. As disaster plans are prepared, steps should be taken to ensure that particular groups are not left out of consideration or discriminated against. Prejudice and bias should not be allowed to influence what gets sent or who receives it. These are deep and difficult issues for every community. Active steps should be taken to ensure that limited resources are distributed fairly according to need, not partiality or favoritism. The global distribution of COVID-19 vaccines was an ethical failure in this regard. Disasters have a way of exposing social injustices and bringing prejudices to the surface. These can pile further problems onto a disaster; or they can be grasped as an opportunity to address underlying ethical problems and work toward become people and communities who are more ethical and more caring toward our world and everyone living here.

Box 2 Climate change as a social justice problem

By Melissa Burt, Department of Atmospheric Science, College of Engineering, Colorado State University.

Climate change is complex and is oftentimes discussed as a scientific problem without the connection to or acknowledgment of the social implications. However, climate change is inherently a social issue, and the impacts are not created or distributed equally. The effects of climate change are mediated through social, cultural, and economic structures and processes, and these issues are explored in this essay through a lens of gender and race.

Climate justice acknowledges that climate change can have differing social, economic, public health, and other adverse impacts on population.[27] As the climate continues to change, more and more people across the world will face greater challenges in terms of weather-related disasters and extremes, vulnerabilities, and insecurities due to changing ecosystems, human health, and food security. Gender, racial, and other social inequities compound and exasperate vulnerability to climate change[28] all of which depend on an individual's position in power and privilege structures based on these categorizations.[29]

Given the urgent nature of the climate issue, gender impacts the ways that people experience and relate to climate change. Women and men can perceive and experience climate change in various ways due to socially constructed gender roles and responsibilities, status, and their intersectional identities.[30] Due to traditional (and historical) gendered roles of being a caretaker and provider, women across the world will continue to experience significant impacts and consequences due to climate change. As an example, women in Senegal are responsible for collecting water for the household. Rainfall in Senegal has declined significantly making it harder for women to collect water, especially in areas where there are no wells or connections to a water distribution network. Due to this, women resort to walking longer distances to find potable water because of clean water shortages.[31] Many women are also responsible for providing food for their households through crop production and selling of crops such as pictured in Malawi (Fig. 6). Increased drought conditions have led to soil that is infertile, and flooding in other areas has resulted in damaged crops and declines in agriculture production.[32]

---

[27] Simmons (2020).
[28] Pörtner et al. (2022).
[29] Kaijser & Kronsell (2014).
[30] Same as above.
[31] Oxfam (2018). *Climate Change and Women Fact Sheet*. Oxfam. https://www.oxfamamerica.org/explore/research-publications/climate-change-and-women-fact-sheet/.
[32] Same as above.

Fig. 6 Malawi women work to support their family by selling produce and other crops, which can be impacted by climate change. Photo by Dónal O'Mathúna

It is also important to note these differences (specifically gender roles) between women and men are socially constructed and context-specific and may shift in the reality of climate change. It is equally important to note that it excludes those who do not fit in these binary categories and excludes the complexity and fluidity of individual identities.[33]

The ways that people experience and relate to climate change are also influenced by social inequities and inequalities that stem from strong-rooted structural challenges such as systemic racism. Consider the devastation that took place in 2005 in the aftermath of Hurricane Katrina in New Orleans, Louisiana, on the Gulf Coast of the United States. The engineered levee system was designed as a flood protection storm surge system for New Orleans to protect the city during storms.[34] During Katrina, the levee system failed and resulted in vastly differing consequences for various groups of people, even though they were in a range of a few miles.[35] This failure affected residents who were poor, mostly black, and some with disabilities more than any other

---

[33] Alaimo (2009).

[34] Lucena & Leydens (2017).

[35] Same as above.

social groups. Much of which was due to these residents living in the lowest-lying floodplain areas. The more affluent areas were able to rebuild and return to life as usual within a matter of weeks to months. But the destruction that disproportionately affected disadvantaged and underserved communities took away property, access to resources, and exposed them to more harm and risks. Overall people were less likely to be able to evacuate and to afford to live somewhere else and had poorer prospects if displaced. This is an example of an intersectional justice issue connected to weather, climate, and engineering.

A clear finding from the IPCC is that vulnerable communities who have historically contributed the least to current climate change are disproportionately affected by climate change.[36] For example, between 2010 and 2020, human mortality from floods, droughts and storms was 15 times higher in highly vulnerable regions, compared to regions with very low vulnerability.[37] The effects can be devastating and widespread. For example, in 2022, Pakistan, a developing nation in south Asia that emits less than 1% of the world's greenhouse gases, experienced especially heavy and prolonged monsoon rains. As a result, one-third of the country was flooded, affecting 33 million people, including the displacement of nearly 8 million people and 1700 fatalities; the regions hardest hit included many of the poorest regions in the country.[38]

In other regions of the world, indigenous peoples, who depend heavily on local resources and live in parts of the world where the climate is changing quickly, are generally at greater risk of economic losses and poor health. Studies of the Inuit people, for example, show that rapid warming of the Canadian Arctic is jeopardizing hunting and many other day-to-day activities, with implications for livelihoods and well-being.[39] This is another example of how impacts of climate change interact with social structures.

---

[36] From Headline Statements in the reporting on the Synthesis Report of the IPCC Sixth Assessment Report (AR6), https://www.ipcc.ch/report/ar6/syr/resources/spm-headline-statements. In Lee, H., and Romero, J. (Eds.) (in press). *Climate Change 2023: Synthesis Report. A Report of the Intergovernmental Panel on Climate Change. Contribution of Working Groups I, II and III to the Sixth Assessment Report of the Intergovernmental Panel on Climate Change*. IPCC.

[37] Hersher (2023).

[38] The Government of Pakistan (2022). *Pakistan floods 2022: Post-Disaster Needs Assessment*. The Government of Pakistan (72 p.). https://thedocs.worldbank.org/en/doc/4a0114eb7d1cecbbbf2f65c5ce0789db-0310012022/original/Pakistan-Floods-2022-PDNA-Main-Report.pdf.

[39] Ford (2009).

> Climate change is expected to increase exposure to environmental health risk factors, including extreme temperatures, air pollution, and air-borne allergens.[40,41] Black and LatinX communities have been reported to be more vulnerable to heat-related deaths than other racial groups in the United States.[42] Indicators have found that these racial groups have access to lower-quality health care,[43] a higher prevalence of chronic health conditions (e.g., obesity and diabetes),[44] outdoor employment in agriculture and construction,[45] or other socio-economic circumstances.[46] Black and Hispanic or LatinX communities in the United States are exposed to far more air pollution than they produce through actions like driving and using electricity as compared to white Americans who experience better air quality than the national average, even though their activities are the source of most pollutants.[47] This further exemplifies the point that climate change and environmental concerns continue to have disproportionate impacts on historically marginalized communities.
>
> Acknowledging that these issues exist builds awareness as an important first step, but it cannot be the last step if we are to right these wrongs. Achieving climate justice will require commitment to ethical responsibilities as described in Box 1 and later explored in Chap. 12. It also means that we must work across disciplines so these social factors are elevated in all decisions that are made as it pertains to climate change. Policymakers, government organizations, and the academic community need to pay closer attention to the nature of climate change adaptation and impacts to ensure that women and people from historically marginalized backgrounds are centered in the research, policy, and action to ensure that these disproportionate risks are avoided. This will lead to justice.

Of course, the impacts of global warming in the future depend on how extreme the temperature rise ultimately is, and on the vulnerability and adaptive readiness of individuals, communities, and nations. For this reason, the member countries of the United Nations (U.N.) reached a landmark agreement in 2015 to address climate change by increasing actions and investments needed to mitigate risk and better adapt.[48] The primary goal of the U.N. Paris Climate Agreement is to hold the global average rise in temperatures to below 2 °C compared to pre-industrial levels, and

---

[40] Fiore et al. (2015).

[41] Kinney (2018).

[42] Colon-Rivera & Plata (2021).

[43] Pörtner et al. (2022).

[44] EPA, 2021. *Climate Change and Social Vulnerability in the United States: A Focus on Six Impacts.* U.S. Environmental Protection Agency, 430-R-21-003.

[45] Colon-Rivera & Plata (2021).

[46] Ostro et al. (2011).

[47] Tessum et al. (2021).

[48] The United Nations Climate Change (accessed 2023, May 22). The Paris Agreement. United Nations. https://unfccc.int/process-and-meetings/the-paris-agreement/what-is-the-paris-agreement

preferably below 1.5 °C, by the end of this century. Two hundred countries pledged limits on carbon emissions to meet this aspirational goal.

Let's take a moment to think about limiting future global average temperature rise to only 1.5–2 °C above pre-industrial levels. These numbers may not seem that consequential, and not even that different from each other (for example, if you were rounding to whole numbers 1.5 °C would round up to 2 °C). However, even if the global average temperature rise were to be kept within the U.N. Paris Climate Agreement targets the expected consequences are serious. A 2018 IPCC report[49] evaluating the expected impacts on people and the environment of global average temperatures rising 1.5 °C compared to 2.0 °C helps illustrate this point. They found that an additional rise of 0.5 °C (the difference between 1.5 and 2.0 °C) makes a big difference. To illustrate this difference we will look at three factors assessed by the IPCC and highlighted in The New York Times:[50] the potentials for heat waves, freshwater scarcity, and sea level rise. Extreme heat on land will be much more common worldwide under 2 °C of warming compared to 1.5 °C, with greatest increases in the tropics. At 1.5 °C, 14% of the world population would be experiencing periods of severe heat waves at least once every five years, but at 2 °C such heat waves are projected to affect 37% of the world's population. At 1.5 °C, more than 350 million people worldwide living in urban settings would be expected to be exposed to severe drought; at 2 °C, this number increases to more than 411 million people. Increased risk of drought is particularly likely for the Mediterranean region, southwestern United States, and southern Africa. Lastly, the world population exposed to flooding from sea level rise in 2100 (assuming no adaptation measures are taken) is expected to be between 31 and 69 million if mean temperatures rise by 1.5 °C, and between 32 and 80 million if mean temperatures rise by 2 °C. Low lying coastal regions, deltas, and small island nations are particularly vulnerable to sea level rise and other climate change impacts.

---

[49] IPCC, 2018: Summary for Policymakers. In Masson-Delmotte, V., Zhai, P. Pörtner, H.-O., Roberts, D. Skea, J., Shukla, P.R., Pirani, A., Moufouma-Okia, W., Péan, C., Pidcock, R., Connors, S., Matthews, J.B.R., Chen, Y., Zhou, X., Gomis, M.I., Lonnoy, E., Maycock, T., Tignor, M., & Waterfield, T., (Eds.) (2018). *Global Warming of 1.5 °C. An IPCC Special Report on the impacts of global warming of 1.5 °C above pre-industrial levels and related global greenhouse gas emission pathways, in the context of strengthening the global response to the threat of climate change, sustainable development, and efforts to eradicate poverty* (pp.3-24). Cambridge University Press. 10.1017/9781009157940.001. https://www.ipcc.ch/sr15/chapter/spm/ and Hoegh-Guldberg, O., Jacob, D., Taylor, M., Bindi, M., Brown, S., Camilloni, I., Diedhiou, A., Djalante, R., Ebi, K.L., Engelbrecht, F., Guiot, J., Hijioka, Y., Mehrotra, S., Payne, A., Seneviratne, S.I., Thomas, A., Warren, R., & Zhou, G. (2018). Impacts of 1.5°C Global Warming on Natural and Human Systems. In Masson-Delmotte, V., Zhai, P. Pörtner, H.-O., Roberts, D. Skea, J., Shukla, P.R., Pirani, A., Moufouma-Okia, W., Péan, C., Pidcock, R., Connors, S., Matthews, J.B.R., Chen, Y., Zhou, X., Gomis, M.I., Lonnoy, E., Maycock, T., Tignor, M., & Waterfield, T., (Eds.) (2018). *Global Warming of 1.5 °C. An IPCC Special Report on the impacts of global warming of 1.5 °C above pre-industrial levels and related global greenhouse gas emission pathways, in the context of strengthening the global response to the threat of climate change, sustainable development, and efforts to eradicate poverty* (pp.175-312). Cambridge University Press. 10.1017/9781009157940.005. https://www.ipcc.ch/sr15/chapter/chapter-3/.

[50] Plumer & Popovich (2018).

Clearly, it is desirable to work toward the lower end of the U.N. Paris Climate Agreement goal in order to minimize the negative effects of climate change. However, is the world on that path? Unfortunately, the answer is 'not yet.' The original pledges were voluntary and non-binding, and progress toward these goals has fallen far short due to socio-economic struggles that result in implementation challenges, as well as changes in political will that affect national priorities. However, recent assessments indicate that even if all the pledges were met, it is not enough to meet the 1.5–2 °C cap on global average temperature rise by the end of the century.[51] Instead, IPCC climate models project that global average surface temperatures are expected to rise 1.5 °C (2.7 °F) above pre-industrial levels within two decades, and much higher by the end of this century[52] (Fig. 7; you can see this in a longer time context in Fig. 2). The different future warming pathways in Fig. 7 reflect different greenhouse gas emissions scenarios, which in turn depend on how policies and practices change (or not) now and in the near future. The pathways that lead to minimal temperature rise and align best to the goal (but more aggressive than the actual pledges) of the U.N. Paris Climate Agreement would require net-zero greenhouse gas emissions by mid-to-late century as well as active carbon removal (a climate intervention effort described in Chap. 11). At the other end of the spectrum, we see that the pathway of maximum temperature rise is expected if there are high greenhouse gas emissions and no mitigation efforts. When the Sixth Assessment Report (AR6) from the IPCC was released in 2021, the pathway that the world was on was probably closest to the high emission scenario in Fig. 7, with temperature rising ~ 4 °C (7.2 °F), which is double the U.N. Paris Climate Agreement goal. Since that time substantial changes in U.S. climate and energy policy (e.g., the Inflation Reduction Act passed by the U.S. Congress in 2022) helped place the U.S. on a more desirable pathway, although the long-term viability of that trajectory is in question due to changing political priorities.[53]

---

[51] Dennis, B. & Mooney C. (2018, February 19). Countries made only modest climate-change promises in Paris. They're falling short anyway. *The Washington Post*. https://www.washingtonpost.com/national/health-science/its-not-fast-enough-its-not-big-enough-theres-not-enough-action/2018/02/19/5cf0a7d4-015a-11e8-9d31-d72cf78dbeee_story.html and Duncombe, J. (2021, August 9). What Five Graphs from the U.N. Climate Report Reveal about our Path to Halting Climate Change. *EOS*. https://eos.org/articles/what-five-graphs-from-the-u-n-climate-report-reveal-about-our-path-to-halting-climate-change.

[52] IPCC, (2021). Summary for Policymakers. In Masson-Delmotte, V., Zhai, P., Pirani, A., Connors, S.L., Péan, C., Berger, S., Caud, N., Chen, Y., Goldfarb, L., Gomis, M.I., Huang, M., Leitzell, K., Lonnoy, E., Matthews, J.B.R., Maycock, T.K., Waterfield, T., Yelekçi, O., Yu, R., & Zhou, B. (Eds.). Climate Change 2021: The Physical Science Basis. Contribution of Working Group I to the Sixth Assessment Report of the Intergovernmental Panel on Climate Change (pp.3-32). Cambridge University Press. 10.1017/9781009157896.001. https://www.ipcc.ch/report/ar6/wg1/downloads/report/IPCC_AR6_WGI_SPM.pdf; and IPCC (in press). Climate Change 2023: Synthesis Report, Summary for Policy Makers. In Lee, H., and Romero, J. (Eds.). *Climate Change 2023: Synthesis Report. A Report of the Intergovernmental Panel on Climate Change. Contribution of Working Groups I, II and III to the Sixth Assessment Report of the Intergovernmental Panel on Climate Change*. IPCC. https://www.ipcc.ch/report/ar6/syr/.

[53] See an independent assessment of the potential climate and energy impact of the Inflation Reduction Act at https://repeatproject.org/.

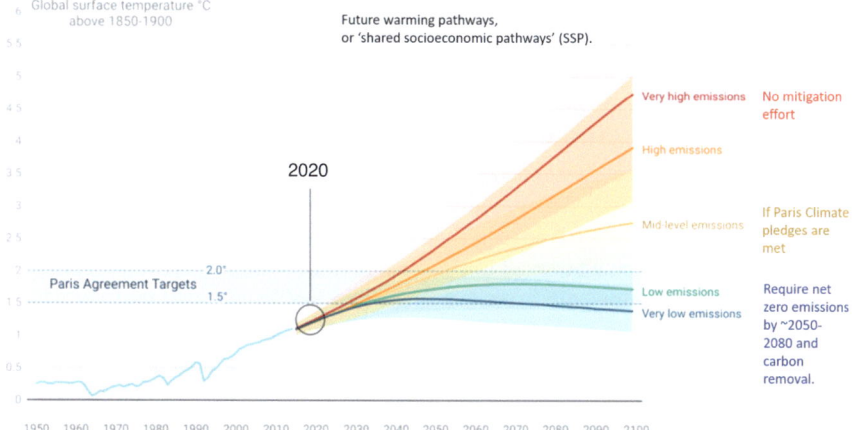

Fig. 7 Historical and projected changes in global average surface temperature above the 1850–1900 average. The global average temperature at the end of the century will be determined by the amount of greenhouse gas emissions over the next several decades. Five different future warming pathways are shown: very low emissions (SSP1-1.9), low emissions (SSP1-2.6), mid-level emissions (SSP2-4.5), high emissions (SSP3-7.0), and very high emissions (SSP5-8.5).[54] Shading shows the 5% and 95% confidence intervals. The two pathways that stay below 2 °C (very low emissions and low emissions) require net-zero emissions by mid- to late century and carbon removal. Annotations on far right provide context based on global policies and practices as of September 2021. The 2015 Paris Climate Agreement goal to hold the rise in temperature to below 2 °C (compared to pre-industrial levels), preferably below 1.5 °C, by the end of the century is indicated with gray dotted lines. Figure modified from Duncombe (2021)[55] based on data from IPCC (2021).[56] Reprinted with permission of Eos and the American Geophysical Union. This figure is excluded from Creative Commons license

---

[54] The five illustrative scenarios in Fig. 7 are referred to as SSPx-y, where 'SSPx' refers to the Shared Socio-economic Pathway or 'SSP' describing the socio-economic trends underlying the scenario, and 'y' refers to the approximate level of radiative forcing (in watts per square meter, or W m^{-2}) resulting from the scenario in the year 2100. SSP definition from IPCC (2021). Summary for Policymakers. In Masson-Delmotte, V., Zhai, P., Pirani, A., Connors, S.L., Péan, C., Berger, S., Caud, N., Chen, Y., Goldfarb, L., Gomis, M.I., Huang, M., Leitzell, K., Lonnoy, E., Matthews, J.B.R., Maycock, T.K., Waterfield, T., Yelekçi, O., Yu, R., & Zhou B. (2021). *Climate Change 2021: The Physical Science Basis. Contribution of Working Group I to the Sixth Assessment Report of the Intergovernmental Panel on Climate Change* (pp. 3-32). Cambridge University Press. 10.1017/9781009157896.001. https://www.ipcc.ch/report/ar6/wg1/downloads/report/IPCC_AR6_WGI_SPM.pdf.

[55] Duncombe, J. (2021, August 9), What five graphs from the U.N. climate report reveal about our path to halting climate change, *Eos* 102, https://doi.org/10.1029/2021EO161811. Reprinted with permission of Eos and the American Geophysical Union.

[56] IPCC (2021). Summary for Policymakers. In Masson-Delmotte, V., Zhai, P., Pirani, A., Connors, S.L., Péan, C., Berger, S., Caud, N., Chen, Y., Goldfarb, L., Gomis, M.I., Huang, M., Leitzell, K., Lonnoy, E., Matthews, J.B.R., Maycock, T.K., Waterfield, T., Yelekçi, O., Yu, R., & Zhou B. (2021). *Climate Change 2021: The Physical Science Basis. Contribution of Working Group I to the Sixth Assessment Report of the Intergovernmental Panel on Climate Change* (pp. 3-32).

While graphs showing global average temperature rise are sobering, it is important to also recognize that warming, and the impacts of it, have not been—and will not be—globally uniform. We saw regional variability within the U.S. in the temperature and precipitation data in Fig. 4 and can see regional variability at a global scale by examining projected future changes in temperature and precipitation in Fig. 8. Nearly everywhere, land is expected to undergo greater warming than the surface ocean at the same latitude. Additionally, the Arctic region is expected to experience two to four times[57] greater warming than the global average. (Why such regional differences exist will be addressed in Chaps. 2 and 3.) With respect to precipitation changes, greater extremes are expected: areas that are usually wet will get wetter, and areas that are usually dry will get drier. Polar latitudes, the equatorial Pacific, and regions affected by monsoons are expected to experience an increase in precipitation; whereas, parts of the subtropics and tropics are expected to experience increased drought.

The seriousness of global warming has been recognized by many industries, governments, communities, and individuals (maybe you). For example, the reinsurance industry (i.e., insurance companies for insurance companies) pays a great deal of attention to climate change models because a lot of money is at stake.[58] Climate-related heat waves, droughts, and coastal flooding all have financial costs. The connection between climate change and economic stability is not lost on private corporations and governmental agencies. For example, in 2021 the U.S. Department of the Treasury's Financial Stability Oversight Council, which is tasked with identifying and addressing vulnerabilities in the U.S. financial system, concluded that climate change is an emerging and increasing threat to financial stability.[59] There are also security risks of climate change. In fact, the U.S. Department of Defense labels climate change as a '**threat multiplier**' because consequences such as extreme weather and water shortages increase the risk of civil unrest, political

---

Cambridge University Press. 10.1017/9781009157896.001. https://www.ipcc.ch/report/ar6/wg1/downloads/report/IPCC_AR6_WGI_SPM.pdf.

[57] Two recent studies point to the Arctic warming four times greater than global averages. These are: Chhlek, P., Folland, C., Klett, J.D., Wang, M., Hengartner, N., Lesins, G., & Fubey, M.K. (2022), Annual Mean Arctic Amplification 1970-2020: Observed and Simulated by CMIP6 Climate Models, *Geophysical Research Letters, 49*(13) https://doi.org/10.1029/2022GL099371, and Rantanmen, M., Karpechko, A.-Y., Lipponen, A., Nordling, K., Hyvärinen, O., Ruosteenoja, K., Vihma, T., & Laaksonen, A. (2022). The Arctic has Warmed Nearly Four Times Faster than the Globe Since 1979, *Communications Earth and Environment, 3*(168), https://doi.org/10.1038/s43247-022-00498-3.

[58] Koerth-Baker, M., (2013, August 27). Mutually Insured Destruction. *The New York Times Magazine*. https://www.nytimes.com/2013/09/01/magazine/mutually-insured-destruction.html?ref=magazine and Rajbhandari, A. (2022, October 23). Climate Change Tops Insurers' Worries With War Close Behind. *Bloomburg*. https://www.bloomberg.com/news/articles/2022-10-23/climate-and-war-risks-top-insurer-s-worries#xj4y7vzkg.

[59] FSOC (2021). Financial Stability Oversight Council, Report on Climate-Related Financial Risk, 2021. US Department of the Treasury. https://home.treasury.gov/system/files/261/FSOC-Climate-Report.pdf. US Department of the Treasury press release (2021, October 21) on this report: https://home.treasury.gov/news/press-releases/jy0426.

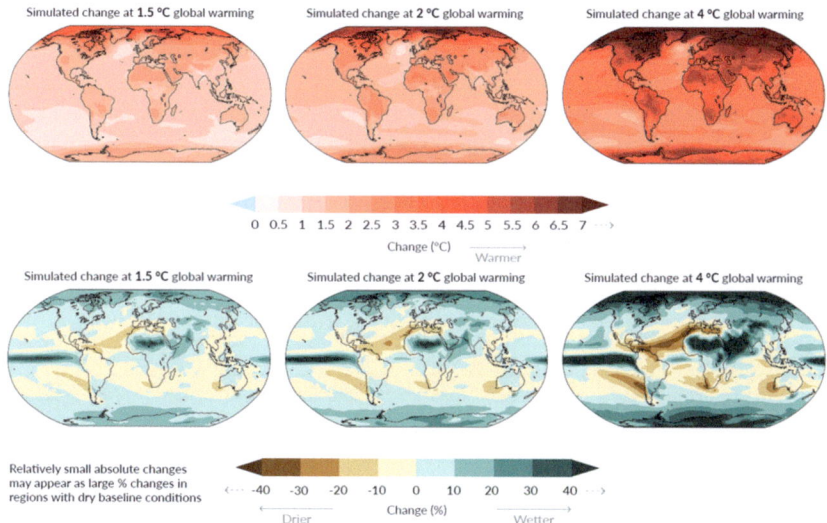

Fig. 8 Upper panel: Simulated annual mean temperature change (°C). Lower panel: Simulated annual mean precipitation change (%). These maps show changes in mean temperature (°C) and precipitation (%) under different global warming scenarios (1.5 °C, 2 °C, and 4 °C). Changes are relative to average temperature and precipitation in 1850–1900. Figure used with permission from the IPCC. Full IPCC caption for this figure can be found in the footnote. From SPM.5 (Panels a and b) in IPCC, 2021.[60] This figure is excluded from Creative Commons license

instability, terrorism, and war.[61] The recognition that climate change poses security

---

[60] Source: SPM.5 (Panels a and b) from IPCC (2021). Summary for Policymakers. In Masson-Delmotte, V., Zhai, P., Pirani, A., Connors, S.L., Péan, C., Berger, S., Caud, N., Chen, Y., Goldfarb, L., Gomis, M.I., Huang, M., Leitzell, K., Lonnoy, E., Matthews, J.B.R., Maycock, T.K., Waterfield, T., Yelekçi, O., Yu, R., & Zhou B. (2021). *Climate Change 2021: The Physical Science Basis. Contribution of Working Group I to the Sixth Assessment Report of the Intergovernmental Panel on Climate Change* (pp. 3-32). Cambridge University Press. 10.1017/9781009157896.001.
https://www.ipcc.ch/report/ar6/wg1/downloads/report/IPCC_AR6_WGI_SPM.pdf. Original caption from IPCC (2021): Figure SPM.5 (b) Simulated annual mean temperature change (°C), panel (c) precipitation change (%) at global warming levels of 1.5 °C, 2 °C and 4 °C (20-year mean global surface temperature change relative to 1850–1900). Simulated changes correspond to Coupled Model Intercomparison Project Phase 6 (CMIP6) multi-model mean change (median change for soil moisture) at the corresponding global warming level. In panel (c), high positive percentage changes in dry regions may correspond to small absolute changes.

[61] YCC Team (2019, June 20). Why climate change is a 'threat' multiplier'. Yale Climate Connections. https://yaleclimateconnections.org/2019/06/why-climate-change-is-a-threat-multiplier/ and Goodman, S. (2021, October, 26). The Pentagon has to include climate risk in all of its plans and budgets. *Defense News.* https://www.defensenews.com/opinion/commentary/2021/10/26/the-pentagon-has-to-include-climate-risk-in-all-of-its-plans-and-budgets/ and Department of Defense, Office of the Undersecretary for Policy (Strategy, Plans, and Capabilities) (2021). Department of Defense Climate Risk Analysis. Report Submitted to National Security Council.

risks was further examined in a 2021 multi-agency U.S. federal report;[62] it was the first time that the U.S. security agencies collectively communicated the climate risks they face. The report projects increased geopolitical tensions over needed climate responses including the demand for financing and technological assistance, highlights the risk of increasing geopolitical flashpoints such as cross-border migration due to climate impacts, and projects a greater demand for aid and humanitarian relief in the next two decades (Table 2).

Table 2 Climate change and international responses that increase risks to US interests through 2040. From: US National Intelligence Council report NIC-NIE-2021–10,030-A

Risks		2021	2030	2040
Geopolitical tensions over climate responses	Perception of insufficient contributions to reduce emissions	L	M	H
	CO_2 removal not at scale for countries' net-zero pledges	–	L	M
	Developing country demands for financing and technological assistance	M	H	H
	Petro-states resisting clean energy transition away from fossil fuels	L	M	H
	Competition with China over key minerals and clean energy technologies	L	M	H
	Contention over use of economic tools to advance climate interests	–	L	M
Climate exacerbated geopolitical flashpoints	Miscalculation over strategic competition in the Arctic leading to conflict	L	M	H
	Cross-border water tension and conflict	L	M	H
	Cross-border migration attributed to climate impacts	M	H	H
	Ungoverned unilateral geoengineering	–	L	M
Climate effects impacting country–level stability	Strain on energy and food systems	L	M	H
	Negative health consequences	L	M	M
	Internal insecurity and conflict	L	L	M
	Greater demand for aid and humanitarian relief	M	H	H
	Strain on military readiness	–	L	M

[63] Risk is rated as none (–), low (L), medium (M), high (H)

[62] Climate Change and International Responses Increasing Challenges to US National Security Through 2040 (NIC-NIE-2021-10030-A). https://www.dni.gov/index.php/newsroom/reports-publications/reports-publications-2021/item/2253-national-intelligence-estimate-on-climate-change. The report was put forth by the U.S. Departments of Homeland Security and Defense, the National Security Council, and the Director of National Intelligence.

[63] Same as above.

Ultimately, the seriousness of climate change outcomes depends on both Earth systems and human systems factors (Fig. 1). The amount and rate of warming, the geographic location, the level of economic development, the stability of governments, political will of leaders, and the cultural norms and values of (and inequalities in) society all play roles in the options that are available and the decisions that will be made about climate change. Therefore, climate change is not just a scientific problem, it is also a social-political-economic one. Furthermore, the public perception of climate change influences the likelihood of taking mitigating and/or adaptive actions. But what is the public perception of climate change? And how does it compare to the scientific perspective? These questions will be addressed in the next section.

4 What the Public Thinks About Climate Change

Compounding the problem and stalling possible solutions is a general disconnect between the public and scientific understandings of climate change. We can examine this by looking at results from Gallup polling data. Gallup is an independent analytics company based in the U.S. that has been conducting polls on a range of topics for over 80 years. Gallup polls of Americans' views on global warming began in 2001 and have continued since. Their polling results through 2022 are shown in Fig. 9. The data show that the adult Americans surveyed in 2022 were very confident they understand global warming well (82%), but were less sure that its effects have begun (60%), that human activity is the main cause of the warming in last century (65%), or whether global warming is something to worry about (65%). These data are consistent with public opinions since 2016, with more varied responses prior to that time. Similar results are observed in polling data from the Yale Program on Climate Change Communication.[64]

At the international level, a 2019 Gallup poll[65] (Fig. 10) of adults in 142 countries and territories shows that majorities in every region of the world perceive climate change as a threat to the people in their country. Education was the most significant factor that shaped attitudes toward climate change risk; people with a college education (i.e., 16 or more years of education) were more likely to say climate change is a serious risk than people who had fewer years of education. However, the study also

---

[64] Yale Program on Climate Change Communication (YPCCC) & George Mason University Center for Climate Change Communication (Mason 4C). (2022). Climate Change in the American Mind: National survey data on public opinion (2008-2022) [Data file and codebook]. 10.17605/OSF.IO/JW79P and Marlon, J., Ballew, M., Rosenthal, S., Maibach E., & Leiserowitz, A. (2022, December 15). *Explore Climate Change in the American Mind.* Yale Program on Climate Change Communication. https://climatecommunication.yale.edu/visualizations-data/americans-climate-views/.

[65] Gallup global survey data from Rzepa, A., & Ray, J. (2020). *World risk poll reveals global threat from climate change.* Gallup. https://news.gallup.com/opinion/gallup/321635/world-risk-poll-reveals-global-threat-climate-change.aspx.

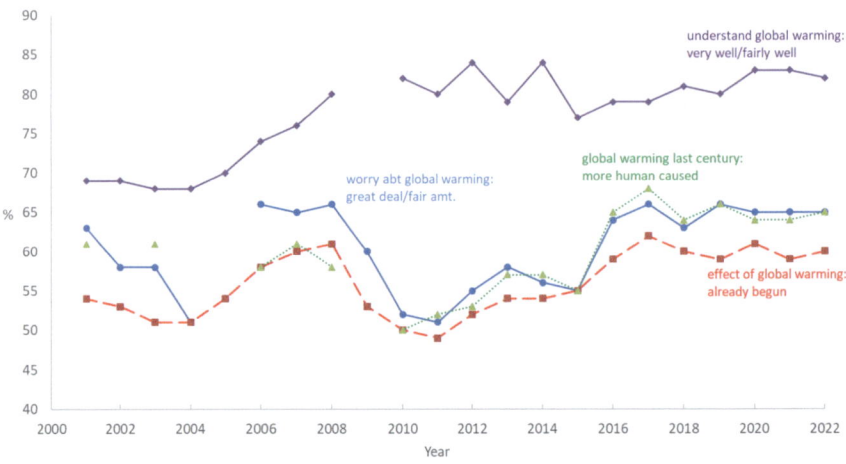

Fig. 9 U.S. public opinions on climate change based on data collected by Gallup[66]

found that people in countries that are top carbon emitters and high oil producers are more skeptical about the threat of climate change. In particular, in 2019 the U.S. was the second-largest carbon emitter and was among the top seven oil producing countries, yet according to the Gallup survey, the U.S. had the highest percentage of climate change skeptics among high-income countries (21% of people in the U.S. viewed climate change as not a threat at all). These results are consistent with a 2018 international Pew Research Center study.[67]

Taken together, the U.S. and international survey results are a mixed bag. It is encouraging that approximately two-thirds (69%) of the international populations surveyed recognize that climate change is a serious problem, and that a similar proportion of people in the U.S. see recent climate change as largely human-caused (65%). However, these public views are still far below the strong scientific consensus (97% in 2013 to > 99% in 2021) about these same issues.[68] Also concerning is

---

[66] Gallup U.S. survey data compiled from Jones, J. M. (2022). *Extreme weather has affected one in three Americans.* Gallup. https://news.gallup.com/poll/391508/extreme-weather-affected-one-three-americans.aspx and prior year survey reports.

[67] From a 2019 PEW Research Center survey: Fagan, M., & Huang, C. (2019, April, 18). *A look at how people around the world view climate change.* Pew Research Center. https://www.pewresearch.org/fact-tank/2019/04/18/a-look-at-how-people-around-the-world-view-climate-change/.

[68] Cook, J., Nuccitelli, D., Green, S.A., Richardson, M., Winkler, B., Painting, R., Way, R., Jacobs, P., & Skuce, A., (2013). Quantifying the consensus on anthropogenic global warming in the scientific literature. *Environmental Research Letters 8*(2) 024024, https://doi.org/10.1088/1748-9326/8/2/024024; and Cook, J., Oreskes, N., Doran, P.T., Anderegg, W.R.L., Verheggen, B., Maibach, E.W., Carlton, J.S., Lewandowsky, S., Skuce, A.G., Green, S.A., Nuccitelli, D., Jacobs, P., Richardson, M., Winkler, B., Painting, R., & Rice, K., (2016) Consensus on consensus: a synthesis of consensus estimates on human-caused global warming. *Environmental Research Letters 11*(4) 048002, https://doi.org/10.1088/1748-9326/11/4/048002; and Lynas, M., Houlton, B.Z., & Perry, S. (2021). Greater than 99% consensus on human caused climate change in the peer-reviewed scientific literature. *Environmental Research Letters 16*(11), 0.1088/1748-9326/ac2966.

Climate Change in Geoscience and Social Contexts

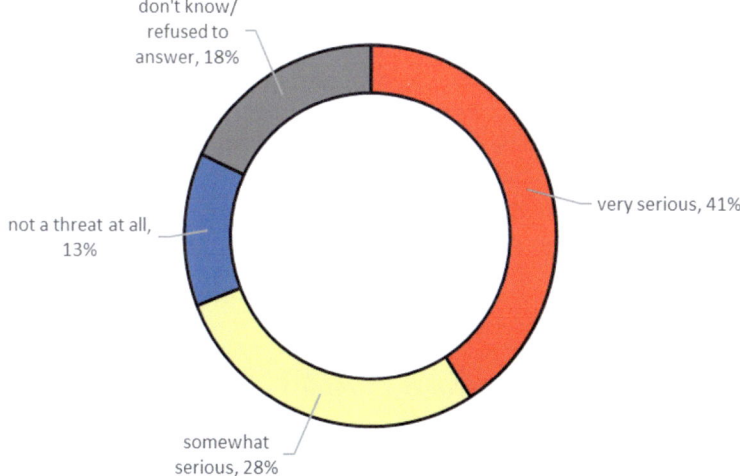

Fig. 10 Global perception of the threat of climate change. Survey participants were asked: Do you think that climate change is a very serious threat, a somewhat serious threat, or not a threat at all to the people in this country in the next 20 years? Based on data collected by Gallup, 2019[69]

the mismatch between the U.S. public's self-confidence in their understanding of global warming issues and the relatively lower level of concern (worry) about it. As laid out by arguments earlier in this chapter, there is much to be concerned about as temperatures rise and precipitation patterns shift (e.g., Boxes 1 and 2). Informed, evidence-based decisions about how to mitigate and adapt will depend on overcoming misconceptions, and on gaining a scientifically sound understanding of climate change and the intersectionality of Earth systems and human systems.

5 Why a Disconnect Exists Between the Public and Scientific Understandings of Climate Change

Bridging the gap between the public and scientific understandings of climate change is a primary goal of this introductory climate change textbook. However, as noted earlier in this chapter, achieving this goal can't depend solely on presenting the science of climate change. It also depends on becoming aware of social science factors that influence how people acquire and process information to make sense of the natural world and our roles in it. Therefore, this last section of Chap. 1 focuses on these topics by drawing on research from the fields of social science and educational

---

[69] Gallup global survey data from Rzepa, A., & Ray, J. (2020). *World risk poll reveals global threat from climate change*. Gallup. https://news.gallup.com/opinion/gallup/321635/world-risk-poll-reveals-global-threat-climate-change.aspx.

psychology. An analysis by psychologists Elke Weber and Paul Stern[70] provides useful framing for the discussion; they posit that there are physical, psychological, and social reasons for the disconnect between the public and scientific understandings of climate change. Physically, the climate system is complex which makes it intrinsically difficult to understand (this book will help break down that complexity). Psychologically, the potential for systemic misconception is greater when people rely on nonscientific approaches to understand climate change. Lastly, there are challenging social influences that frame and shape the way people view climate change.

5.1 Why the Climate System is Complex

One of the reasons climate change is a challenging topic to understand is that it is intrinsically complex. The climate system contains multiple interconnected parts (Fig. 11). It includes the atmosphere, ocean, ice, the solid Earth, and life (the biosphere). Processes within these system components control the global movement of energy, water, and carbon. Energy in the climate system is fueled primarily by the amount and distribution of solar radiation that reaches the Earth's surface, but also by the longer wavelength radiation emitted by Earth itself and interactions with gases in the atmosphere, as well as the reflectivity of Earth's surface and atmosphere. Natural changes to the climate system are caused by **external drivers** (also called **forcings**), such as changes in the Sun's strength, changes in Earth's orbit around the Sun, and the occasional occurrence of large explosive volcanic eruptions. Natural changes to the climate system are further influenced by **internal variabilities,** such as El Niño-Southern Oscillation (ENSO), and **internal interactions** (known as **feedbacks**) among internal climate system components. Human system interactions with the climate system, especially related to the emission of carbon-rich greenhouse gases, have become an additional driver of climate change. The relative importance of the many drivers and feedbacks that influence climate change depends on the temporal (time) and spatial (geographic) scales under consideration. For example, the drivers and feedbacks that are important for climate events occurring over a few months to a few years, or affecting an area of a few 100 km^2, are different from the important drivers and feedbacks for events occurring over millions to 10 s of millions of years or affecting the entire globe.

To illustrate the complexity and interconnectedness of the climate system let's look at an example that focuses on the Arctic. Given its high latitude setting, the Arctic receives less solar radiation over an entire year than the temperate (mid latitude) and tropical (low latitude) regions, which explains why very cold temperatures prevail for much of the year, snow and glaciers cover much of the Arctic landscape, and sea ice covers large parts of the Arctic Ocean. However, recall that the Arctic is experiencing two to four times greater warming than the global average (Fig. 8). Why

---

[70] Weber & Stern (2011).

Climate Change in Geoscience and Social Contexts 33

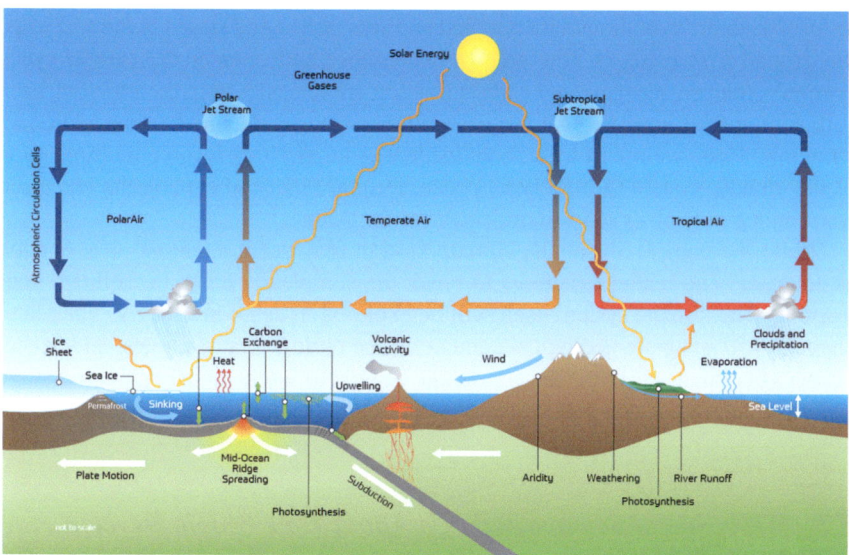

Fig. 11 Representation of the Earth's climate system, which includes five subsystems: the atmosphere, hydrosphere, cryosphere, biosphere, and geosphere (sometimes also called the lithosphere). The Earth's climate system also includes processes and interactions among these systems. The cycling of energy, water, and carbon are three processes that are particularly important in Earth's climate system. This representation of the Earth's climate system shows one hemisphere; the equator would be on the right margin of the diagram. Illustration by Geo Prose. From Koppers and Coggon (2020)[71] Figure used with permission from Geo Prose

is that? The answer lies in system feedbacks that amplify warming in this region. The increase in temperature causes more snow and ice to melt during the Arctic summer. As these bright white surfaces disappear, they expose the darker land and ocean surfaces beneath them. Darker surfaces absorb more heat energy from the Sun than do lighter surfaces (you may have experienced this yourself—think about how hot dark pavement can be on your, or your pets,' feet). Temperature will go up as more heat is absorbed. As temperature goes up, even more snow and ice will melt, which in turn causes temperature to go up even more—amplifying warming. In this example air temperature was affected by the changing presence of ice—a different part of the climate system. There is more to the story though that involves the ocean. As the thin lid of sea ice melts away from the Arctic Ocean surface (sea ice—unlike glacial ice on land—rarely is more than a few meters thick), heat energy stored in the ocean can now be transferred to the atmosphere, amplifying atmospheric warming even more. We can even add a third influence that is important in the Arctic: changes in permafrost (i.e., soil that's remained frozen for a long time, in some places frozen for hundreds of thousands of years). As air and land surface temperatures increase, permafrost destabilizes—it melts (i.e., defrosts). Organic carbon, produced by photosynthesis

---

[71] Koppers & Coggon (2020).

to form ancient vegetation that later died and was buried and stored in the soil when the ground first froze, is now able to decompose and re-enter the carbon cycle. Formerly trapped methane and carbon dioxide are released from the now-unfrozen soil, adding additional greenhouse gases to the atmosphere which cause additional warming. Thus, we see at least three feedbacks at play that involve multiple parts of the climate system (atmosphere, ocean, ice, life) that can explain the enhanced warming already observed and projected for the Arctic.

While the complexity of the climate system may seem daunting, it is our job as educators and textbook authors to break the complexity down to facilitate your learning. The climate system itself will be explored in detail in Chap. 2. Then later chapters will integrate geological context on the multiple causes, rates, and consequences of climate change. This is essential context to assess modern climate change and our role in it, as well as to forecast future climate conditions.

5.2 What Social Science Tells Us About How People Develop Their Understandings of the Natural World

Scientists and non-scientists often develop their understandings of the natural world in different ways. Ideally, scientists base scientific conclusions and recommendations on data, evidence-based reasoning, and ethical commitments. These approaches help reduce error and fraud. Non-scientific approaches—ones that tend to be driven more by personal experiences, feelings, and social influences than by systematic observations and measurements—can be more vulnerable to misconceptions. And as noted in Sect. 2.4, once incorrect mental models are formed, they are difficult to correct. The goal of this last section of Chap. 1 is to build awareness of the obstacles that can trip-up the public (and scientists) in trying to develop a scientifically sound understanding of the natural world. To set the stage, let's first examine the nature of scientific research.

5.2.1 How Science Works: The Nature of Scientific Research

When you were in grade school you likely were taught some version of 'the scientific method' that involved a set sequence of steps starting with asking a question, forming a hypothesis, conducting an experiment, collecting data, and then drawing conclusions. This very linear checklist approach oversimplifies the actual process of scientific investigation. While it is generally sufficient for describing a simple laboratory experiment (e.g., for a grade-school science fair), it is insufficient to describe scientific practices that involve observations of the natural world. A more realistic depiction (Fig. 12) of the processes used to conduct scientific research was put forth by a team of collaborators led by the University of California Museum of Paleontology of the University of California at Berkeley.[72] Asking questions and making

---

[72] Understanding Science 101. (Accessed 2023, May 22). *How science works.* University of California Museum of Paleontology. https://undsci.berkeley.edu/article/howscienceworks_01.

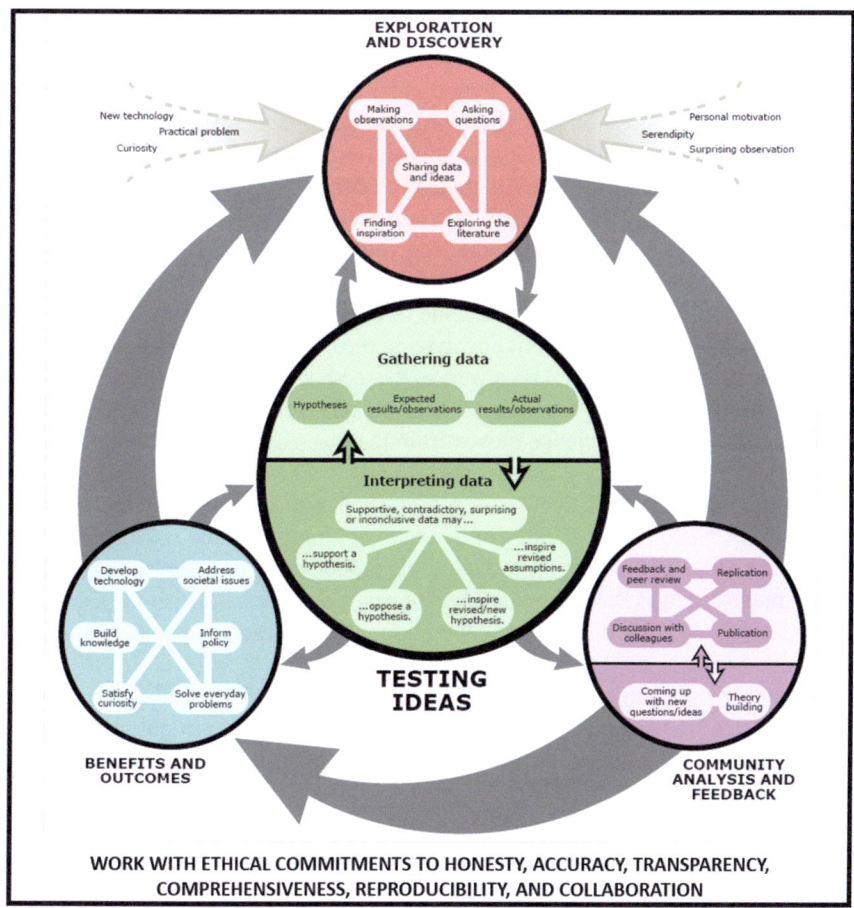

Fig. 12 A model of the scientific research process. Modified from: The University of California Museum of Paleontology, Understanding Science—how science really works. © UC Museum of Paleontology Understanding Science www.understandingscience.org.[73] This figure is excluded from Creative Commons license

evidence-based (data-based) conclusions are at the core of both the simple model of the scientific method from grade school, and the more realistic model that is used in practice. The more realistic model also shows how dynamic the scientific research process actually is. It's not a straight line from question to answer. It's one question leading to other questions leading to answers and more questions to reach a deeper understanding. Therefore, scientists engage in many different activities during the research process. They explore and make discoveries. They make

---

[73] Modified "Complex Science Flowchart." From How science works Complex Flow Handout. (Accessed 2022, August 12). Understanding Science 101. University of California Museum of Paleontology. https://undsci.berkeley.edu/lessons/pdfs/complex_flow_handout.pdf

systematic observations and measurements in the natural world and in laboratories. They use mathematical models that incorporate theories and observational data, while testing tentative explanations (i.e., hypotheses) against new data. They are trained to base their conclusions on analytical processing of large amounts of data, while also dealing with incomplete data sets and uncertainty. Scientists are also OK being wrong and revising their ideas based on new evidence.

From this more realistic model, we also see how collaborative the research process is. Rarely is science done by a single individual in isolation. Instead, teams of researchers, sometimes from the same location and sometimes from many places around the world, work together to tackle complex, puzzling scientific questions. Researchers who were not directly involved in the data collection and analysis also play important roles—as expert reviewers of the methods and of the interpretations of research results. Constructive feedback from the peer-review process is an essential quality control check as science moves from the investigation stage to the dissemination (publication) stage. Rigorous peer review is a key reason why publications in scientific journals and in synthesis reports (such as those conducted by the IPCC and the U.S. National Academies of Science) are highly trusted sources of scientific information. Importantly, underlying all sound scientific practices are ethical commitments, such as the value placed on honesty, accuracy, transparency, comprehensiveness, reproducibility, and collaboration which impact the trustworthiness of science.

Ultimately, outcomes of the scientific research process have the potential to address societal challenges (such as climate change) and inform policy—if those outcomes are effectively disseminated and valued by decision-makers and the public-at-large. Unfortunately, major hurdles exist outside the realm of science that need to be overcome in order to bridge the disconnect between the public and scientific understandings of climate change. Some of the most challenging hurdles are explored in the next section.

5.2.2 Why Rigorous Science Isn't Enough to Change Public Thinking About Climate Change

Social science research tells us that changing opinions and behaviors—about climate change or other challenging issues—requires learners to overcome serious obstacles. Three of these obstacles are groupthink, media bias, and the Internet bubble. Let's consider each of these in turn before closing-out Chap. 1. A fourth factor, personal values, is closely related to ethics and will be addressed in detail in Chap. 12.

In Sect. 4 of this chapter, we looked at data concerning public opinion about climate change. The U.S. data were presented in aggregate; in other words, the results were not broken down by any demographic groupings. Do different groups (people with different lived or shared identities) have different opinions about climate change? Most definitely. Perspectives differ widely by demographic group, including by age, education, race, gender, urban, and rural. But some of the biggest differences correspond to political party affiliation, as shown by results compiled from annual

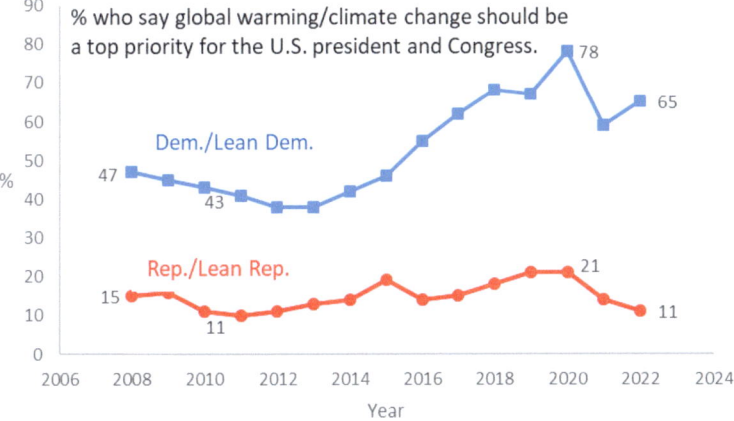

Fig. 13 U.S. survey data showing the percent of respondents who identify as Democrat/Lean Democrat and who identify as Republican/Lean Republican who said that addressing global warming/climate change should be a top priority for the president and Congress. The term 'global warming' was used 2008–2014. The term 'climate change' was used starting in 2015. Figure compiled from data collected and reported by the Pew Research Center (2008–2022)[77]

U.S. public surveys conducted by the Pew Research Center[74] (Fig. 13). People's views on climate change have become polarized along political party lines (as have many other complex issues). Why? One explanation is that people's views about climate change are heavily influenced by feelings of political or cultural group identity. As lawyer and psychologist Dan Kahan noted—'When people are asked for their views on climate change…they translate this into a broader question: whose side are you on?'.[75] This is a consequence of **groupthink**—when our individual opinions and rational thinking take back seats to the desire for maintaining a sense of belonging and cohesion of a group that we identify with. When this happens, the influence that group leaders have on framing opinions and mental models for the in-group members grows. According to psychologists Gordon Pennycook and David Rand, groupthink puts at risk our critical-thinking ability, as facts are rationalized to 'fit' the values of our group.[76] Thus, objective data are explained (interpreted) through the lens of a political or social group.

---

[74] Tyson, A., Funk, C., & Kennedy, B. (2023, April 18). *What the data say about Americans' views of climate change*. Pew Research Center. https://www.pewresearch.org/fact-tank/2022/04/22/for-earth-day-key-facts-about-americans-views-of-climate-change-and-renewable-energy/.

[75] The Economist. (2015, November 26). *Groupthink*. The Economist. https://www.economist.com/special-report/2015/11/26/groupthink. Another useful resource on this topic is: Guber, D.L., (2017, June 28). Partisan Cueing and Polarization in Public Opinion About Climate Change. Climate Science, Oxford University Encyclopedia of Climate Science. https://doi.org/10.1093/acrefore/9780190228620.013.306.

[76] Pennycook, G., & Rand, D. (2019, January 19). *Why Do People Fall for Fake News?* The New York Times. https://www.nytimes.com/2019/01/19/opinion/sunday/fake-news.html.

[77] Data compiled from Tyson, A., Funk, C., & Kennedy, B. (2023, April 18). *What the data say about Americans' views of climate change*. Pew Research Center.

While groupthink can influence individual opinions and behaviors in ways that are counter to a scientifically sound understanding of climate change, so can where and how people get information. Articles in peer-reviewed scientific journals are written for a particular expert audience; they are not written for general public consumption. Science communication specialists recognize that a gap exists between these primary sources of climate research information and the broader population. Certainly, dissemination outlets exist that bridge this communication gap. For example, each IPCC report include a 'summary for policy-makers'.[78] These are slimmed-down executive summaries in text and diagrams of the most important takeaway points of the much longer data-rich assessment report. They are specifically written for non-experts by science and science communication experts. There are also news media outlets that are reliable, accurate sources of climate information and analysis in forms accessible to the general public. However, being able to identify reliable sources of information can be particularly challenging when there are so many options to choose from. Let's look at this further.

The media world plays by different rules than the science world, and those rules also vary across the media landscape (e.g., 'mainstream' mass media, social media), so learning the rules of the game can help us be more discerning in choosing where we get our information on climate change (or any other topic) and evaluating its efficacy. Both the mass media news industry and social media seek to maximize audience share and conform to the interests and values of that media outlet's owners or advertisers. Therefore, it is important to identify and move away from news and information sources affected by extreme **bias**.

But how do we get our news about climate change or other topics? Primarily in two ways: by habit and by algorithm.[79] A habit is a pattern of regular (essentially automatic) behavior. For example, a 2014 Pew Research Center[80] study on political polarization and media habits in the U.S. showed that ideological conservatives and liberals had different media habits. At that time, respondents who identified as consistently conservatives primarily turned to one mass media news source (Fox News). In contrast, respondents who identified as consistently liberal turned to a different and broader set of mass media news sources (with CNN, National Public Radio [NPR],

---

https://www.pewresearch.org/fact-tank/2022/04/22/for-earth-day-key-facts-about-americans-views-of-climate-change-and-renewable-energy/ and prior years reports on the PEW Research Center site https://www.pewresearch.org/.

[78] For example, the Summary for Policymakers of the 2022 IPCC Sixth Assessment (AR6) Report on Impacts, Adaptations, and Vulnerability can be accessed here: https://www.ipcc.ch/report/ar6/wg2/. Additionally, the 2021 IPCC report also includes a 'Summary for All' as an outreach component of their work: IPCC (2022). Climate Change 2021: Summary for All. https://www.ipcc.ch/report/ar6/wg1/downloads/outreach/IPCC_AR6_WGI_SummaryForAll.pdf.

[79] Point made by Vanessa Otero, Founder of Ad Fontes Media (https://adfontesmedia.com/) in an interview with Sen, P. (2021, May 6). MarTech Interview with Vanessa Otero, Founder and CEO at Ad Fontes Media. Marketing Technology Insights. https://martechseries.com/mts-insights/interviews/martech-interview-with-vanessa-otero-founder-and-ceo-ad-fontes-media/.

[80] Mitchell, A., Gottfried, J., Kiley, J., & Matsa, K.A., (2014, October 21). *Political Polarization & Media Habits*. Pew Research Center. https://www.pewresearch.org/journalism/2014/10/21/political-polarization-media-habits/.

MSNBC and The New York Times receiving the most mention). More recent (2022) results from analyses by the Reuters Institute for the Study of Journalism at the University of Oxford are consistent with these findings.[81] The polarization in news habits also extends to information accessed via social media. For example, a 2023 peer-reviewed study published in the journal *Science* examined 280 million Facebook users. They determined that there's little overlap between political news consumption of ideological conservatives and liberals. The researchers also found that the separation increases as a news link moves from being selected by the algorithm, to being seen by a user, to being interacted with (e.g., clicked, reacted, liked, reshared, or commented on the post).[82]

Why might media habits among different ideological groups make a difference in people's views on topics such as climate change? If we get different information, we perceive different realities. Influencing this perception challenge is the relationship between media bias and reliability. One way to visualize this relationship is through work done by Ad Fontes Media, which ranks bias and reliability of articles from a wide range of web/print, podcast, and television media sources. Their analytical methods[83] use factors of language (i.e., is it infused with opponent characterization and inflammatory terminology?), political position (i.e., is the article taking a left, right, or neutral position?), and a comparison to other articles (to identify important omissions) to characterize bias. Factors of expression (i.e., is information presented as fact, analysis, or opinion?), veracity (i.e., is it provable and accepted?), and the relevance and fairness of the headline and supporting graphics are used to characterize reliability. Their findings indicate a clear correlation: the more extreme the media bias the more unreliable the information that the article contains (Fig. 14). The most extreme category (red at the bottom of the charts) is essentially direct **disinformation** (i.e., propaganda, fabrication). In contrast, centrist perspectives correlate (with some exceptions) to higher reliability, thereby providing the media consumer with more fact-based, trustworthy analysis of information (green at the top of the charts). A 2022 study by Saumya Bhadani and collaborators reinforces and expands on these observations about bias and reliability. In that case, the researchers focused on news media sources and the ideological biases of their U.S. *audiences*. They found news media sources with more extreme and less politically diverse audiences have lower journalistic standards (i.e., lower reliability), whereas news media sources that

---

[81] Newman, N., Fletcher, R., Robertson, C.T., Eddy, K., & Nielson, R.K., (2022). *Reuters Institute Digital News Report 2022*, University of Oxford, p. 39, data on cross-platform news audience polarisation for selected countries (including the US), https://reutersinstitute.politics.ox.ac.uk/digital-news-report/2022.

[82] González-Bailón et al. (2023).

[83] Otero, V. (2021). Ad Fontes Media's Multi-Analyst Content Analysis White Paper https://adfontesmedia.com/white-paper-2021/; Summary methodology videos available at: https://adfontesmedia.com/how-ad-fontes-ranks-news-sources/?utm_source=Ad+Fontes+Media&utm_campaign=1bcedceb06-EMAIL_CAMPAIGN_2022_08_11_04_08&utm_medium=email&utm_term=0_7489d5d067-1bcedceb06-141690115&goal=0_7489d5d067-1bcedceb06-141690115&mc_cid=1bcedceb06&mc_eid=5b62156ea4 (accessed 2023, May 22).

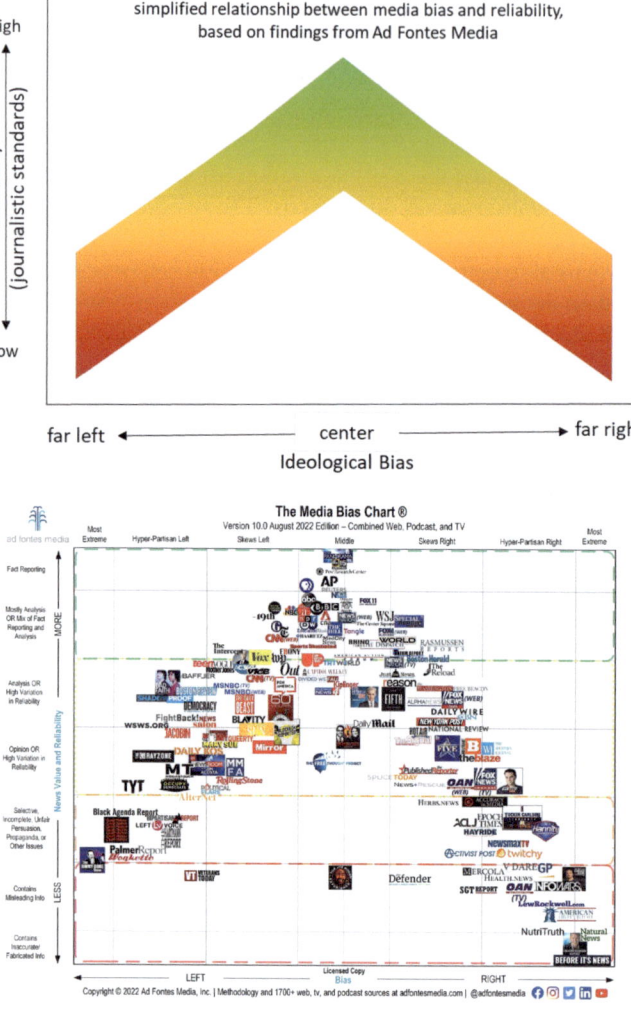

Fig. 14 (upper) A simplified representation of the relationship between ideological bias and reliability (i.e., journalistic standards) of media sources based on the Media Bias Chart by Ad Fontes Media. (lower) Media Bias Chart 10.0, August 2022 Edition, based on Ad Fontes Media ratings conducted in 2022. Ad Fontes ® and the Media Bias Chart ™ are trademarks of Ad Fontes Media Inc. Copyright permission granted from Ad Fontes Media 2022. A detailed interactive chart with ~ 1700 news sources plotted can be accessed from their website, https://adfontesmedia.com/. These charts are updated approximately every 6 months. Figure 14 (lower) is excluded from Creative Commons license

attract more centrist and more politically diverse audiences have higher journalistic standards (higher reliability).[84]

Choosing reliable media sources is important for closing the gap between the public and scientific understandings of climate change. The good news is that it has

---

[84] Bhadani et al. (2022).

been demonstrated that people across the political ideology spectrum rate mainstream news sources as far more trustworthy than either hyperpartisan or fake news sources[85] and regularly updated rating systems (e.g., Ad Fontes Media, NewsGuard,[86] among others) provide publicly available measures of reliability. However, we must overcome the human tendency to gravitate toward information that validates our attitudes, beliefs, behaviors, and feelings rather than seeking out information that is accurate.[87] Media sources on the far left and far right exploit this human tendency and so does social media.

When we get our news from 'side-door' pathways, such as internet search engines and social media, the technology of internet searches becomes an additional factor that can affect the scientific literacy (including climate change knowledge) of the general public. Traditional internet search design uses algorithms to tailor the range of results we get based on personal information it has collected about each of us. Our individual past browsing and search histories are used to make selected information more prominent in our search results. This customization of information has been coined by Eli Pariser[88] as **the filter bubble**. In some ways, the filter bubble has parallels to groupthink; however, the technology is primarily doing the filtering rather than the groups in which individuals belong and interact. In both cases, people can be unaware that these factors influence our perceptions about the world, including our understanding about climate change. As search engine design evolves to integrate more artificial intelligence (AI) tools, new challenges to ensuring awareness of information reliability are emerging as well. For example, AI-based responses to internet queries may not necessarily include sources (e.g., a list of links such as provided in a traditional Internet search), making it difficult for a user to evaluate the quality of the information provided. In addition, early research on people's perceptions of AI-based interactions with consumer internet chatbots shows that human-like chatbot 'conversations' enhance people's trust.[89] The risk of accessing, spreading, and trusting inaccurate, unreliable information is both increasing and concerning.

Social media is now among the most common pathways for people, especially young adults, to get their news and information. The filter bubble and habits of social media use, where news articles are often skimmed or headlines just glanced at, have consequences—they help extend the possibility of misconceptions. A 2019–2020 Pew Research Center study shows that 'Americans who rely on social media as their pathway to news are more ignorant and more misinformed than those who come

---

[85] Pennycook & Rand (2019).

[86] NewsGuard is another source of news ratings: https://www.newsguardtech.com/.

[87] Hart et al. (2009).

[88] Pariser, E., (2012). *The Filter Bubble – How the New Personalization Web is Changing What We Read and How We Think.* Penguin Random House. ISBN 9780143121237.

[89] Stokel-Walker, C. (2023, February 13). AI chatbots are coming to search engines — can you trust the results? *Nature*, https://www.nature.com/articles/d41586-023-00423-4; and Lu, L., McDonald, C., Kelleher, T., Lee, S., Chung, Y.J., Mueller, S., Vielledent, M. Yue, C.A. (2022). Measuring consumer-perceived humanness of online organizational agents. *Computers in Human Behavior 128*, 107,092. https://doi.org/10.1016/j.chb.2021.107092.

to news through print, a news app on their phones or network TV.'[90] While the survey items in the Pew study specifically addressed political and pandemic news, the implications are expected to be equally disheartening for other complex topics, including climate change. A 2018 study[91] further highlighted the challenges that social media pose for getting reliable information. To understand how false news spreads, Soroush Vosoughi and co-authors used a data set of rumor cascades on a popular social media platform, Twitter, from 2006 to 2017. They found that false news spread six times faster than real news, and that false news reached up to one hundred times more people than real news. They hypothesized that the degree of novelty and the emotional reactions of recipients may be responsible for differences observed in the spread of false versus real news. This hypothesis is consistent with the concept of **cognitive biases** described by informatics expert Filippo Menczer and psychologist Thomas Hills.[92] Cognitive biases include our tendency to prefer information from people we identify with, our tendency to pay attention to and share information about risks or fears, and our tendency to seek and better remember information that already fits well with our established world views. These cognitive biases explain why we gravitate to attention-getting information on social media (and mass media) that ultimately goes viral. The obvious challenge is that popular posts can be misunderstood as accurate posts.

Think back to the process of science (Fig. 12). In order for the societal benefits of scientific research on climate change to be fully realized, the public and policy-makers must successfully overcome obstacles of social media filter bubbles, media biases, and cognitive biases. Becoming aware of these navigational hazards is an important first step. What must follow is for each of us to apply that awareness to enhance our critical thinking and modify our behaviors. We need to make wise choices about where we get our information, and then use scientifically sound information to make evidence-based conclusions about climate change. Those conclusions can then spur action, so that at the individual to international levels we work to mitigate and adapt to the negative effects of climate change on Earth systems and human systems.

---

[90] Quote from Sullivan, M. (2020, July 30). This was the week America lost the war on misinformation. The Washington Post. https://www.washingtonpost.com/lifestyle/media/this-was-the-week-america-lost-the-war-on-misinformation/2020/07/30/d8359e2e-d257-11ea-9038-af089b63ac21_story.html reporting on a study from: Mitchell, A., Jurkowitz, M., Oliphant, J.B., & Shearer, E. (2020, July 30). Americans Who Mainly Get Their News on Social Media Are Less Engaged, Less Knowledgeable. Pew Research Center. https://www.journalism.org/2020/07/30/americans-who-mainly-get-their-news-on-social-media-are-less-engaged-less-knowledgeable/.

[91] Vosoughi et al. (2018).

[92] Menczer & Hills (2020).

6 Conclusions

This chapter sought to address two guiding questions: Why does climate change matter? How does examining climate change in geoscience and social contexts help our understanding? In answering these questions, you explored a wide range of natural science and social science concepts which serve as a foundation for the rest of the book. You explored examples about recent and projected future climate change and impacts on humanity. You were introduced to concepts of vulnerability and readiness to deal with the effects of climate change, and through Chapter Boxes connected these concepts to humanitarian crises and to social injustices. You learned from U.S. and international surveys what the publics' views are on the seriousness of climate change, and recognized that public perception is not fully aligned with the scientific understanding. With that realization, you examined reasons why this disconnect exists: that there are physical, psychological, and social reasons for the disconnect between the public and scientific understandings of climate change. Systematic observations, in-depth data analysis, models, and peer review help provide information for evidence-based decisions; whereas media biases, the social media bubble, groupthink and other cognitive biases create opportunities for misinformation (and deliberate disinformation) and thus misunderstandings. In addition, this chapter made the case that in order to understand climate change today, the scenarios of future warming pathways, and the risks those pose, we must understand the processes, patterns, events, rates, and consequences of climate change in the past. The geoscience perspective of climate change is therefore the focus of this textbook. You will learn how the climate system works in Chap. 2 and explore how climate has changed in the past (and how we know it) in Chaps. 3, 4, 5, 6, 7, 8 and 9. You will learn how climate models work in Chap. 10 and learn in Chap. 11 about the role that climate intervention may need to play now and in the future to mitigate the negative effects of climate change. However, we know that addressing the challenges of climate change (just as addressing the challenges of a global pandemic) will take more than understanding the science. Therefore, in Chap. 12 you will delve into the field of environmental humanities to see how ethical reasoning can help us make personal and policy decisions that determine climate change mitigation and adaptation.

> **Box 3 Questions for introspection and discussion**
>
> In this chapter, you learned that the public perceptions of climate change are not fully aligned with the scientific understanding of climate change. While most of the rest of this book explains and explores the climate system and climate change from a geoscience perspective, it is worthwhile to pause here for some introspection on what *you* think about climate change, and *why* you think it. Consider the following questions, and after your own personal reflection, discuss your responses with your family or classmates:

- Reexamine Table 1. What misconceptions have you or your family held? What do you want to know more about?
- Examine current reports of surveyed public perceptions of climate change. How does what you find compare to the recent and historical public opinion described in Chap. 1? Are there any distinct changes in the trends of public opinion? What might be the reason for such changes?
- Where do you get your information about climate change? Where do those sources lie in the most current media bias and reliability chart? If you primarily get your information on climate change from social media, do you know its source? How can you change your social media habits to best differentiate between reliable and unreliable information?
- Identify a specific climate change topic that you want to explore. Pick one outlet each from the far left, center, and far right (Fig. 14; or more recent versions of the media bias and reliability chart produced by Ad Fontes Media, https://adfontesmedia.com/, or from another news rating resource, such as NewsGuard, https://www.newsguardtech.com/) and compare the reporting on that topic from each source. What differences do you find?

References

Alaimo, S. (2009). Insurgent vulnerability and the carbon footprint of gender. *Women, Gender and Research, 18*(3–4), 22–35.

Allan, R. P. et al. (2021). IPCC, 2021: Summary for Policymakers. In Masson-Delmotte, V., Zhai, P., Pirani, A. et al. (Eds.), *Climate Change 2021: The Physical Science Basis. Contribution of Working Group I to the Sixth Assessment Report of the Intergovernmental Panel on Climate Change* (pp. 3–32). Cambridge University Press. https://doi.org/10.1017/9781009157896.001 https://www.ipcc.ch/report/ar6/wg1/downloads/report/IPCC_AR6_WGI_SPM.pdf

Bhadani, S., Yamaya, S., Flammini, A., Menczer, F., Ciampaglia, G. L., & Nyhan, B. (2022). Political audience diversity and reliability in algorithmic ranking. *Nature Human Behavior, 6*, 495–505. https://doi.org/10.1038/s41562-021-01276-5

Burke, K. D., Williams, J. D., Chandler, M. A., Haywood, A. M., Lunt, D. J., & Otto-Bliesner, B. L. (2018). Pliocene and Eocene provide best analogs for near-future climates. *PNAS, 115*(52), 13288–13293. https://doi.org/10.1073/pnas.1809600115

Colon-Rivera, H. A., & Plata, G. (2021). The effects of climate change on hispanic and latinx communities. Pyychiatric Times. https://www.psychiatrictimes.com/view/the-effects-of-climate-change-on-hispanic-and-latinx-communities

Fiore, A. M., Naik, V., & Leibensperger, E. M. (2015). Air quality and climate connections. *Journal of the Air & Waste Management Association, 65*(6), 645–85.

Ford, J. D. (2009). Dangerous climate change and the importance of adaptation for the Arctic's Inuit population. *Environmental Research Letters 4*, 024006. https://doi.org/10.1088/1748-9326/4/2/024006/pdf

González-Bailón, S., Lazer, D., Barberá, P. S, Zhang, M., Allcott, H., Brown, T., Crespo-Tenorio, A., Freelon, D., Gentzkow, M., Guess, A. M., Iyengar, S., Mie Kim, Y. , Malhotra, N., Moehler, D., Nyhan, B., Pan, J., Velasco Rivera, C., Settle, J., Thorson, E., Tromble, R. Wilkins, A.,

Wojcieszak, M., Kiewiet De Jonge, C., Franco, A., Mason, W., Jomini Stroud, N., & Tucker, J. A. (2023). Asymmetric ideological segregation in exposure to political news on Facebook. *Science 381*, 392–398. https://doi.org/10.1126/science.ade7138

Hart, W., Albarracín, D., Eagly, A. H., Brechan, I., Lindberg, M. J., & Merrill, L. (2009). Feeling validated versus being correct: A meta-analysis of selective exposure to Information. *Psychological Bulletin, 135*(4), 555–588. https://doi.org/10.1037/a0015701

Hersher, R. (2023). Cut emissions quickly to save lives, scientists warn in a new U.N. report. NPR. https://www.npr.org/2023/03/20/1162711459/cut-emissions-quickly-to-save-lives-scientists-warn-in-a-new-u-n-report (reporting on the Summary for Policy Makers Synthesis Report of the IPCC Sixth Assessment Report (AR6), https://www.ipcc.ch/report/ar6/syr/)

Kaijser, A., & Kronsell., A. (2014). Climate change through the lens of intersectionality. *Environmental Politics 23* (3) 417–433. https://doi.org/10.1080/09644016.2013.835203

Kinney, P. L. (2018). Interactions of climate change, air pollution, and human health. *Current Environmental Health Report, 5*(1), 179–86.

Klein, R. J. T., Huq, S., Denton, F., Downing, T. E., Richels, R. G., Robinson, J. B., & Toth, F. L. (2007). Inter-relationships between adaptation and mitigation. In M. L. Parry, O. F. Canziani, J. P. Palutikof, P. J. van der Linden & C. E. Hanson (Eds.), *Climate Change 2007: Impacts, Adaptation and Vulnerability. Contribution of Working Group II to the Fourth Assessment Report of the Intergovernmental Panel on Climate Change* (pp. 745–777). Cambridge University Press. https://www.ipcc.ch/site/assets/uploads/2018/02/ar4-wg2-chapter18-1.pdf

Koppers, A., & Coggon, R. (Eds.) (2020). *Exploring Earth by Scientific Ocean Drilling: 2050 Science Framework*. Integrated Ocean Discovery Program, 132 pages, https://www.iodp.org/2050-science-framework

Kossin, J. P., Knapp, K. R., Olander, T. L., & Velden, C. S. (2020). Global increase in major tropical cyclone exceedance probability over the past four decades. *PNAS 117* (22), 11975–11980, https://www.pnas.org/content/117/22/11975

Lucena, J. C., & Leydens, J. A. (2017). *Engineering Justice: Transforming Engineering Education and Practice*. Wiley. ISBN: 978-1-118-75730-7

Lynas, M., Houlton, B. Z., & Perry, S. (2021). Greater than 99% consensus on human caused climate change in the peer-reviewed scientific literature. *Environmental Research Letters, 16*(11), 114005. https://doi.org/10.1088/1748-9326/ac2966

Madhav, N., Oppenheim, B., Gallivan, M., Mulembakani, P., Rubin, E., & Wolfe, N. (2017). Pandemics: Risks, Impacts, and Mitigation. In D. T. Jamison, H., Gelband, S. Horton, P. Jha, R. Laxminarayan, C. N. Mock, & R. Nugent (Eds.), *Disease Control Priorities: Improving Health and Reducing Poverty* (3rd ed., Chapter 17), The International Bank for Reconstruction and Development, The World Bank. https://doi.org/10.1596/978-1-4648-0527-1_ch17

Menczer, F., & Hills, T. (2020). Information overload helps fake news spread, and social media knows it; originally published with the title "The Attention Economy" in *Scientific American 323*, 6, 54–61. https://doi.org/10.1038/scientificamerican1220-54, https://www.scientificamerican.com/article/information-overload-helps-fake-news-spread-and-social-media-knows-it/

Ostro, B., Lipsett, M., Reynolds, P., Goldberg, D., Hertz, A., Garcia, C., Henderson, K. D., & Bernstein, L. (2011). Assessing long-term exposure in the California teachers study. *Environmental Health Perspectives, 119*(6). https://doi.org/10.1289/ehp.119-3114832 and Ostro, B., Lipsett, M., Reynolds, P., Goldberg, D., Hertz, A., Garcia, C., Henderson, K. D., & Bernstein, L. (2010). Long-term exposure to constituents of fine particulate air pollution and mortality: Results from the California Teachers Study. *Environmental Health Perspectives, 118*(3):363–9.

Pennycook, G., & Rand, D. (2019). Fighting misinformation on social media using crowdsourced judgements of news source quality, *PNAS 116*(97) 2521–2526. https://doi.org/10.1073/pnas.1806781116

Perkins-Kirpatrick, S. E., & Lewis, S. C. (2020). Increasing trends in regional heatwaves, *Nature Communications 1*, 3357. https://www.nature.com/articles/s41467-020-16970-7

Plumer, B. & Popovich, N. (2018). *Why Half a Degree of Global Warming is a Big Deal*. New York Times. https://www.nytimes.com/interactive/2018/10/07/climate/ipcc-report-half-degree.html

Pörtner, H.-O., Roberts, D. C., & Tignor, M. et al. (Eds.). (2022). *Climate Change 2022: Impacts, Adaptation, and Vulnerability.* Contribution of Working Group II to the Sixth Assessment Report of the Intergovernmental Panel on Climate Change Cambridge University Press, 3056 pp. https://doi.org/10.1017/9781009325844

Simmons, D. (2020). What is 'climate justice'? Yale Climate Connections. https://yaleclimateconnections.org/2020/07/what-is-climate-justice/

Tessum, C. W., Paolella, D. A., Chambliss, S. E., Apte, J. S., Hill, J. D., & Marshall, J. D. (2021). PM2.5 polluters disproportionately and systemically affect people of color in the United States. *Science Advances 7,* https://doi.org/10.1126/sciadv.abf44

Tippet, C. D. (2010). Refutation text in science education: a review of two decades of research. *International Journal of Science and Mathematics Education 8,* 951–970; and Cook, J., Bedford, D., & Mandia, S. (2014). Raising climate literacy through addressing misinformation: case studies in agnotology-based learning, *Journal of Geoscience Education 62,* 296–306. https://doi.org/10.5408/13-071.1

Vosoughi, S., Roy, D., & Aral, S. (2018). The spread of true and false news online, *Science 359*(6380), 1146–1151. https://doi.org/10.1126/science.aap9559

Weber, E., & Stern, P. (2011). Public understanding of climate change in the United States. *American Psychologist, 66*(4), 315–328. https://doi.org/10.1037/a0023253

Kristen St. John Intent on being a political science major in college, Kristen's path changed after taking physical and historical geology as her science requirements. She's found that piecing together the stories that are preserved in climate archives of Earth's history is like working on a 10,000+ piece jigsaw puzzle. The challenges and rewards of figuring things out are much more fun - and solutions more attainable - when working with others. Synergistic research collaborations, and seeking ways to make geoscience more accessible and meaningful for students, have been guideposts for her career path. Her research focuses on reconstructing the history of land ice and sea ice by looking at clues in marine sediments. Kristen's degrees are from Furman University and The Ohio State University; she is a Professor at James Madison University. Kristen is the primary author of Chap. 1, a primary author of Chap. 2, and the author of Box 1 in Chap. 8.

Dónal O'Mathúna Dónal grew up in Ireland and developed an interest in geosciences because of the deeper appreciation it gave him of the mountains he loved to hike and the waters he kayaked. Course selection in college brought him to other natural products – the ones pharmacy uses to develop medicinal agents. This brought him into uncharted territories as he grappled with the ethical issues related to medicinal products and then to healthcare ethics more broadly. Things have come full circle as now he conducts research on the ethical issues at the intersection of human, animal and environmental factors in global health, aka One Health. Dónal is an Associate Professor at The Ohio State University College of Nursing and the Center for Bioethics. He has spoken and published widely in bioethics, especially disaster ethics, and has contributed to ethics initiatives with the World Health Organization, UNICEF and other international agencies. Dónal is the author of Box 1 in Chap. 1.

Melissa Burt Melissa's interest in weather started from her fear of tornadoes, which sparked a curiosity to learn more about the environment, weather, and climate. Melissa A. Burt is an Associate Professor in Atmospheric Science and Associate Dean for Diversity and Inclusion at Colorado State University. Her research spans the intersection of atmospheric science and social justice issues, climate change, and science communication. She also leads and facilitates diversity, equity, and inclusion efforts to strengthen a culture of inclusion and belonging. Melissa is the Vice President for the non-profit organization, the Earth Science Women's Network, and co-founder of Science Moms, a non-partisan group of climate scientists working to demystify climate science and solutions that preserve the planet. Melissa has a B.S. degree in Meteorology from Millersville

University and a M.S. and Ph.D. in Atmospheric Science from Colorado State University. Melissa is the author of Box 2 in Chap. 1.

Open Access This chapter is licensed under the terms of the Creative Commons Attribution 4.0 International License (http://creativecommons.org/licenses/by/4.0/), which permits use, sharing, adaptation, distribution and reproduction in any medium or format, as long as you give appropriate credit to the original author(s) and the source, provide a link to the Creative Commons license and indicate if changes were made.

The images or other third party material in this chapter are included in the chapter's Creative Commons license, unless indicated otherwise in a credit line to the material. If material is not included in the chapter's Creative Commons license and your intended use is not permitted by statutory regulation or exceeds the permitted use, you will need to obtain permission directly from the copyright holder.

The Earth's Climate System

Lawrence Krissek and Kristen St. John

Guiding Question: What are the parts of the climate system and how do they interact?

1 Key Take-Away Points

- The climate system includes an interconnected group of subsystems, or spheres: the atmosphere, the hydrosphere, the cryosphere, the biosphere, and the lithosphere (sometimes also called the geosphere).
- The compositions, structures, and circulation processes of the subsystems, as well as the interactions between subsystems, are factors that affect how Earth's climate system responds to external climate forcings over different timescales and spatial scales. These and other characteristics and internal processes of each subsystem affect the storage or transfer of energy, water, and carbon in the climate system.
- The lower two layers of the atmosphere—the troposphere and the stratosphere—are most directly involved in Earth's climate system. Carbon dioxide and water vapor as greenhouse gases, and in the carbon and water cycles, affect the habitability of the planet and climate variability. Aerosols and greenhouse gases affect the energy balance and therefore global temperatures. Atmospheric circulation redistributes heat, drives large-scale precipitation patterns, and is a primary factor influencing surface ocean circulation.

L. Krissek (✉)
School of Earth Sciences, The Ohio State University, Columbus, Ohio, USA
e-mail: krissek.1@osu.edu

K. St. John
Department of Geology and Environmental Sciences, James Madison University, Harrisonburg, USA
e-mail: stjohnke@jmu.edu

© The Author(s) 2025
K. St. John and L. Krissek (eds.), *Climate Change*,
https://doi.org/10.1007/978-3-031-82869-0_2

- The hydrosphere is dominated by the global ocean, which is a major player in the transfer and storage of heat, carbon and other nutrients, and oxygen within the Earth system. Physical and chemical properties of sea water, especially temperature, salinity, and dissolved gases, affect the ocean's structure, circulation, and habitability for marine life. The currents of the surface ocean and deep ocean redistribute heat across the planet and influence the acidity and oxygenation of the deep ocean. High-latitude ocean regions are particularly sensitive areas that affect overturning circulation of the entire ocean.
- The cryosphere is composed of the various forms of frozen water (ice) present at, or near, Earth's surface, including ice sheets, mountain glaciers, snow and ice fields, sea ice, and permafrost (frozen ground). The cryosphere affects the climate system in several ways: the size of glaciers and ice sheets affects global sea level; the high reflectivity of snow and ice impacts the planet's energy balance; the volume of water in ice sheets affects the global water cycle; and carbon stored in permafrost affects the carbon cycle and thus global temperatures.
- The geosphere is composed of the nonliving materials that form the "solid Earth": soil, sediment, and rock near the Earth's surface as well as geologic materials in Earth's interior. The composition of the geosphere, as well as tectonic and weathering processes that move the materials through the rock cycle, impacts the climate system over short and long-time periods especially by influencing boundary conditions and the carbon cycle, which in turn affect the energy balance of the planet.
- The biosphere is composed of all living organisms on Earth. The biosphere plays a role in Earth's surface energy budget primarily by affecting albedo and on the carbon cycle through photosynthesis and respiration, which are evident in the seasonal cycles of CO_2 in the atmosphere.
- Energy in Earth's climate system controls the surface temperature of the planet. When the incoming solar radiation (insolation) is not balanced by the output of Earth's thermal radiation, the temperature on Earth's surface changes. Earth's surface and atmospheric albedo and the abundance of greenhouse gases in the atmosphere affect this energy budget. Geological and biological changes affecting these radiative forcing factors caused past changes in global climate; however, anthropogenic activities are the dominant controls on the recent energy budget imbalance causing modern global warming.
- The movement of water within and between the various components of Earth's climate system is driven primarily by the interplay of incoming solar energy and gravity. The movement of water around Earth's surface and near-surface is tied to the effects of insolation and global average temperatures, which vary on timescales ranging from a few months to 10 s of thousands of years. Because the atmosphere has one of the shortest residence times for water, its behavior during evaporation and precipitation is observed to be responding relatively quickly to the increasing global temperatures of the past several decades.
- The redistribution of carbon among reservoirs is a major cause of climate change, now and in the geologic past. Some climate system processes transfer carbon quickly (e.g., biosphere-driven seasonal cycles) and others much more slowly. The

contrast in rates between the very slow processes that transfer photosynthetically produced organic carbon from land and the surface ocean into the geosphere, where it is transformed into fossil fuels, and the rapid processes by which modern society returns this carbon back to the atmosphere is the primary cause of the rise of atmospheric CO_2 today and consequently modern global warming.

2 Introduction

Any complex system involves a number of components, processes and interactions among them. For example, the human body contains several systems, such as the circulatory system or the digestive system, each of which is composed of multiple parts of the human anatomy and involves organs, processes and interactions to produce a specific outcome (e.g., distributing oxygen throughout the body or extracting nutrition from food). The body itself therefore is a system of systems. Similarly, Earth's climate system is also a system of systems.

In this book, we are exploring Earth's **climate system**, its history, and its potential future. To do this, we are examining key components and processes that interact to control Earth's climate, how Earth's climate has changed at times in the past due to changes in those various components and processes (and the evidence used to interpret those changes and their causes), and predictions for how Earth's climate system will respond to present and future changes in those factors and processes. Because the climate system is complex, though, it often is discussed in terms of an interconnected group of subsystems, or **spheres**: the atmosphere, the hydrosphere, the cryosphere, the biosphere, and the lithosphere (sometimes also called the geosphere). In this chapter, we will begin by introducing these spheres and their characteristics that are particularly important for affecting their interactions with energy, water, carbon, and each other (Fig. 1). We will then integrate this introduction to explore, in turn, the movements of energy, water, and carbon through the climate system. This focus on the flows of energy, water, and carbon identifies three threads of discussion used in later chapters.

Before we consider the various spheres of Earth's climate system, it is beneficial to become familiar with a few concepts and some terminology that are fundamental to this discussion, as well as to the content of subsequent chapters. These concepts and terms are described below:

- **Climate versus Weather**: **Weather** is the short-term condition of the atmosphere at a location. Weather conditions (temperature, precipitation, wind speed and direction, cloud cover) can—and often do—change over intervals of 24 h or less. **Climate**, in contrast, is the "long-term" average set of conditions at that location, usually determined over a period of at least 30 years. Given this difference, it is not correct to talk about the difference in conditions from one day to the next—or even one year to the next—as "climate change", unless that difference is consistent with a pattern of change over 30 years or more.

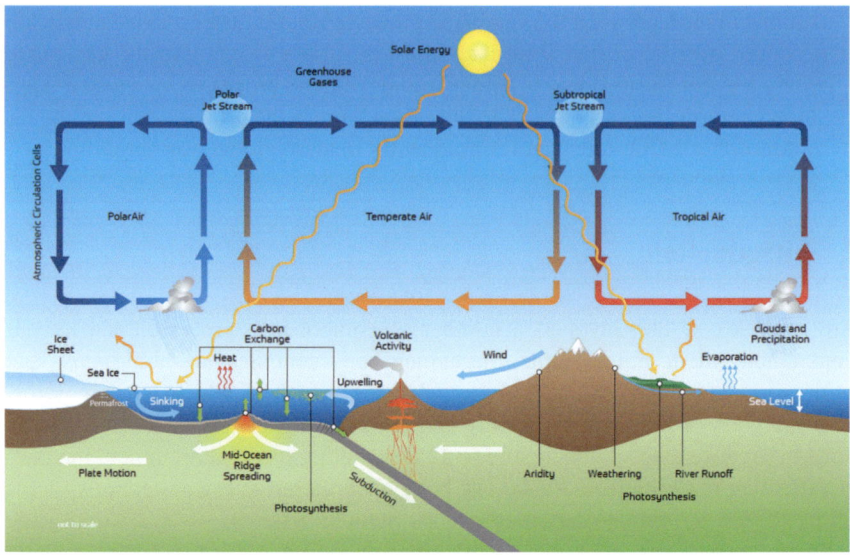

Fig. 1 Representation of the Earth's climate system, which includes five subsystems: the atmosphere, hydrosphere, cryosphere, biosphere, and geosphere (sometimes also called the lithosphere). The Earth's climate system also includes processes and interactions among these systems. The cycling of energy, water, and carbon are three processes that are particularly important in Earth's climate system. This representation of the Earth's climate system shows one hemisphere; the equator would be on the right margin of the diagram. Illustration by Geo Prose. From Koppers and Coggon (2020)[1] Figure used with permission from Geo Prose

- **Forcings**: A **forcing** is an external input of some sort that causes a response in the system overall, or in one of its spheres. If we use a coiled spring as an example of a simple system, then the forcing would be the energy added by compressing or stretching the spring. A forcing of primary importance in our discussions of Earth's climate system will be the incoming solar radiation (i.e., **insolation**) to the Earth system. As we'll see soon, Earth's responses to insolation include heating of Earth's atmosphere and hydrosphere, as well as energy uptake through photosynthesis.
- **Feedbacks**: You may be familiar with audio feedback, where a sound that's initially low volume is increased (or amplified), perhaps because a speaker is too close to a microphone. That's an example of a **positive feedback**, where the interaction increases the effect of a change that's already underway (i.e., an original increase is increased further, or an original decrease is decreased further). Positive feedbacks do occur in Earth's climate system; one example, which was introduced in Chap. 1, described how a decrease in Arctic snow or ice cover due to an initial small amount of warming leads to a greater loss of snow or ice cover

---

[1] Koppers, A., & Coggon, R. (Eds.) (2020). *Exploring Earth by Scientific Ocean Drilling: 2050 Science Framework*. Integrated Ocean Discovery Program, 132 pages, https://www.iodp.org/2050-science-framework.

and additional warming. Some feedbacks can have the opposite effect, where the interaction reduces the effect of the change that's already underway; this is called a **negative feedback**. As will be discussed in Chap. 7, an example of a negative feedback in the climate system is the chemical weathering of silicate rocks, which extracts CO_2 from the atmosphere, thereby reducing atmospheric temperatures and slowing the rate of additional chemical weathering.

- **Timescales and Spatial Scales**: Processes acting within Earth's climate system take place over intervals of time ranging from a few seconds or less (e.g., the interaction of a photon of solar energy with a molecule of gas in the atmosphere) to hundreds of thousands of years or more (e.g., variations in the position and orientation of Earth in its orbit relative to the sun). As a result, we use the term **timescale** to allow us to differentiate across this wide range of durations. Similarly, processes and their effects within Earth's climate system take place over distances ranging from the size of a single molecule to the entire Earth, so the term **spatial scale** is used to differentiate across this wide range of distances.
- **Response Time**: **Response time** provides information about how quickly a given part of Earth's climate system responds to a specific forcing, or change in forcing, and is closely linked to the concept of **inertia** in some parts of the climate system. Response time is controlled by the size of the affected part of the climate system and its composition (e.g., air, liquid water, ice, or rock; Table 1). For example, we have all experienced the fairly rapid response of the atmosphere to a change in insolation, as illustrated by the change from a nighttime low temperature to a daytime high temperature (assuming that a strong cold front of weather doesn't pass through the area during the day). In contrast, the response time to warming of a large ice sheet, such as those found in Antarctica and Greenland today, is much longer and requires that warming be maintained longer before we see a response in the system. In this example, the large ice sheet is said to have much more inertia within the climate system.
- **Reservoirs and Fluxes**: Many discussions of Earth's climate system and its spheres are framed as considerations of "budgets" and consider **reservoirs** (which are the various "holding bins" of the component of interest) and **fluxes** (which are the transfers between the various reservoirs during a specific interval of time, such as a year). An example that you're probably familiar with is a bank account of some sort, where the reservoir is the account balance (the money in the account) and the fluxes are the rates of income and expense. Changes in that reservoir (i.e., changes in the account balance) are determined by the relative sizes of the income flux and the expense flux. For example, as we will see later in this Chapter, and in Chaps. 6 and 7, the distribution of carbon (C) on Earth often is considered in terms of a budget with reservoirs and fluxes, and with very different effects if more carbon is stored in the reservoir of Earth's atmosphere or in the reservoir of Earth's oceans (the hydrosphere).
- **Boundary Conditions**: **Boundary conditions** are important characteristics that affect the processes and outcomes of a system, but are not necessarily part of the system itself. That is why these are said to be located along the "boundary" of the system. An example of a boundary condition you might imagine would be a

Table 1 General response times of the various climate system "spheres"

Climate system sphere	Area responding	Response time
Atmosphere	Local to regional, few 100 km x few 100 km	Hours
	Continent-scale	Days
	Hemisphere-scale	Few weeks to few months
Hydrosphere	Surface ocean, few 100 km x few 100 km	Days
	Surface ocean, subtropical gyre	Months to years
	Surface ocean, global	Decades to centuries
	Deep ocean	1000 s of years
Cryosphere	Sea ice	Months to decades
	Mountain glaciers and icefields	Decades
	Continental-scale icesheets	Decades to 10,000 s of years
Geosphere	Continent-scale, surficial processes	Decades to 1000 s of years
	Global, plate tectonic processes	Millions of years
Biosphere	Local to regional, few 100 km x few 100 km	Months to decades
	Continent-scale	Months to 1000 s of years
	Global	Months to 1000 s of years

For global-scale responses, the atmosphere responds most quickly, followed by slower responses of the surface ocean and the deep ocean, an even slower response of the continental ice sheets of the cryosphere, and the extremely slow plate tectonic response of the geosphere

sports team whose best player is injured and not available for a game. The absence of that player does not affect the rules or duration of the game itself (e.g., if this is U.S. men's college basketball, the game still consists of two 20-min halves, each team still has five players on the court at a time, each free throw is still worth one point, etc.), but it's very possible that the outcome of the game will be different than it would have been if that team's best player had played (i.e., if the boundary condition had been different). In terms of Earth's climate system, one very important boundary condition is the location and elevation of continental landmasses on the globe at a particular time, as well as their changes through time (Chaps. 6, 7, and 10), since continental position and elevation affect circulation in the atmosphere and ocean, as well as the details of the cryosphere.

- **Density and Convection**: **Density** is the mass of a material divided by its volume (i.e., units of grams per cubic centimeter, g/cm^3), and is an essential physical characteristic for understanding how air in the atmosphere and water in the ocean move (i.e., circulation in these two spheres of the climate system). Density is also an essential control on plate tectonics and Earth's internal structure, including the lithosphere, with influences on Earth's climate expressed over timescales of millions of years. Changes in the temperature of air and/or liquid water also change the density of those materials; the details of these changes will be discussed in the next section, but some of these changes can produce situations where more-dense air or liquid water is sitting on top of less-dense air or liquid water. For any fluid, such as air or liquid water, a configuration with more-dense material sitting on top of less-dense material is unstable, and the fluid will respond by mixing vertically until the less-dense material is at the top and the more-dense material is at the bottom. This process is called **convection**. You've probably seen convection taking place in a pot of water on a stove. In that pot, the convection is driven by the heating at the bottom of the pot, which produces warmer, lower density water, and the cooling at the top of the pot, which produces cooler, more-dense water. The lower density warm water rises while the more-dense cooler water sinks, in an attempt to establish a "stable" profile where water density increases from the top of the pot to its bottom.

- **Water's Unusual Properties**: Water (H_2O) molecules have a dipolar molecular configuration, meaning they each have a positively charged end and a negatively charged end. As a result, neighboring hydrogen and oxygen atoms of adjacent water molecules are linked together via a weak attracting force (hydrogen bonds). This extra attraction between water molecules gives water a number of its unusual properties that are crucial to Earth's climate system. An essential property is the elevated melting and boiling temperatures of water; without this effect all water on Earth would be in the form of water vapor, and Earth's climate and habitability would be fundamentally different. A second important property is high **heat capacity and heats of phase changes**, so that water absorbs or releases large quantities of heat without major changes in temperature or phase (i.e., solid, liquid, or gas). These properties are important in heat transfer as water changes state and moves through the hydrological cycle (e.g., evaporating from the ocean to the atmosphere). The underlying molecular structure that produces hydrogen bonds, as well as the hydrogen bonds themselves, also greatly increase water's ability as a solvent—a property that explains why seawater is salty.

3 Components of Earth's Climate System

Earth's climate system is often discussed as being composed of five subsystems, or spheres: the atmosphere, hydrosphere, cryosphere, biosphere, and lithosphere (or geosphere). In this section we will consider key characteristics of each sphere and highlight the characteristics and/or internal processes of each that affect the storage or transfer of energy, water, and carbon.

Fig. 2 Layered structure of Earth's atmosphere. The bottom two layers, the troposphere and stratosphere, are most important when considering the Earth's climate system. From NOAA, https://www.noaa.gov/jetstream/atmosphere/layers-of-atmosphere

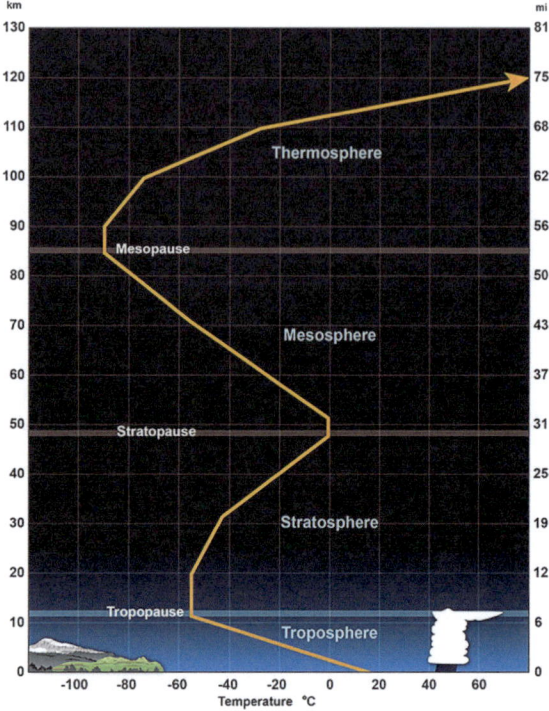

3.1 Earth's Atmosphere

Compared to the diameter of Earth, its atmosphere is a relatively thin layer of gases and very small suspended particles, but with an importance to life and climate that far outweighs its size. This importance arises from the composition of Earth's atmosphere, which is quite different from the atmospheric compositions known or interpreted for other planets in and beyond our solar system.

3.1.1 Atmosphere Structure and Composition

The layered structure of Earth's atmosphere is defined on the basis of patterns of temperature change with increasing altitude (Fig. 2). This classification identifies four atmospheric layers.[2] Of these layers, the lower two—the troposphere and the stratosphere—are most directly involved in Earth's climate system and will be discussed further here. The upper layers of the atmosphere do play a role, however, in absorbing incoming solar radiation.

---

[2] Some atmospheric classifications include a 5th layer, the exosphere, as an outermost layer.

The troposphere is the lowest layer in the atmosphere and is defined by a decrease in temperature and air density as elevation increases; these decreases are the reasons why mountaintops are colder than the adjacent valleys, and why there is less oxygen to breathe at higher elevations. Within this general pattern of decreasing density upward, however, local-to-regional conditions of concentrated heating near Earth's surface can sometimes create the opposite pattern, with lower density air near the surface; when this pattern occurs, convection takes place within the troposphere. You will have experienced this if you're familiar with afternoon thunderstorms on warm humid days, as are common in the US Midwest. A similar scenario takes place on a larger scale to drive Earth's major winds belts, as will be discussed in Sect. 3.1.2.

The troposphere averages approximately 10 km (6.2 miles) thick, but its thickness varies systematically from 18 to 20 km (11.2–12.4 miles) at the equator to ~ 6 km (3.7 miles) at the poles. Because the density of air in the troposphere is greater than air density in the overlying layers of the atmosphere, the troposphere contains ~ 75% of the total mass of the atmosphere and more than 99% of the atmosphere's water. As an illustration of the decrease in air density at higher elevations, humans enter the so-called "death zone" at an elevation of 8 km (5 miles, or ~ 26,000 ft), where the density of air and its constituents (including the oxygen needed by humans) is ~ 1/3rd its value at sea level and is insufficient to sustain human life for an extended period of time.

The stratosphere extends from the top of the troposphere to an altitude of ~ 50 km (31 miles) and contains ~ 19% of the atmosphere's mass but very little water. In contrast to the troposphere, air temperature in the stratosphere increases with increasing altitude, due to energy released during the formation of ozone (O_3) by natural processes there. Stratospheric ozone provides protection to life on Earth by absorbing ultraviolet (UV) solar radiation, which is harmful to life on Earth; you may be aware of concerns about the so-called "ozone hole" in the stratosphere in the high southern latitudes, and the potential impact of increased exposure of life in that region to cellular and genetic damage due to UV radiation.

The fact that air temperature increases and air density decreases with increasing altitude in the stratosphere means that little to no convection takes place in the stratosphere, so any convection originating in the underlying troposphere stops when it reaches the base of the stratosphere. This loss of convection causes the "anvil-shaped" tops of cumulonimbus clouds, with the cloud tops located at the base of the stratosphere. Because of the stable density configuration of the stratosphere, aerosols injected there—such as volcanic emissions or, potentially, aerosols intentionally injected by humans (see Chap. 11)—tend to remain in the stratosphere for several years and be transported over a large area by its circulation.

The composition of Earth's atmosphere below ~ 80 km (50 miles) is dominated by three gases—nitrogen (N_2), oxygen (O_2), and argon (Ar)—as shown in Table 2. These gases are well-mixed, so their relative proportions remain constant throughout the atmosphere and through time. The remaining gases are present in much lower (or "trace") concentrations, which do vary through time and across locations. Among these lower-abundance gases are several of the important "greenhouse gases", including water vapor (H_2O), carbon dioxide (CO_2), methane (CH_4), nitrous

Table 2 Relative abundance of key gases in Earth's atmosphere. Except for water vapor, all reported percentages are for a completely dry atmosphere

Atmospheric gas composition	Relative abundance (%)
Nitrogen (N_2)	78
Oxygen (O_2)	21
Water vapor (H_2O)	from ~ 0 to ~ 4
Argon (Ar)	0.9
Carbon dioxide (CO_2)	0.035
Methane (CH_4)	0.00017
Nitrous oxide (N_2O)	0.000031
Ozone (O_3)	0.000007

Data from NOAA (https://www.noaa.gov/jetstream/atmosphere)

oxide (N_2O), and ozone (O_3). Water vapor abundance varies tremendously between locations and through time. The role of water vapor and the other greenhouse gases in creating the "greenhouse effect" will be discussed in more detail in Sect. 4.1, but it's worth emphasizing here that Earth's surface would be uninhabitable—at least by most forms of life we know—without the activity of CO_2 and water vapor as greenhouse gases and in the carbon and water cycles.

3.1.2 Atmospheric Circulation

Circulation in the atmosphere that impacts the Earth's climate system primarily takes place within the troposphere, and this circulation varies over a wide range of timescales and spatial scales. Here we will focus on the large-scale (i.e., global) patterns of atmospheric circulation, which are represented in Figs. 1 and 3. Notice that there are three atmospheric circulation cells in each hemisphere (Northern and Southern). Each circulation cell contains a limb of rising air and a limb of descending air which are connected by lateral wind belts, one close to the Earth's surface and one at higher elevation, located near the top of the troposphere. The lateral wind belts close to the Earth's surface are additionally represented in Fig. 3 with curved arrows that indicate the direction of lateral motion in more detail as the winds move across the latitudes. The three most important causes of these large-scale circulation patterns are: (1) uneven heating and cooling of the atmosphere, (2) the addition and subtraction of water vapor in the atmosphere, and (3) the Coriolis effect. Each of these factors varies by location on the Earth. Heating vs. cooling and water vapor content also varies over time (e.g., seasonally).

The heating and cooling of the troposphere involves both the gains and losses of heat energy through interactions with the Earth's surface and the stratosphere, and the gain or loss of energy associated with changes of state of water as water is added to, or removed from, the atmosphere (i.e., through evaporation and precipitation). These differences in locations of heating and cooling mean that the atmosphere has different temperatures, and therefore different densities, at different locations around

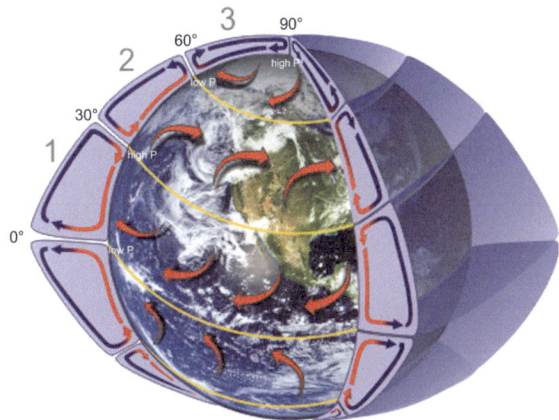

Fig. 3 Representation of the Northern Hemisphere tropospheric circulation cells (connected loops of red and blue arrows, labeled 1–2–3) and resulting wind belts (red curved arrows across Earth's surface). The trade winds blow from the northeast to the southwest between 30° and the equator (0°). The westerlies blow from the southwest to the northeast between 30° and 60°N. The polar easterlies blow from the northeast to the southwest between 90° and 60°N. Latitudes of low vs. high air pressure at Earth's surface are noted at 0°, 30°, 60° and 90° N. Modified from NOAA, https://www.noaa.gov/jetstream/global/global-atmospheric-circulations

Earth. Consider the air at the equator, which is being heated by a relatively large amount of insolation, and compare that to air at the North Pole, which receives much less insolation (and no insolation at all during the Northern Hemisphere winter). The relatively warmer air at the Equator will have a lower density and will tend to rise, whereas the relatively colder air near the pole will be more dense and will sink. Consequently, these separate areas of rising and sinking air create regional differences of **atmospheric pressure**, which is the pressure caused by the atmosphere's weight pushing down on the Earth's surface at a location. Atmospheric pressure is lower at a location where the atmosphere is being warmed and low-density air is rising than it is at a location where the atmosphere is being cooled and dense air is sinking. As a result, there are regions of high atmospheric pressure at the Earth's surface at the poles and a region of low atmospheric pressure at the Earth's surface at the equator. Since air is a fluid (in the sense that it "flows" continuously in response to a pressure difference), the air at the Earth's surface would flow laterally from the areas of high atmospheric pressure toward the area of low atmospheric pressure; in other words, laterally away from each pole toward the equator.

If differences in heating and cooling were the only factor influencing atmospheric circulation, there would be a single circulation cell in each hemisphere, rather than the three circulation cells depicted in Figs. 1 and 3. The additional effects of adding/removing water vapor to/from the atmosphere, and the **Coriolis effect** interact to break a single convection cell in the atmosphere of each hemisphere into a system with three smaller convection cells in each hemisphere. The water vapor effect on atmospheric circulation comes from the reservoir of heat energy associated with water as it changes state, termed **latent heat**. As water evaporates from the Earth's land and ocean surfaces, it draws in heat energy from its surroundings. This heat energy is used to change liquid water to water vapor and has no effect on the temperature of the water. Later, the heat energy absorbed by the water is released to the atmosphere when the water vapor condenses and precipitates as rain or snow. The Coriolis effect is an apparent deflection, or "turning", of masses of air as they move laterally across the Earth. This effect results from us observing the movement of the air while we sit on a planet that is rotating. In the Northern Hemisphere, the "turning" is to the right of the original direction of motion (e.g., imagine driving a car with steering that constantly pulls to the right), whereas the "turning" is to the left in the Southern Hemisphere.

The expressions of these convection cells at the Earth's surface are—in each hemisphere—a zone of low atmospheric pressure near the Equator, a zone of high atmospheric pressure centered at a latitude of ~ 30°, another zone of low atmospheric pressure at a latitude of ~ 60°, and a zone of high atmospheric pressure at the Pole. Because the descending air at ~ 30° is relatively cold and dry, these regions experience little rainfall; note that this latitude on the continents is marked today by many of the world's major deserts. In contrast, the warm rising air near the Equator tends to carry abundant water vapor to altitudes where much of the water vapor condenses and falls as rain; note the relationship between this latitudinal belt and the world's tropical rainforests. This region of warming and rising air, with associated high precipitation, is known as the **Intertropical Convergence Zone, or ITCZ.**

The large-scale global surface wind patterns are formed as a result of air motion responding to these pressure differences. Air is pushed laterally across the Earth's surface from areas of high surface pressure to areas of low surface pressure, and its direction of flow is modified by the Coriolis effect. For example, the high surface pressure at ~ 30° pushes air toward the low surface pressure near the Equator, but the moving air is turned by the Coriolis deflection. As a result, the so-called **prevailing winds** primarily flow from the east (more accurately, from the northeast in the Northern Hemisphere and from the southeast in the Southern Hemisphere, due to the difference in direction of the Coriolis effect in the two hemispheres), and commonly are known as the **trade winds**. Air at mid latitudes is pushed from the surface high pressure at ~ 30° toward the low pressure at ~ 60°, but is also turned by the Coriolis effect, producing the mid-latitude prevailing winds that blow from the west (more accurately, from the southwest in the Northern Hemisphere and the northwest in the Southern Hemisphere) and are known as the **westerlies**. Finally, air at high latitudes is pushed from the surface high pressure near the pole toward the low pressure at ~ 60°, with turning due to the Coriolis effect. Similar to the interaction at low latitudes, this produces winds that blow predominantly from the east—the so-called **polar easterlies**.

The pattern of tropospheric circulation that produces three major wind belts in each hemisphere—the low-latitude trade winds, the mid-latitude westerlies, and the high-latitude polar easterlies—is the long-term average atmospheric circulation; as such, it can be considered the "climate" of atmospheric circulation and plays an essential role in transferring excess heat from low latitudes to high latitudes. It also plays an essential role in transporting water—in the form of water vapor—through the atmosphere and from the oceans to the continents. The primary factors that cause this pattern of atmospheric circulation—the changes in insolation from low latitudes to high latitudes on Earth, the energy transfer effects associated with changing the water vapor content of the atmosphere, and the Coriolis effect—are interpreted to have remained relatively constant through geologic time, suggesting that the present pattern of atmospheric circulation can be a guide to understanding Earth's past climate history. In applying such a uniformitarian approach (i.e., "the present is the key to the past"), however, we must recognize that some very important climate-influencing boundary conditions were significantly different at times in the past, and that the influences of those changed boundary conditions may have produced atmospheric circulation patterns that were quite different from those of today. Examples of boundary conditions that are known to have the potential to significantly modify global atmospheric circulation patterns include the locations and sizes of continental landmasses on the Earth (e.g., evidence that "megamonsoons" were a significant feature of atmospheric circulation during times when a supercontinent such as Pangaea existed; see Box 1 on monsoons) and the location, areal extent, and elevation of major mountain ranges. Some climatic implications of these changed boundary conditions will be discussed in Chaps. 6, 7, 9, and 10.

> **Box 1 Example of the interconnected climate system: monsoons**
>
> When you hear the word "monsoon", you may think of heavy rains—but a scientific definition of the word often places at least as much emphasis on the characteristic of a seasonally reversing pattern of winds as on the amount of rainfall. In many places, the seasonal reversal of the winds is associated with a wet season and a dry season, so that heavy rains are only characteristic of one of the monsoon seasons. In many people's minds the concept of a monsoon also is tied closely to southern Asia, but a monsoon can occur anywhere in the tropics and subtropics (i.e., between 35 °N and 35 °S latitudes) when a relatively strong temperature difference develops between an adjacent continent and ocean. Areas of strong monsoon systems are shown in Fig. 4. The tropics and subtropics are also areas of high population density today, with ~ 2/3rds of the global human population living within monsoon-influenced regions (see Fig. 4).

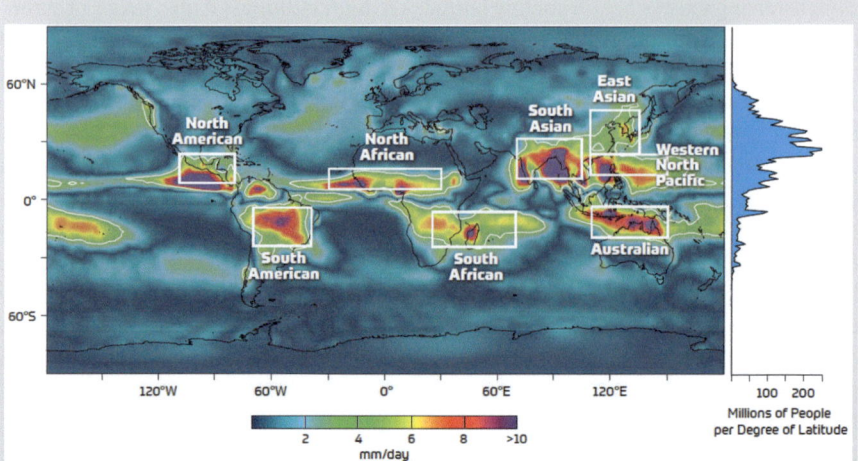

Fig. 4 Global monsoon regions (white rectangles and labels), illustrated using the average difference in precipitation (mm/day) between Northern Hemisphere summer (June–August) and winter (December–February) between 1979 and 2018. Illustration by Geo Prose, from Koppers and Coggin (2020) and based on precipitation data from the Climate Prediction Center Merged Analysis of Precipitation 1979–2018 and population data from the Center for International Earth Science Information Network (CIESIN), Columbia University (2018). Figure used with permission from Geo Prose

The development of this temperature difference is favored if the continent has a relatively large area located at high elevation, where summertime heating and wintertime cooling are enhanced; examples in the world today include the Himalayan Mountains/Tibetan Plateau of Asia and the Colorado Plateau of southwestern North America. The geologic record of Earth's past climate suggests that monsoons were stronger and more widespread during the existence of the supercontinent Pangaea (to the point that those have been described as "megamonsoons on a megacontinent"), and that the "modern" monsoon system of southern Asia only began as the Himalayan Mountains and the Tibetan Plateau were uplifted above a critical elevation by plate tectonic interactions during the last ~ 15 million years. In other words, the development and characteristics of monsoon systems are very sensitive to geologic and geographic "boundary conditions".

A monsoonal wind pattern develops because of differences in the rates and amounts of heating or cooling of continental surfaces (rock) and the ocean (water). Because rock and soil have lower heat capacities than water (i.e., each gram of rock requires fewer calories to heat up or cool down by 1° C than water does), continental surfaces heat up more—and faster—than the adjacent ocean during the summer and cool down more—and faster—than the adjacent ocean during the winter. During the summer, air overlying the relatively warmer continent is warmed, becomes less dense, and rises, producing a region of low atmospheric pressure over the land. At the same time, air overlying the relatively cooler ocean is cooled, becomes more dense, and sinks, producing an offshore region of high atmospheric pressure. Following the same principles we used earlier when discussing global atmospheric circulation, air in the lower troposphere responds to this pressure difference by flowing from the offshore high pressure toward the onshore low; this onshore flow also transports water vapor evaporated from the tropical and subtropical ocean onto land, resulting in a link between the summer onshore wind and the rainy season. In winter the locations of atmospheric heating and cooling are reversed, because the rocks of the continent cool more—and more quickly—than the water of the adjacent ocean. This produces atmospheric high pressure over the relatively cooler continent and low pressure over the relatively warmer ocean, which drives winds from the continent to the ocean (also known as "offshore winds"). These offshore winds reduce the amount of water vapor that is transported to the continent, producing the association of offshore winter winds and the dry season.

At this point, you can understand that onshore monsoonal winds and the associated rainy season are driven by the increased insolation of summer in the respective hemispheres. The geographically larger ITCZ, which also is driven by strong insolation, also responds to the seasonal shift in stronger insolation (i.e., summer) by migrating northward a few degrees of latitude during the Northern Hemisphere summer, and a few degrees of latitude southward during the Southern Hemisphere summer. Additionally, while the monsoons are a pattern that generally repeats during the summer and winter of each year, larger-scale variations within the climate system can interrupt this pattern for one or several years; this is described as a "failure" of the monsoon, and generally is used to emphasize a delay and/or lack of precipitation during the rainy season. The underlying physical causes for such failures are not fully understood, but may have connections to changes in Northern Hemisphere temperatures, quasi-periodic ocean–atmosphere oscillations in other parts of the tropics, and regional river discharges.[3] The impacts of a "failed" monsoon season on humans can be significant; for example, a drought in India from failed summer monsoons in south Asia between 2015 and 2018 affected crop production, freshwater availability, and hydroelectric power production.[4]

Another example of the interconnected climate system is the phenomenon known as **El Nino-Southern Oscillation, or ENSO**, which is a quasi-periodic (3 to 7 year) ocean–atmosphere oscillation in the tropical Pacific, with effects extending elsewhere in the global climate system. Like the monsoons, ENSO variability affects hundreds of millions of people. ENSO is described further

---

[3] Gayatri et al. (2017); Mishra (2020); and Mishra, Aadhar (2021).
[4] Mishra (2020)

in Chap. 3. Understanding how the monsoons and ENSO will respond in a warming future is essential for assessing their potential impacts on physical and biological systems, and also for predicting and mitigating their impacts on people and their social, political, and economic infrastructures.

3.2 Earth's Hydrosphere

Earth's hydrosphere is composed of the liquid water at Earth's surface and in the shallow subsurface. This includes water in the oceans, lakes and ponds, rivers and streams, and groundwater, each of which can be considered a **reservoir**, or "holding tank" of a particular size. Of these reservoirs, the oceans are the largest by far, containing more than 99% of Earth's liquid water and covering ~ 70% of Earth's surface. This introduction to the hydrosphere will focus on the oceans, because the oceans contain the vast majority of Earth's liquid water and because the oceans are such a major player in Earth's overall climate system (Fig. 1). Glaciers and ice sheets contain ~ 2.5% of Earth's total surface and near-surface water and are the largest reservoir of potential freshwater, so some people consider that frozen water also to be part of the hydrosphere. Because frozen water behaves in such a different way than liquid water in the Earth's climate system—and at such different timescales—we will consider the frozen water separately as the cryosphere.

3.2.1 Ocean Composition and Physical Properties

Seawater is salty. Even though salt forms only a very small portion of seawater, these dissolved inorganic materials give seawater some physical properties that are important for influencing the hydrosphere's role in Earth's climate system; unfortunately for humans, this salt content also makes seawater difficult and expensive for humans to use as a water resource. Present-day seawater has a global average salt content of ~ 3.5%, more often stated as a **salinity** of ~ 35‰ ("parts per thousand"). In the open ocean (i.e., away from coastal zones with major river input or zones with restricted circulation where excessive evaporation can take place), seawater salinity ranges from ~ 30‰ to ~ 38‰. The salinity of open-ocean surface waters varies with latitude over this entire range, reflecting changes in the relative amounts of precipitation and evaporation across latitudes.[5] The salinity of seawaters below ~ 500–1000 m depth varies over a smaller range of values, and is controlled by the salinity of the surface waters that sank to form that subsurface water.

---

[5] For a map of sea surface salinities see https://www.ncei.noaa.gov/access/world-ocean-atlas-2023f/bin/woa23f.pl.

Although most naturally occurring elements and a large number of compounds have been detected in seawater, > 99% of the salt in seawater is contributed by only six components (sodium, chloride, sulfate, magnesium, calcium, and potassium; Na^+, Cl^-, SO_4^{2-}, Mg^{2+}, Ca^{2+}, and K^+, respectively). In addition, the proportional contribution of each of these six major salts to the total salinity is always the same, regardless of the water's location (e.g., Atlantic Ocean or Pacific Ocean, surface water or subsurface water); for example, chloride always forms 55.04% of the total salinity in seawater. The chemical breakdown, or **weathering**, of rocks on land is the primary source of the dissolved salts that enter the ocean, as will be discussed in more detail in Chaps. 5 and 7. The fact that these six components dominate seawater salinity tells us that the input rates of these components far exceed the rates for any processes that remove them from seawater; for comparison, silicon (Si) and iron (Fe) also are released by chemical weathering of rocks on land, but their abundances in seawater are low because both Si and Fe are taken up relatively quickly by a variety of inorganic and biological processes in the ocean.

The salt in seawater has several important effects on ocean water in terms of understanding the ocean as a component of the climate system. The presence of salt in seawater lowers the freezing point (i.e., to a temperature below 0° C), and the freezing point continues to decrease as the salinity increases. When seawater does freeze, however, the freezing involves a process that preferentially excludes (or "expels") salt ions and the water molecules bonded to them from relatively low-salinity ice. This process occurs because the salt ions and their clusters of associated water molecules (linked because of the presence of hydrogen bonds) are too large to fit in the hexagonal crystal structure of ice (which you may have seen reflected in the sixfold symmetry of snowflakes). The result of this freezing process is the formation of relatively low-salinity sea ice (some salt is present as small pockets of brine trapped within the ice) and the residual high-salinity water (often referred to as a "brine", because of its high salinity). These two products have very different densities, so the low-density sea ice remains at the ocean surface while the high-salinity brines are more dense and sink into the deeper ocean. Both of these products play important roles in Earth's climate system; we will explore the importance of sea ice in our discussion of the Cryosphere (Sect. 3.3), and the importance of the high-salinity brines in our discussion of subsurface ocean circulation (Sect. 3.2.2). The presence of salt in seawater also decreases the rate of evaporation from the ocean surface, because the bonds between water molecules and salt ions are stronger than the hydrogen bonds between water molecules themselves. Those stronger salt-water bonds must be broken for water molecules to evaporate, which slows the rate of evaporation. Although the climatic effects of changes in salinity on evaporation rates are less obvious than the effects of sea ice formation, the amounts and rates of evaporation are important controls on Earth's water cycle, as we'll discuss in Sect. 4.2.

In addition to its dissolved inorganic solids (i.e., salts), seawater also contains a number of dissolved gases and organic components that are important in the climate system and to the marine biosphere. Two of the important dissolved gases are CO_2 (and products of its interaction with water) and O_2. The CO_2 in the ocean originates from two sources: uptake from the atmosphere at the ocean surface, and production by the biological process of respiration, which has a major impact on CO_2 contents in the deep ocean. The ocean can hold a tremendous amount of CO_2, primarily because the CO_2 interacts with water to form carbonic acid (H_sCO_3), the bicarbonate ion (HCO_3^-), and the carbonate ion (CO_3^{2-}), as will be discussed in more detail when discussing rock weathering in Chap. 7. As a result, the ocean is estimated to have absorbed 30–50% of the CO_2 that has been released by the burning of fossil fuels in the last ~ 175 years. The natural transfer of CO_2 between the ocean and the atmosphere has been a major cause of past changes in Earth's climate system, as will be explored in a number of later chapters in this book. Dissolved O_2 gas in the ocean also has two sources: uptake from the atmosphere at the ocean surface, and production during the biological process of photosynthesis, which requires sunlight. Since both of these sources are most active near the ocean surface, dissolved O_2 contents generally are high in the upper few 100 m of the ocean and are lower in the subsurface, where the biological process of respiration consumes dissolved O_2.

The important major nutrients for marine plant growth are carbon (C), hydrogen (H), oxygen (O), nitrogen (N), and phosphorus (P), all of which are derived from water molecules (for H and O) or from compounds of C, N, and P dissolved in seawater. The forms of N and P used directly in photosynthesis are nitrate (NO_3^{2-}) and phosphate (PO_4^{3-}), respectively; each passes through its own nutrient cycle within the ocean, with limited additions of new N and P from river runoff and aerosol deposition on the ocean surface (except for coastal regions affected by human activities that supply nutrients in agricultural fertilizer or wastewater). As nutrients, nitrate and phosphate are used during photosynthesis in the sunlit surface waters of the ocean, so concentrations of these nutrients tend to be low near the surface. When the near-surface marine plants (almost entirely microscopic phytoplankton) and the animals that consume them (mostly microscopic zooplankton and higher-level consumers) excrete solids or die themselves, the resulting solid organic detritus settles deeper in the ocean. Respiration at those depths releases dissolved N and P, as well as the other nutrients, back into the seawater, so that concentrations of dissolved N and P generally are low in the surface waters and higher in the deep ocean. This transfer of nutrients (including carbon) from the surface to the deep ocean is sometimes termed the "biological pump", as will be discussed in Sect. 4.4.

Temperature is another physical property of the ocean that is important to characterize for understanding the role of the ocean in Earth's climate system. Ocean temperatures are primarily controlled by solar heating and heat exchange with the atmosphere. The distribution of modern surface ocean temperatures ranges from ~ 30 °C (86°F) in the tropics to < 0 °C (<32°F) in the polar ocean.[6] Ocean temperatures decrease rapidly with increasing water depth; for example, even at the equator where the maximum annual average solar heating occurs, temperatures are only 13 °C

---

[6] For a map of sea surface temperatures see https://www.ncei.noaa.gov/access/world-ocean-atlas-2023f/bin/woa23f.pl.

(55 °F) at 200 m (660 ft) below the ocean surface, and the deeper ocean water is only about 3.5 °C (38 °F).

The large amount of energy stored in the surface ocean acts as a modulator of coastal climates. In Box 1 on the monsoons, the points were made that water has a higher heat capacity than rock or soil, and that this difference in heat "storage-ability" plays an important role in the differential heating (and cooling) of the ocean and continents, which in turn is a primary factor in generating the monsoons. Another important climatic effect of water's high heat capacity is that a large amount of energy can be added to the ocean from the atmosphere without changing the ocean's temperature significantly; conversely, the ocean can release a large amount of energy to an overlying cool atmosphere without cooling the ocean water significantly. As a result, the ocean is often described as a "thermal buffer", helping to moderate temperature changes on adjacent coastlines. On a large scale, this effect is recognized by the distinction between "maritime climates", found along coastlines and characterized by relatively cool summers and relatively warm winters, and "continental climates", found in continental interiors and characterized by hot summers and cold winters. A similar but smaller moderating effect also can occur along the coastlines of large lakes, as discussed in Box 2.

Temperature, like salinity, also plays a role in controlling the density of ocean water. As seawater cools (e.g., seasonally, or as ocean currents flow from low to higher latitudes), it becomes more dense. Physical processes in the ocean can change both seawater temperature and seawater salinity, so the interactions of these two controls on density are important to the Earth's climate system. Those interactions are complex but quantifiable. In general, changes in seawater temperature are a stronger control on seawater density than changes in seawater salinity, so scientists often focus on seawater temperature and its changes—for example, with increasing depth into the ocean or across a geographic area of the surface ocean—as a reasonable approximation of the distribution of seawater densities. We will see in the next section that water density is a key factor in deep ocean circulation.

> **Box 2 Weather and climate effects of large lakes**
>
> The same important properties and behaviors of liquid water that make the oceans such a major player in Earth's overall climate system also allow smaller bodies of water to exert a significant influence on the climates of their immediate surroundings. One often-publicized example of this type of influence is the "lake effect" snow in the region of Buffalo, New York, which occurs when a combination of cold air temperatures and winds blowing from the west across the unfrozen surface of Lake Erie evaporates large amounts of water from the lake, transports that water vapor eastward onto land, and deposits that condensed water as snow on the colder land. Another example of the Lake Erie "lake effect" on local climate is not publicized as widely as the "lake effect" snows but has a significant and more widespread impact on human

activities and economies than the heavy snows; this effect is a moderation of air temperatures in the autumn as heat is released from the warm waters of the lake into the atmosphere on cool nights. This temperature moderation means that the average date of first frost near Lake Erie (i.e., within a few 10 s of kilometers of the lake) is one to three weeks later than the average date of first frost farther away (Fig. 5). As a result, industries based on a longer autumnal growing season for frost-sensitive plants—such as vineyards and nurseries for landscaping plants—are concentrated close to the Lake Erie shore.

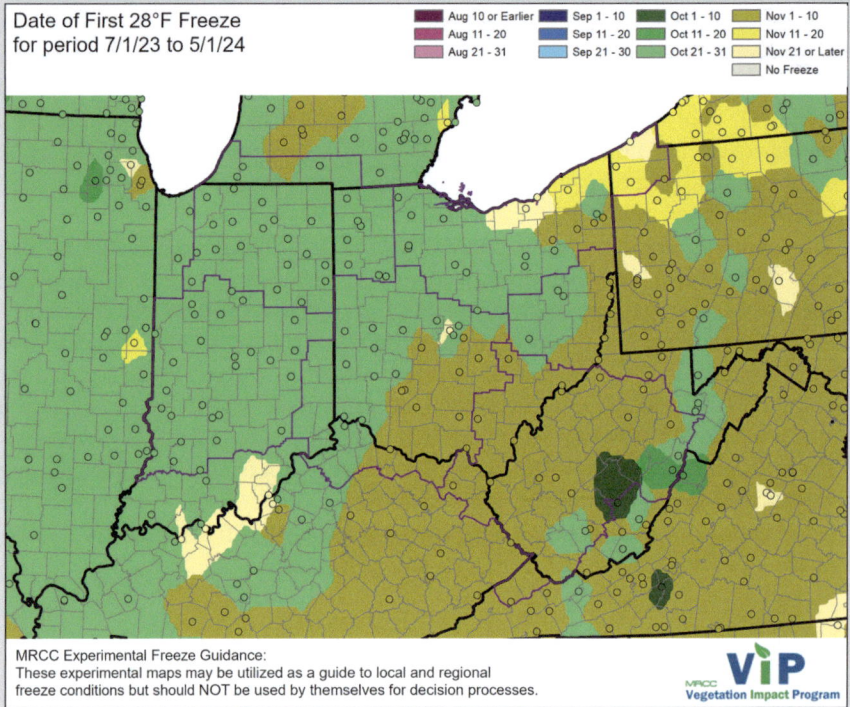

Fig. 5 Map of first freeze date in the region ranging from central Illinois (west side of map) to east-central Pennsylvania, 2023–2024. Counties within each state are outlined. Note that the first freeze dates for counties along the southeastern (downwind) shore of Lake Erie generally are one to three weeks later than the first freeze dates for counties farther from the lake. *Source* Vegetation Impact Program (VIP), Midwestern Regional Climate Center (MRCC). Purdue University, https://mrcc.purdue.edu/VIP/frz_maps/area_160#Layer_poly_dtf_28 accessed on May 1 2024. Figure used with permission from MRCC. This figure is excluded from Creative Commons license

3.2.2 Ocean Circulation

Working together, circulation in the ocean and the atmosphere is the major mechanism for redistributing heat and water within Earth's climate system. At the same time, ocean circulation also plays a major role in transporting and storing carbon dioxide (CO_2) and other greenhouse gases, as well as other nutrients and dissolved oxygen that are essential to the biosphere. In this section we will explore the fundamentals and large-scale patterns of ocean circulation, with brief discussions about their impacts on the distributions of heat, water, carbon, and other biologically important components. More focused discussions of Earth's energy budget and its water and carbon cycles will take place in Sect. 4 of this chapter.

The patterns of ocean circulation, as well as the important drivers of that circulation, are very different for the "surface ocean", which generally is the uppermost several 10–100 s of meters (depending on location) and the "deep ocean", which generally is at depths below 500–1000 m (again, depending on location). As a result, we will discuss those two layers separately. But first let's explore the characteristics of, and the reasons for, the ocean's layered structure, while fully realizing that this generalized model of a two-layer ocean does not incorporate important dynamic processes that cause variations at a range of spatial scales and timescales.

Ocean Layering Defined by Temperature, Salinity and Density

The vertical distributions (i.e., vertical cross-sections) of seawater temperature and salinity measured from the ocean surface to the deep ocean along a north–south transect in the Atlantic Ocean are shown in Fig. 6 and the resulting seawater density is shown in Fig. 7. Remember from our earlier consideration of seawater density that temperature and salinity are the dominant controls on the density of a parcel of seawater, with temperature generally being the more important of those two.[7] That strong influence of temperature on seawater density is illustrated by the fact that the temperature cross-section (Fig. 6a) and the density (Fig. 7) cross-section in the Atlantic Ocean show a strong inverse relationship; temperature decreases are accompanied by seawater density increases, whereas salinity changes (Fig. 6b) are not always accompanied by the expected direction of change in seawater density. The patterns of the temperature and salinity cross-sections also give us an opportunity to recognize three important facts about today's oceans: 1) the ocean basins are deep (with an average depth of ~ 3800 m), 2) ocean water exhibits a layered structure,

---

[7] As a detail that's sometimes important, the density of seawater is also influenced to a much smaller extent by the pressure exerted on it—in other words, the depth at which that seawater is located. Pressure increases relatively quickly with increasing water depth, with one "atmosphere" of pressure added for ~ every 10 m of depth. However, water is difficult to compress, so the effect of pressure changes on seawater density generally only becomes important enough to consider if a circulation scenario involves moving a parcel of water up or down in the ocean by 1000s of meters.

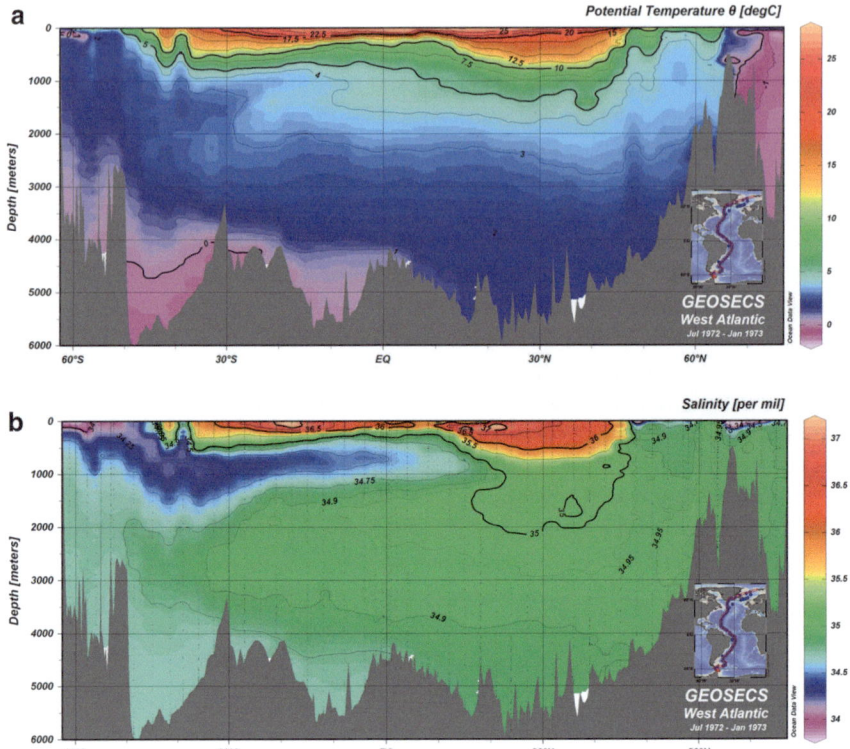

Fig. 6 **a** Temperature (°C) and **b** salinity (per mil, or parts per thousand, ‰) distributions measured from the ocean surface to the deep ocean along a north–south transect in the Atlantic Ocean. The GEOSECS temperature cross-section uses potential temperatures, which are the in situ temperatures corrected for heating due to pressure effects. Those effects are small and don't change the stable temperature (or density) distributions, even for waters at the greatest depths. *Source* Courtesy Reiner Schlitzer, Alfred Wegener Institute (AWI); https://odv.awi.de/data/ocean/geosecs/ Ocean Data View figure used with permission from AWI. This figure is excluded from Creative Commons license

especially at the low and mid latitudes, and 3) the vast majority of the ocean basins is filled with waters that are cold (an average temperature of ~ 3.5 °C) and salty (an average salinity of 35‰, or 3.5%).

In the low to mid-latitude ocean (0 to ~20–35° N or S), the surface waters are warm and have high salinity; these characteristics can be explained by the high insolation at low latitudes and the enhanced evaporation of surface waters by the relatively dry descending air at those latitudes. Notice that the salinity pattern is different very close to the Equator, with slightly lower salinities under the influence of the increased rainfall in the Intertropical Convergence Zone. The surface waters in the low to mid latitudes have a relatively uniform but low density; the low density indicates the influence of warm seawater temperatures on density. The density (as well as the temperature and the salinity) is relatively uniform **within** the surface waters of a

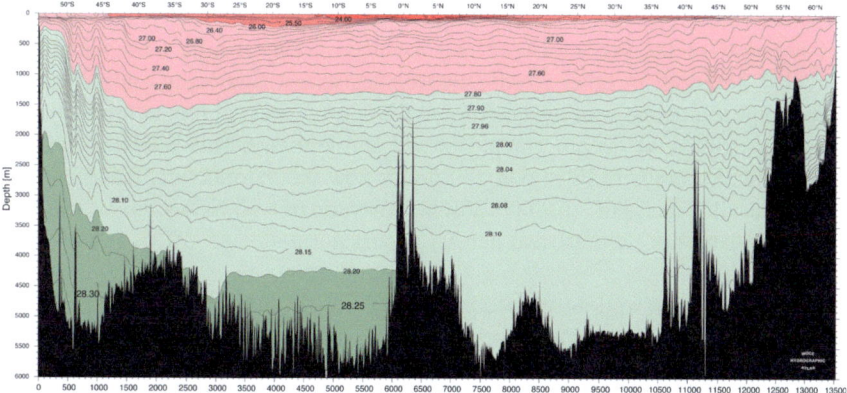

Fig. 7 Profile of seawater density anomaly (i.e., density of the seawater minus density of freshwater, multiplied by 1000, in kg/m^3. This means that a density anomaly of 24.00 indicates seawater with a density of 1.024 g/cm^3.) of the Atlantic Ocean. This profile is from a slightly different north–south transect location in the Atlantic Ocean than the temperature and salinity data shown in Fig. 6. From the WOCE Atlantic Ocean Atlas[8]

location because those waters are homogenized (i.e., "mixed") by the effects of the wind and waves. These are the characteristics of the **surface ocean** at low to mid latitudes. All three parameters begin to change significantly with increasing depth below a level of ~ 100–200 m, and those changes continue to a depth of 500–1000 m. These zones of change (or "-cline", which means a gradation or change) are named the thermocline, halocline, and pycnocline, depending on which property of seawater is being considered: "thermocline" for the change in temperature, "halocline" for the change in salinity, and "pycnocline" for the change in density. The halocline in the North Atlantic is not as well defined as in the South Atlantic; this is because salty (and warm) water from the Mediterranean Sea flows into the North Atlantic, mixing and disrupting its layered structure. The thermocline is defined by a marked temperature decrease with increasing water depth, so the accompanying pycnocline is defined by a marked increase in seawater density. As a result, these vertical cross-sections indicate that the water column at low to mid latitudes has a stable density distribution, with less-dense water positioned on top of more-dense water; this stable density layering inhibits vertical mixing, unless some other process comes into play to physically force water to move vertically. The **deep ocean** lies below the pycnocline, and is characterized by cold temperatures and high seawater densities. Temperature, salinity, and density do change as depth increases within the deep ocean, but those changes are much smaller than the changes from the surface ocean to the deep ocean. As we will discuss soon, though, even those small changes are important controls on the movement of water in the deep ocean.

---

[8] From Koltermann et al. (2011)

In contrast to the low and mid latitudes, the seawater temperatures, salinities and densities of the surface ocean and the deep ocean at high latitudes are not markedly different; in other words, this region lacks a significant and permanent thermocline, halocline, and pycnocline, so the entire water column at high latitudes closely resembles the deep ocean at low and mid latitudes. This is best seen in the South Atlantic waters in Figs. 6 and 7; notice how the thermocline, halocline, and pycnocline all "pinch-out" (become absent) poleward of ~ 50°S. A similar "pinching out" of temperature, salinity, and density occurs north of ~ 45–50 °N in the North Atlantic.

When there is no well-developed pycnocline only a small change in temperature and/or salinity is needed to produce an unstable density profile in the high-latitude ocean, with high-density seawater positioned on top of low-density seawater. This situation could develop, for example, if the surface waters cooled by losing heat to an overlying cold atmosphere during the winter, or if the salinity of the surface waters increased because of sea ice formation during the winter (also a result of surface cooling). Although we will not explore this third possibility in detail, the mixing of two bodies of water (known as "water masses") with different temperatures and salinities (but the same original densities) can also produce a new water mass that is more dense than the original water masses. Regardless of what causes an unstable density profile, the response is the same; the more-dense water sinks while the less-dense water is displaced upward, with this response continuing until a stable density profile is reestablished. This response is a form of **convection**, driven primarily by processes at the ocean surface, rather than by heating at the bottom of the system (the latter is the driver of convection in a pot of water on a hot stove). As a result, this style of top-driven convection is often described as "**overturning**".

To summarize, we now understand that the generalized distribution of ocean seawater properties defines three zones: a surface layer of warm, relatively low-density water that is thickest at low to mid latitudes (with a maximum thickness of ~ 100 m) and is thin or non-existent at high latitudes; an underlying transition zone, marked by decreasing seawater temperatures and increasing seawater densities as water depth increases, that is 500–1000 m thick at low latitudes but thin to non-existent at high latitudes; and the deep ocean of cold high-density water, which underlies the transition zone at low to mid latitudes but extends to the surface at high latitudes. Based on these depth ranges you can understand that the "surface ocean" is a small fraction of the total ocean volume; the vast majority of the ocean can be classified as the "deep ocean". The causes of circulation are very different for the surface ocean and the deep ocean, as are the resulting patterns and rates of flow. Circulation in the surface ocean is considered to be driven by the winds through a multi-step process, whereas circulation in the deep ocean is a response to the density differences between various bodies of water. We'll now turn our attention to those causes and patterns.

Surface Circulation

We've already discussed some of the important interactions between the surface ocean and the overlying atmosphere—specifically, the exchange of heat and moisture—but moving air also exerts a direct physical influence on the water surface through friction. You will have seen one indication of this effect if you've ever been on or near the ocean (or a lake) when conditions change from calm to windy, and waves begin to build on the water surface. Those waves—which are disturbances of the air/water boundary—form because energy is being transferred by friction from the gas molecules in the moving air to the water molecules on the surface of the lake or ocean. The details of this transfer are complicated, and those details change as the wind speed increases and the size of the waves already present on the water surface change, but for our purposes the important point is the illustration that energy can be transferred directly from the moving air to the water surface. Winds, and wind patterns, operate over larger areas and longer timescales to drive surface ocean currents, and current patterns, than are needed to produce surface ocean waves by local winds; these differences illustrate a climatically important contrast between the atmosphere and the ocean. This climatically important difference is a contrast in the "response time" of the atmosphere compared to that of the ocean; in other words, how long a change of some sort (i.e., the cause) must be maintained in order to generate a response (i.e., the effect). A similar idea is sometimes embodied in the concept of the "memory" of the atmosphere compared to that of the ocean; in other words, how long the "effect" persists after the "cause" has been removed. As you've probably noticed from your surroundings, the direction and strength of the winds can change very quickly on a gusty day—on timescales of a few seconds or less—and the predominant wind can change from one day to the next (or even within a single day, if a strong weather front moves through). As a result, the response time of the atmosphere to the pressure differences that cause the winds is on timescales ranging from seconds (or less) to a few hours to a few days. In comparison, generating wind-driven currents in the surface ocean—or changing those current patterns—requires relatively consistent forcing by the wind for periods of hours to days (for currents in an area of a few 100 km by a few 100 km) to months or longer (for currents extending over an entire ocean basin). In other words, the surface ocean has a much longer response time, and correspondingly a much longer "memory", than the atmosphere (Table 1); this difference arises primarily because water is both more dense than air and has a much higher internal resistance to flow (also known as viscosity).

Figure 8 shows the large-scale average pattern of surface currents in today's ocean; in other words, this is the equivalent of the climate of ocean surface circulation. The current systems shown in Fig. 8 are important players in Earth's climate system, as we'll soon discuss, but it's also important to recognize that variations from these patterns take place over a range of timescales and spatial scales.

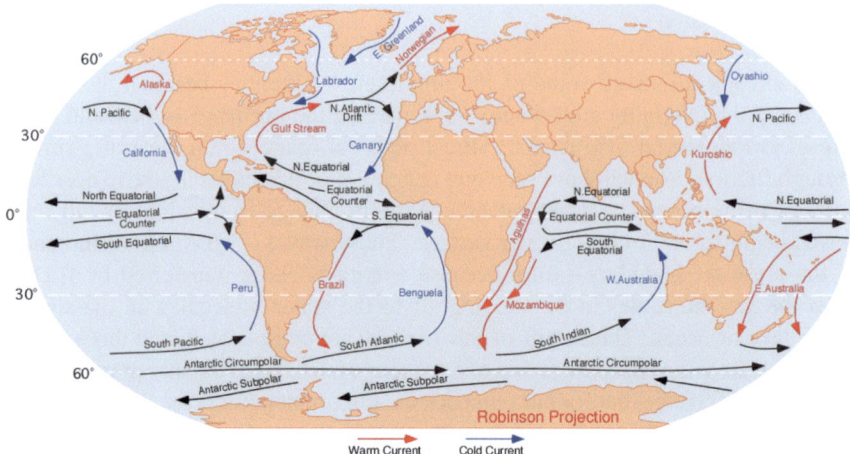

Fig. 8 Large-scale surface ocean currents. The relative temperatures of western and eastern boundary currents are shown using red (warmer) versus blue (cooler) arrows. *Source* Michael Pidwirny, public domain via https://commons.wikimedia.org/wiki/File:Corrientes-oceanicas.png?uselang=en#Licensing

The Subtropical Gyres

Look carefully at the currents shown in Fig. 8 in the low to mid-latitude portions of the North Atlantic, the North Pacific, the South Atlantic, the South Pacific, and the southern Indian Ocean. Notice that the currents in the low to mid-latitude portions of each of these five basins connect to form a more-or-less closed loop of circulation; these are known as **subtropical gyres**. Each subtropical gyre has four components:

- A westward-flowing current near the Equator. This current is generally named the North Equatorial current or the South Equatorial Current, depending on the hemisphere where the current is located.
- A poleward-flowing current along the western edge of the basin (which is the eastern edge of the adjacent continent). This current is known as a western boundary current, and generally is given a name related to its geographic location.
- An eastward-flowing current at the mid latitudes. This current is generally termed the West Wind Drift, although it is given a more specific name based on its geographic location in some basins.
- An equatorward-flowing current along the eastern edge of the basin (which is the western edge of the other adjacent continent). This current is known as an eastern boundary current, again generally given a name related to its geographic location.

A fundamental difference between the subtropical gyres of the Northern Hemisphere and those of the Southern Hemisphere is their direction of circulation, with the Northern Hemisphere gyres circulating in a clockwise direction while the Southern Hemisphere gyres circulate counterclockwise.

The particular pattern of the subtropical gyres raises the question: what causes the currents to flow in this way, (i.e., in this connected system) and why do they flow in opposite directions in the Northern and Southern Hemispheres? As a starting point to answering these questions, remember that wind blowing across the ocean surface can transfer energy via friction to the water. As a result, it's informative to compare the pattern of major wind belts at the Earth's surface (Fig. 3) to the map of surface currents (Fig. 8). What you likely notice from comparing these two figures is that the equatorial currents and the trade winds are located at the same latitudes and have the same direction of flow, while a similar relationship exists between the West Wind Drift and the atmospheric westerlies. This might suggest to you that the winds "pull" the surface waters into motion through the influence of friction; this is a good starting point, but only partially correct. While the frictional pull of the wind on the ocean surface causes the initial motion, two other factors influence the surface ocean circulation: the boundary effects of landmasses, which confine the ocean water flow pathways, and the Coriolis effect, which causes the deviation from the original direction of motion (a deflection to the right in the Northern Hemisphere and a deflection to the left in the Southern Hemisphere). Collectively (and through a complex set of steps important for an oceanography text but beyond the scope of this chapter[9]), these factors generate and maintain the surface current systems of the global ocean.

The climatic impacts of the subtropical gyres are contributed especially by the boundary currents, which play an important role in transporting heat and moisture across latitudes. **Western boundary currents** (WBC), such as the Gulf Stream in the North Atlantic and the Kuroshio Current in the North Pacific, are especially important in poleward heat transport, because WBCs are fast and large. For example, WBCs can flow at speeds up to 2 m/sec, which is very fast in the world of ocean circulation, so they can transport warm water from the tropics to the mid latitudes in approximately one month. And as an example of its large volume, the Gulf Stream is estimated to transport ~ 30×10^6 m^3 of water past one point of its flow every second. A back-of-the-envelope estimate is that the average volume of large U.S. university football stadiums (the entire stadium, not just the playing field) is ~ 1×10^6 m^3, so the Gulf Stream is transporting ~ 30 football stadiums worth of warm water past an observation point each second. That's a lot of water, carrying a lot of heat poleward!

Note that the present-day locations of WBCs are determined by both the wind patterns and the locations of the continents that form the ocean boundaries. As we try to evaluate ocean surface circulation and its potential contribution to poleward heat transport during Earth's past, it's essential to consider the interplay between the atmosphere, the surface ocean, and continental geography as determined by plate tectonics (see Chap. 6).

---

[9] For more information on the factors that set-up and maintain surface ocean gyres see https://oceanservice.noaa.gov/education/tutorial_currents/ and https://oceanmotion.org/html/background/ocean-in-motion.htm, https://oceanmotion.org/html/background/geostrophic-flow.htm and related https://oceanmotion.org pages.

The High-Latitude Surface Ocean

Let's look again at the map of general surface ocean circulation (Fig. 8), focusing now on the regions poleward of the subtropical gyres. The present-day current patterns are very different in the Northern Hemisphere compared to the Southern Hemisphere; the North Atlantic and North Pacific each are characterized by a subpolar gyre of currents, which covers a smaller area than the subtropical gyres and circulates in a counterclockwise direction, whereas the high-latitude Southern Ocean is dominated by the Antarctic Circumpolar Current, which flows uninterrupted around the globe (thus the reason it is called "circumpolar"). This difference in circulation patterns highlights the importance of geography on factors that play a major role in Earth's climate system; in this case, the important geographic difference is the present-day locations of landmasses and ocean basins. In the high-latitude Northern Hemisphere, the continental landmasses of North America, Europe, and Asia separate the North Atlantic and the North Pacific, so that ocean water cannot flow uninterrupted around the globe. Instead, the east–west currents, which are driven by the interactions of the westerlies and the polar easterlies and Coriolis, are deflected by the continents to form southward-flowing western boundary currents and northward-flowing eastern boundary currents. Although their respective flow directions are the opposite of the boundary currents in the subtropical gyres, the boundary currents of the subpolar gyres still are important in the transfer of heat across latitudes as we noted previously for the boundary currents of the subtropical gyres. (As an interesting aside, the iceberg that sank the *Titanic* probably was calved from a glacier in southwest Greenland and was transported south by the Labrador Current, the western boundary current of the North Atlantic subpolar gyre.)

In contrast to the high-latitude Northern Hemisphere, the southern sides of South America, Africa, and Australia do not connect to Antarctica, thereby providing a conduit for continuous west-to-east current flow around the globe (i.e., crossing all lines of longitude) in the Southern Hemisphere, poleward of the subtropical gyres. As a consequence, oceanic heat transport to Antarctica is significantly reduced. As a result, the Antarctic Circumpolar Current has been shown to play a crucial role in thermally isolating Antarctica from the heat stored in the rest of the ocean, producing the uniquely cold and dry climate of that continent. The continuous ocean connection presently occupied by the Antarctic Circumpolar Current is, geologically speaking, a relatively young feature, which has developed over the last ~ 100 million years during the plate tectonic-driven breakup of the former supercontinent Gondwanaland (see Chap. 6 for more discussion of plate tectonics and the development of present-day ocean and continent distributions). The last part of this circum-Antarctic passageway to open was the present-day Drake Passage, which connects the Pacific and Atlantic Oceans. An important connection such as the Drake Passage is often termed a **gateway**, and the process of plate tectonics has acted to both open and close oceanographic gateways through Earth history. Geologic evidence indicates that this opening of the Drake Passage involved multiple steps, so a single date cannot be given for its opening; instead, the opening may have begun as long ago as ~ 50 Ma and appears to have been completed (in other words, an unrestricted deepwater connection had formed) by ~ 15 Ma. Evidence from Late Cretaceous-age (~100 to ~ 65 Ma) geologic records from Antarctica clearly indicates that its climate was warmer and more humid than today, with abundant terrestrial plant and animal life; in some cases, the environment at that time would be considered to have been a temperate rainforest. This dramatic difference between Antarctica's climate during the Cretaceous and today clearly demonstrates the important role of the Antarctic Circumpolar Current in limiting southward heat transport.

In contrast to the polar waters in the Southern Hemisphere, the polar waters in the Northern Hemisphere are only narrowly connected to its neighboring oceans, via the shallow Bering Strait between Asia and Alaska and the deeper Fram Strait between Greenland and Svalbard. In addition to inflow of ocean water through these two straits, the Arctic receives large volumes of freshwater inflow from major rivers in Siberia, Alaska, and Arctic Canada, so that the salinity of Arctic seawater (generally 30–34 ‰) is lower than the salinity of other ocean basins. The circulation pattern in the Arctic Ocean, driven by the polar easterlies and Coriolis and constricted by landmasses and wide continental shelves, is dominated by a clockwise gyre and a transpolar current that ultimately feeds cold, low-salinity waters to the East Greenland Current (Fig. 8).

As shown in Figs. 6 and 7, the seawater temperature, salinity, and density distributions in the high-latitude oceans exhibit little or no thermocline or halocline, and therefore correspondingly have little to no pycnocline. These characteristics apply to both the high-latitude North Atlantic—especially in the regions where water flowing southward from the Arctic Ocean mixes with North Atlantic waters—and to regions within, and south of, the Antarctic Circumpolar Current (especially near the large embayments in the Antarctic coast known as the Weddell Sea and the Ross Sea). In these regions a combination of surface cooling, increase in salinity as sea ice forms, and mixing with other water masses reduces or eliminates the density difference between the surface ocean and the deep ocean, so that surface waters readily sink and deeper waters rise. This convective overturning, which is driven by density increases at the surface rather than by density decreases at the seafloor, supplies water to the ocean's deep circulation system, which will be discussed in more detail in the next section.

Deep Ocean Circulation

In preceding sections, you've been introduced to the important point that circulation in the deep ocean (i.e., below the thermocline and pycnocline) is driven by small differences in the densities of distinctive bodies of water (called **water masses**). These density differences are caused by small differences in their temperatures or salinities, which is why circulation in the deep ocean is called "**thermohaline**" (thermo = temperature, haline = salt). You've also been introduced to the idea that subsurface water masses are formed when surface waters in a region—usually in the high latitudes—become more dense and sink below the surface. This increase in density can occur as a result of cooling, of an increase in salinity, or of mixing two

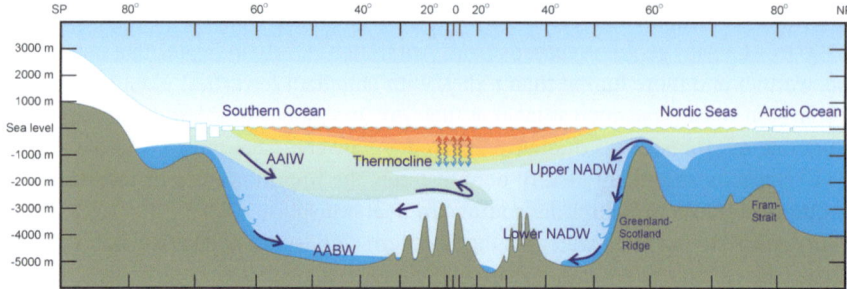

Fig. 9 Schematic depiction of subsurface water masses in the Atlantic Ocean. Arrows indicate the locations of formation and flow of subsurface water masses. AAIW = Antarctica Intermediate Water; NADW = North Atlantic Deep Water, AABW = Antarctic Bottom Water. Gray-green shading indicates the underlying seafloor. White is used to indicate an Antarctic ice sheet and floating iceshelf with icebergs calving off and for sea ice in the Arctic Ocean. Near its area of formation, the NADW occupies approximately the entire depth of the ocean (hence, the upper and lower NADW boundaries are shown here); as it flows southward, however, the NADW encounters northward-flowing AABW, which "wedges in" below the NADW due to the AABW's greater density. Figure used with permission from Iiker Fer and Svein Østerhus. Figure by Ilker Fer, University of Bergen, https://www.uib.no/en/rg/fysos/52636/polar-oceanography

surface water masses with different temperatures and salinity. As a result, the temperature and salinity of each subsurface water mass is determined by the temperature and salinity characteristics of its surface ocean "parent" or "parents", and can be used as a "fingerprint" to track that water mass as it moves through the deep ocean. As a demonstration of this fingerprint concept, compare the subsurface temperature and salinity distributions in Fig. 6 to the schematic representation of subsurface water masses in Fig. 9.

Two of the most widespread and largest-volume water masses in today's ocean are the North Atlantic Deep Water (NADW) and Antarctic Bottom Water (AABW); we will discuss these two water masses as examples of deep ocean circulation, because they play an important role in Earth's climate system. These (and other) subsurface water masses of the Atlantic Ocean are shown in Fig. 9. Note that the name of a subsurface water mass generally carries two important pieces of information: 1) a geographic name that indicates where that water mass formed (i.e., where it sank from the surface into the subsurface), and 2) a term that indicates the level within the deep ocean where that water mass is located. As an example, the name "NADW" indicates that this water mass left the surface in the North Atlantic and is located "deep" in the subsurface, whereas the AABW left the surface somewhere near Antarctica and is found at the "bottom" of the ocean. As a result of these names, we can infer that the NADW will be positioned above the AABW anywhere these two water masses are found together, because "bottom" implies a deeper depth that "deep". In addition, knowing that NADW will be located above AABW tells us that the AABW is more dense than the NADW, due to a colder temperature and/or a higher salinity.

The density of surface waters will increase if that water cools and/or if its salinity increases, both of which are changes common at high latitudes today. Large-scale cooling of surface waters is most likely at high latitudes because of the large temperature difference on the Earth today between the tropics and the poles (often described as a large "thermal gradient"). These cold conditions also enhance the formation of sea ice at high latitudes, producing "brines" that contain the salt excluded as sea ice forms. As a result, the vast majority of today's subsurface water is formed in the high latitudes, with that water subsequently flowing predominantly north–south, as discussed further below. The locations where subsurface water masses formed may have been quite different in the past, however, especially during times when large ice sheets were not present at high latitudes and the global thermal gradient was reduced. During those globally warmer times, salinity differences may have been more important than temperature differences as the primary control on water mass density differences; the differences in salinity would have been driven more by the geographic distribution of zones of high evaporation, rather than areas of sea ice formation. Under those conditions, low-latitude continental shelves and restricted basins (analogous to the present-day Mediterranean Sea, for example) may have been the most important sites of deepwater formation, and the patterns and effects of deepwater circulation may have been quite different than those of today.

As the NADW and the AABW sink into the subsurface ocean, their subsequent paths of flow are driven by pressure differences arising from density differences, modified slightly by the Coriolis force and—for the deepest waters—the shape of the ocean floor. Average speeds of deepwater flow are a few centimeters per second (cm/sec), although short-lived "benthic storms" have been recorded with current speeds of a few 10s of cm/sec. As a result, current speeds in the deep ocean are slower—often much slower—than the speeds of surface currents. The strength of the "turning" due to the Coriolis effect is affected by the speed of motion, so the Coriolis effect on these very slow deepwater currents is much less than its effect on the faster surface currents. As a result, general patterns of deepwater flow can be visualized as direct paths, rather than the gyre pattern observed in the subtropical surface ocean.

NADW is formed primarily in regions around southeast and southwest Greenland, where warm and salty surface waters carried northward from the Gulf Stream and its extensions cool and mix with cold water (e.g., the East Greenland and Labrador Currents in Fig. 8) flowing south from the Arctic. The formation rate of NADW is estimated at $20-50 \times 10^6$ m^3/sec (or enough water to fill a large college football stadium in the U.S. 20–50 times each second), although recent studies have suggested that this formation rate has slowed in the past decade or two due to increased freshwater input from melting in Greenland. The NADW sinks and flows southward (Fig. 9), in a pattern termed **Atlantic Meridional Overturning Circulation**, or **AMOC** ("Meridional" means flow along meridians of longitude, which are aligned north–south; "overturning" reflects the origin of the NADW by density-driven sinking). Near its area of formation, the NADW occupies approximately the entire depth of the ocean; as it flows southward, however, the NADW encounters northward-flowing AABW, which "wedges in" below the NADW due to the AABW's greater density.

The majority of AABW forms in the Weddell Sea, which is the large embayment in Antarctica south of the Atlantic Ocean. The AABW is the coldest major subsurface water mass in today's ocean, with salt added during the extensive formation of sea ice in the Weddell Sea. After overturning, some of the AABW flows northward at the bottom of the Atlantic; this is termed Antarctic Meridional Overturning Circulation. As it flows northward, the AABW interacts with the southward-flowing NADW. Other AABW is entrained in the Antarctic Circumpolar Current, which extends throughout the Southern Ocean water column, and flows eastward and then northward into the deep Indian and Pacific Oceans.

As a result of these two meridional overturning locations, the global view of deep ocean circulation visualizes flow southward through the Atlantic, eastward in the Southern Ocean, and northward into the deep North Pacific, where the deep waters move toward the surface in a very diffuse process. This upward flow of subsurface water is termed "upwelling", although the diffuse upwelling across the North Pacific is very different from the more rapid and geographically restricted wind-driven upwelling that takes place in other parts of the ocean.[10] Studies over the past 50 years have estimated that the average travel time for a parcel of water to move from the Atlantic source regions (i.e., where it was last at the surface) to the deep North Pacific is ~ 1000–2000 years, so this value is used as an estimate of the "mixing time" of the deep ocean (or correspondingly, the "residence time" of water in the deep ocean). A very generalized view of circulation in the total ocean then connects the water moving to the surface in the North Pacific back to the surface Atlantic through several routes; some water flows through the Bering Strait, across the Arctic Ocean, and into the North Atlantic through the Fram Strait, but most of the return flow occurs through interconnections of the various surface ocean subtropical gyres—in sum forming an **ocean conveyor**. The return flow of surface water from the North Pacific to the Atlantic is estimated to take a few decades to a century, given the faster speeds of surface currents. The ocean conveyor concept especially emphasizes the importance of the AMOC, with the deepwater circulation being the outgoing cold lower limb of the conveyor and the surface flow being the returning warm upper limb. This model is also sometimes known as the **salt conveyor** because of the important role of increased salinity in driving AMOC. Note that meridional overturning circulation (MOC) presently does not take place in the North Pacific, because sea surface salinities are too low there. The situation may have been different in the past, however, and searching for evidence of past episodes of North Pacific MOC remains an area of active research.

Thermohaline circulation plays an important role in the transfer and storage of heat, carbon and other nutrients, and oxygen within the Earth system. As part of the oceanic conveyor, AMOC is estimated to carry approximately half of the total global

---

[10] Upwelling, not from the deep ocean as described here, but from thermocline depths also occurs in the ocean, especially in coastal ocean settings where the winds and Coriolis drive water away from the coast. In such cases, ocean water upwells from below, bringing with it nutrients to support the surface ocean food web. Areas of coastal upwelling are therefore typically biologically productive, supporting a strong fisheries economy. Examples of these locations include the coastal regions offshore Peru, northern California, northwest Africa, and southwest Africa.

north–south (or meridional) heat transfer; in other words, the low latitudes would become warmer and the northern high latitudes would become cooler if the AMOC slowed or stopped. A reduction in AMOC likely was responsible for a dramatic cooling event in the Northern Hemisphere—if not globally—approximately 12 to 11 ka. This event is termed the Younger Dryas. The possibility that AMOC has weakened in the past 10–20 years has raised concerns that a similar cooling event might be triggered by increased glacial melting in Greenland; however, the precise "tipping point" that would initiate such cooling is still unknown, although the subject of much research. As described in Chap. 10, climate modelers focus on the Younger Dryas as one paleoclimate event that can inform predictive models of climate and ocean change in future.

In addition to transporting heat, thermohaline circulation is also the primary process supplying dissolved oxygen to the deep ocean (also known as "ventilating" the deep ocean; Fig. 10). The surface waters that overturn are saturated—or nearly so—with dissolved oxygen, due to interaction with the oxygen-rich atmosphere plus the production of dissolved oxygen during photosynthesis by marine plant plankton (phytoplankton). The overturned waters carry this dissolved oxygen into the deep ocean and along their flow paths (this type of transport is called **advection**, in contrast to **diffusion**, which is transport by molecular processes and is much slower). Animals in the deep ocean use this dissolved oxygen during respiration, so the average dissolved oxygen content of the deep waters gradually decreases along the lower limb of the global ocean conveyor, from the North Atlantic to the North Pacific. In today's ocean the total demand for dissolved oxygen in the deep ocean rarely exceeds the amount that is supplied by thermohaline circulation, so that most of the deep ocean remains well-oxygenated. If the rate of supply decreases, due to less-vigorous thermohaline circulation, or if the rate of biological consumption increases significantly, however, then large portions of the deep ocean could become depleted in dissolved oxygen (this depleted condition is known as **hypoxia** for low-oxygen conditions, or **anoxia** if no oxygen is present). This would produce oceanic "dead zones", with major impacts on the marine biosphere. (You may be familiar with the term "dead zone" from news reports of modern conditions off the mouth of the Mississippi River, especially during the summer.) Ocean sediments carry a record of multiple episodes of widespread low-oxygen conditions in Earth's past, especially during times of reduced global temperature gradient and sluggish deepwater circulation.

Thermohaline circulation also transports dissolved CO_2 from the high-latitude surface waters into the deep ocean, where that CO_2 is isolated from the atmosphere for 1000–2000 years (i.e., the "mixing time" of the deep ocean). Studies estimate that 30–50% of the CO_2 released by the burning of fossil fuels since the start of the Industrial Revolution (~1850 C.E.) has been stored in the deep ocean by thermohaline circulation. The benefit of this process is that the stored CO_2 has not been available to act as a greenhouse gas in the atmosphere; the drawback is that the increased CO_2 content has increased the acidity of the ocean by an average of ~ 0.1 pH units, representing an increase of ~ 30% in ocean acidity. This effect is known as **ocean acidification**, and has been observed to be affecting a wide range of

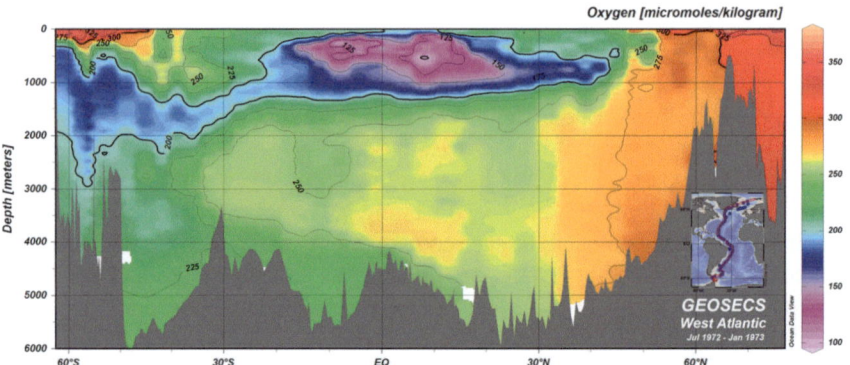

Fig. 10 Profile of dissolved oxygen levels in the Atlantic Ocean. Depth is in meters; dissolved oxygen is in micromoles per kilogram (μmol/kg). Notice the lowest dissolved oxygen levels are in intermediate depths of the low to mid latitudes, far from sources of deepwater formation. *Source* Courtesy Reiner Schlitzer, Alfred Wegener Institute (AWI); https://odv.awi.de/data/ocean/geosecs/ Ocean Data View figure used with permission from AWI. This figure is excluded from Creative Commons license

marine organisms—especially those that secrete skeletal material made of calcium carbonate. Evidence of widespread ocean acidification during a time of extreme global warming in the geologic past is explored in Chap. 8.

3.3 Earth's Cryosphere

The cryosphere is composed of the various forms of frozen water (ice) present at, or near, Earth's surface (Figs. 1 and 11). This includes large ice sheets, such as those present today in Greenland and Antarctica, mountain glaciers, snow and ice fields, sea ice, and permafrost (frozen ground). Taken together, the cryosphere today contains ~ 2% of the world's water and ~ 70% of the world's freshwater[11]; melting of mountain glaciers is the primary source of freshwater for human consumption in some parts of the world (e.g., regions in and near the Andes, and regions in and near the Himalayas). The large ice sheets hold more than 90% of all ice on Earth, with maximum thicknesses of 3 km or more. It is estimated that sea level would rise approximately 7.4 m (23 ft) if the entire Greenland Ice Sheet melted, and approximately 60 m (200 ft) if all the ice in Antarctica melted. Conversely, sea level was lower by ~ 130 m (420 ft) at the Last Glacial Maximum, approximately 18 ka, when the high latitude ice sheets were somewhat larger, similar large ice sheets covered much of northern North America and Europe, and mountain glaciers were more abundant and larger. These numbers make it clear that Earth's cryosphere both responds to, and is an important driver of, Earth's climate system.

---

[11] McConnell (2005)

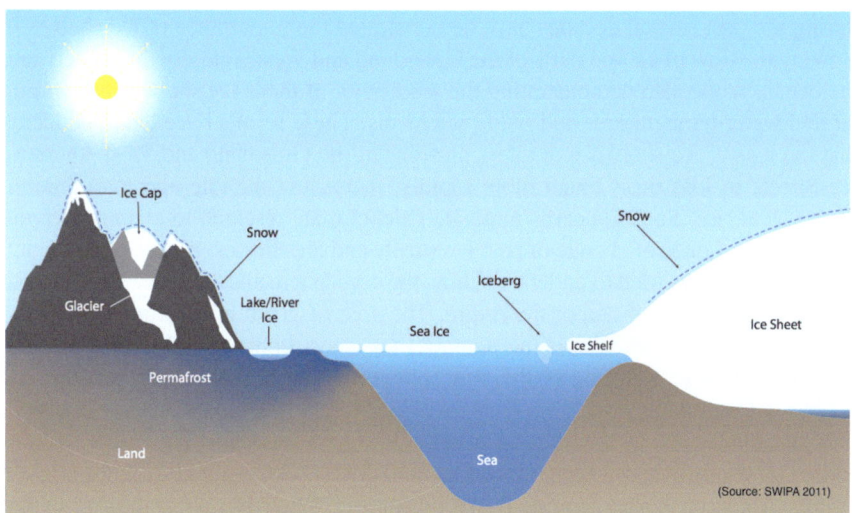

Fig. 11 Illustration of the parts of the Earth's cryosphere, including large ice sheets, smaller mountain glaciers, permafrost and floating ice shelves, sea ice and lake/river ice. Figured used with permission from Arctic Monitoring and Assessment Programme (AMAP; 2012)[12] This figure is excluded from Creative Commons license

Both large ice sheets and mountain glaciers form when the amount of snowfall each year exceeds the amount of melting; this is said to be a positive **mass balance**. Under these conditions, each year's net accumulation is buried by the net accumulation of subsequent years. As snow accumulates and the depth of burial increases, the increased pressure transforms the buried snow to glacial ice through a predictable sequence of changes. Once formed, the ice in both mountain glaciers and large ice sheets flows like a very viscous ("sticky") fluid away from its region of accumulation. This flow is in response to preexisting topography (i.e., downhill for a mountain glacier) or pressure differences (i.e., radially outward for a dome-shaped ice sheet). You may have seen a demonstration of this style of flow using "flubber", which is made from glue, water, and cornstarch. Cores of ice recovered from the areas where the ice is thickest and least disturbed by flow provide excellent archives of a range of past climatic conditions and are discussed further in Chap. 4.

At some distance away from its area of accumulation, the glacier or ice sheet begins to lose ice mass. This loss may occur as: melting on land, producing glacial outwash streams and lakes; as the calving of icebergs if the glacier or ice sheet ends in a lake or the ocean; or as the melting of ice shelves, which are floating extensions of a glacier or ice sheet into the ocean. Recall from Table 1 that mountain glaciers generally have a faster response time than ice sheets to climate forcings. Studies

---

[12] AMAP, 2012. Climate Change in the Arctic–A Hot Topic. SWIPA 2011: Snow, Water, Ice and Permafrost in the Arctic. Arctic Monitoring and Assessment Programme (AMAP), Oslo. 17 pp, 978–82-7971–072-1, https://www.amap.no/documents/doc/climate-change-in-the-arctic-a-hot-topic/101.

during the past several decades have demonstrated the importance of ice shelves in slowing the flow of certain parts of the Greenland and West Antarctic Ice Sheets into the ocean, while also demonstrating the sensitivity of those ice shelves to warming air and water temperatures and rising sea levels. These results have led to concerns that the loss of ice shelves will allow portions of the Greenland and West Antarctic Ice Sheets to loss mass much more rapidly, increasing the rate and total amount of sea level rise. Such concerns motivate paleoclimate research to characterize the behaviors and rates of change of past ice sheets and ice shelves, the results of which can inform models used to understand how the cryosphere and sea level may respond to rapid warming now and in the future. The area of greatest uncertainty in future projections of sea level rise is the potential for catastrophic response of ice sheets and ice shelves to warming of the atmosphere and ocean. This topic is addressed further in Chap. 10.

Sea ice is formed by the freezing of sea water, as discussed in Sect. 3.2.2, accompanied by the formation of the brines that help drive overturning in the high-latitude oceans. Much of the sea ice that forms during the winter season in the Arctic and around Antarctica (December/January/February for the Arctic and June/July/August for the Antarctic) melts the following summer, so is termed "first year ice". Some of the sea ice persists for several years and forms "multi-year ice", however, and is transported around its respective ocean basin by surface currents. It is important to note that the formation and melting of sea ice does not affect sea level, because the water involved is already in the ocean. In contrast, adding new water to the ocean by melting land-based glaciers or ice sheets does affect sea level. Changing sea ice conditions in recent decades is one of the areas of focus in Chap. 3.

The cryosphere plays several important roles in the climate system. In addition to the effects that glaciers and ice sheets have on global sea level change, the cryosphere impacts the global energy balance of Earth's climate system by affecting the proportions of incoming solar radiation (**insolation**) that are absorbed vs. reflected at Earth's surface. In general, snow and ice form a light-colored and reflective surface, which reflects more incoming solar energy than a dark-colored and non-reflective surface like land vegetation or open water. As a result, components of the cryosphere generally have a high **albedo** (i.e., reflectivity), which reduces the amount of insolation absorbed by Earth's surface. In an example of a positive feedback loop, the high albedo of snow and ice reduces the amount of solar energy absorbed by Earth's surface, thereby cooling the surface and enhancing the preservation of additional snow and ice. Another positive feedback in the Earth's climate involving the cryosphere is specific to sea ice: when sea ice is present it forms an insulating layer on the ocean surface, limiting the heat transfer from the ocean to the atmosphere; however, during times of global warming, the insulating sea ice cover melts, allowing more heat to be transferred from the ocean to the atmosphere, causing temperatures to further rise. Changes in the cryosphere also have a large impact on Earth's water cycle, since the cryosphere is the second-largest reservoir of water at Earth's surface. In addition, the formation of sea ice—or lake ice, to a much smaller extent—significantly reduces the amount of water transferred to the atmosphere from that region. Finally, the cryosphere even plays indirect roles in Earth's carbon cycle:

the areal extent of ice sheets and glaciers affects the amount of land available for terrestrial vegetation to grow. Perhaps more importantly, when ice sheets, glaciers, and permafrost melt, buried organic carbon stored in the frozen soil (permafrost) can re-enter the carbon cycle, decomposing to form CO_2 and CH_4 and causing further warming (once again, a positive feedback).

3.4 Earth's Geosphere

The geosphere is composed of the nonliving materials that form the "solid Earth"—both its surface and its interior (Fig. 1). These include soil, unconsolidated sediment, such as sand and mud, and rock found in Earth's outermost layer, or **crust**, as well as geologic materials in Earth's deeper **mantle** and **core** that cannot be observed directly. The composition and properties of the materials present in the mantle and core can be interpreted, however, from indirect evidence such as their densities, their behavior during the passage of earthquake waves, the rare instances where unusual geologic processes have brought pieces of mantle material to the surface, and the compositions of meteorites (which formed from the same material as the early Earth). The composition of crustal materials, as well as important processes that move the materials of the geosphere through the **rock cycle** and impact the climate system over short and long-time periods, are discussed further in Chaps. 5, 6, 7, and 8.

The geosphere most directly impacts the energy budget of Earth's surface by influencing global albedo. Areas of exposed rock have an albedo that is lower than the albedo of snow or ice; in other words, the areas of exposed rock absorb a larger proportion of the incoming solar radiation than areas of snow or ice. In detail, though, the albedo of exposed rock does vary depending on the rock's color, with dark-colored rocks (such as **basalts**, which are discussed further in Chap. 6) having a lower albedo than light-colored rocks (such as **granites**, also discussed further in Chap. 6). Through interactions with the atmosphere, the geosphere also can affect Earth's energy budget through processes associated with **plate tectonics**, such as explosive volcanic eruptions that inject particles and gases into the upper troposphere and the stratosphere and help block incoming solar radiation. Earth's energy budget is also affected by the tectonically driven uplift of large portions of continents, such as the present-day Tibetan Plateau of Asia and Colorado Plateau of southwestern North America; the uplifted landmasses are subjected to more intense insolation because the incoming solar radiation has passed through a thinner atmosphere.

Over timespans of millions of years or more, the geosphere primarily interacts with the Earth's carbon cycle through processes associated with plate tectonics, as discussed in Chap. 6; this set of interactions is considered the "long-term" carbon cycle. The geosphere's impact on the long-term carbon cycle also involves interactions between CO_2 in the atmosphere and processes of rock breakdown (called "**weathering**") and/or formation in the Earth's crust. This set of interactions is discussed in Chap. 7, and is—as you will learn—crucial as a "thermostat" for the maintenance of conditions on Earth that are favorable to life as we know it.

The geosphere's interaction with Earth's water cycle primarily takes place at or near the surface, beginning with the potential control of topography produced by plate tectonic processes on the distribution of precipitation. This difference is illustrated by the "rain shadow" effect in the northwestern U.S., where regions west of the Cascade Mountains generally receive abundant precipitation, while regions east of the Cascades lie in the arid "rain shadow" of those mountains. Once precipitation occurs, the characteristics of the surficial material influence how much water moves along the surface as runoff, and how much moves into the ground to form groundwater (often described as "percolating" or "infiltrating" into the ground). Both surface water (e.g., rivers, natural lakes, and human-built reservoirs) and groundwater are important sources of freshwater for humans, and the surficial geosphere plays an important role in determining how much of the precipitation in an area is available for human use, and from what form of storage. A very small fraction of the precipitation in an area also interacts with, and may be taken up by, the surficial geosphere during processes of **chemical weathering**, as will be discussed further in Chaps. 5, 6, and 7.

3.5 Earth's Biosphere

The biosphere is composed of all living organisms on Earth, ranging from the highest-altitude vegetation in mountain ranges to the unusual forms of life found in the deepest ocean trenches, and from the single-celled primitive bacteria and Archea to the largest trees and vertebrate mammals. The largest amount of biosphere mass, however, is composed of **primary producers**, which are the organisms that produce their own energy stores via photosynthesis (using light) or chemosynthesis (using chemical reactions in unusual environments). In the oceans, the most abundant primary producers are the phytoplankton—free-floating microscopic photosynthesizers. Some groups of phytoplankton produce mineralized skeletal material that is abundant in seafloor sediments; these are discussed further in Chap. 5. On land, primary producers tend to be macroscopic (i.e., visible with the naked eye) and linked to the ground by roots, which provide the water and nutrients needed for photosynthesis and growth.

Some components of the biosphere can survive across a very wide range of climatic conditions; these generalists can be found from the tropics to the high latitudes (Fig. 1). Many components of the biosphere, however, are more specialized in their environmental tolerances, so their geographic distribution is more limited. This type of specialization leads us to classify some organisms as "subtropical" and others as "polar", for example. Identifying analogous distributions for fossilized groups of organisms means that the record of Earth's biosphere in the past can provide strong evidence of Earth's climate history, as will be discussed in Chaps. 5, 8, and 9.

The biosphere plays a role in Earth's surface energy budget primarily by affecting albedo and the greenhouse gas (especially CO_2) levels in the atmosphere. Although the exact albedo value for a vegetated area will depend on the types and amount of vegetation present, vegetated land has a much lower albedo than snow or ice and a slightly lower albedo than unvegetated land. The solar energy absorbed by vegetation is used to drive photosynthesis, which has a direct effect on the carbon and water cycles.

Effects of the biosphere on Earth's carbon and water cycles arise from the fact that primary producers use CO_2 and water to produce organic matter as an energy store, and that plants, animals, and some microbes release CO_2 and water during respiration (the breakdown of organic matter to extract its energy). The overall effect of today's global biosphere on the carbon cycle is illustrated in the so-called Keeling curve (Fig. 12), which is based on the CO_2 of air samples collected every day at the top of Moana Loa, Hawaii for more than 60 years. For now we'll ignore the long-term trend shown by these data; instead, let's focus on the shorter-term "sawtooth" pattern illustrated by the Keeling curve. Note the duration of each "tooth"; the time from one peak to the next peak—or from one valley to the next valley—is one year. The decrease from a peak toward the following valley consistently begins in April or May, whereas the increase from the valley to the following peak begins in about October.[13] These changes are interpreted as a large-scale recorder of the increase in plant growth in the Northern Hemisphere spring, which draws down atmospheric CO_2 by ~ 3%, and the reduction in Northern Hemisphere photosynthesis in the fall, when ongoing respiration releases the stored CO_2 back to the atmosphere. A similar long-term record is not available to show the biosphere's influence on the global water cycle, because water content in the atmosphere is also affected by a number of other variables (e.g., temperature, wind direction and strength). Studies at a regional scale, however, do give some indication of how large the biosphere's influence can be in an area; the Amazon Basin, for example, is estimated to contain approximately 390 billion trees, with each tree releasing ~ 100 gallons of water to the atmosphere each day through the process of evapotranspiration. That's enough water to fill approximately 80 million Olympic-sized swimming pools each day!

4 Climate System Interconnections: Cycles of Energy, Water, and Carbon

In the previous Section you were introduced to the 5 major subsystems (i.e., spheres) of Earth's climate system: the atmosphere, the hydrosphere (with special emphasis on the global ocean), the cryosphere, the geosphere, and the biosphere. Interconnections among these spheres naturally emerged in the discussion as none of these systems operate in isolation. Each has characteristics important in the transfer of energy and the movement of matter throughout the climate system. The cycling of energy and of matter—in particular water and carbon—are fundamental concepts to understand how the climate system operates, as these involve physical and chemical processes that can determine climate system boundary conditions, forcings, and feedbacks.

---

[13] Such details can be observed from the more detailed monthly graphs available at https://gml.noaa.gov/ccgg/trends/.

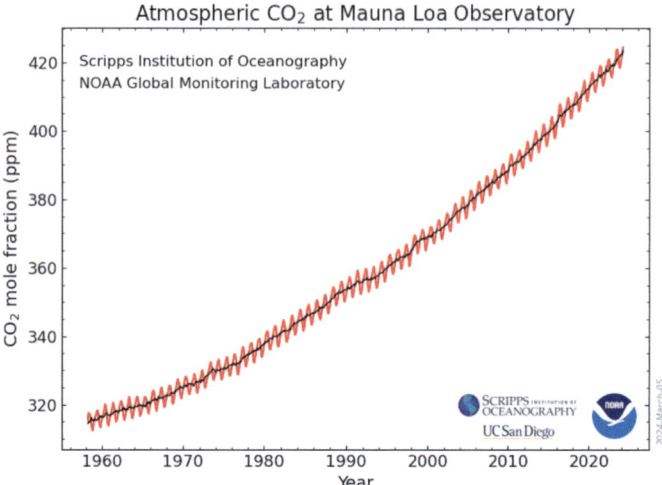

Fig. 12 Carbon dioxide (CO_2) data on Mauna Loa, Hawaii, directly measured from the atmosphere. From NOAA, https://gml.noaa.gov/ccgg/trends/

Note that the cycling of the solid Earth itself (the geosphere) via plate tectonics is another critical cycle in the Earth's climate system, which is especially important for understanding boundary conditions, forcings, and feedbacks over long-time scales. A thorough description of plate tectonics and its direct connections to past climate change are addressed in Chap. 6.

Processes that are part of the cycling of energy, water, and carbon occur at different rates and occur across a range of timescales and spatial scales. Here we will focus on identifying major reservoirs (i.e., natural storage places) and global-scale pathways (i.e., processes) by which energy, water, and carbon move among and between reservoirs. As such, the concept of natural "**budgets**" will also be introduced for each to characterize (i.e., quantify) the amounts and rates of exchange (i.e., the fluxes) which will highlight the variations in how quickly or slowly different processes operate.

4.1 *Energy Transfer and the Global Energy Budget*

Energy in Earth's climate system controls the surface temperature of the planet. As you learned in Sect. 2, incoming solar radiation (i.e., insolation) provides the heat energy that drives circulation in (and between) the atmosphere and hydrosphere. Incoming solar radiation also provides the energy needed for biochemical reactions of photosynthesis, which serves as the base of a complex planetary food web that involves nearly the entire biosphere. Energy also moves to the surface from Earth's hot interior. This geothermal energy drives convection in the geosphere which creates, moves, and destroys tectonic plates—and vents heat, gases, and molten rock at Earth's surface in the process. However, of these two end member sources of energy (the Sun vs. Earth's interior), it is Sun that contributes by far the most energy available to the Earth's surface, contributing over 99.9% of the heat flux[14] to Earth's surface environment. For this reason, our discussion of energy in the climate system will focus on insolation as the primary input, and we can disregard the input of heat energy from Earth's interior until it is discussed in Chap. 6 with reference to plate tectonic-drivers of climate change.

Energy is not only added to Earth's climate system, driving a wide range of processes, but energy also leaves Earth's climate system. Therefore, it is customary to consider energy in Earth's climate system as being part of a "**budget**" of inputs and outputs. Like a financial budget, if the inputs are equal to the outputs, the budget remains steady and is in balance. Contrastingly, if the inputs and outputs are not in balance, then there will be either a net gain or loss; in the case of Earth's energy budget this would results in Earth's global average temperature increasing or decreasing, respectively.

In the following discussion we will examine the global energy budget of the planet as a way to understand the input and output of energy and to characterize the energy flows of relevance to the climate system. Figure 13 is central to this discussion. The colored arrows show the directions and interactions of energy flow, and the numbers quantify the annual magnitudes of energy flux during the early twenty-first century. The numbers are in units of watts per square meter (W/m^2), which can be thought of as the amount of energy striking, or being emitted by, a 1 m by 1 m area each year. Let's break this energy budget down into pieces, by starting with the incoming solar energy and seeing how that interacts with Earth's atmosphere and surface; then turning to the outgoing thermal energy and the particularly important interactions with gases in the atmosphere that result in the greenhouse effect. (Recall you were introduced to the greenhouse effect in Sect. 3.1.)

Approximately 340 W/m^2 of energy from the Sun enters Earth's atmosphere each year (Fig. 13 yellow arrow). This energy is in the form of electromagnetic radiation from the ultraviolet (UV), visible, and a portion of the infrared (IR) parts of the electromagnetic spectrum. Appropriately, this radiation from the Sun is termed "shortwave radiation" as these wavelengths are shorter than those of the outgoing energy in the climate system. Thirty percent (100 W/m^2) of this insolation is reflected by the atmosphere and Earth's surface, whereas 70% is absorbed by the atmosphere (80 W/m^2) and Earth's surface (160 W/m^2). In other words, the globally averaged albedo of Earth is 30%. The albedo of the atmosphere, land, and ocean surface is the major factor in whether the insolation is reflected or absorbed, with higher albedo surfaces (e.g., the upper surface of many clouds, snow and ice) being more reflective. The angle at which the solar radiation strikes the Earth's surface is a secondary factor in determining whether the insolation energy is reflected or absorbed, with lower angles promoting greater reflectivity. Energy that is absorbed contributes to heating the atmosphere, land, and surface ocean.

---

[14] See https://ugc.berkeley.edu/background-content/earths-internal-heat/

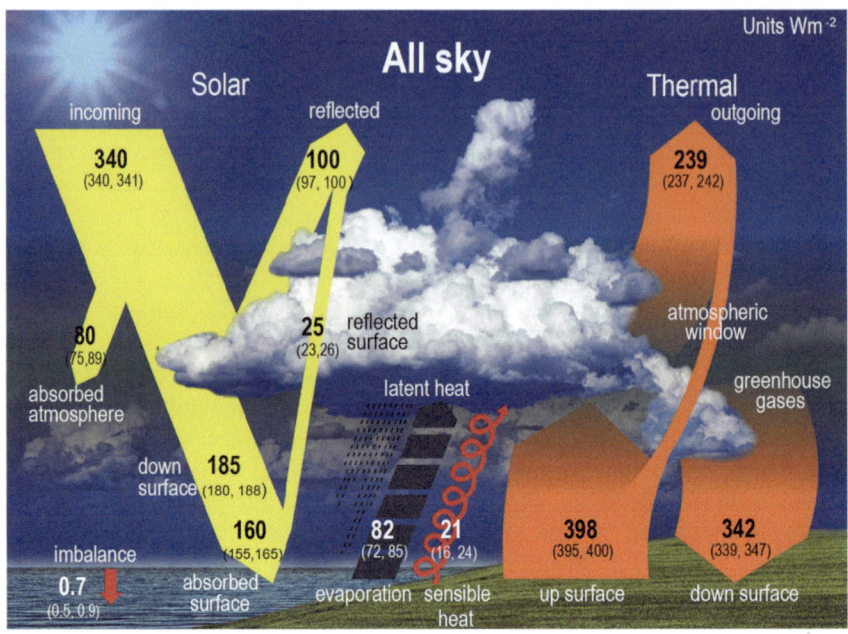

Fig. 13 Schematic representation of the global mean energy budget of the Earth. Numbers indicate best estimates for the magnitudes of the globally averaged energy balance components in W m^{-2} together with their uncertainty ranges in parentheses (5–95% confidence range), representing climate conditions at the beginning of the twenty-first century. *Source*: IPCC (2021)[15] Figure used with permission from the IPCC. This figure is excluded from Creative Commons license

While the Earth's surface absorbs radiation, it also emits radiation because it is warm (398 W/m^2; Fig. 12, orange arrow); this **thermal energy** is entirely within the IR part of the electromagnetic spectrum and has longer wavelengths than the incoming solar radiation entering Earth's climate system. Greenhouse gases in the upper troposphere interact with a large percentage of this longwave radiation, absorbing the outgoing radiation and re-radiating its thermal energy both up (out to space) and down (back to Earth's surface). The portion radiated back down to Earth's surface (342 W/m^2) creates the **greenhouse effect**, and is critical to keeping the average global temperature in moderation. Without a greenhouse effect Earth's

---

[15] *Source* Fig. 2 in Chap. 7 in IPCC, 2021: Chapter . In: Climate Change 2021: The Physical Science Basis. Contribution of Working Group I to the Sixth Assessment Report of the Intergovernmental Panel on Climate Change [Forster, P., T. Storelvmo, K. Armour, W. Collins, J.-L. Dufresne, D. Frame, D.J. Lunt, T. Mauritsen, M.D. Palmer, M. Watanabe, M. Wild, and H. Zhang, 2021: The Earth's Energy Budget, Climate Feedbacks, and Climate Sensitivity. In Climate Change 2021: The Physical Science Basis. Contribution of Working Group I to the Sixth Assessment Report of the Intergovernmental Panel on Climate Change [Masson-Delmotte, V., P. Zhai, A. Pirani, S.L. Connors, C. Péan, S. Berger, N. Caud, Y. Chen, L. Goldfarb, M.I. Gomis, M. Huang, K. Leitzell, E. Lonnoy, J.B.R. Matthews, T.K. Maycock, T. Waterfield, O. Yelekçi, R. Yu, and B. Zhou (eds.)]. Cambridge University Press, Cambridge, United Kingdom and New York, NY, USA, pp. 923–1054].

average surface temperature would plunge to a very cold $-16\,°C$. This is an important point: while the primary input of energy to the Earth's climate system is solar energy, the Sun's radiation is not sufficient on its own to keep global temperatures above freezing. The augmented heating from a moderate greenhouse effect is one of the key factors that keeps our planet habitable (Chap. 7 explores this point further).

There are also two non-radiative components of the Earth's energy budget; (1) the processes of convection and conduction, collectively labeled as "sensible heat" transfer (represented by a red spiral arrow in Fig. 13), and (2) the process of evaporative latent heat transfer (represented in Fig. 13 by the dark gray arrow). The transfer of heat via molecule-to-molecule conduction and the vertical movement of heat as hot air rises (convection) transfers 21 W/m^2 of heat energy from the Earth's surface to the atmosphere and changes the temperature in these parts of the system. Latent heat transfer is a more significant process, redistributing 82 W/m^2 as water absorbs heat energy from its surroundings when it evaporates from the Earth's surface and subsequently releases this heat energy to the atmosphere when it condenses as liquid water droplets.

Notice that the energy budget diagram (Fig. 13) shows that the inputs and outputs are currently not in balance; the input of solar radiation is 340 W/m^2 and the output from a combination of reflected solar (short wave) radiation (100 W/m^2) and Earth emitted thermal (longwave) radiation (239 W/m^2) are not equal. There is an imbalance of $+$ 1 W/m^2 (or more specifically $+$ 0.7 W/m^2 as shown in the bottom left corner of the diagram). This net gain of energy in the climate system is the reason why global average temperatures have been increasing in recent decades; it is the reason for modern global warming. Variations in the energy balance can occur over short and long-time periods. Processes that increase or decrease the amount of greenhouse gases impact the Earth's energy balance, as do changes in solar radiation and changes in Earth's albedo. The current imbalance is driven by anthropogenic greenhouse gas forcing.

One way to see this impact is to consider "**radiative forcing**", which is a measure of the direct effect a forcing factor has on the Earth's energy budget. Figure 14 shows the change in radiative forcing of the most important factors impacting Earth's energy budget in the last ~ 250 years. The vertical axis is in the same units (W/m^2) as the energy budget diagram (Fig. 13). Positive values add energy to the climate system, and negative values reduce energy in the climate system. All but two factors plotted are anthropogenic in origin; the two natural system factors are solar and volcanic forcings. Both of these factors will be explored more in Chap. 3. Anthropogenic CO_2 is clearly the primary driver of the total change in radiative forcing. It has caused the multi-decadal rise in total CO_2 in the atmosphere (Fig. 12) and produced the positive radiative forcing (Fig. 14). Notice too that short-lived explosive volcanic eruptions and more persistent tropospheric aerosols from human activities have negative radiative forcings; this is because the aerosols from these sources increase the planet's albedo, blocking some of the incoming solar radiation. The potential intentional use of human-introduced aerosols as a way to reduce greenhouse gas warming will be discussed in Chap. 11.

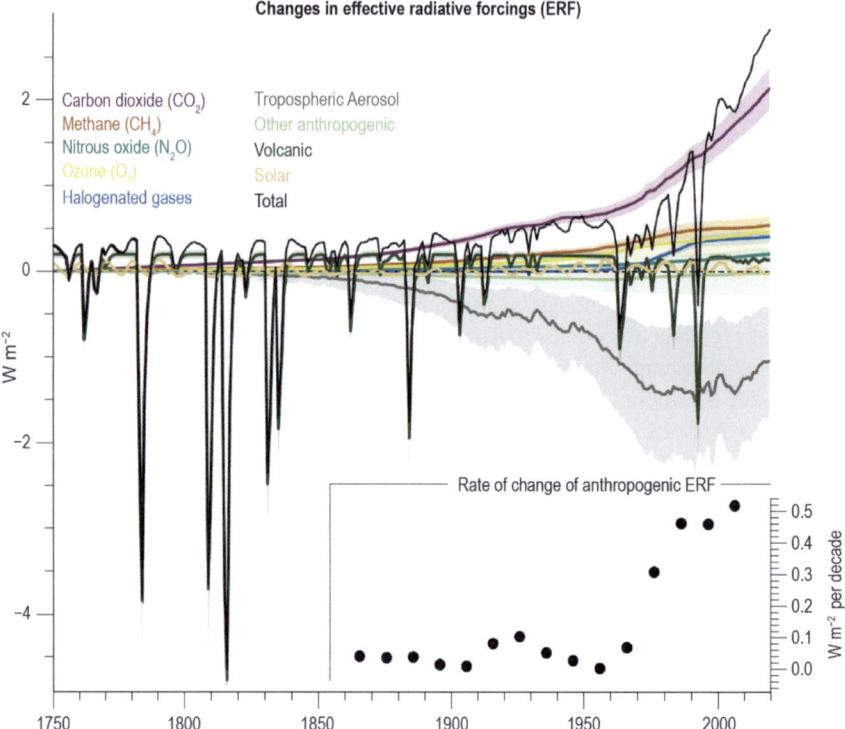

Fig. 14 Measure of the effect that natural (volcanic and solar) and anthropogenic forcing factors (all others) have had on Earth's energy budget between 1750 and 2021. The effective radiative forcing of these factors is in Watts per square meter (W/m^2), the same unit of the energy budget displayed in Fig. 13. The inset graph shows the rate of change of anthropogenic effective radiative forcing (W/m^2 per decade) and indicates a steep increase in the rate of anthropogenic forcing after 1950. *Source*: IPCC (2021)[16] Figure used with permission from the IPCC. This figure is excluded from Creative Commons license https://www.ipcc.ch/report/ar6/wg1/figures/chapter-2/figure-2-10. Original caption and full source reference in footnote

---

[16] *Source*: Fig. 10 in IPCC, 2021: Chap. 2. In: Climate Change 2021: The Physical Science Basis. Contribution of Working Group I to the Sixth Assessment Report of the Intergovernmental Panel on Climate Change [Gulev, S.K., P.W. Thorne, J. Ahn, F.J. Dentener, C.M. Domingues, S. Gerland, D. Gong, D.S. Kaufman, H.C. Nnamchi, J. Quaas, J.A. Rivera, S. Sathyendranath, S.L. Smith, B. Trewin, K. von Schuckmann, and R.S. Vose, 2021: Changing State of the Climate System. In Climate Change 2021: The Physical Science Basis. Contribution of Working Group I to the Sixth Assessment Report of the Intergovernmental Panel on Climate Change [Masson-Delmotte, V., P. Zhai, A. Pirani, S.L. Connors, C. Péan, S. Berger, N. Caud, Y. Chen, L. Goldfarb, M.I. Gomis, M. Huang, K. Leitzell, E. Lonnoy, J.B.R. Matthews, T.K. Maycock, T. Waterfield, O. Yelekçi, R. Yu, and B. Zhou (eds.)]. Cambridge University Press, Cambridge, United Kingdom and New York, NY, USA, pp. 287–422, https://doi.org/10.1017/9781009157896.004.] Original caption for Fig. 10 in IPCC, (2021): Temporal evolution of effective radiative forcing (ERF) related to the drivers assessed in Sect. 2.2. ERFs are based upon the calculations described in Chapter , of which the global annual mean, central assessment values are shown as lines and the 5–95% uncertainty range as shading

Additionally, processes in the climate system can indirectly affect the Earth's energy budget—through feedbacks, which either amplify or diminish the impact of existing forcings. Some examples include the snow and ice albedo feedback (described in Sect. 3.3), a positive feedback affecting the albedo in the energy budget. The sea ice insulation feedback (also described in Sect. 3.3) is also a positive feedback, affecting the energy budget by changing sensible and latent heat transfer from the surface ocean to the atmosphere. A third example is the water vapor feedback; this too is a positive feedback and is triggered when natural or anthropogenic forcings cause a temperature change that affect the rate of evaporation and thus the transfer of water vapor to the atmosphere. Any change in the amount of water vapor in the atmosphere will impact the energy balance because water vapor is a greenhouse gas.

In sum, changes in Earth's temperature result from a combination of direct and indirect processes that affect Earth's energy budget. The directions, magnitudes, and rates of change vary over time, and a deep time geologic perspective on the processes that cause warming and cooling helps put the modern climate in perspective. Chaps. 3 through 9 will provide you with a greater appreciation for that geologic perspective. Understanding and quantifying the processes that affect the energy budget are also essential for projecting future climate change using global climate models (the focus of Chap. 10), and for designing/engineering climate intervention approaches (the focus of Chap. 11).

4.2 Water Cycle Reservoirs and Transfer Fluxes

The movement of water within and between the various components of Earth's climate system is driven primarily by the interplay of incoming solar energy and gravity. Processes driven by Earth's internal heat, such as plate tectonics and associated volcanic eruptions, processes in the upper atmosphere, and the arrival of meteoric material on Earth can also influence the amount and movement of water in Earth's climate system, but the influence of surficial heating by insolation generally is much more important than those other influences. As a result, the amount of water present at Earth's surface and moving through the water cycle is considered to have been essentially constant through all but Earth's earliest history (i.e., approximately its first 800 million years of existence). This discussion will focus on the natural water cycle at and close to Earth's surface, including its major natural **reservoirs**, the rates of transfer between those reservoirs (i.e., the **fluxes**), and a reminder of the role and impact of water transfer on the various subsystems of Earth's climate system. The role that human activities play in the water cycle will also be pointed out.

(Sect. 3 in Chap. 7, see Figs. 6, 7 and 8 in Chap. 7 for more detail on uncertainties). The inset plot shows the rate of change (linear trend) in total anthropogenic ERF (total without TSI and volcanic ERF) for 30-year periods centered at each dot. Further details on data sources and processing are available in the chapter data table (Table 2.SM.1).

The movement of water around Earth's surface and near-surface is tied fundamentally to the effects of insolation and global average temperatures, and the geographic variations in these factors. Regions receiving high insolation are provided the energy needed to warm water, to melt ice, and to evaporate water, whereas regions receiving low insolation are areas where water cools, condenses from water vapor and precipitates, and freezes. Water has a high heat capacity, so a large amount of heat is needed to change its temperature. In addition, the amount of heat needed to change water's phase is also very high. As a result, water acts as a "thermal buffer" in Earth's climate system, absorbing or releasing large amounts of energy (i.e., both sensible and latent heats) without undergoing extreme or rapid changes in temperature or phase. In addition, the movement of large quantities of water by circulation in the atmosphere and the ocean is an effective mechanism for the meridional transfer (i.e., north–south transfer) of excess heat from the low latitudes to the high latitudes.

The large-scale geographic variations in insolation have some temporal variability as well, as observed by seasonal shifts in solar heating and the response to that heating by the climate system. One large-scale observable example of seasonally paced variability in the water cycle is the north and south shifting of the Intertropical Convergence Zone (ITCZ) and its associated rainfall, which on a first order aligns to the latitudes of maximum solar heating and is secondarily influenced by the differential heating of the continents and the ocean. A second example of seasonal changes in the water cycle involves the seasonal formation and melting of sea ice in the Arctic. Between 1981 and 2010, late winter sea ice covered ~ 15.5 million square kilometers (6 million square miles) of the Arctic Ocean, whereas late summer sea ice covered ~ 6.5 million square kilometers (2.5 million square miles) of the Arctic Ocean; a seasonal reduction (and regrowth) of ~ 58%.[17] Changes in water cycle reservoir sizes and fluxes also occur over longer time scales. For example, regional multi-year to multi-decade variations in heating and cooling, such as the El Nino-Southern Oscillation, affect precipitation and drought patterns in the equatorial Pacific and beyond (Chap. 3). Planetary-scale orbital-driven changes affecting insolation occur over many thousands of years, and cause cyclic redistributions of water between its major reservoirs—expanding water in the cryosphere during cool periods and redistributing it back to the oceans during warm periods (Chap. 9). But most immediate are the modern human-driven changes to the Earth's energy budget that impact the water cycle; the extra energy in the system from excess greenhouse gases is accelerating and altering the global water cycle. Anthropogenic activities that generate more CO_2 and CH_4 (and other greenhouse gases; Fig. 14) provide additional heat and cause more evaporation, and thus more water vapor, a greenhouse gas itself, which in turn drives a positive feedback resulting in greater warming. With greater heat energy and water vapor in the atmosphere, shifts in the amount and location of precipitation are occurring, resulting in more extreme weather events, including hurricanes, floods, and droughts depending on geographic location.

---

[17] From the National Snow and Ice Data Center, https://nsidc.org/learn/parts-cryosphere/sea-ice.

Gravity also plays an important role in Earth's water cycle, although gravity's role is mentioned much less often than that of insolation. Gravity is essential for driving precipitation from the atmosphere to Earth's surface, for driving the downhill flow of surface water, groundwater, and ice on land, and for driving flow in the oceans between areas of excess precipitation and areas of excess evaporation. As a result, the cyclic nature of Earth's water cycle depends on the effects of gravity.

Figure 15 is a visual summary of Earth's "natural" water cycle, with additional notation on human influences such as recognition that population growth, urbanization, deforestation and agricultural activities impact the size of the terrestrial biosphere and the rates of both water use and transpiration. While qualitative, there is a lot going on in this diagram. As you might imagine, it is very challenging to measure the fluxes between reservoirs on a global scale, although modern satellite-based instruments have improved our ability to measure some of them (e.g., evaporation from remote portions of the oceans and continents, or discharge from large and remote rivers). As a result, much research continues with the goal of better understanding Earth's water budget today, and how that budget—and its components—are likely to change in the future.

The results of recent research on reservoir volumes and the magnitudes of important water fluxes are summarized in Table 3. The dominant role of the ocean and its fluxes—loss of water due to evaporation and gain of water due to precipitation—are evident when the size of this reservoir and its fluxes is compared to the other entries in Table 3. Note that the majority of water evaporated from the ocean is returned to the ocean by precipitation, although likely at a different latitude after being transported in the atmosphere. Evaporation from the surface ocean is concentrated in the regions of the trade winds, whereas precipitation to the surface ocean is enhanced in the low-latitude ITCZ and in regions poleward of the trade winds. These geographic differences are an important control on the salinity of the surface ocean, producing highest sea surface salinities in the regions subject to enhanced evaporation and lower sea surface salinities in the regions receiving enhanced precipitation.

Precipitation over land is sourced from a combination of moisture that was evaporated over the ocean and moisture supplied by transpiration. The relative proportions of ocean-supplied vs. land-supplied moisture for precipitation vary significantly across the continents, and also can vary significantly at a specific location across the seasons. The dramatic seasonal variations in precipitation in areas that experience summer monsoonal rains (e.g., the Himalayas and the Colorado Plateau) are a result of such variability, with ocean-supplied moisture contributing more during the wet season than during the dry season.

An estimate of the "residence time"—or the average stay of a water molecule in a reservoir—can be calculated from the volume of each reservoir and its input flux or output flux. The estimated residence times are listed in Table 3 and vary from a few hours or days for the terrestrial biosphere, to a few days or weeks for the atmosphere, to several thousand years for the oceans, and to as much as hundreds of thousands of years for the large ice sheets. Although the timescales involved are not exactly

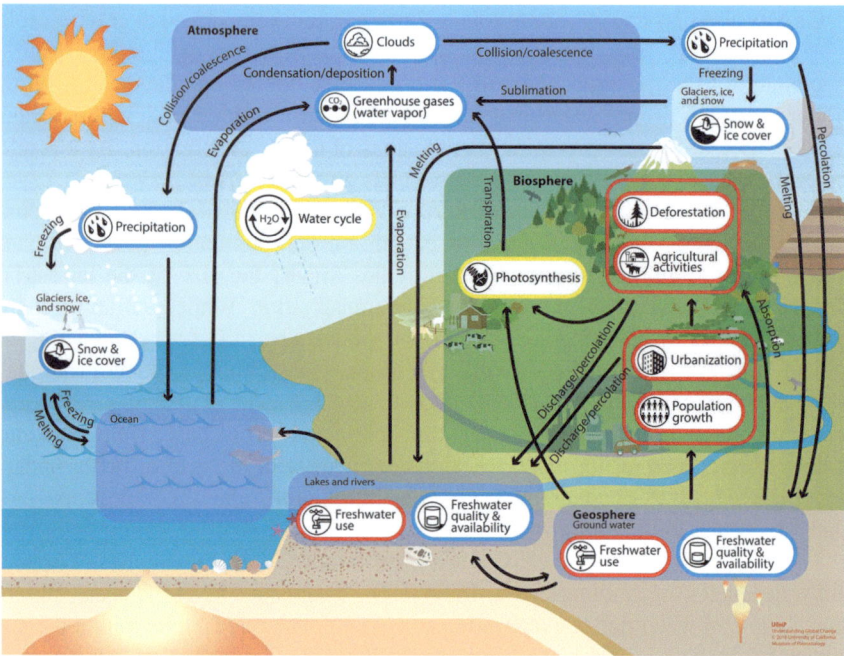

Fig. 15 Representation of Earth's water cycle, including major biosphere, geosphere, hydrosphere, cryosphere, and atmosphere reservoirs and the transfer processes (indicated by black arrows). Human activities that play roles in the water cycle are noted in red. Used with permission from the University of California Museum of Paleontology, https://ugc.berkeley.edu/background-content/water-cycle/. © UC Museum of Paleontology Understanding Global Change, www.understandingglobalchange.org This figure is excluded from Creative Commons license

the same values, these residence times for water also provide a general indication of how quickly the respective parts of the climate system respond to climate change. The atmosphere and biosphere respond most quickly, followed successively by the non-ocean hydrosphere, the oceans, and the cryosphere.

Because the atmosphere has one of the shortest residence times for water, its behavior during evaporation and precipitation has been predicted to respond relatively quickly to the increasing global temperatures of the past several decades. This prediction is based on the physical fact that a warm atmosphere can hold much more water vapor than a cold atmosphere. Under present conditions, the atmosphere can hold 7–10% more moisture for each 1° C increase in temperature, increasing the probability of more-frequent and more-intense rainfall events, with accompanying risks for floods, other natural hazards, and risks to agriculture. Recent research has found evidence for this acceleration of the water cycle, which has important societal and socioeconomic implications. For example, city planners are realizing that the capacities of urban stormwater drainage systems have to be increased, and large insurance companies are recalculating the risks from more-frequent large storms and floods.

Table 3 Natural water cycle reservoirs and transfer fluxes

Reservoir	Volume (km^3)	Output flux to (km^3/yr)	Input flux from (km^3/yr)	Residence time
Ocean	1.3 billion	Evaporation to Atmosphere 0.4 million	Precipitation from Atmosphere 0.38 million	Several 1000 years
			Inflow from Streams and Groundwater 0.05 million	
Ice Sheets and Glaciers	26 million	Melting and Iceberg Calving to Rivers, Lakes, and Ocean Poorly known	Precipitation from Atmosphere Poorly known	Decades to hundreds of 1000s of years
Groundwater	15 million	Discharge to Ocean 0.004 million	Precipitation from Atmosphere 0.013 million	Several years to tens of 1000s of years
Lakes and Rivers	0.202 million	Evaporation to Atmosphere and Discharge to Ocean Poorly known	Precipitation from Atmosphere Poorly known	Several months to several 100s of years
Soil Moisture (Not shown in Fig. 14)	0.05 million	Evaporation to Atmosphere and Uptake by Biosphere 0.073 million	Precipitation from Atmosphere Poorly known	Several months
Atmosphere	0.013 million	Precipitation to Ocean and Land 0.5 million	Evaporation from Land and Ocean and Transpiration from Biosphere 0.5 million	A few days to several weeks
Terrestrial Biosphere	0.001 million	Transpiration from Plants < 0.07 million	Precipitation from Atmosphere and Uptake from Soil Moisture 0.1 million	Hours to days

4.3 Carbon Cycle Reservoirs and Transfer Fluxes

The importance of carbon in the climate system has already been made clear in our discussion of Earth's energy budget; the rise in atmospheric CO_2 from fossil fuel-sourced human activities is the major driver of the net gain in radiative forcing, creating an energy budget imbalance and causing the rise of global average temperatures. Fossil fuels are not the only reservoir of carbon in the Earth's climate system, nor are human activities the only means of transferring carbon between reservoirs. The redistribution of carbon among reservoirs is a major cause of climate change, now and in the geologic past. Some processes transfer carbon quickly and others slowly; thus, carbon cycles among the reservoirs at different rates—from biosphere-driven seasonal cycles (e.g., Fig. 12) to geosphere-driven million-year shifts in global climate state. Understanding the reservoirs, transfer pathways, and rates of carbon exchange in the global carbon cycle is the focus of this final section of Chap. 2.

Figure 16 shows the main reservoirs (i.e., storage bins) of carbon in the Earth's climate system. The numbers in the figure are the sizes of reservoirs, in gigatons (1 gigaton = 1 billion metric tons). The same reservoirs are listed in Table 4, which has added information on the form of carbon in each reservoir, the input and output processes and rates, and the residence times of carbon for these reservoirs. The largest carbon reservoir by far is the geosphere, most abundantly in sedimentary materials composed either of microscopic $CaCO_3$ fossil shells or of muds rich in organic carbon; these sedimentary materials are described further in Chap. 5. Fossil fuels comprise a much smaller portion of the geologically stored carbon, and are derived—over very long-timescales—from the concentrated, buried remains of organic carbon produced through photosynthesis. The oceans are the second-largest reservoir of carbon, and include carbon in many forms: as dissolved carbonate (CO_3^{2-}) and bicarbonate (HCO_3^-) ions, dissolved organic C, dissolved CO_2 gas, and as marine biota. Together, carbon stored in the geosphere and ocean makes up 99.9% of all carbon in the Earth system. The smaller reservoirs of the carbon cycle are likely the ones that we interact with the most: the terrestrial biosphere, the soil and the atmosphere. Carbon in the terrestrial biosphere includes all living land-based biota. Soil, being at the interface between the underlying bedrock and the overlying terrestrial biosphere and atmosphere, is a mix of carbon in rock fragments, living and decomposing organic matter, and gases. Finally, carbon in the atmosphere is primarily in the form of CO_2 and CH_4 gas.

Many different natural processes are involved in the transfer of carbon among the reservoirs. Some of these processes move large amounts of carbon relatively quickly and others not so. The input and output fluxes (in gigatons per year (Gt/yr; Table 4) make this clear. For example, large fluxes of carbon are exchanged between the atmosphere and the terrestrial biosphere via photosynthesis (142 Gt/yr) and respiration (111 Gt/yr), with the remainder being transferred from plants to soil via decomposition and burial. Similarly, gas exchange from the atmosphere to the ocean moves 80 Gt/yr, and a return from the ocean to the atmosphere of 78 Gt/yr. All of the other natural transfer processes among the various reservoirs move much smaller amounts on an annual basis. For this reason, we can consider there being fast pathways within the carbon cycle—the exchanges between the terrestrial biosphere and the atmosphere, and between the surface ocean and the atmosphere—and slow pathways (all of the rest). As a result, the residence time of a carbon atom in the atmosphere and the terrestrial biosphere is very short—only ~ 10 years. A carbon-containing molecule may stay in unfrozen soil for ~ 50 years (if permafrost it would be much longer), and in the ocean for ~ 500 years. In contrast, a carbon-bearing molecule that is transferred to the geosphere typically resides in the geosphere for

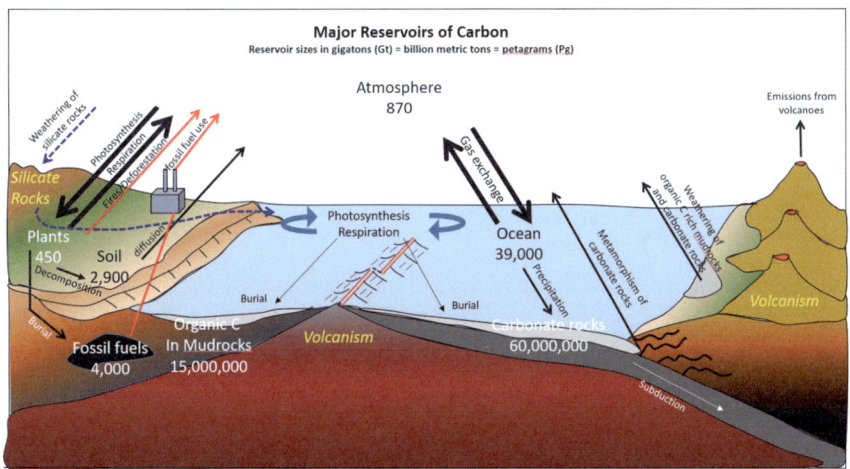

Fig. 16 Representation of the global carbon cycle. The sizes of reservoirs are in gigatons (1 gigaton = a billion metric tons = 1 petagram). For example, the atmosphere reservoir is 870 gigatons. The arrows in the figure mark the flow of carbon (also called carbon flux). Arrows show the pathways carbon can take among and between the different reservoirs. Black and blue arrows are used for natural processes, whereas red arrows are used for processes related to human activities. Heavy black arrows indicate the processes that occur most quickly. Figure by Kristen St John and Lee Kump

much longer—from ~ 5000 (for organic C-rich mudstones) to 200,000,000 years (for fossil fuels and carbonate rock). That is unless human activities intercede by transferring geologically-stored carbon to the atmosphere much more quickly by extracting and combusting fossil fuels.

To illustrate this point let's take a special look at the rates and process by which molecules containing carbon in the atmosphere (CO_2) are transferred via the biosphere to the geosphere to create fossil fuels, and contrast that with the rate and process by which the CO_2 in fossil fuels is returned to the atmosphere. Two terms are associated with the processes that move carbon through the carbon cycle to ultimately create fossils fuels: the **terrestrial biological pump** and the **marine biological pump**. In the terrestrial biological pump, molecules of atmospheric CO_2 are transferred to the terrestrial biosphere via photosynthesis. When the plant dies a small fraction of that carbon may be buried in soil or sediment. Over time this organic-rich material (peat) gets buried even more, heated and compressed, creating the fossil fuel sedimentary rock coal. In the marine biological pump, gas exchange moves CO_2 from the atmosphere to the ocean, where photosynthetic microscopic life synthesizes the dissolved CO_2 into organic carbon. When marine life in the ocean dies and sinks, the organic carbon is moved to the deeper ocean. A small fraction of it will be buried on the seafloor forming organic-rich mudrock, which over time, and with additional burial and heat, can form the fossil fuels oil and natural gas. In either case, these biological pumps take many millions of years to move CO_2 from the atmosphere into fossil fuels in the geosphere. The transfer rate of carbon from

Table 4 Carbon cycle reservoirs and transfer fluxes. Anthropogenic processes are in italic[18]

Reservoir	Form of carbon	Mass of carbon stored (billion metric tons = "gigatons" $Gt = 10^9$ tons)	Output flux (Gt/yr)	Input flux (Gt/yr)	Residence time (years)
Atmosphere	As carbon dioxide (CO_2) and methane (CH_4)	870	To plants via photosynthesis 142	From plants via respiration 111	10
				From plants via fires 26	
				From land use changes (deforestation, urbanization) 1.6	
				From fossil fuels use and cement production 9.4	
			To ocean via gas exchange 80	From freshwater via gas exchange 1.5	
				From ocean via gas exchange 78	
			To silicate rocks via weathering reactions 0.3	From organic-rick C mudrocks and carbonate rocks via weathering reactions < 0.1	
				Form carbonate rocks via degassing during metamorphism < 0.1	
				From volcanic emissions 0.1	

(continued)

---

[18] Sources for Table 2.4: IPCC https://www.ipcc.ch/report/ar6/wg1/figures/chapter-5/figure-5-12; and NNEP/GRID-Arendal https://www.grida.no/resources/6453.

Table 4 (continued)

Reservoir	Form of carbon	Mass of carbon stored (billion metric tons = "gigatons" Gt = 10^9 tons)	Output flux (Gt/yr)	Input flux (Gt/yr)	Residence time (years)
Fossil fuels	As coal, oil, natural gas (all are organic carbon molecules)	4000	*To atmosphere via fossil fuels use and cement production 9.4*	From plants via the burial and conversion of organic C to fossil fuels < 0.1	200,000,000
Oceans	As dissolved carbonate (CO_3^{2-}) and bicarbonate (HCO_3^-) ions, dissolved organic C, dissolved CO_2 gas, and marine biota	39,000	To atmosphere via gas exchange 78	From river transport of weathered rocks and soils 0.8	500
			To ocean floor sediment via burial 0.2		
Organic C in Mudrocks	For example, black, organic-rich mudstone and shales	15,000,000	To fossil fuels via burial, heat and pressure < 0.1	From ocean via burial 0.2	5000
			To atmosphere or ocean via weathering < 0.1		

(continued)

Table 4 (continued)

Reservoir	Form of carbon	Mass of carbon stored (billion metric tons = "gigatons" Gt = 10^9 tons)	Output flux (Gt/yr)	Input flux (Gt/yr)	Residence time (years)
Terrestrial Vegetation	Plants on land	450	To atmosphere via plant respiration 111	From atmosphere via photosynthesis 142	10
			To atmosphere via fires 26		
			To soil via burial 0.2		
			To organic C in mudrock via burial < 0.1		
			To atmosphere from land use changes (deforestation, urbanization) 1.6		
Carbonate Rocks	As limestone and marble ($CaCO_3$)	66,000,000	To atmosphere or ocean via weathering < 0.1	From ocean via rock precipitation < 0.1	200,000,000
			To atmosphere via degassing during metamorphism < 0.1		
Soil, including permafrost (frozen soil)	As organic matter, CO_2, and methane (CH_4)	2900	To atmosphere via diffusion < 0.1	From plants via burial 0.2	50 (for unfrozen soil)
			To rivers 2.5		

natural weathering of exposed organic-rich sedimentary rocks (e.g., organic C-rich mudrock and fossil fuels) back to the atmosphere is < 0.1 Gt/yr. In contrast, the extraction and combustion of fossil fuels can occur within months to years, adding almost two orders of magnitude more CO_2 back to the atmosphere than the natural weathering rates of organic-rich sedimentary rocks. The result is an imbalance in the carbon cycle caused by anthropogenic activities; and because CO_2 is a greenhouse gas, this creates an imbalance in the Earth's energy budget, resulting in the increase in global average temperatures in recent decades (Sect. 4.2). Other anthropogenic activities that add carbon to the atmosphere and increase this modern imbalance include the production of cement (which is similar to the metamorphism of carbonate rock, but occurs on a much faster timescale), changes in land use such as deforestation, and intentional or unintentional forest fires.

While we know that anthropogenic activities are responsible for the increase in carbon-based atmospheric greenhouse gases in recent decades, paleoclimate proxies of the geologic past (including several you will learn about in future chapters) indicate that natural processes in the carbon cycle that impact atmospheric CO_2 have sometimes changed as well. Changing rates of volcanism, of carbon burial and rock formation, and of rock weathering, as well as feedbacks in the climate system (e.g., a warm ocean can't store as much dissolved CO_2 as a cold one), have redistributed carbon among the different reservoirs. At different times in the geologic past—long before people were influencing the carbon cycle—the atmospheric carbon reservoir was sometimes smaller than today, and at other times much larger than today; at times the atmospheric carbon reservoir was so much larger that global temperatures prevented ice sheets from existing anywhere on the planet. Understanding the natural climate variability, the climatic forcings and feedbacks, and the sometimes-interrelated processes that affect the global carbon cycle, the water cycle and the Earth's energy budget will be explored throughout the rest of the book.

References

Gayatri, K., Cheng, H., Sinha, A., Yi, L., Li, X., Zhang, H., Li, H., Ning, Y., and Edwards, L.A. (2017). The Indian monsoon variability and civilization changes in the Indian subcontinent. *Science Advances, 3*, e1701296. https://doi.org/10.1126/sciadv.1701296

Koltermann, K. P., Gouretski, V. V., Jancke K. (2011). Hydrographic atlas of the world ocean circulation experiment (WOCE). Volume 3: Atlantic Ocean (eds.) Sparrow, M., Chapman P., Gould J. International WOCE Project Office, Southampton, UK, ISBN 090417557X. https://doi.org/10.21976/C6RP4Z, specific image from: http://whp-atlas.ucsd.edu/whp_atlas/atlantic/a16/sections/printatlas/printatlas.htm

Koppers, A., & Coggon, R. (Eds.) (2020). *Exploring Earth by Scientific Ocean Drilling: 2050 Science Framework*. Integrated Ocean Discovery Program, p 132, https://www.iodp.org/2050-science-framework

McConnell, J. R. (2005). 198 Role and importance of cryospheric processes in climate system. In Anderson, M. G., (Ed.), *Encyclopedia of Hydrological Sciences*. https://doi.org/10.1002/0470848944.hsa208

Mishra, V., Aadhar, S. (2021). Famines and likelihood of consecutive megadroughts in India. *Nature Climate and Atmospheric Science, 4*(59). https://doi.org/10.1038/s41612-021-00219-1

Mishra, V. (2020). Long-term (1870–2018) drought reconstruction in context of surface water security in India. *Journal of Hydrogeology, 580*, 124228. https://doi.org/10.1016/j.jhydrol.2019.124228

Lawrence Krissek Larry Krissek became intrigued with the oceans while completing the Boy Scouts' Oceanography merit badge and a middle school science fair project, even though he lived far from the coast. He has been fortunate to be able to pursue that interest throughout his career, combining his initial focus on scientific research with more-recent efforts to improve earth- and climate-science literacy across the K-16 spectrum. Larry's research interests focus on the information about past climates carried by land-derived sediments deposited in the ocean, especially the history of past ice sheets. Larry has spent his academic career in the School of Earth Sciences (and its predecessors) at Ohio State University. Larry is a primary author of Chaps. 2 and 9.

Kristen St. John Intent on being a political science major in college, Kristen's path changed after taking physical and historical geology as her science requirements. She's found that piecing together the stories that are preserved in climate archives of Earth's history is like working on a 10,000 + piece jigsaw puzzle. The challenges and rewards of figuring things out are much more fun–and solutions more attainable–when working with others. Synergistic research collaborations, and seeking ways to make geoscience more accessible and meaningful for students, have been guideposts for her career path. Her research focuses on reconstructing the history of land ice and sea ice by looking at clues in marine sediments. Kristen's degrees are from Furman University and The Ohio State University; she is a Professor at James Madison University. Kristen is the primary author of Chap. 1, a primary author of Chap. 2, and the author of Box 1 in Chap. 8.

Open Access This chapter is licensed under the terms of the Creative Commons Attribution 4.0 International License (http://creativecommons.org/licenses/by/4.0/), which permits use, sharing, adaptation, distribution and reproduction in any medium or format, as long as you give appropriate credit to the original author(s) and the source, provide a link to the Creative Commons license and indicate if changes were made.

The images or other third party material in this chapter are included in the chapter's Creative Commons license, unless indicated otherwise in a credit line to the material. If material is not included in the chapter's Creative Commons license and your intended use is not permitted by statutory regulation or exceeds the permitted use, you will need to obtain permission directly from the copyright holder.

Modern Climate Change in the Context of the Last Two Millennia

Thomas M. Cronin

Guiding Question: How are instrumental and paleoclimate records used to document and interpret recent climate changes?

1 Key Take–Away Points

- Determining the relative contributions of external "forcing" factors, including solar output, volcanic activity, and greenhouse gas concentrations, to observed changes in the global climate system is important for understanding how the climate system operates.
- In addition to external forcing, Earth's atmosphere, oceans, biosphere, and cryosphere interact with one another to produce internal climate variability, which has large impacts on hemispheric and regional climate patterns.
- Some patterns of internal climate variability take place at timescales of decades to centuries, which complicates our understanding of the warming and precipitation changes of the last century.
- An example of internal climate variability is El Niño-Southern Oscillation (ENSO), a quasi-periodic (3–7 year) oscillation originating in the tropical Pacific, that has huge impacts on seasonal climate patterns in the tropics and higher latitudes as well.
- Instrumental records of Earth's climate system obtained from satellites, thermometers, tide gauges, and other methods show major changes in Earth's climate system but are limited to the last few decades to centuries and often limited in their geographic distribution.

T. M. Cronin (✉)
United States Geological Survey, Reston, USA
e-mail: tcronin@usgs.gov

© The Author(s) 2025
K. St. John and L. Krissek (eds.), *Climate Change*,
https://doi.org/10.1007/978-3-031-82869-0_3

- Archives of paleoclimate from tree rings, coral skeletons, lake and ocean sediments, glaciers and ice sheets, among others preserve records of both external and internal modes of variability over all timescales.
- A historical example of the value of paleoclimate records is our effort to understand the Little Ice Age (LIA), cool periods in many regions from about 1550 to 1880 CE that preceded major industrialization and the human-caused rise in greenhouse gas concentrations. Most instrumental records do not extend back through the LIA, and thus paleoclimatic proxy reconstructions are required to understand its causes.
- Paleoclimate archives can be spliced (connected and combined) with instrumental records to construct long-term continuous records of key climate parameters, such as temperature, sea ice, sea level, glaciers and ice sheets, ocean circulation, and ocean chemistry.
- An example of the benefits of spliced instrumental and proxy records comes from the Arctic. Satellite measurements of Arctic sea-ice cover began only in 1979, however sea-ice proxies from sediment records (e.g., molecules produced by ice algae and microfossils that live in sub-sea-ice environments) let us extend sea-ice records back centuries to millennia.
- The most studied aspects of climate using spliced records are Northern Hemisphere and global atmospheric temperatures. Paleotemperature reconstructions going back 2000 years and longer can be compared to climate model simulations of temperature changes due to radiative forcing from volcanic, solar and greenhouse gas influences. These comparisons show unequivocally that forcing by increased greenhouse gas abundances has caused warming over the last century.
- An important focus of present and future work using spliced instrumental and proxy records is to better understand the natural modes of internal climate variability prior to significant human impact on the climate system. By understanding this natural variability better, researchers will improve their ability to identify more recent climate variations influenced by human activities.

2 Introduction

Understanding the patterns, causes and impacts of changes in Earth's climate system is a daunting challenge. It requires a multifaceted approach to complex, partially understood processes in the atmosphere, oceans, cryosphere and biosphere involving diverse specialists from a variety of disciplines. One of the most important applications of paleoclimatology is the study of modern climate change, including human-induced changes, in the context of climate changes over the last few millennia. In this sense, understanding modern climate change helps us to predict what is likely to happen in the future—the next year, decade or century, and what those future changes are likely to do to our physical and biological worlds and to human society. This chapter describes what we know about climate in the last 2000 years and how we know it. It will involve examination of several types of data depicted in graphs.

If you are relatively new to learning about climate change, take some time now to read Box 1, which will give you some tips and explain the different ways you may encounter climate and geologic data in graphical forms.

Box 1 Tips, conventions, and tools used in the study of geological and historical climate data

Figures in this chapter adopt several conventions used in many studies splicing instrumental and paleo-proxy records. First let's look at the issue of plotting time series of data. In figures, you will see the x-axes represent time, going from older on the left to younger [usually up to the present] on the right. The y-axes represent the particular climate parameter, such as temperature, or the proxy record of paleotemperature. In the literature, other conventions for graphical depiction of time are also used. For example, some studies plot the time scale in the opposite direction [older to the right]. Others adopt a traditional "stratigraphic" convention by plotting time on the y-axis, with oldest data at the bottom and youngest at the top, and the climate parameters along the x-axis. Stratigraphic approaches are used commonly for longer-term paleo-records of the Quaternary [the last 2.5 million years, Myr], or the Cenozoic Era (the last 66 Myr). Keep in mind that there is no universally accepted format for plotting time, so pay careful attention when comparing records from different studies (including records throughout this book, which draw on studies from a wide range of sources).

Another choice pertains to how age is depicted. There are many options depending on the type and quality of chronostratigraphic and age data available in paleo-proxy records. If radiocarbon dating is used, a method useful back to a maximum of about 50 kilo-annum (ka, or "thousand years"), results are typically presented in years before present (yr BP, which means years before 1950). In studies of the last 2 kyr (i.e., the last 2 thousand years), the "Common Era" (CE), the timescale is usually converted to calendar years. Common Era paleo-records often use annually resolved time series of proxy data—that is, data in which the value for each year can be determined. These annually resolved records come mostly from tree rings and coral skeletons, and less commonly from glacial ice and sediments. Annual resolution is preferred when comparing paleo and instrumental records, but decadal scale data (i.e., data where average values are determined decade-by-decade) are also used to examine low frequency climate variability.

The study of trends in climate parameters, such as temperature, is an extremely complex process. Imagine if you were asked what was the Earth's mean "temperature" for the year 2022. It would require many decisions and statistical approaches. Would you use daytime temperature? Nighttime? Seasonal averages? Summer only? How would you fill geographic gaps where no data are available? In order to obtain long instrumental records, researchers use temperatures measured with a variety of methods. Pre-satellite atmospheric records (~pre-1970s) are of various quality and geographic coverage, but go back to the mid-nineteenth century. The oceans, which cover 70% of Earth's surface, were not reliably monitored until the 1950s. Beginning about 20 years ago, an international program began that now has a fleet of ~ 4000 robotic instruments that drift with the ocean currents and move up and down between the surface and mid-water levels. These "ARGO" floats are providing continuous information about surface to sub-surface-ocean temperature and salinity.[1] ARGO data is then combined with prior ocean monitoring data to extend the record farther back in time.

According to NASA, instrumental records show that since 1880 total warming has been about 1.02 °C. This represents a rate of 0.07° C per decade, but, importantly, warming has accelerated as the rate rose to 0.18 °C per decade since 1981. Total human-induced warming since the industrial revolution of about 1.0° C is of concern reflected in highly publicized serious environmental implications if warming continues to 1.5 or 2.0 °C (see Chap. 1). We note that one convention is to report global temperature in terms of **temperature anomalies**. These are the difference of a temperature relative to a mean value for a baseline period (e.g., 1951–1980), rather than absolute planetary temperature. Such an approach is used for example in Fig. 4.

One important tool used by researchers is an approach called "**reanalysis**" in which temperature records of various spatial resolution are subdivided into geographic grid cells and subjected to rigorous statistical and modeling efforts to verify the instrumental observations. One effort, run by the National Centers for Environmental Prediction (NCEP) and the National Center for Atmospheric Research, exemplifies the scrutiny that atmospheric and oceanic temperatures are subjected to and the interested reader can find more information at the NCEP site.[2] The latest reanalysis (version 3), a 1.0 degree latitude × 1.0 degree longitude global grid, extends back to the year 1836. Some studies attempt to "hindcast" climate patterns using models and compare model output with climate observations to assess the strength of the models (see Chap. 10).

---

[1] Roemmich et al. (2019).

[2] National Centers for Environmental Prediction (NCEP), https://psl.noaa.gov/data/gridded/data.ncep.reanalysis2.html.

> Paleo-proxies are subjected to equally rigorous methods to establish how well they represent a particular climate parameter. Specifically, the proxy methods must be tested in lab or field experiments, clearly identifying their efficacy in recording a parameter with quantitative statistics and error bars. Using multiple proxies from sediment, ice, and coral cores is preferred. Tree-ring dendroclimatology adopts its own rigorous, quantitative methodology for each study region, type of tree, and proxy used. Paleoclimatology continues to evolve with the development of large hemispheric or global databases of proxy records, including many international efforts aimed at the Common Era of the last 2 kyr[3] or the entire Holocene.[4]

The basic approach to learning about climate change for this timeframe is to **splice** (i.e., connect and combine) instrumental records covering the last few decades to centuries with pre-instrumental paleoclimate "proxy" records covering previous centuries that come from various archives like sediments, coral reefs, speleothems (mineral deposits formed underground in caverns), tree rings, and ice cores. The fundamental goals of combining instrumental and proxy records are to assess natural variability and changes in the climate system due to natural and anthropogenic causes. This knowledge can then be applied to improving climatic, ocean, carbon cycling, glaciological, and other models that help predict future changes.

Combining instrumental and proxy records is needed because instrumental records, herein meant to include all types of observations from monitoring systems, including satellite measurements and imagery over recent decades, are used largely to understand weather and climate over short timescales (years to decades). This limitation applies to all human-observed records—both data from instruments (i.e., thermometers, complex earth-sensing satellites), and observations made by humans (i.e., qualitative reports from diaries or correspondence, or extracted from cultural artifacts like paintings). Therefore, instrumental and historical records are inherently insufficient to fully understand the longer-term patterns and causes of climate changes.

In this chapter, you will explore several examples of the value of splicing instrumental and paleoclimate records to document and understand climate change, but let's start with two simple and highly relevant examples: The **Little Ice Age (LIA)** describes cool periods in many regions from 1550 to as late as 1880 Common Era (CE). The LIA was the final and most intense of a series of millennial-scale cooling events and glacial advances that took place within the otherwise generally warm Holocene epoch (11.7 ka-present). Industrialization and the human-caused rise in greenhouse gas concentrations coincided with the end of the LIA. Most instrumental records do not extend back through the LIA, and thus paleoclimatic proxy reconstructions are required to understand its causes (e.g., a Peruvian Andes ice core record of the LIA is explored in Chap. 4).

---

[3] PAGES2k Consortium (2017), PAGES2k Consortium (2019).

[4] Kaufman et al. (2020).

More generally, splicing instrumental and geological records has helped geoscientists consider whether an Anthropocene Epoch should be defined on the geologic timescale. The possibility of defining an Anthropocene has been widely discussed, based on the global impact of *Homo sapiens* on Earth's environment (you will revisit the discussion and recent decision about the Anthropocene in Chap. 12). Splicing instrumental and paleo-records helps establish what criteria—industrialization (nineteenth century), radioactivity (1950), plastic pollution, land clearance, fertilizers, or other human activities—might be used as a "golden spike" to define the start of the Anthropocene as a unique, chronostratigraphic (the branch of geoscience dealing with ages of strata) geological epoch. Although opinions have differed widely over whether and on what criteria the Anthropocene should be defined—and a recent decision has been made to *not* define an Anthropocene Epoch on the geologic timescale[5]—the impacts of population growth, industrialization, greenhouse gas emissions, and climate change are clearly evident in combined paleo-instrumental records.

Together, instrumental measurements, historical archives, and paleoclimate proxy records of pre-instrumental climate variables provide baseline information about natural climate variability due to **external forcing** (e.g., volcanic activity, solar irradiance variability, and greenhouse gases [GHG]) and climate variability from **internal processes** (e.g., El Niño-Southern Oscillation [ENSO]). In particular, spliced instrumental and proxy records, when compared to climate model simulations and used to assess contributions of external and internal climate forcing, constitute an important and rapidly growing research field.

This chapter focuses on reconstructing the climate changes during the **Common Era (CE)**, the last 2000 years, focusing on records of temperature, precipitation, atmospheric greenhouse gas levels, and other climate parameters. We begin with a brief review of external forcing mechanisms and internal climate variability recognized from instrumental records (e.g., satellites, buoys, tide gauges). Then, we present paleoclimate records, some overlapping the instrumental time series, from faunal, floral, geochemical, and physical proxies recovered from a variety of archives (tree rings, ice cores, lake and ocean sediments, coral skeletons, and others). Finally, the chapter concludes with an exploration of spliced instrumental-paleo-proxy records. You will see that proxies can be calibrated by comparing instrumental and paleo-proxy records during an overlapping time period, verifying their ability to record climate change. Such efforts are the foundation for developing quantitative paleoclimate reconstructions of the late Holocene and, as discussed in other chapters, the deeper geological past. They are also a means of testing climate models used to hindcast past and predict future climate changes out to the year 2300 and beyond.

---

[5] Voosen, P. (2024). The Anthropocene is dead. Long live the Anthropocene Panel rejects a proposed geologic time division reflecting human influence, but the concept is here to stay. Science, https://doi.org/10.1126/science.z3wcw7b

The most important and well-studied combined paleo/instrumental reconstructions will be discussed here.

3 Climate Forcing

You learned about external factors that cause climate change, defined as those that influence Earth's radiative budget and global and regional temperatures, in Chap. 2. Climate forcing by volcanic activity, changes in solar activity, and changes in greenhouse gas concentrations are each important when we consider climate change in the Common Era. Therefore, it is important to collect and examine instrumental and proxy records of these factors. Let's explore Common Era records of each of these climate forcing factors (Fig. 1).

3.1 Volcanic Activity

Explosive volcanic eruptions produce large quantities of sulfate aerosols that are ejected into the troposphere (the lower level of the atmosphere, in which all of our weather occurs) and into the stratosphere (the atmospheric layer above the troposphere). Because the sulfate aerosols are highly reflective of sunlight, their presence in the stratosphere exerts a cooling effect on Earth's atmosphere for several years to decades. We can reconstruct explosive volcanic activity of the past based on the record of sulfate aerosols deposited and preserved in glacial ice. The sulfate aerosol records obtained from Northern and Southern Hemisphere ice cores provide a quantitative measure of volcanic forcing, called the Global Volcanic Forcing (GVF) index (Fig. 1a). Volcanic input from the atmosphere is a major source of extra sulfate (non-sea-salt sulfur, NSSS) to the ocean where surface-dwelling organisms take up extra sulfate and later release some to the atmosphere, which in turn is incorporated into snow and glacial ice.

Figure 1a shows a clear correlation between the GVF and NSSS proxies. Notice in Fig. 1a that the downward spikes (i.e., large negative forcing on the Earth's radiation [heat] budget) in GVF indicate times of large volcanic activity, such as major historical events in Krakatau, Indonesia (thirteenth and fifteenth centuries) and Tambora (1815 CE), the latter of which produced the 1816 "year without a summer" in Europe.

3.2 Solar Activity

Solar activity, measured by Total Solar Irradiance and sunspot number (Fig. 1b), is extremely variable over interannual (i.e., two or more years) to centennial timescales

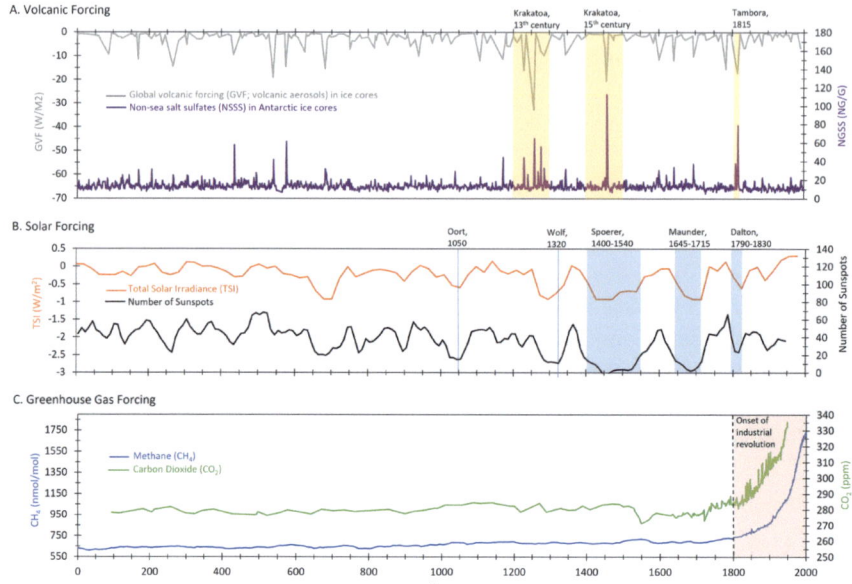

Fig. 1 Climate forcing mechanisms affecting Earth's radiative balance and climate system for the past ~ 2000 years. For each climate forcing mechanism, two types of proxy data are shown.[6] **A** Volcanic Forcing data: Global volcanic forcing (GVF; left Y-axis), a measure of volcanic aerosols from ice cores in the Northern and Southern Hemispheres. Also plotted (right axis) is a measure of the average annual non-sea-salt sulfur (nanograms per gram, ng/g; which is the same as 1 part per billion [ppb]) in the West Antarctic Ice Sheet Divide Core ice core. Significant volcanic eruption events are labeled. **B** Solar Forcing data: indicated by the Total Solar Irradiance (TSI, in watts per square meter, W/m^2; left axis), a proxy of solar activity measured from cosmic isotopic records. TSI values are plotted relative to the solar activity in 1986. Also plotted are data on the number of sunspots (right axis). Significant periods of sunspot minima are labeled (e.g., Maunder, 1645–1715). **C** Greenhouse gas data: methane (CH_4) and carbon dioxide (CO_2) concentrations measured from ice cores over the last ~ 2000 years. The atmospheric CH_4 (left axis) axis is in nanomoles per mole (nmol/mol, which is the same as 1 ppb), the atmospheric CO_2 is in parts per million (ppm). Approximate onset of Industrial Revolution labeled. Figure by Thomas M. Cronin and Anna Golub, USGS

and thus also influences the Earth's weather and climate. For example, modern observations dating back to the nineteenth century have established an 11-year sunspot cycle. Sunspots appear as darker regions on the surface of the Sun caused by partially understood processes of the sun's magnetic field. Modern observations of solar activity show that high solar activity corresponds to high sunspot activity, making

---

[6] Buizert et al. 2015a, Buizert et al. 2015b, Dlugokencky & Tans (2016), Fröhlich (2009), Garcia-Munoz et al. (1975), Knudsen et al. (2008), Köhler et al. (2017a), Köhler et al. (2017b), MacFarling-Meure et al. (2006), Marcott et al. (2014), Masarik & Beer (2009), Mitchell et al. (2013), Mitchell et al. (2011), Rubino et al. (2019), Rubino et al. (2013), Sigl et al. (2016), Sigl et al. (2015), Steinhilber et al. (2009), Steinhilber et al. (2012), Usoskin et al. (2014), NOAA Paleoclimatology Program, National Centers for Environmental Information, accessed 2019. https://www.ncdc.noaa.gov/paleo-search/study/25830.

sunspot abundance a reliable proxy for solar activity. Researchers also reconstruct longer-term solar activity using records of cosmogenic isotopes such as beryllium-10, chlorine-36 and carbon-14, which are produced by cosmic ray bombardment of molecules in Earth's upper atmosphere and signify changes in both solar output and the Earth's magnetic field. These measurements are reported as Total Solar Irradiance (TSI) in watts per square meter (W/m^2). Thus, both the TSI and the number of sunspots are proxies of solar activity.

By examining TSI and variability in the number of sunspots for the late Holocene, it is clear that the two related proxies generally correlate over decadal and centennial timescales (Fig. 1b). This comparison also shows the value of using multiple proxies—and even multiple proxies for the same climate parameter—in paleoclimatology. The well-known Maunder, Dalton, Spoerer, Oort, and Wolf periods (marked on Fig. 1b) represent periods of sunspot minima, and additionally, based on beryllium-10 and carbon-14, represent periods of lower TSI.

3.3 Greenhouse Gases

The third major category of external forcing mechanisms is atmospheric greenhouse gases (GHGs), especially carbon dioxide (CO_2) and methane (CH_4). Due to their ability to "trap" long wave radiation re-emitted from Earth's surface (as discussed in Chap. 2), GHGs have a major impact on Earth's atmospheric temperature over decadal timescales, such as today's anthropogenic climate warming, and throughout much of geological history over longer timescales (10^5 to 10^7 years). Figure 1c shows records of CH_4 and CO_2 reconstructed for the last ~ 2000 years from several Antarctic ice cores. Notice how the atmospheric CO_2 and CH_4 curves show a stark rise in GHG concentrations beginning in the early nineteenth century associated with human activity (i.e., fossil-fuel based industrialization and land-use changes).

Carefully measured and detailed records of the volcanic, solar, and GHG changes during the Common Era are essential components for understanding and evaluating the cause(s) of recent rising atmospheric temperatures and other recent changes in the climate system. In particular, comparing the timing, trends, and magnitudes of changes in these drivers help us determine the role of natural vs anthropogenic causes of climate change in the Common Era. Also important are natural patterns of short-term *internal* climate variability as an additional backdrop.

Table 1 Examples of regional internal climate oscillations

Name	Semi-regular periodicity
El Niño-Southern oscillation (ENSO)	3–7 years
North Atlantic oscillation (NAO)	~ 10 years
Pacific-North American Pattern (PNA)	2–3, 19, and 74 years
Atlantic multidecadal oscillation (AMO)	60–90 years

4 Short-Term Climate Variability

Internal Climate Variability refers to quasi-cyclic changes in Earth's regional or global weather and climate caused by processes internal to the climate system. These processes include interactions among the ocean, the atmosphere, and ice over inter-annual to multidecadal timescales. Internal variability has large effects on weather, including regional seasonal temperature and precipitation patterns.

The most well-studied examples of internal climate variability are listed in Table 1. These features are called **oscillations** because they switch back-and-forth between two opposing modes of regional climate in an irregular pattern (unlike a cycle, which in the strict sense switches between modes in a regular pattern, as discussed in Chap. 9). The oscillations represent changes in meridional (i.e., latitudinal, north–south) or zonal (i.e., longitudinal, east–west) patterns of heat and moisture. Quantitative measurements of atmospheric pressure and/or ocean surface temperature document the semi-regular temporal patterns of these oscillations. Let's look at the instrumental record of El Niño-Southern Oscillation (ENSO) as one example of climate oscillations over the last ~ 200 years. The discovery of ENSO is described in Box 2.

ESNO is a pattern of changes in regional ocean temperatures and precipitation across the tropical Pacific Ocean, which often affects regional patterns of seasonal weather in other parts of the globe. The "state" of an oscillation like ENSO can be presented as a quantitative index. Figure 2 shows one of the most commonly used ENSO indices, which is the average sea-surface temperature (SST) anomaly between 5°S and 5°N latitude and 170°W and 120°W longitude (located in the tropical Pacific). Atmospheric indices of the Southern Oscillation include the atmospheric pressure difference between the opposing high pressure and low-pressure systems in Tahiti and Darwin, Northern Australia. The two opposing modes of ENSO variability are referred to as the El Niño (warm phase) and the La Niña (cool phase). Three strong warm phase El Niño events in 1982–83, 1997–98, and 2015–2016 are highlighted in Fig. 2. Note too that although all these modes of climate variability are based on regional, usually seasonal patterns, these processes have ripple-effect impacts on weather patterns that extend to other regions. These impacts notably affect temperature, precipitation, and hurricane activity and are called **teleconnections** (Fig. 3). For example, in winter, the strong 1997–98 El Niño event caused a disruption to the jet stream and focused unusually heavy precipitation on California, resulting in devastating floods and mudslides in a region far from the equatorial Pacific.

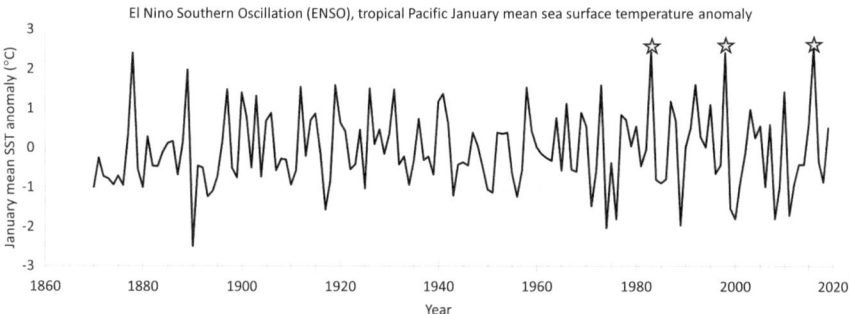

Fig. 2 Internal climate variability over the last ~ 150 years of El Niño-Southern Oscillation (ENSO), an atmospheric-oceanic variation in the tropical Pacific Ocean. The ENSO index plotted as January average sea-surface temperature (SST) anomalies between 5°S-5°N and 170°-120°W since 1870 (data from Rayner et al. 1996, 2003; HadISST1 dataset; Parker et al. 1995a).[7] These data are in degrees C relative to the 1971-2000 mean. Three significant El Niño Warm Events are labeled with stars: 1982–1983, 1997–1998, and 2015–2016

---

[7] Parker, D.E., Folland, C.K., Bevan, A., Ward, M.N., Jackson, M. & Maskell, K. (1995a). Marine surface data for analysis of climatic fluctuations on interannual to century timescales. In Martinson, D.G., et al. (Eds.), *Natural Climate Variability on Decade-to-Century Time Scales*, 241–250 and color figures on 222–228, National Academies Press; Rayner, N.A., Horton, E.B., Parker, D.E., Folland, C. K., & Hackett, R.B., (1996). Version 2.2 of the global sea ice and sea-surface temperature dataset, 1903–1994, *Climate Research Technical Note CRTN74*, Hadley Centre; Rayner, A., Parker, D.E., Horton, E.B., Folland, C.K., Alexander, L.V., Rowell, D.P., Kent, E.C., & Kaplan, A., (2003). Global analyses of sea-surface temperature, sea ice, and night marine air temperature since the late nineteenth century, *Journal of Geophysical Research 108* (D14), 4407. https://doi.org/10.1029/2002JD002670.

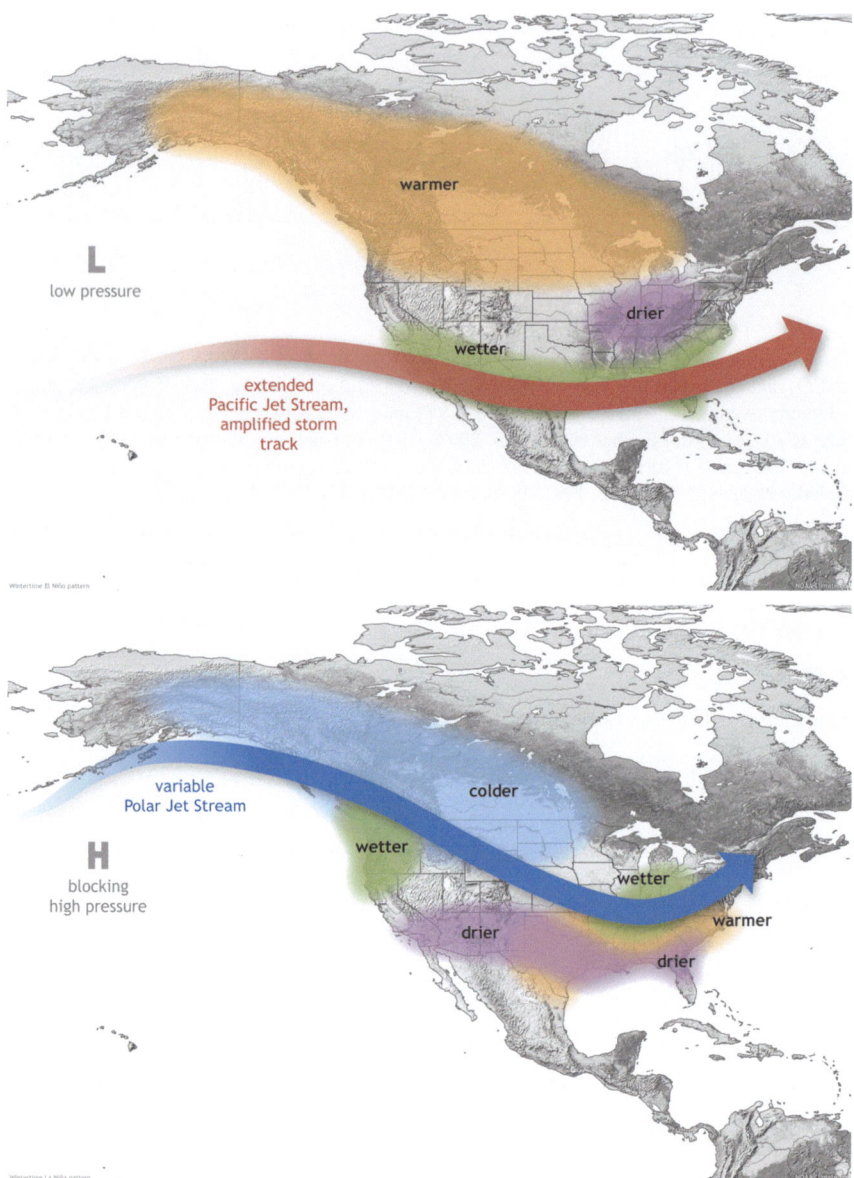

Fig. 3 ENSO teleconnections in North America during El Niño (upper panel) and La Niña (lower panel). From the National Weather Service[8]

---

[8] National Weather Service (accessed 2023, June 10). ENSO Teleconnections. National Weather Service. NOAA. https://www.weather.gov/fwd/teleconnections. Public Domain.

Box 2 The discovery of ENSO

The scientific discovery of ENSO came from two lines of investigation. Sir Gilbert Walker, a British mathematician, was prompted by a major drought in 1877 due to seasonal monsoon failure on the Indian subcontinent to pioneer the study of atmospheric oscillations in large-scale weather. Walker discovered periodic oscillations in atmospheric pressure in the tropical Indo-Pacific by comparing pressure records from Darwin, Australia, and Tahiti. Approximately every few years, the typical pattern of low pressure over Darwin and high pressure over Tahiti would reverse, leading to near-drought in northern Australia and parts of Asia and anomalously heavy rain in Tahiti. Walker had identified the atmospheric phenomenon known as the Southern Oscillation (SO) and its connection between Asian monsoon patterns and the zonal [west–east] atmospheric circulation pattern now bearing his name, the Walker Circulation.

Separately, the term "El Niño" (Spanish for "Christ Child") was coined by Spanish colonists for an ocean phenomenon—the near-annual Christmas-time warming of surface waters off the Peruvian coast. Typically, surface waters off northwestern South America are relatively cool due the northward-flowing Peru (Humboldt) Current and the upwelling of nutrient-rich colder deep water. Some years, however, the surface warming was especially severe leading to disastrous impacts on local fishing industries.

In the 1960s in Sweden, meteorologist Jacob Bjerknes linked the atmospheric Southern Oscillation to the ocean's El Niño in studies of the 1957–58 ENSO event. He found that the atmosphere–ocean link involved a feedback loop such that Walker Circulation weakened and warm surface water dominated the normally cool eastern tropical Pacific. This eastern tropical position of the water mass, known today as the "warm pool", is a characteristic feature of El Niño events. During El Niño events, due to weakened easterly winds, the thickness of the warm surface-ocean layer decreases in the central and western equatorial Pacific and increases in the eastern equatorial Pacific (this involves changes in the depth of the thermocline, the depth with the most rapid ocean temperature change from surface to deeper water). Bjerknes coined the term ENSO and also recognized its global "teleconnections" outside the tropical Pacific. The opposite of El Niño "warm" events are La Niña "cool" events.

The El Niño events in 1982–83 and 1997–98 had large global economic impacts, which helped researchers identify and begin to model regional temperature and rainfall anomalies and other weather events. Besides the importance of predicting an ENSO event months in advance for its effects on agriculture, hurricane activity, economics, and society in general, one practical aspect of understanding ENSO is seen in our interpretation of recent human-induced GHG warming. In 2015–2016, a strong El Niño event was partially responsible for one of the warmest years on record in terms of global temperature. Conversely, late in the year 2020, we experienced the inception of a new La

Niña event, which typically would diminish mean annual global temperatures. Despite that La Niña, however, Earth still experienced a warm global average temperature equal to that reached in 2016. As we see in this chapter (Sects. 4 and 7), paleo-reconstructions of ENSO, the Asian Monsoon and related phenomena indicate possible perturbations to the normal late Holocene pattern due to human activities. Excellent graphics of the ENSO phenomenon can be found at NOAA's Pacific Marine Environmental Laboratory[9] and NASA's visualization studio.[10]

5 Instrumental Records of Climate

In this section, we focus on a set of instrumental data: surface atmospheric temperature, sea-surface temperature, Arctic Ocean sea ice, and global mean sea level. How have each of these climatic parameters changed in recent decades? Tracking changes in these parameters helps us to examine the amount of recent climate change, and better distinguish the natural and anthropogenic contributions to that climate change.

5.1 Instrumental Records of Temperature

Surface air temperature anomalies for both the Northern and Southern Hemispheres are shown in Fig. 4a for the past ~ 150 years. They are plotted relative to the mean temperature for 1971–2000. A clear warming trend beginning at the turn of the twentieth century is evident in both hemispheres with the warming accelerating in the Northern Hemisphere during the last few decades. A plateau in the temperature curve occurred during the years 1940–1970, followed by a rapid warming since the 1970s.

In Fig. 4b, sea-surface temperature data obtained from ship records, oceanographic stations, and buoys since 1850 are combined to form a single surface-ocean temperature curve. Temperature data used in these reconstructions have been corrected for any biases due to different sampling methods. As seen in the air temperature graphs, a clear warming trend totaling 0.7–0.8° C began at the turn of the twentieth century and, following a plateau from 1940 to 1970, has continued during the last few decades. These parallel trends in air and ocean temperature indicate that a large portion of planetary warming over the last century (mainly since the 1970s) is taking place in

---

[9] NOAA's Pacific Marine Environmental Laboratory, https://www.pmel.noaa.gov/elnino/animations-graphics.

[10] NASA's visualization studio, https://svs.gsfc.nasa.gov/search/?chunk=0&studio=svs&series=31.

Modern Climate Change in the Context of the Last Two Millennia

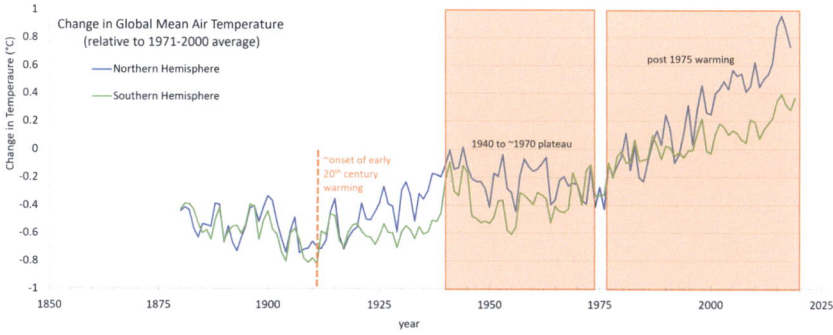

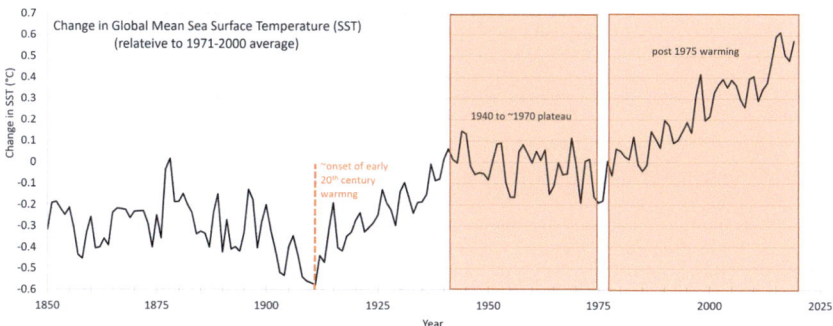

Fig. 4 Upper: Global mean surface air temperature anomalies for the Northern and Southern Hemispheres over the past ~ 150 years. **Lower:** Global mean sea-surface temperature anomalies from ships, oceanographic stations, and buoys from 1850–2006 CE corrected for methodological biases. The onset of early twentieth century warming is labeled, as well as the distinct phases of the ~ 1940–1970 plateau and post-1975 warming. Temperature anomalies are relative to 1971–2000 average, and are based on several sources[11]

---

[11] HadCrut4.3 dataset, Met Office Hadley Centre observations datasets, accessed 2019. https://www.metoffice.gov.uk/hadobs/hadcrut4/, HADISST1 dataset, Met Office Hadley Centre observations datasets, accessed 2019. https://www.metoffice.gov.uk/hadobs/hadisst/data/download.html, HADISST2 dataset, Met Office Hadley Centre observations datasets, accessed 2019. https://www.metoffice.gov.uk/hadobs/hadsst2/, HADISST2 dataset, Met Office Hadley Centre observations datasets, accessed 2019.

https://www.metoffice.gov.uk/hadobs/hadsst3/, Huang et al. (2017a), Huang et al. (2017b), International Comprehensive Ocean Atmosphere Data Set (ICOADS), accessed 2019, Kennedy et al. (2011), Menne et al. (2018), NOAAGlobalTemp V5 dataset, accessed 2019, Smith et al. (2008), Woodruff et al. (2011), Zhang et al. (2019), Zhang, H.-M., Huang, B., Lawrimore, J., Menne, M., Smith, T.M., NOAA Global Surface Temperature Dataset (NOAAGlobalTemp), Version 4.0 [indicate subset used], NOAA National Centers for Environmental Information.

Fig. 5 Arctic Ocean sea ice from northern Greenland's Sherard Osborn Fjord, represents frozen surface-ocean water, often reaching several meters thick depending on the age and region of the Arctic. Most regions of the Arctic Ocean featured perennial sea ice until recent decades when many areas experienced major decline in summer sea-ice cover. Sea-ice thickness is also on the decline. Sea-ice decline influences ocean circulation, solar heat absorption, and marine ecosystems but, because the sea ice is floating, its changes do not affect global sea level. Photo courtesy of Dr. Thomas M. Cronin, U. S. Geological Survey, and the Swedish Polar Secretariat

the ocean. In fact, more than 90% of the total heat trapped by greenhouse gases in the atmosphere is transferred to the world's oceans, with a much smaller percentage being absorbed by land and ice sheet/glacial surfaces. In addition, warming in deeper layers of the ocean, reflecting the progressive transfer of total planetary surface heat into the deep ocean, has been recorded since the 1950s; this warming has accelerated since the 1990s, and has been well-documented by oceanographic studies. This evidence of recent surface and deep-ocean warming emphasizes the importance of the world's oceans in understanding recent climate change.

5.2 Instrumental Records of Arctic Ocean Sea Ice

Until the last few decades, Arctic Ocean sea ice (Fig. 5) was considered to be "perennial" (i.e., present year-round) in nature, with summer seasonal declines only around the margins of the Arctic. Arctic Ocean summer sea ice, which historically covered 14–16 million square km, has recently shown some of the most rapid changes of any part of Earth's climate system. Arctic sea-ice variability, usually measured in sea-ice extent (in km^2) or concentration (defined as ocean area covered by > 15% ice), is related to changes in surface air temperature and atmospheric patterns caused

Fig. 6 Arctic Ocean sea-ice extent in square kilometers on the 15th of September back to 1850 (data from Walsh et al. 2015; Walsh et al. 2019).[12] This curve shows major rapid loss of sea ice over the last few decades relative to previous recent sea-ice variability[13]; satellite passive microwave data since 1979, pre-satellite data reconstructed from historical data (ships logs and national ice services). The onset of major decline in summer Arctic sea ice is evident since the 1960s

by two regional internal climate oscillations: the North Atlantic Oscillation and the related Arctic Oscillation. Figure 6 shows sea-ice area in square kilometers in the Arctic, measured for September, the month of minimal sea-ice cover in most Arctic regions, extending back to the year 1850. The curve is constructed from satellite passive microwave records since 1979 and historical data (e.g., ice charts) back to the year 1850. This graph illustrates the drastic nature of the sea-ice loss, which was about 4% per decade from 1958 to 1997 and almost 13% per decade since the 1990s. Massive downward spikes in sea-ice cover are most evident beginning in the year 2007. Summer sea-ice extent declined from about 7 million km^2 in the 1990s to only 3.5–5 million km^2 since 2007. Although estimates vary, and depend on the time period covered, approximately 50% of the total sea-ice decline has been attributed to greenhouse gas forcing, with the other half due to internal climate variability of the North Atlantic and Arctic Oscillations.

In addition to large decreases in summer sea-ice cover, other studies show that the age of the Arctic sea ice has decreased. Sea-ice age is regionally variable, usually less than 1 year to several years in age, but in recent decades, sea ice > 4 years old has virtually disappeared. Related to the age of sea ice, its thickness has also declined (e.g., up to 20% since the period 2003–2009). Decreasing sea-ice thickness means that the total *volume* of Arctic sea ice has decreased significantly in the last

---

[12] Walsh, J. E., Chapman, W.L., Fetterer, F., & Stewart, S., (2019). Gridded Monthly Sea Ice Extent and Concentration, 1850 Onward, Version 2, Boulder, Colorado USA, NSIDC: National Snow and Ice Data Center. https://nsidc.org/data/g10010; Walsh, J. E., Chapman, W.L., & Fetterer, F., (2015, updated 2016). Gridded Monthly Sea Ice Extent and Concentration, 1850 Onward, Version 1, Boulder, Colorado USA, NSIDC: National Snow and Ice Data Center. https://nsidc.org/data/g10010/versions/1.

[13] In addition to Walsh et al. (2019) and Walsh et al. (2015) (references provided above), previous sea ice variability also available from: Fetterer, F., Knowles, K., Meier, W.N., Savoie, M., & Windnagel, A.K., (2017, updated daily). Sea Ice Index, Version 3. Boulder, Colorado USA. NSIDC: National Snow and Ice Data Center.

few decades. In summary, then, the instrumental records show marked decreases in Arctic sea-ice extent, sea-ice age, sea-ice thickness, and total sea-ice volume, especially in the past several decades.

Using a variety of proxy methods, sea-ice variability has been documented for the last 11.7 kyr, that is, during the Holocene interglacial. Although more research is needed to provide a better understanding, the decline in Arctic sea ice in the past few decades generally represents the disruption of near-perennial sea-ice cover that existed for much of the Holocene interglacial. These patterns support the idea of **Arctic Amplification** (also called Polar Amplification), a hypothesis that holds that warming climates involve greater degrees of warming at high latitudes than those seen in mid and low latitudes. Arctic Amplification is supported by paleoclimatic records of high-latitude temperature history and, to a certain degree, long-term paleo-CO_2 reconstructions, which show amplified high-latitude warming over all timescales including parts of the Cretaceous Period, the Paleocene-Eocene Thermal Maximum (PETM), the Eocene thermal maxima, the Pliocene warm period, several Quaternary interglacial periods, and the early mid Holocene (several of these will be discussed in greater detail in later chapters).

The implications of declining Arctic sea ice are significant. These include altered hemispheric weather patterns; changes in ocean circulation both within the Arctic and in the exchange of ocean water between the Arctic and the Nordic Seas and North Atlantic Ocean; decreased albedo and associated feedbacks (the open ocean absorbs more short-wave solar radiation than the more reflective ice and snow do); and large changes to Arctic biological productivity, marine ecosystems, and species distributions and biodiversity.

5.3 *Instrumental Records of Global Sea Level*

Global sea-level rise is caused by several factors, the two largest being thermal expansion of the ocean as it warms and melting land ice from glaciers and ice sheets. Other factors also affect regional sea level. These include the subsidence or uplift of land surfaces due to changing ice sheet mass during and since the end of the last deglaciation (i.e., glacio-isostatic adjustment, GIA); other vertical land movements such as mountain building from tectonic activity and subsidence from ground water withdrawal; and regional oceanographic changes due to Earth's rotation and gravitational effects (notably linked to melting land ice impacts on ocean circulation).

We can measure historical changes in sea level through a global network of tide gauge stations. Figure 7 shows a sea-level curve (blue line) representing Global Mean Sea Level (GMSL) based on such records from 1880 to 2013. Although relatively few tide gauge stations were available during the nineteenth and early twentieth centuries, this curve nonetheless clearly captures a rise in global sea level beginning in the late nineteenth century and steadily continuing over the last century.

Since 1993, the tide gauge record has been augmented and improved by satellite altimeter records (orange line) of global sea level. The satellite record shows a steep

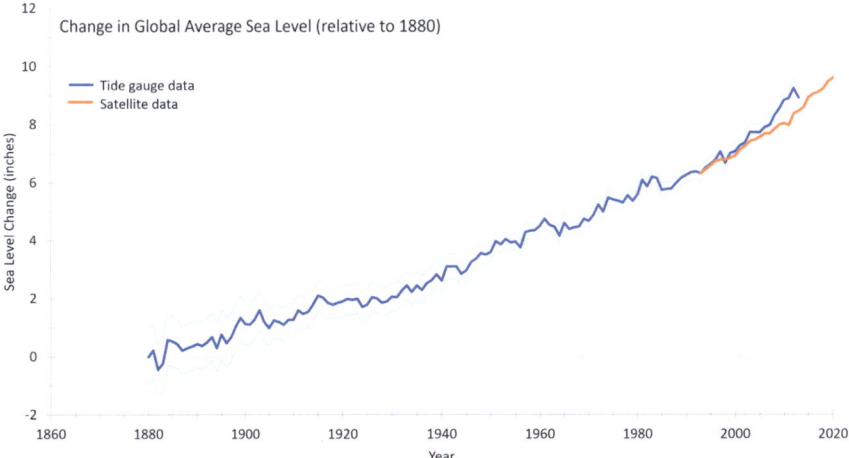

Fig. 7 The indicator shows an increase in global average sea level since 1880, in inches; the blue line becomes steeper in more recent decades, indicating an increased rate of change. The blue line shows sea level as measured by tide gauges (1880–2013); the surrounding light blue-shaded area shows upper and lower 95% confidence intervals; the orange line shows sea level as measured by satellites for comparison (1993–2021). Data from the U.S. Global Change Research Program[14]

rise in sea level the last few decades supporting the overlapping tide gauge record. These data indicate a rise in GMSL of over 8 inches (20 cm) since 1880, with 2.5 of those inches (6 cm) occurring since 1993. The rate of global average sea-level rise has accelerated from 0.06 inches (1.4 mm) per year throughout most of the twentieth century to 0.14 inches (3.6 mm) per year from 2006 to 2015,[15] which is consistent with the trends in global temperature and sea-ice changes described above. The causes of the acceleration during recent decades have been recently recognized as due to melting glaciers and greater contributions from the melting of the Greenland and Antarctic Ice Sheets, a trend confirmed by glaciological studies of ice sheet mass

---

[14] US Global Change Research Program (accessed 2023, June 10). Sea Level Rise. U.S. Global Change Research Program, https://www.globalchange.gov/browse/indicators/global-sea-level-rise, from EPA, 2022: Climate Change Indicators: Sea Level. https://www.epa.gov/climate-indicators/climate-change-indicators-sea-level.

[15] Lindsay, R., (2022). Climate Change: Global Sea Level. Climate.gov. https://www.climate.gov/news-features/understanding-climate/climate-change-global-sea-level.

balances.[16] It is expected that melting land ice will make an even greater contribution to global sea-level rise in the future.

6 Splicing Instrumental and Paleoclimate Proxy Data

At this point in the chapter, you have considered important external climate forcing and short-term internal variability, and you have tracked recent changes in climate based on a variety of instrumental records. How do we expand our climate knowledge to older time periods for which no instrumental records exist? Paleoclimate archives can provide us with proxy (i.e., indirect or surrogate) data that we can collect and measure. But how do we connect and combine the instrumental records with paleoclimate records so that we are confident in the information the proxies provide and in the continuity of the records? In other words, how do we **splice** the instrumental and paleoclimate proxy data?

6.1 Generalized Method of Splicing

There are a variety of approaches to calibration of climate proxies and verification of their ability to record climate changes. Most are carried out by comparing instrumental and paleo-proxy records during an overlapping time period. One example of splicing instrumental and proxy records involves extending instrumental sea-surface temperature records by using tropical corals.[17] Some tropical corals build aragonite skeletons that often have annual banding (Fig. 8) that allows paleo-reconstruction of seasonal and interannual sea-surface temperatures, similar to the use of tree rings to reconstruct past temperature and precipitation. Paleotemperatures obtained from geochemical proxies from coral skeletons are spliced with instrumental temperatures to provide long-term records of tropical climate variability (see Sect. 7.2).

---

[16] There are about 200,000 glaciers today (estimates vary depending on definitions) covering more than 726×10^3 square kilometers and, should they all melt, it would cause less than a 0.5 m sea-level rise. There are two ice sheets today, the Greenland and Antarctic Ice Sheets, covering about 1.7 $\times 10^6$ and 14×10^6 square kilometers, respectively. Melting glacial land ice mass measured in gigatons (one Gt = billion [1×10^9] tons) per year can be converted to sea level, measured in mm per year, for a global ocean area = 361×106 square kilometers as follows: 1 mm of global sea-level rise results from about 360 Gt of melted ice. To give concrete examples linking ice sheet mass balance loss to observed sea-level rise, the total contribution of the Greenland Ice Sheet melting since 1972 is about 13.7 mm of sea-level equivalent, with increasing mass loss occurring since the 1990s. Similar acceleration of Antarctic Ice Sheet mass loss has occurred rising from 40 Gt per year from 1979 to 2000 to 252 Gt per year from 2009 to 2017.

[17] See Felis (2020).

Fig. 8 *Orbicella*, a coral record with annual banding from St. Croix, US Virgin Islands used to reconstruct tropical climate variability. Image courtesy Dr Casey Saenger, Western Washington University. This figure is excluded from Creative Commons license

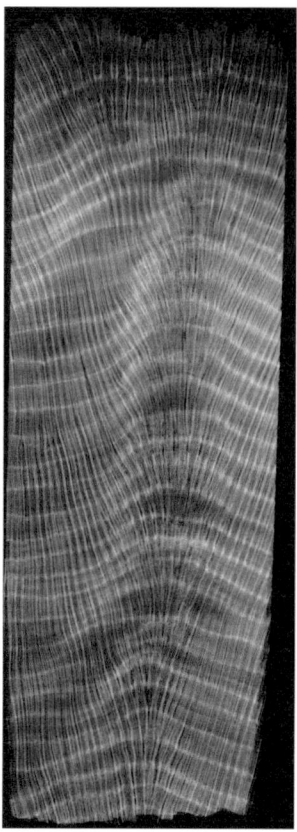

6.2 What Spliced Records Allow Us to Do

Splicing instrumental and paleoclimate proxy data provides the foundation for developing *quantitative* paleoclimate reconstructions of the late Holocene and, as discussed in other chapters, the deeper geological past (Chaps. 4–9). Such efforts also provide a means of testing climate models (Chap. 10) used to hindcast past and to forecast future climate changes.

Connecting and combining instrumental and paleoclimate proxy records also make possible research efforts that aim to distinguish natural versus human-induced climate trends. Such work is essential to providing the public and policy makers with confident conclusions about the relative impact humans have on the climate system. There are two general approaches to identifying the human influence. One approach is the **detection and attribution (D&A)** of climate patterns where the main objectives are to detect statistically significant global or regional climate patterns and attempt to attribute causes, including human-induced changes in greenhouse gases.

D&A analyses typically use paleo-proxy records to extend the instrumental record back in time.

A second related approach to identifying human influence involves studies that attempt to identify the **timing of the emergence (ToE)** of an anthropogenic signal above the background of natural variability. For example, although much attention is given to hemispheric and global temperature trends (e.g., Fig. 4), ToE estimates for other parts of the climate system suggest that human influence appears to be "emerging" even earlier than its signal in temperature trends. To cite one example, Earth system modeling has established that changing trends in ocean pCO_2 and pH (ocean acidification) emerge much faster (ToE of about 12 years) than the ToE for sea-surface temperature (SST; ToE of 45–90 years). Other examples of apparent early anthropogenic influence include rates of Arctic summer sea-ice decline and increases in global mean sea level due to warming oceans and melting glacial ice.

While the details of splicing, including statistical aspects of D&A and ToE studies, are beyond the scope of this chapter, connecting paleo and instrumental records is important to extending our understanding of the climate system. The spliced results described below are among the most important and well-studied combined paleo/instrumental reconstructions of climate change from the Common Era.

7 Paleoclimate Reconstructions from the Common Era

In this section, we present several paleoclimate reconstructions that are composites of multiple individual records and are spliced into the instrumental records described in Sect. 5. The paleoclimate reconstructions (which cover centuries to a few millennia, depending on the record) depict mean annual temperature, the tropical ocean–atmosphere system, global mean sea level, Arctic Ocean sea-surface temperature, and summer Arctic Ocean sea ice.

7.1 Global Air Temperature During the Common Era

Figure 9 shows Northern Hemisphere decadal surface Mean Annual Temperature (MAT) anomalies (dark blue, with respect to the mean temperature for 1961–1990 CE) plotted for the last ~ 2000 years and reconstructed from a number of original datasets and collaborative efforts of dozens of researchers. Although several different paleotemperature proxies are used to construct the curve in Fig. 9, the majority are from tree-ring records, specifically compiled from tree-ring characteristics such as ring width, maximum latewood density, or stable isotopes used as temperature proxies. These tree-ring measurements have undergone a variety of rigorous, statistical analyses of tree-ring/temperature relationships to establish the "skill" for tree-ring paleotemperature estimation. First, proxy and instrumental data for a calibration time period—such as beginning in the year 1850 CE up to the

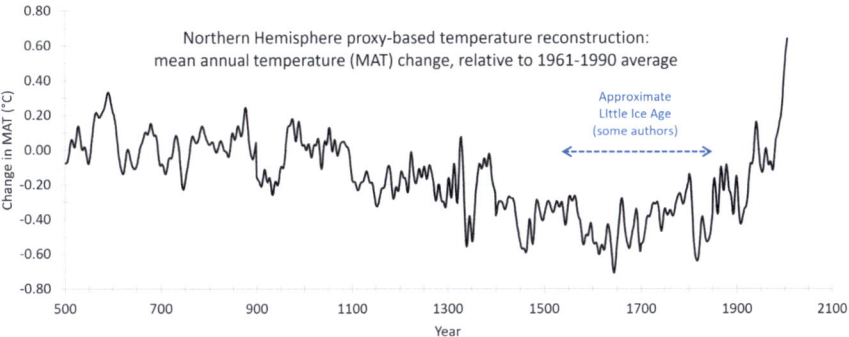

Fig. 9 Northern Hemisphere decadal surface mean annual temperature (MAT) anomalies for the last ~ 1500 years reconstructed from multiple different proxies, including tree rings.[18] Warming during the last ~ 100 years illustrates a rapid rise in Northern Hemisphere temperatures. Temperature anomalies are shown as differences from the average MAT for 1961–1990

twenty-first century for temperature—are used to determine statistical relationships between the tree-ring features and temperature for a historical period. The second step involves verification, in a procedure called cross-validation, which typically involves a shorter, more recent time period aimed at testing the calibration statistics. We emphasize that there are a variety of approaches that have been applied to the statistical analyses of proxy data using cross-validation methods, but they all produce temperature patterns similar to those shown in Fig. 9.

There are several important features of the graph in Fig. 9. The most noteworthy is the steep temperature increase of the last ~ 100 years, which illustrates a rapid rise in Northern Hemisphere temperatures. This rise in temperatures is often referred to as the "hockey stick" curve due to its shape. In addition, periods of relatively warm and cool MATs are seen during the Medieval Climate Anomaly (900–1300 CE, also called the Medieval Warm Period) and the Little Ice Age (1550–1880 CE), respectively. This 2000-year temperature curve, used in conjunction with the volcanic, solar, and greenhouse gas forcing curves in Fig. 1, and accounting for climate variability shown in Fig. 2, leads to the conclusion that GHG forcing explains most of the observed warming over the last century.

7.2 Tropical Sea-Surface Temperature During the Common Era

The study of shallow water hermatypic (reef-building) corals (the order Scleractinia, stony corals), which secrete an aragonite ($CaCO_3$) skeleton, is known as sclerocli-matology. Modern reef environments provide archives of tropical paleoceanography

---

[18] *Data sources*: Brohan et al. (2006), Kerr (2000), Mann et al. (2009), Mann (2008), Mantua et al. (1997).

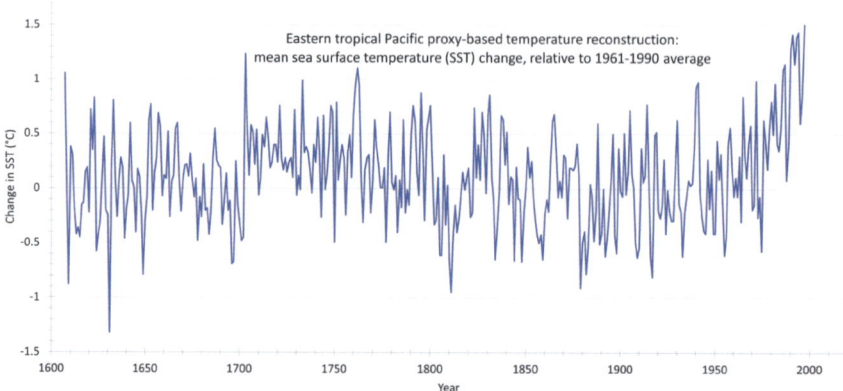

Fig. 10 Tropical paleoclimate record of changing April–May sea-surface temperatures (SST) for the past ~ 400 years in degrees Celsius from coral proxies, expressed as anomalies relative to the 1961–1990 reference period. El Niño–Southern Oscillation (ENSO) influence and modern global warming are evident. Data from Tierney et al. (2015)[19]

using several paleotemperature proxy methods (i.e., skeletal trace element concentrations, stable isotopes). A number of shallow water coral genera have been used to reconstruct tropical region variability to assess changes in ENSO patterns, anomalous sea-surface temperatures (including coral bleaching events), prehistorical ENSO events, Intertropical Convergence Zone (ITCZ) patterns, and ENSO teleconnections.

The study of ENSO variability requires sampling the skeleton at either seasonally and/or interannually resolved timescales, which can be achieved by using coral genera such as *Porites* in which the growth pattern of skeletons is rapid (up to ~ 1 cm per year). However, large, older coral colonies are difficult to find, and as a result, composite, spatially robust tropical sea-surface temperature (SST) reconstructions (Fig. 10) do not extend back as far in time as air temperature reconstructions from tree rings (Fig. 9) and ice cores. Nevertheless, the coral-based SST reconstructions are quite detailed for the timeframe they represent. For example, the composite SST anomaly record for the East Pacific Ocean (a region of strong seasonal and ENSO interannual variability) was reconstructed from multiple regional coral datasets. It shows strong interannual variability, as well as a warming trend that began during the early twentieth century (Sect. 4).

7.3 Arctic Temperature During the Common Era

Climate change in the Arctic region, especially Arctic air and ocean temperatures and sea-ice cover, has been intensely studied using spiced instrumental and paleoceanographic records. The composite curve in Fig. 11, constructed by a consortium of

---

[19] Tierney et al. (2015).

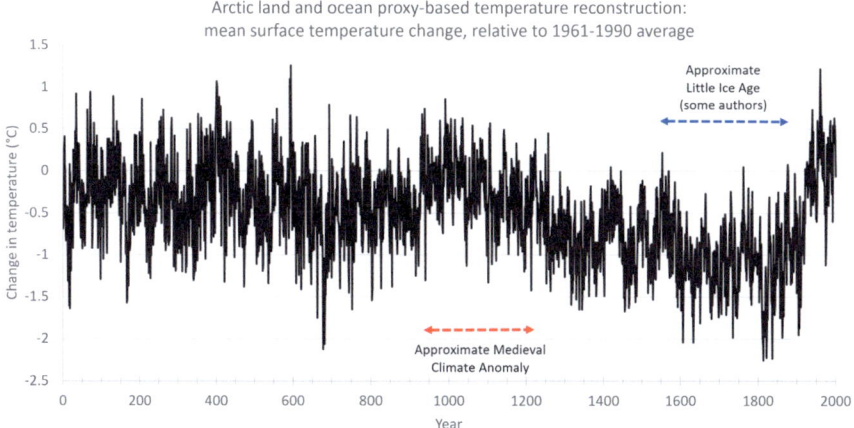

Fig. 11 Arctic temperature reconstructed over the past 2000 years using PAGES 2k Consortium Arctic database (data from McKay and Kaufman 2014)[20]. The curve illustrates surface temperature anomalies relative to the reference time 1961–1990. The record clearly displays the modern climate warming beginning ~ 1900 CE in the Arctic

researchers, is based on tree ring, marine sediment, glacial ice, lake sediment, historical, and speleothem proxies of temperature from locations north of 60° latitude. The graph shows clearly both pre-anthropogenic centennial and multidecadal temperature variability (measured by temperature anomalies) and the emergence of recent warming in the early twentieth century. Multiple relatively cool periods are notable during the Little Ice Age. The onset of recent warming is similar to trends discussed in previous sections; however, relative to the tropical ocean and global atmospheric temperatures, the Arctic trend of temperature increase is steeper, reflecting Arctic Amplification.

7.4 Arctic Ocean Sea Ice During the Common Era

A ~ 1450-year record of reconstructed late-summer Arctic sea-ice extent (red curve) and the observed late-summer Arctic sea-ice extent over the last ~ 150 years (yellow curve) are shown in Fig. 12. The sixty-nine proxy records (mostly ice cores and tree rings) used to reconstruct the pre-instrumental period come from a circum-Arctic network with excellent spatial coverage and temporal resolution (<10-yr). Due to the scarcity of marine sediment records with high temporal resolution, which potentially could provide direct proxies of sea-ice extent, each proxy method used in this compilation reconstructs indirect climatic factors rather than direct indicators of sea-ice cover. Nevertheless, these proxies are calibrated to instrumental late-summer sea-ice records from the historical period and as such, they serve as indicators of atmospheric

---

[20] McKay and Kaufman (2014).

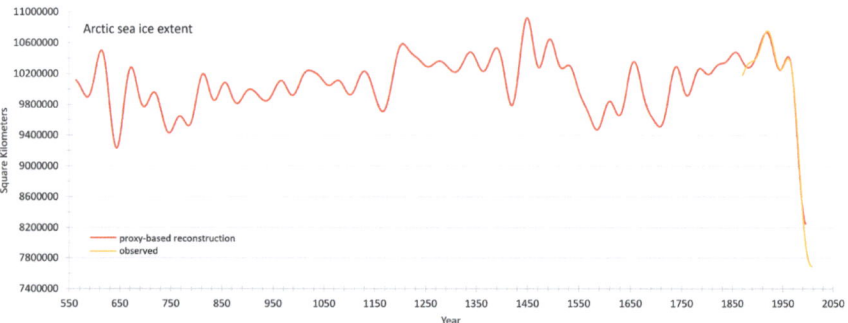

Fig. 12 A ~ 1500-year record of reconstructed late-summer Arctic sea-ice extent (red), interpreted from multiple terrestrial climate proxies (ice cores, tree rings, lake sediments, and historical data), and observed Arctic sea-ice extent over ~ 150 years (yellow). The curve shows a two million square kilometer reduction in Arctic sea ice in the last few decades. Data from Kinnard et al. (2011) and references therein[21]

forcing and atmospheric response (temperature, moisture etc.) to changes in sea-ice cover. They are also supported by a few but growing number of low-resolution marine sediment records using a sea-ice proxy called IP25, a lipid molecule with 25 carbon atoms. IP25 and related organic molecules are organic biomarkers produced by a sea-ice dwelling diatom species that lives near the sea-ice edge.

Arctic sea-ice variability is linked to the strength of a flow of warm Atlantic water, called the North Atlantic Drift, that moves from low to high latitudes and enters the Arctic Ocean Basin through the Barents Sea and the eastern Fram Strait between Greenland and Svalbard. The 1500-year reconstruction shown in Fig. 12 supports the hypothesis that variability in Atlantic water inflow (linked to the North Atlantic Oscillation) is an important forcing factor in the Arctic. In addition, the record shows relatively lower sea-ice extent during the Medieval Climate Anomaly, relatively higher values during the early part of the Little Ice Age, and finally, an unprecedented decline of 2×10^6 square kilometers during the last few decades, providing additional evidence of anthropogenic-driven climate change.

7.5 Sea-Level Change During the Common Era

Variations in global sea level have occurred over all timescales and paleo-sea-level history is reconstructed using several types of proxy records, depending on the timescale of interest. During the last glacial maximum (LGM, 21–19 ka), global sea level was about 130 m below current sea level, a result of the growth of land-based ice sheet mass in North and South America, Antarctica, Greenland and Eurasia.

---

[21] Kinnard et al. (2011), de Vernal et al. (2008), Fisher et al. (2006), Grumet et al. (2001), Jones et al. (2009), Jacoby et al. (1982), Masse' et al. (2008), Macias-Fauria et al. (2010), Polyak et al. (2010).

Fig. 13 Outcrop of uplifted and exposed paleo-reef in Applehall, Barbados. During the peak warmth of the last interglacial, about 125,000 years ago, global sea level was 6 to 9 m above current sea level. This reef was below sea level at that time. The Barbados sea-level record is famous for documenting glacial-interglacial sea-level oscillations. In the foreground, a Holocene tidal notch cuts into the 125,000 year old exposed reef. Photo courtesy of Robert Poirier (U. S. Geological Survey). This figure is excluded from Creative Commons license

Sea level rose during the last deglaciation (19 to 11.7 ka) at varying rates up to 40–50 mm yr^{-1} (~15 times today's rate), until reaching a level near present-day mean sea level about 7 ka. Global deglacial and early Holocene interglacial (11.7–7 ka) sea level is reconstructed from tropical coral reef stratigraphic records (New Guinea, Barbados, Kiritimati, Tahiti, and many other tropical sites, Fig. 13) and marsh and coastal sediment records from continental margins at all latitudes.

While coral records provide a first-order record of the last deglacial sea-level rise, marsh records from modern day coasts are considered the best available archives for detecting relatively small relative sea-level changes for the last 7–8 ka for three main reasons. First, glacio-isostatic adjustment rates (GIA) are relatively well known for most coastlines, allowing correction for this important process. Many mid-latitude regions, such as the mid-Atlantic coast of the eastern US, are undergoing gradual subsidence due to their location[22] outside the areas covered by LGM ice sheet margins. Coastal records in high-latitude settings that were covered by LGM ice sheets are, conversely, undergoing GIA-induced uplift. Second, the radiocarbon chronology of marsh sediments that accumulated in non-glaciated coastal regions and paleo-shorelines that have been uplifted in glaciated regions is excellent, often providing a near-continuous chronology for Holocene sea level change. Third, the proxies of past sea level for mid and low latitudes are based on marsh (i.e., *Spartina*

---

[22] Glacial isostatic adjustment is the ongoing vertical movement of land after an ice sheet melts. Land that was once under an ice sheet rises up (or rebounds). Land that was outside the margin of former ice sheets subsides (sinks). A good analogy is getting up from a soft sofa. Your body mass pressed the sofa cushion down when you were sitting on it (as an ice sheet would to underlying land) and created a puffed-up area peripheral to where you sat (as land does adjacent to an ice sheet). When you get up it takes time for the sofa to relax back to its original shape. Land takes time to adjust as well after an ice sheet melts.

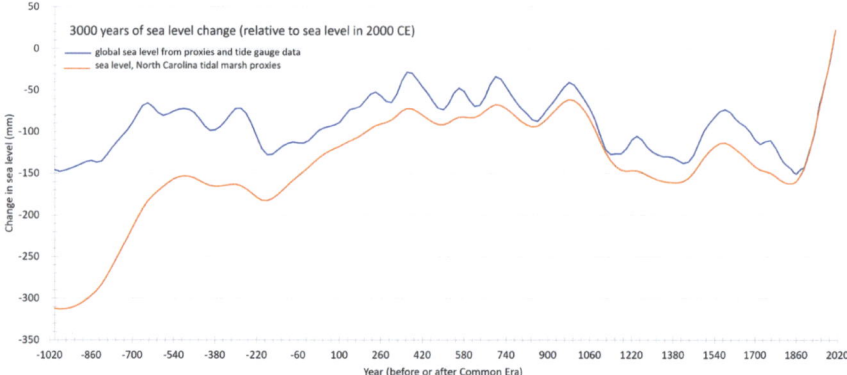

Fig. 14 Change in global sea level during the past ~ 3000 years, relative to sea level in 2000 CE. Global paleo-sea level (blue) from a global database of relative sea level reconstructions and tide gauge data.[23] Rise in global sea level during the last ~ 100 years has been more rapid than during any time in the last 2700 years (data from Kopp et al. 2016). Tidal marsh paleo-sea-level curve from North Carolina (orange line), reconstructed from foraminifera and marsh sediments and corrected for local glacial isostatic adjustments and paleo-elevation above sea level (data from Kemp et al. 2011).[24] Horizontal axis is the year: positive numbers are years common era (CE), whereas negative numbers are years before common era (BCE)

mid-latitude marshes, and mangrove environments in low latitudes) paleoecological data from foraminifera, diatoms and marsh vegetation. These proxies typically have a relatively small error bar as sea-level datums (from 5 to 20 cm) due to the narrow ecological ranges of coastal zone plant and animal species.

Focusing on the last 3000 years of mean global Holocene sea-level variability, Fig. 14 shows global paleo-sea level (blue line) created from a global database of relative sea level reconstructions spliced in with tide gauge data. The sea-level curve is the culmination of decades of local and regional sea-level studies, geophysical modeling of glaciological and oceanographic processes, and statistical analyses of the results.

Relatively high sea-level peaks occurred during the "Roman Warm Period" (~2000 years ago) and parts of the Medieval Climate Anomaly, with relatively low sea level during the Little Ice Age. These natural centennial-scale oscillations are followed by a steep increase in global sea level during the last ~ 100 years. This rapid sea-level rise (roughly 2.1 mm yr^{-1} on average),[25] also shown in Fig. 7, has been rigorously and statistically analyzed and is interpreted by experts as signifying an

---

[23] Grinsted et al. (2009), Hay (2015), Jevrejeva et al. (2009), Kemp et al. (2011), Kopp et al. (2016), Lambeck et al. (2004), Lambeck et al. (2014), Mann et al. (2019), Marcott et al. (2013), Masson-Delmotte et al. (2013), PAGES 2k Consortium (2013), Sivan et al. (2004), Woodroffe et al. (2012).

[24] Kemp et al. (2011).

[25] Kemp et al. (2011).

abrupt, anomalous, human-induced sea-level rise more rapid than any seen during the preceding three millennia.

The paleo-sea-level record of marshes in modern North Carolina is also shown in Fig. 14 (orange line), reconstructed from foraminifera and marsh sediments and corrected for local glacial isostatic adjustments. This record is typical of marsh records from the mid-Atlantic region of the eastern United States. The shape is similar to the global sea-level curve (blue line) with clearly defined sea-level maxima during the MCA and minima during the LIA. Although no single sea-level curve can be viewed as a "global" curve due to local/regional tectonic, glacio-isostatic and ocean circulation processes, records from marsh sediments have highly resolved sea-level datums (like foraminiferal assemblages), excellent chronology (radiocarbon dating), and solid glacio-isostatic corrections. They thus reveal important first-order patterns over centennial timescales and they also can exhibit an obvious and significant emergence of human-induced influences on sea level.

8 Closing Thoughts on the Use of Instrumental and Proxy Records to Assess Climate Change in the Common Era

This chapter provided you an overview of climate change in the last 2000 years. It introduced both what we know and how we know it. Connecting and combining late Holocene paleoclimate proxy records with instrumental records of various parts of the climate system enables us to recognize and better understand natural modes of internal variability, as well as the anthropogenic-driven climate changes. The following considerations are important to keep in mind as future research expands and improves the splicing of paleo and instrumental records:

- Spatial and temporal gaps and uncertainty exist is both paleo and instrumental datasets. It is therefore critical to improve spatial coverage, especially in paleo datasets, through more sediment core, coral, ice core, and tree-ring paleoclimatology. In addition, data-climate modeling simulations can provide a more robust spatial coverage and point toward a need for improved coverage. Statistical and probability analyses can improve interpretations of what external and internal factors are driving regional climate changes.
- Although many types of well-tested proxy methods exist for specific climatic parameters, multiproxy records are preferred. This requires multidisciplinary teams of researchers addressing common goals of climate reconstruction and interpretation. Moreover, proxy method development can be improved by using both field-based approaches in natural environments and lab analyses under controlled environmental conditions.
- Chronology is most crucial in all paleoclimate reconstructions. Although it is not always possible to achieve the temporal resolution reached in instrumental time series, there is still often sufficient overlap—e.g., in the case of annual mean

temperature—for climate modelers to use paleo datasets to test and improve model simulations.
- It must be recognized that multiple forcing mechanisms operate over various time scales. In fact, distinguishing between natural and anthropogenic climate changes and variability requires acknowledging such complexity. A prerequisite is understanding unforced climate variability—those interannual to centennial-scale changes unrelated to greenhouse gases, solar variability and volcanic activity—by integrating paleo and instrumental records.
- Neither the paleo nor the instrumental approach is perfect—we need both paleo and instrumental records. The former has the advantage of longer timescales and more extreme climate states, including those with elevated pre-industrial GHG concentrations, found in the geological record. Instrumental records, notably those extending to the nineteenth century, capture Earth's climate changes during growing human activity including the rise in GHG concentrations.
- Searching for the "emergence of trends" in different parts of the climate system (e.g., changes in the oceans, sea ice, glacier, and ice sheets), while not as highly publicized as hemispheric or global mean temperature, should continue to be a major focus in the detection and attribution of climate trends.

References

Brohan, P., Kennedy, J. J., Harris, I., Tett, S. F. B., & Jones, P. D. (2006). Uncertainty estimates in regional and global observed temperature changes: A new data set from 1850. *Journal of Geophysical Research Atmosphere, 111*, D12106.

Buizert, C., Adrian, B., Ahn, J., Albert, M., Alley, R. B., Baggenstos, D., Bauska, T. K., Bay, R. C., Bencivengo, B. B., Bentley, C. R., Brook, E. J., Chellman, N. J., Clow, G. D., Cole-Dai, J., Conway, H., Cravens, E., Cuffey, K. M., Dunbar, N. W., Edwards, J. S., Fegyveresi, J. M., Ferris, D. G., Fitzpatrick, J. J., Fudge, T. J., Gibson, C. J., Gkinis, V., Goetz, J. J., Gregory, S., Hargreaves, G. M., Iverson, N., Johnson, J. A., Jones, T. R., Kalk, M. L., Kippenhan, M. J., Koffman, B. G., Kreutz, K., Kuhl, T. W., Lebar, D. A., Lee, J. E., Marcott, S. A., Markle, B. R., Maselli, O. J., McConnell, J. R., McGwire, K. C., Mitchell, L. E., Mortensen, N. B., Neff, P. D., Nishiizumi, K., Nunn, R. M., Orsi, A. J., Pasteris, D. R., Pedro, J. B., Pettit, E. C., Price, P. B., Priscu, J. C., Rhodes, R. H., Rosen, J. L., Schauer, A. J., Schoenemann, S. W., Sendelbach, P. J., Severinghaus, J. P., Shturmakov, A. J., Sigl, Slawny, K. R., Souney, J. M., Sowers, T. A., Spencer, M. K., Steig, E. J., Taylor, K. C., Twickler, M. S., Vaughn, B. H., Voigt, D. E., Waddington, E. D., Welten, K. C., Wendricks, A. W., White, J. W. C., Winstrup, M., Wong, G. J., & Woodruff, T. E. (2015a). Precise interpolar phasing of abrupt climate change during last ice age. *Nature 520*, 661–665.

Buizert, C., Cuffey, K. M., Severinghaus, J. P., Baggenstos, D., Fudge, T. J., Steig, E. J., Markle, B. R., Winstrup, M., Rhodes, R. H., Brook, E. J., Sowers, T. A., Clow, G. D., Cheng, H., Edwards, R. L., Sigl, M., McConnell, J. R., & Taylor, K. C. (2015). The WAIS Divide deep ice core WD2014 chronology - Part 1: Methane synchronization (68–31 ka BP) and the gas age–ice age difference. *Climate of the past, 11*(2), 153–173.

de Vernal, A., Hillaire-Marcel, C., Solignac, S., Radi, T. and Rochon, A. (2008). Reconstructing sea ice conditions in the arctic and sub-arctic prior to human observations. In DeWeaver, E.

T., Bitz, C. M., & Tremblay, L.-B. (Eds.), Arctic Sea Ice Decline: Observations, Projections, Mechanisms, and Implications. *Geophysical Monograph Series 80*, 27–45. Wiley.

Dlugokencky, E., & Tans, P. (2016). Trends in Atmospheric Carbon Dioxide, NOAA Global Monitoring Library, Earth System Research Laboratories, Available at www.esrl.noaa.gov/gmd/ccgg/trends/

Felis, T. (2020). Extending the instrumental record of ocean-atmosphere variability into the last interglacial using tropical corals. *Oceanography, 33*(2), 68–79. https://doi.org/10.5670/oceanog.2020.209

Fisher, D., Dyke, A., Koerner, R., Bourgeois, J., Kinnard, C., Zdanowicz, A., de Vernal, C., Hillaire-Marcel, S., & Rochon, A. (2006). Natural variability of Arctic sea ice over the Holocene. *Eos, 87*, 273–275.

Fröhlich, C. (2009). Evidence of a long-term trend in total solar irradiance. *Astronomy and Astrophysics, 501*, L27–L30.

Garcia-Munoz, M., Mason, G. M., & Simpson, J. A. (1975). The anomalous He-4 component in the cosmic-ray spectrum at <MeV per nucleon during 1972–1974. *The Astrophysical Journal, 202*, 265–275.

Grinsted, A., Moore, J. C., & Jevrejeva, S. (2009). Reconstructing sea level from paleo and projected temperatures 200 to 2100 AD. *Climate Dynamics, 34*(4), 461–472.

Grumet, N. S., Wake, C. P., Mayewski, P. A., Zielinski, G. A., Whitlow, S. I., Koerner, R. N., Fisher, D. A., & Woollett, J. M. (2001). Variability of sea-ice extent in Baffin Bay over the last millennium. *Climate Change, 49*, 129–145.

Hay, C. C., Morrow, E., Kopp, R. E., & Mitrovica, J. X. (2015). Probabilistic reanalysis of twentieth-century sea-level rise. *Nature, 517*(7535), 481–484.

Huang, B., Thorne, P. W., Banzon, V. F., Boyer, T., Chepurin, G., Lawrimore, J. H., Menne, M.J., Smith, T. M., Vose, R. S., & Zhang, H.-M. (2017a). Extended reconstructed sea surface temperature, version 5 (ERSSTv5). *NOAA National Centers for Environmental Information.* https://doi.org/10.7289/V5T72FNM, Accessed 2019

Huang, B., Thorne, P. W., Banzon, V. F., Boyer, T., Chepurin, G., Lawrimore, J. H., Menne, M. J., Smith, T. M., Vose, R. S., & Zhang, H.-M. (2017). Extended Reconstructed Sea Surface Temperature, Version 5 (ERSSTv5): Upgrades, Validations, and Intercomparisons. *American Meteorological Society, 30*, 8179–8205. https://doi.org/10.1175/JCLI-D-16-0836.1

Jacoby, G. C., & Ulan, L. D. (1982). Reconstruction of past ice conditions in a Hudson Bay estuary using tree rings. *Nature, 298*, 637–639.

Jevrejeva, S., Grinsted, A., & Moore, J. (2009). Anthropogenic forcing dominates sea level rise since 1850. *Geophysical Research Letters, 36*(20), L20706.

Jones, P. D., Briffa, K. R., Osborn, T. J., Lough, J. M., van Ommen, T. D., Vinther, B. M., Luterbacher, J., Wahl, E. R., Zwiers, F. W., Mann, M. E., Schmidt, G. A., Ammann, C. M., Buckley, M., Cobb, K. M., Esper, J., Goosse, H., Graham, N., Jansen, E., Kiefer, T., … Xoplaki, E. (2009). High-resolution palaeoclimatology of the last millennium: A review of current status and future prospects. *Holocene, 19*, 3–49.

Kaufman, D., McKay, N., Routson, C., Erb, M., Dätwyler, C., Sommer, P. S., Heiri, O., & Davis, B. (2020). Holocene global mean surface temperature, a multi-method reconstruction approach. *Nature Scientific Data, 7*, 201. https://doi.org/10.1038/s41597-020-0530-7

Kemp, A. C., Horton, B. P., Donnelly, J. P., Mann, M. E., Vermeer, M., & Rahmstorf, S. (2011). Climate related sea-level variations over the past two millennia. *Proceedings of the National Academies of Science, 108*(27), 11017–11022.

Kennedy, J. J., Rayner, N. A., Atkinson, C. P., & Killick, R. E. (2011). An ensemble data set of sea surface temperature change from 1850: The met office Hadley Centre HadSST.4.0.0.0 Data Set, *Journal of Geophysical Research–Atmospheres*, 7719–7763.

Kerr, R. A. (2000). A North Atlantic climate pacemaker for the centuries. *Science, 288*, 1984–1985.

Kinnard, C., Zdanowicz, C. M., Fisher, D. A., Isaksson, E., de Vernal, A., & Thompson, L. G. (2011). Reconstructed changes in Arctic sea ice over the past 1,450 years. *Nature, 479*, 509.

Knudsen, M. F., Riisager, P., Donadini, F., Snowball, I., Muscheler, R., Korhonen, K., & Pesonen, L. J. (2008). Variations in the geomagnetic dipole moment during the Holocene and the past 50 kyr. *Earth Planetary Science Letters, 272*, 319–329.

Köhler, P., Nehrbass-Ahles, C., Schmitt, J., Stocker, T. F., & Fischer, H. (2017). Comment on "Changes in atmospheric CO_2 levels recorded by the isotopic signature of n-alkanes from plants", from K. S. Machado and S. Froehner. *Global and Planetary Change, 156*, 24–25.

Köhler, P., Nehrbass-Ahles, C., Schmitt, J., Stocker, T. F., & Fischer, H. (2017). A 156 kyr smoothed history of the atmospheric greenhouse gases CO_2, CH_4, and N_2O and their radiative forcing. *Earth System Science Data, 9*, 363–387.

Kopp, R. E., Kemp, A. C., Bittermann, K., Horton, B. P., Donnelly, J. P., Gehrels, W. R., Hay, C. C., Mitrovic, J. X., Morrow, E. D., & Rahmstorf, S. (2016). Temperature-driven global sea-level variability in the Common Era. *PNAS, 113*(11), E1434–E1441.

Lambeck, K., Anzidei, M., Antonioli, F., Benini, A., & Esposito, A. (2004). Sea level in Roman time in the Central Mediterranean and implications for recent change. *Earth and Planetary Science Letters, 224*(3–4), 563–575.

Lambeck, K., Rouby, H., Purcell, A., Sun, Y., & Sambridge, M. (2014). Sea level and global ice volumes from the Last Glacial Maximum to the Holocene. *Proceedings of the National Academies of Science, 111*(43), 15296–15303.

MacFarling-Meure, C., Etheridge, D., Trudinger, C., Steele, P., Langenfelds, R., van Ommen, T., Smith, A., & Elkins, J. (2006). Law Dome CO_2, CH_4 and N_2O ice core records extended to 2000 years BP. *Geophysical Research Letters, 33*, L14810.

Macias-Fauria, M., Grinsted, A., Helama, S., Moore, J., Timonen, M., Martma, T., Isaksson, E., & Eronen, M. (2010). Unprecedented low twentieth century winter sea ice extent in the Western Nordic Seas since AD 1200. *Climate Dynamics, 34*, 781–795.

Mann, M. E., Zhang, Z., Hughes, M. K., Bradley, R. S., Miller, S. K., Rutherford, S., & Ni, F. (2008). *Proceeding of the National Academy of Sciences, 105*, 13252.

Mann, M., Zhang, Z., Rutherford, S., Bradley, R. S., Hughes, M. K., Shindell, D., Ammann, C., Faluvegi, G., & Ni, F. (2009). Global signatures and dynamical origins of the Little Ice Age and Medieval Climate Anomaly. *Science, 326*, 1256–1260.

Mantua, N. J., Hare, S. R., Zhang, Y., Wallace, J. M., & Francis, R. C. (1997). A Pacific interdecadal climate oscillation with impacts on salmon production. *Bulletin of the American Meteorology Society, 78*, 1069.

Marcott, S. A., Bauska, T. K., Buizert, C., Steig, E. J., Rosen, J. L., Cuffey, K. M., Fudge, T. J., Severinghaus, J. P., Ahn, J., Kalk, M. L., McConnell, J. R., Sowers, T., Taylor, K. C., White, J. W. C., & Brook, E. J. (2014). Centennial-scale changes in the global carbon cycle during the last deglaciation. *Nature, 514*, 616–619.

Marcott, S. A., Shakun, J. D., Clark, P. U., & Mix, A. C. (2013). A reconstruction of regional and global temperature for the past 11,300 years. *Science, 339*(6124), 1198–1201.

Masarik, J., & Beer, J. (2009). An updated simulation of particle fluxes and cosmogenic nuclide production in the Earth's atmosphere. *Journal of Geophysical Research–atmosphere, 114*, D11103.

Masse', G., Rowland, S. J., Alexandrine Sicre, M., Jacob, J., Jansen, E., & Belt, S. T. (2008). Abrupt climate changes for Iceland during the last millennium: evidence from high resolution sea ice reconstructions. *Earth Planetary Science Letters 269*, 565–569

Masson-Delmotte V., Schulz, M., Abe-Ouchi, A., Beer, J., Ganopolski, A., Gonzalez Rouco, J. F., Jansen, E., Lambeck, K., Luterbacher, J., Naish, T., Osborn, T., Otto-Bliesner, B., Quinn, T., Ramesh, R., Rojas, M., Shao, X., & Timmermann, A. (2013). Information from paleoclimate archives. Climate Change 2013: The Physical Science Basis. In Stocker TF, et al. (Eds.) *Contribution of Working Group I to the Fifth Assessment Report of the Intergovernmental Panel on Climate Change.* Cambridge University Press. 383–464

McKay, N. P., & Kaufman, D. S. (2014). An extended Arctic proxy temperature database for the past 2,000 years. *Sci. Data, 1*(140026), 2014. https://doi.org/10.1038/sdata.2014.26

Menne, M. J., Williams, C. N., Gleason, B. E., Rennie, J. J., & Lawrimore, J. H. (2018). The global historical climatology network monthly temperature dataset, version 4. *J. Climate,* 9835–9854. https://doi.org/10.1175/JCLI-D-18-0094.1

Mitchell, L. E., Brook, E., Lee, J. E., Buizert, C., & Sowers, T. (2013). Constraints on the late Holocene atmospheric methane budget. *Science, 342,* 964–967.

Mitchell, L. E., Brook, E. J., Sowers, T., McConnell, J. R., & Taylor, K. (2011). Multi-decadal variability of atmospheric methane, 1000–1800 C.E. *Journal of Geophysical Research-Biogeoscience, 116,* 1–16. https://doi.org/10.1029/2010JG001441

PAGES 2k Consortium. (2013). Continental-scale temperature variability during the past two millennia. *Nature Geoscience, 6,* 339–346.

PAGES2k Consortium. (2017). A global multiproxy database for temperature reconstructions of the Common Era. *Nature Scientific Data, 4,* 170088. https://doi.org/10.1038/sdata.2017.88

PAGES2k Consortium. (2019). Consistent multi-decadal variability in global temperature reconstructions and simulations over the Common Era. *Nature Geoscience, 12,* 643–649. https://doi.org/10.1038/s41561-019-0400-0

Polyak, L., Alley, R. B., Andrews, J. T., Brigham-Grette, J., Cronin, T. M., Darby, D. A., Dyke, A. S., Fitzpatrick, J., Funder, S., Holland, M., Jennings, A. E., Miller, G. H., O'Regan, M., Savelle, J., Serreze, M., St. John, K., White, J. W. C., & Wolff, E. (2010). History of sea ice in the Arctic. *Quaternary Science Reviews 29,* 1757–1778.

Roemmich, D. et al. (2019). On the future of Argo: A global, full-depth, multi-disciplinary array. *Frontiers in Marine Science 6,* 439, https://doi.org/10.3389/fmars.2019.00439. See also https://argo.ucsd.edu/

Rubino, M., Etheridge, D. M., Thornton, D. P., Howden, R., Allison, C. E., Francey, R. J., Langenfelds, R. L., Steele, P. L., Trudinger, C. M., Spencer, D. A., Curran, M. A. J., Van Ommen, T. D., & Smith, A. M. (2019). Revised records of atmospheric trace gases CO_2, CH_4, N_2O and $\delta^{13}C$-CO_2 over the last 2000 years from Law Dome, Antarctica. *Earth System Science Data, 11,* 473–492. https://doi.org/10.5194/essd-11-473-2019

Rubino, M., Etheridge, D. M., Trudinger, C. M., Allison, C. E., Battle, M. O., Langenfelds, R. L., Steele, L. P., Curran, M., Bender, M., White, J. W. C., Jenk, T. M., Blunier, T., & Francey, R. J. (2013). A revised 1000 year atmospheric $\delta^{13}C$-CO_2 record from Law Dome and South Pole, Antarctica. *Journal of Geophysical Research - Atmosphere, 118,* 8482–8499.

Sigl, M, Winstrup, M, McConnell, J. R., Welten, K. C., Plunkett, G, Ludlow, F., Büntgen, U., Caffee, M., Chellman, N., Dahl-Jensen, D., Fischer, H., Kipfstuhl, S., Kostick, C., Maselli, O. J., Mekhaldi, F., Mulvaney, R., Muscheler, R., Pasteris, D. R., Pilcher, J. R., Salzer, M., Schüpbach, S., Steffensen, J. P., Vinther, B. M., & Woodruff, T. E. (2015). Timing and climate forcing of volcanic eruptions for the past 2,500 years. *Nature 523,* 543–549

Sigl, M., Fudge, T. J., Winstrup, M., Cole-Dai, J., Ferris, D., McConnell, J. R., Taylor, K. C., Welten, K. C., Woodruff, T. E., Adolphi, F., Bisiaux, M., Brook, E. J., Buizert, C., Caffee, M. W., Dunbar, N. W., Edwards, R., Geng, L., Iverson, N., Koffman, B., … Sowers, T. A. (2016). The WAIS Divide deep ice core WD2014 chronology—Part 2: Annual-layer counting (0–31 ka BP). *Climate of the past, 12*(3), 769–786.

Sivan, D., Lambeck, K., Toueg, R., Raban, A., Porath, Y., & Shirman, B. (2004). Ancient coastal wells of Caesarea Maritima, Israel, an indicator for relative sea level changes during the last 2000 years. *Earth and Planetary Science Letters, 222*(1), 315–330.

Smith, T. M., Reynolds, R. W., Peterson, T. C., & Lawrimore, J. (2008). Improvements to NOAA's historical merged land-ocean surface temperature analysis (1880–2006). *Journal of Climate, 21,* 2283–2296.

Steinhilber, F., Abreu, J. A., Beer, J., Brunner, I., Christl, M., Fischer, H., Heikkilä, U., Kubik, P. W., Manna, M., McCracken, K. G., Miller, H., Miyahara, H., Oerter, H., & Wilhelms, F. (2012). 9,400 years of cosmic radiation and solar activity from ice cores and tree rings. *PNAS, 109*(16), 5967–5971.

Steinhilber, F., Beer, J., & Frohlich, C. (2009). Total solar irradiance during the Holocene. *Geophysical Research Letters, 36,* L19704.

Tierney, J. E., Abram, N. J., Anchukaitis, K. J., Evans, M. N., Giry, C., Kilbourne, K. H., Saenger, C. P., Wu, H. C., & Zinke, J. (2015). Tropical sea-surface temperatures for the past four centuries reconstructed from coral archives. *Paleoceanography, 30*, 226–252. https://doi.org/10.1002/2014PA002717

Usoskin, I. G., Hulot, G., Gallet, Y., Roth, R., Licht, A., Joos, F., Kovaltsov, G. A., Thébault, E., & Khokhlov, A. (2014). Evidence for distinct modes of solar activity. *Astronomy and Astrophysics 562*, L10. World Data Center for Paleoclimatology, Accessed 2019

Woodroffe, C. D., McGregor, H. V., Lambeck, K., Smithers, S. G., & Fink, D. (2012). Mid-Pacific micro-atolls record sea-level stability over the past 5000 yr. *Geology, 40*(10), 951–954.

Woodruff, S. D., Worley, S. J., Lubker, S. J., Ji, Z., Freeman, J. E., Berry, D. I., et al. (2011). ICOADS release 2.5: Extensions and enhancements to the surface marine meteorological archive. *International Journal of Climatology 31*(7), 951–967.

Zhang, H.-M., J. H. Lawrimore, B. Huang, M. J. Menne, X. Yin, A. Sanchez-Lugo, B. E. Gleason, R. Vose, D. Arndt, J. J. Rennie, and C. N. Williams. (2019). Updated temperature data give a sharper view of climate trends. *EOS*, 100, https://doi.org/10.1029/2019EO128229

Thomas M. Cronin [USGS emeritus research geologist] became interested in glacial geology and climate change growing up in Connecticut where evidence of the last great ice sheets is everywhere. He was lucky to go to Colgate University to play college basketball where he met excellent professors who encouraged him to pursue a geology major. After getting his BA from Colgate he again was lucky to pursue an MA and PhD in Geology from Harvard University and then a National Research Council Post-doc at the USGS and Museum of Natural History, Washington DC. His research since at USGS, which coincided with a growing international awareness of the importance of geological records of climate change, has focused on paleoclimatology, sea-level change, biostratigraphy, geochemistry and ecosystems. He was an NSF-sponsored visiting researcher at Shizuoka University, Japan (1991), taught at the Urbino (Italy) Summer School for Paleoclimatology Faculty (2008–2016), was adjunct faculty at Georgetown University's Science Technology International Affairs Program, Walsh School of Foreign Service (2005–2021), and served in the White House Office of Science, Technology and Policy (OSTP) (1996–97). He has participated in sediment coring expeditions including four to the Arctic Ocean. His awards include the Brady Medal (TMS London), Duke of Montefeltro Medal (USSP Urbino), Fellow, American Association for Advancement of Science, Wilmot H Bradley lecture (Geological Society Washington), US Coast Guard Service Medal, AGU Citation for Excellence Reviewing, Bolin Climate Center Annual Lecturer (Stockholm), USGS Leadership, Meritorious Service, and Excellence Award. During his career, he worked extensively with high school, bachelors, masters and PhD students leading to many fruitful careers in the sciences. Tom is the primary author of Chap. 3.

Open Access This chapter is licensed under the terms of the Creative Commons Attribution 4.0 International License (http://creativecommons.org/licenses/by/4.0/), which permits use, sharing, adaptation, distribution and reproduction in any medium or format, as long as you give appropriate credit to the original author(s) and the source, provide a link to the Creative Commons license and indicate if changes were made.

The images or other third party material in this chapter are included in the chapter's Creative Commons license, unless indicated otherwise in a credit line to the material. If material is not included in the chapter's Creative Commons license and your intended use is not permitted by statutory regulation or exceeds the permitted use, you will need to obtain permission directly from the copyright holder.

Icy Secrets Preserved in Earth's Glaciers

Lonnie G. Thompson and Ellen Mosley-Thompson

Guiding Question: How do we read the clues about past climate preserved in Earth's ice?

1 Key Take-Away Points

- It is important to study the past to better understand both natural- and human-driven climate and environmental change.
- Glacier ice is one of the most versatile recorders of Earth's past climate over multiple time scales.
- Ice core records from glaciers and ice caps in the polar regions and from low-latitude, high-elevation glaciers provide valuable information on past regional to global-scale climate variability (e.g., temperature and precipitation) and forcing mechanisms that cool (e.g., volcanic eruptions) and warm (e.g., greenhouse gases) the planet.
- Climate histories are recorded in ice by stable isotopes of water ($\delta^{18}O$, δD) and by impurities in that ice (e.g., dust, major anions and cations, trace elements, gases trapped in the bubbles, black carbon, and microbes).

L. G. Thompson (✉)
School of Earth Sciences and the Byrd Polar and Climate Research Center, The Ohio State University, Columbus, USA
e-mail: thompson.3@osu.edu

E. Mosley-Thompson
Department of Geography, Atmospheric Sciences, and the Byrd Polar and Climate Research Center, The Ohio State University, Columbus, USA
e-mail: thompson.4@osu.edu

© The Author(s) 2025
K. St. John and L. Krissek (eds.), *Climate Change*,
https://doi.org/10.1007/978-3-031-82869-0_4

- The oldest continuous ice core records come from the polar ice sheets, and—until recently—the oldest core (800,000 years) recovered was drilled at Dome C in East Antarctica.
- Ice core records from the low latitudes, although typically encompassing shorter timescales, can provide detailed information about climate and environments where most people live and where most human civilizations developed.
- Glaciers are one of our best recorders of climatic and environmental changes, and for many indigenous people, they are an important part of their cultural history; however, both glaciers and people are at risk due to climate change.
- The world's cryosphere is in retreat as atmospheric and oceanic temperatures warm in response to rapidly increasing concentrations of greenhouse gases, and unfortunately, the rate of increase shows no sign of slowing.
- The shrinking of Earth's cryosphere, including mountain glaciers that have existed for thousands of years, endangers water resources necessary for societies to survive and prosper, especially in mountain regions and areas dependent on rivers that originate in glaciated areas such as the Himalayas.
- Melting of polar ice sheets and mountain glaciers is contributing to rising sea levels which is endangering coastal cities, communities, and ecosystems.

2 Introduction

Climate change is expected to impact many aspects of life in the coming centuries. Predictions of future climate changes are provided by climate models, the development of which has depended in large measure on our basic understanding of the Earth Systems and our knowledge of its climate history. This history extends from the present to millions of years back in time, with time intervals (or temporal resolutions) varying from sub-annual to millennial. It is based on information extracted from many different proxy sources: tree rings, lake and marine cores, cave deposits, corals, ice cores, and even historical records.

Paleoclimatologists, or scientists who study past climate, not only strive to reconstruct past climate variations on regional to global scales, but they conduct research to determine the natural and anthropogenic (human-caused) mechanisms that influence climate changes. This information is essential to place the current climate into both historical and geologic perspectives and to provide data used by computer models to predict future scenarios. It is the role of paleoclimatologists to reconstruct records of climate, including temperature and precipitation variability over hundreds to thousands of years from geological and biological "media" that preserve evidence of these variations. Paleoclimatologists also strive to find evidence of the causes of climate change, or "forcing mechanisms" in these records. Three forcing mechanisms that you have already been introduced to in early chapters include variations in solar energy reaching Earth's atmosphere and surface, volcanic activity, and change in the abundance of greenhouse gases.

This chapter discusses records of climatic and environmental variations preserved in ice cores obtained from ice sheets and glaciers, at locations ranging from Earth's polar regions to the tropics. These ice core-derived paleoclimate histories complement those extracted from other sedimentary systems, such as those presented in Chap. 5. In addition to the histories of regional and global climate variations that glaciers and ice sheets produce, mapping the current retreat of their margins provides some of the most dramatically visible evidence of recent climate changes.

3 Why Is Ice Important in the Earth System?

In addition to glacier ice, Earth's cryosphere is composed of frozen water in all its forms: snow cover, sea ice and other floating ice, and permafrost. Although the cryosphere contains just about 2% of all the water on Earth, it constitutes 70% of the freshwater.[1] The most significant components of the cryosphere are the vast ice sheets of Antarctica and Greenland, the Arctic sea ice, and mountain glaciers that exist at all latitudes. Ice and snow cover is an essential part of Earth's thermostat as its white surface has a high albedo which means that it reflects much of the Sun's incoming radiation that warms the surface which then radiates energy back to space. The polar ice sheets also help maintain the temperature difference (or thermal gradient) between the poles and the tropics. This thermal gradient is essential for maintaining the constant movement of energy (heat) throughout the world via its water (oceans) and air (atmosphere).

Because ice and snow on Earth exist close to their melting point, they are among the first components of the Earth system to respond to climate changes, particularly those involving temperature. It is indisputable that Earth's cryosphere is threatened by the current climate warming. The most publicized consequence of this warming is the melting of the Arctic sea ice (as described in Chap. 3), the Greenland ice sheet, and the large ice shelves circling the Antarctic ice sheet. However, the world's mountain glaciers are also melting and many are now vanishing. The ongoing rapid retreat of the world's glaciers and ice sheets is a major contributor to global sea level rise.[2] More recently, strong evidence indicates an acceleration of the rate of ice loss in the tropics, which presents particular danger to water supplies for at-risk populations in South America and southeast Asia. The loss of mountain glaciers, often viewed as the world's water towers, threatens water resources that are essential for hydroelectric power, crop irrigation, municipal water supplies, and even tourism.[3] Today we live in climatological conditions that are unprecedented in at least the last 800,000 years, which—until recently—was the longest continuous history documented by ice core

---

[1] McConnell (2005).

[2] Zemp et al. (2019).

[3] Carey et al. (2017).

records,[4] and the world's ice cover is responding dramatically. Social science studies suggest that the effects of glacier retreat extend beyond variability in glacier runoff and must be analyzed in the context of their deep historical, socio-economic, cultural, and spiritual settings (see Box 1). Unfortunately, today the future human response to the shrinking of the cryosphere remains ambiguous. Compelling evidence from both data and models confirms that anthropogenic greenhouse gases (GHGs) are largely responsible for the 20th- and 21st-century warming of our planet, and yet efforts to reduce carbon dioxide and methane emissions are insufficient as the atmospheric concentrations of these gases continue to accelerate.[5]

Box 1 Glaciers: home of the gods

In the history of human civilization, people have always traveled to the mountains for inspiration and guidance. Although a significant effort is required to get to these areas, the isolation and the pristine air and water allow for clear and uninterrupted thought. Glaciers atop the highest mountains of the world are often the focus of belief systems by indigenous societies that consider them to be sacred places such as the home of the gods and/or the souls of their ancestors. These glaciers also have immediate secular value as the source of small streams that are vital to the survival of nearby communities. People who live near and have close connections to these glaciers quite understandably have concerns when outsiders arrive to conduct activities on them, even nondestructive scientific research. When scientific expeditions, whether national or international, arrive at these glaciated mountains with multiple team members and perhaps tons of equipment, encounters with local communities can be sudden and unexpected.

Ice core drilling projects on tropical mountains present logistical and political complications that typically are not encountered by scientists drilling in polar locations. Researchers from the Byrd Polar and Climate Research Center at The Ohio State University (OSU) have conducted dozens of expeditions to high and remote mountain glaciers. For 2–3 years before the start of a project, contacts must be made with appropriate agencies within the host country to obtain all the necessary permits and custom clearances and to set up collaborations with scientists within the country, who often join the team and participate in the fieldwork. Although local governments are informed of the expedition's mission to help understand regional climate change, this critical information is not always communicated to the citizens who live near the glaciers.

---

[4] One of the longest ice core records to date comes from Dome C in East Antarctica (Fig. 1) and shows that modern temperature and greenhouse gas concentrations are the highest in 800,000 years. A 2025 drilling campaign on Little Dome C (~35 km away from Dome C) successfully recovered an even longer ice core record, potentially extending ice core climate reconstructions back to 1.2 million years ago.

[5] Allan et al. (2021).

Interactions with local communities in the Bolivian and Peruvian Andes

The ice core research team at OSU under the direction of Dr. Lonnie Thompson takes several steps to establish a rapport with the citizens of communities near the glaciers under study. For example, brochures are produced in both English and local languages describing the goals of the expedition and why it is important to advance our understanding of climate change. Prior to the expedition, Thompson travels to the area to meet with local officials and give lectures so that concerns can be raised and addressed. Nonetheless, these efforts do not guarantee that there will not be issues or that some communities will not be overlooked. Such was the case in 1997, when leaders of the Aymara community living at the base of Nevado Sajama in Bolivia insisted on the sacrifice of an alpaca at the start of the field program. This was described in the first chapter in *Thin Ice, Unlocking the Secrets of Climate in the World's Highest Mountains* by Mark Bowen. In 2019, the drilling of the glacier on Nevado Huascarán in northern Peru, the world's highest tropical mountain, sparked a protest among members of the Quechua community who thought the expedition was acting under the direction of mining interests. This incident was described in a September 17, 2019, news piece in *Nature* entitled "Daring scientists extract ice from Earth's highest tropical glacier" by Barbara Fraser.

Encounter with communities in Papua, New Guinea, Indonesia

The island of New Guinea contains the only ice field between the Himalayas in South Asia and the Andes in South America. In 2010, Thompson's team drilled this ice field located in the shadow of Puncak Jaya, the highest mountain in Oceania. The mountains and remote, isolated ice fields are located on Earth's richest gold deposit in the middle of a tropical rainforest in a setting that is reminiscent of the scenes in the movie *Avatar*. Cloud banks are a persistent feature, and at times enormous waterfalls which spill from the mountains cascade through the clouds. A large mining complex near the ice field is run by Freeport-McMoRan of Phoenix, Arizona and staffed by people (mostly locals) who live in the small city of Tembagapura (pop. ~ 20,000).

The mine and the ice field are surrounded by four tribes that are in conflict with each other, and at times disagreements erupt into violence. While Thompson's team drilled the ice field near Puncak Jaya, about 150 members of the Amungme people, who had not been consulted by the government about the expedition, took exception to the team's activities. In light of these tensions, Thompson was invited by the management of the mine to confer with representatives from the Amungme. An overview of the ice coring procedure was provided, along with the reasons why such an undertaking was so valuable, not only in New Guinea but throughout the world. As the meeting progressed, Thompson gained an understanding of the Amungme culture and beliefs. To them, the ridges and the valleys of the mountains are the arms and legs of their God, and the glacier through which the expedition was drilling is the head. In their words, the expedition was "drilling into the skull of God to steal memories." Thompson realized that in a way that was exactly what his team was doing; by drilling ice cores they were pulling out of the ice the "memories" of past climate changes. During the meeting the realities and hazards of the changing climate were discussed, revealing a generational divide. Older people, who had gazed upon the ice fields during their long lives, asserted that the ice had always been on the mountain and would remain there forever. However, younger people were aware of the changes in the glaciers, and to them, their eventual demise was a reality. Thompson indicated that the time would come very soon when the last piece of ice would melt away, but a small part of it would be preserved in the freezers at OSU. By the end of the meeting, the Amungme granted the team permission to remove the 88 m of ice core they had drilled. To this day, the ice from New Guinea is in storage at $-30\ °C$, although some has been used for various analyses to reconstruct past climate changes in the equatorial West Pacific. Since the events of 2010, aerial observations of the New Guinea ice fields have confirmed the fears expressed by the younger Amungme during that meeting; at the current rate of retreat, all the ice on those mountains will disappear within the next decade.[6]

---

[6] Permana et al. (2019b, 2019a).

The disagreement between the older and younger Amungme serves as one example of the divisions sparked by culture and beliefs among groups throughout the world, and especially between generations. Like the Amungme, the history and traditions of societies, no matter the economic level, form the basis of beliefs that environmental scientists must respect while they try to promote an understanding of how a rapidly changing climate will alter the lives and livelihoods in these societies (Figs. 1, 2 and 3).

Fig. 1 Meeting with representatives of the Amungme on June 16, 2010

Fig. 2 Drilling camp in July (2010) where the OSU team drilled the rapidly disappearing East Northwall Firn ice field, Papua, New Guinea, Indonesia

Fig. 3 Aerial photos show the Northwall Firn ice fields in 2010 (left) and in 2021 (right)

4 How Is Information Stored in Glaciers and Ice Fields Produced, Preserved, and Collected?

Glaciers and ice sheets are among the world's best recorders of natural and anthropogenic climate change and provide a time perspective for current climatic and environmental variations. An ice core drilled through a glacier can produce an extensive variety of unique data about past climates, environments, and atmospheric composition that are not available from other sources. Because these records are long, continuous, of high temporal resolution, and contain information about numerous physical and chemical properties of the atmosphere, they are invaluable. They often provide relatively precise histories of climatic events, and reveal the potential forcing mechanisms that have been active over the history of human development, as well as before humans had significant influences over their environment.

Glacier ice is considered one of nature's best "thermometers"[7] and provides critical evidence of recent warming by its retreat throughout the world. Over the last 60 years, such records have been recovered from the polar regions as well as low-latitude, high-elevation ice fields (Fig. 4). Analyses of these ice cores and of the glaciers from which they have been drilled have provided three lines of evidence for past and present abrupt climate change: (1) the temperature and precipitation histories recorded in the glaciers as revealed by the climate records extracted from the

---

[7] Pollack, H. (2010). A World Without Ice, Avery. ISBN 9781583334072.

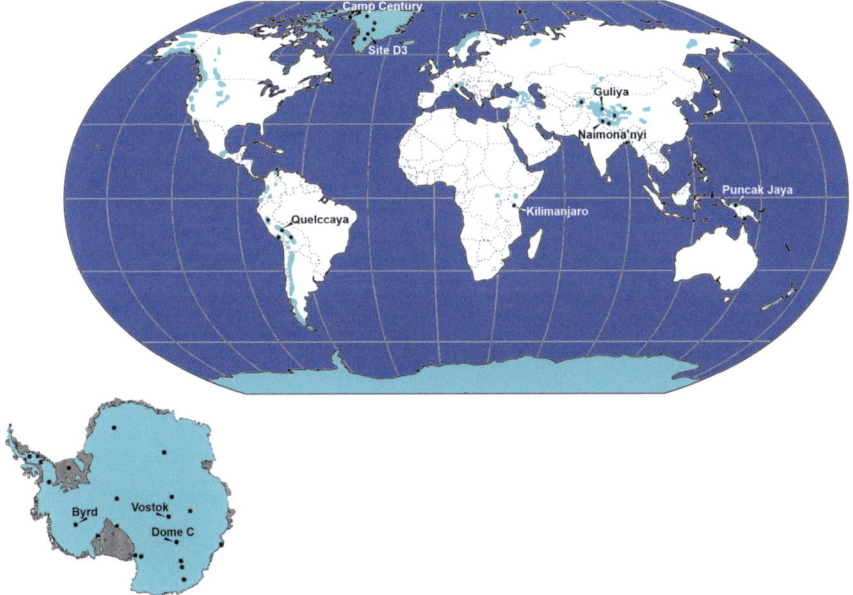

Fig. 4 Global map showing locations of terrestrial ice cover (light blue) and the locations of many ice core drilling sites. Antarctic ice shelves are shown with grey shading. The ice core sites discussed in the text are labeled

ice cores; (2) the accelerating loss of the glaciers themselves, and; (3) the exposure of ancient fauna and flora along the retreating margins of the glaciers as a result of their recent melting and shrinking (see Box 2).

Box 2 Placing the recent glacier retreat into a time perspective

The recent retreat of glaciers around the world is well documented. The real challenge for paleoclimatologists is placing the retreat into a longer term time perspective. The retreat of the Quelccaya ice cap has been documented over the last three decades and allows us to address this question in a unique way. When the margins of the ice cap advanced in the past as the glacier grew, they covered nearby wetland plants in their growth position. These wetland plants have no woody structure and are perfectly preserved as illustrated in Fig. 5. However, the recent retreat of Quelccaya's margins exposed the long-buried plants. Once these plants are exposed to the environment, they decay within a few years. During an expedition to Quelccaya in the austral winter of 2002, some of these freshly exposed plants were spotted for the first time. The margin along which they were found was very familiar to the field team since its retreat had been investigated for several years, ever since a lake began to form in 1985 as the ice melted. In subsequent years additional plant samples were collected, and radiocarbon dating of twenty plants collected between 2004 and 2007 revealed an average calibrated age of ~ 4700 years before present (yrs BP). The radiocarbon dates tell us how long ago the plants were growing and when they were covered by the advancing ice, thus stopping the radiocarbon clock. The ages and positions of these plants also indicated that in 2004–2007 the ice cap was smaller than it had been in almost 5000 years.

Fig. 5 Photographs and photomicrographs of the wetland plants collected along the western margin of the Quelccaya ice cap

After 2007, more plants were exposed as the ice retreated up the valley, and their radiocarbon dates revealed that they were progressively older in age. In 2011, fresh plant remains were uncovered on the eastern side of the expanding lake as the retreat continued. The average radiocarbon date on five samples collected in that year was ~ 6300 yrs before present (BP), or approximately 1600 years older than those collected in 2004–2007. This indicated that in 2011 the ice cap was smaller than it had been in over 6000 years (Fig. 6). Moreover, all plant ages together confirmed that about 6000 years ago the advance of the western margin of the Quelccaya ice cap was much slower (~300 m in ~ 1600 years) than its current rate of retreat (~300 m in 25 years).

Fig. 6 Photograph of the western margin of the Quelccaya ice cap showing the lake that has been growing from the melting ice since 1985 and the collection sites of plants from 2004 to 2011.[8] Figure used with permission from AAAS. This figure is excluded from Creative Commons license

The retreat of glaciers in the polar regions is also exposing old plants which have been collected and radiocarbon dated. For example, newly exposed tundra plants, still in their growth positions, were collected along the margins of 30 ice caps in the Canadian Arctic.[9] Their radiocarbon dates indicate that the plants were entombed in ice over 40,000 years ago. Until recently, these plants were never exposed after their burial, which along with paleotemperature reconstructions suggests that these ice caps have not been smaller since the previous warm interglacial period about 110,000 years ago.

4.1 What Is Analyzed in an Ice Core?

Glacier ice is one of the most versatile recorders of Earth's past climate over multiple timescales. The records extracted from ice cores from around the world range from a few decades to hundreds of millennia. This section discusses the various atmospheric constituents that are deposited within the snowfall, or that fall out on the snow/ice

---

[8] Thompson et al. (2013b, 2013a).
[9] Pendleton et al. (2019).

surface under dry conditions and are trapped within the porous upper layers of snow as it slowly densifies into firn[10] and eventually into solid ice.

Ice can store anything that is found in the atmosphere, such as gases, dust, chemical species (i.e., elements, ions, isotopes, molecules), pollen, and fallout from volcanic eruptions and anthropogenic contaminants (e.g., thermonuclear emissions, lead, sulfate, and nitrate) (Fig. 7, Table 1). Glacier ice can even trap emissions from biological activity, which gives us information about past vegetation patterns. Solar activity is recorded by measuring certain radioactive isotopes preserved in the ice, such as carbon-14 (^{14}C), chlorine-36 (^{36}Cl), and beryllium-10 (^{10}Be) which are produced by the cosmic energy bombardment of chemical species in the atmosphere. Occasionally layers of volcanic ash and/or acids are trapped in ice if the atmospheric conditions are conducive, and the glacier is located close to and/or downwind from an eruption, or if the eruption is large enough to disperse material over very large areas. Temperature records can be derived from stable isotopes[11] (atoms of the same element but of different weights) of oxygen (^{18}O and ^{16}O) and hydrogen (^{1}H and ^{2}H, or D for deuterium), the elements of water and ice. Gases in the atmosphere such as carbon dioxide (CO_2), methane (CH_4), and nitrous oxide (N_2O) move through the snow and firn, and as the firn is compacted into ice, the gases are trapped in air bubbles in the same concentrations that they occur in the atmosphere at the time.

Many glaciated regions receive precipitation on a seasonal basis, i.e., more snow falls in "wet" seasons than "dry" seasons. This seasonality of precipitation is very pronounced in tropical monsoon regions, where high-altitude mountain glaciers receive the majority of their snowfall during the summer when monsoons are active. Dust and other aerosols in the atmosphere are concentrated on the ice during the dry season and are preserved as thin laminae alternating with thick, cleaner wet season ice (Figs. 7, 8 and 9). These alternating dark and light layers constitute yearly cycles that are analogous to tree rings and can be counted back in time. The ice also records the seasonal variations in atmospheric dust sizes and in various chemical species that make up salts (e.g., consisting of cations such as sodium and anions such as chloride), biological emissions (e.g., sulfate, ammonium, nitrate), and other seasonally varying atmospheric aerosols.

Below is a brief discussion of the information that can be obtained from the physical and chemical properties of ice which were introduced above. These can be divided into five types: (a) stable isotopes of oxygen and hydrogen ($\delta^{18}O$ and δD); (b) dissolved and particulate matter preserved in firn and ice; (c) gases extracted from air bubbles trapped in the ice; (d) physical characteristics of the firn and ice; and (e) new areas of research such as extraction and identification of ancient microorganisms such as bacteria and viruses. In the past, the first three types of analyses required that the ice core be cut into discrete samples and melted, or sometimes crushed under a vacuum in the case of gas measurements. More recently, techniques have been

---

[10] Firn is partially compacted granular snow left over from past seasons. It is of intermediate density between that of snow (~400 kg/m^3) and that of glacier ice (~830 kg/m^3).

[11] Stable isotopes, such as those of oxygen, do not undergo radioactive decay while radioactive isotopes undergo decay of their nuclei to form daughter isotopes.

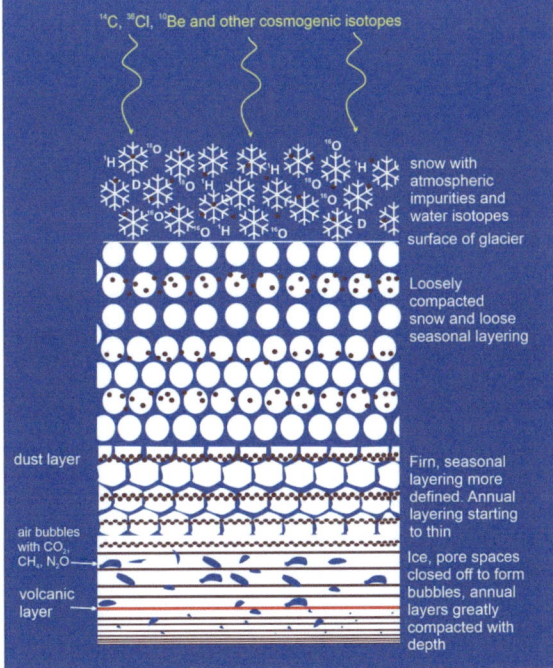

Fig. 7 The formation of ice in a glacier. Snow and impurities (dust, salts, pollen, chemical species) fall on the surface of a glacier and undergo gradual compaction. Impurities are generally concentrated in dry seasons and are interbedded with cleaner wet season snow. After being buried by subsequent snow, compacted snow in the glacier becomes firn, which is the intermediate phase with densities between those of snow and ice. With further compaction, the spaces between snow crystals close off, and ice is formed. Air bubbles in the ice contain entrapped atmospheric gases such as carbon dioxide (CO_2), methane (CH_4), and nitrous oxide (N_2O). The dry season layers are also compacted and thin with depth. If a volcanic eruption occurs in the vicinity, a layer of volcanic ash may be deposited and preserved in the glacier and can be used as a known time horizon to construct or calibrate an ice core's timescale

developed to continuously measure dust and other "impurities" that allow the ice to remain mostly intact.[12]

4.1.1 Stable Isotopes of Water ($\delta^{18}O$, δD)

Water is the most abundant compound on Earth's surface and is the substance of which glaciers are composed, thus the measurement of the isotopes of oxygen and hydrogen is one of the main tools used by glaciologists and paleoclimatologists to

---

[12] This method is known as "continuous flow analysis" and often consists of melting a small channel through the center of a length of ice core and directing the meltwater online through instruments that measure dust and various chemical species.

Table 1 Information gleaned from ice core analyses

Measurement	Source (examples)	What it tells us (examples)
$d^{18}O$, δD	Oceans, lakes, water vapor, precipitation (i.e., the hydrosphere)	Temperature, precipitation amount, precipitation source
Mineral dust concentration	Soils, deserts, dried lake beds, glacial outwash, and stream beds	Aridity, ground exposure, wind strength
Mineral dust sizes	Soils, deserts, dried lake beds, glacial outwash, and stream beds	Wind strength, relative distance between the dust source and glacier
Methanesulfonic acid (MSA)	Oceans	Marine biological activity, it is affected by sea ice coverage in polar waters and upwelling ocean water
Fluoride (F^-)	Soils, volcanic eruptions, groundwater	Salts, volcanic activity, aerosol source, wind direction
Chloride (Cl^-)	Salt flats, oceans, volcanic eruptions, dried lake and stream beds	Aridity, ground exposure, sea ice coverage, volcanic activity, aerosol source, wind direction
Nitrate (NO_3^-)	Salt flats, soils, vegetation, production in the atmosphere by lightning, anthropogenic emissions	Aridity, biological activity in soils, storm activity, aerosol source, wind direction, human impact
Sulfate (SO_4^{2-})	Salt flats, oceans, volcanic eruptions, anthropogenic emissions	Volcanic activity (climate forcing), drought, fire, aerosol source, wind direction, human impact
Sodium (Na^+)	Salt flats, oceans, volcanic eruptions, dried lake and stream beds, loess deposits	Drought, sea water spray, aerosol source, wind direction
Ammonium (NH_4^+)	Salt flats, lakes, soils, anthropogenic emissions	Aridity, ground exposure, biological activity in soils, human impact
Potassium (K^+)	Soils, salt flats, dried lake and stream beds, glacial outwash	Aridity, ground exposure, aerosol source, wind direction
Magnesium (Mg^{2+})	Soils, salt flats, dried lake and stream beds, glacial outwash	Aridity, ground exposure, aerosol source, wind direction

(continued)

Table 1 (continued)

Measurement	Source (examples)	What it tells us (examples)
Calcium (Ca^{2+})	Soils, salt flats, dried lake and stream beds, glacial outwash	Aridity, ground exposure, aerosol source, wind direction
Black Carbon (BC)	Vegetation burning, anthropogenic emissions	Forest and grassland fire history, industrial activity
Trace and Rare Earth elements	Micrometeorites, dust, volcanic eruptions	Extraterrestrial and terrestrial dust, volcanic emissions, dust sources
Microbes	Viruses, bacteria, fungi in the environment	Past climate conditions, ecological diversity in extreme environments, evolution
Gases (CO_2, CH_4, N_2O)	Plant and animal respiration and decay, oceans, agriculture, anthropogenic emissions, volcanic eruptions, permafrost	Atmospheric composition, climate forcing, human impacts
Cosmogenic radionuclides (^{10}Be, ^{36}Cl, ^{14}C)	Produced by interaction of stable nuclides in the atmosphere with cosmic rays	Age determination, changes in solar output and Earth's magnetic field

Listed are the parameters that are measured, examples of their sources, and examples of the information they provide.

study past climate variations. The isotopes of oxygen (^{16}O with 8 neutrons, ^{17}O with 9 neutrons, ^{18}O with 10 neutrons) and hydrogen (^{1}H with 1 neutron, ^{2}H or deuterium (D) with 2 neutrons) combine to form water molecules with nine possible isotopic combinations. In paleoclimate research, the two most important molecules of water are $^{1}HD^{16}O$ (one ^{1}H ion + one D ion + one ^{16}O oxygen ion) and $^{1}H_2^{18}O$ (two ^{1}H ions + one ^{18}O ion). Of the three isotopes of oxygen, ^{16}O ("light" oxygen) is by far the most abundant (99.76%), ^{18}O ("heavy" oxygen) is rarer (0.20%), and ^{17}O is the rarest (0.04%), while ^{1}H has an abundance of 99.98% and D has an abundance of only 0.02%. In precipitation, including snow, the ratio of the heavy to light isotopes of oxygen and hydrogen is determined by many atmospheric factors, but among the most influential is the sea surface temperature at the source of moisture and the air temperature at the site of condensation and deposition.

Traditionally, isotopes of oxygen and hydrogen are determined by mass spectrometry, by which ions of different masses are separated by passage through a magnetic beam and measured. The ratios of $^{18}O/^{16}O$ and $^{2}H/^{1}H$ are denoted as $\delta^{18}O$ and δD after they are calculated with respect to the isotopic ratios of a laboratory standard.[13] In the polar regions, stable isotopes of snow have long been used as a temperature proxy.[14] This is illustrated in a section of an ice core from Greenland in Fig. 8a,

---

[13] One of the most common laboratory standards used for analysis of stable isotopes of water is the Vienna Standard Mean Ocean Water (VSMOW), which is composed of distilled deep offshore ocean water.

[14] Dansgaard (1964); Johnsen et al. (1972).

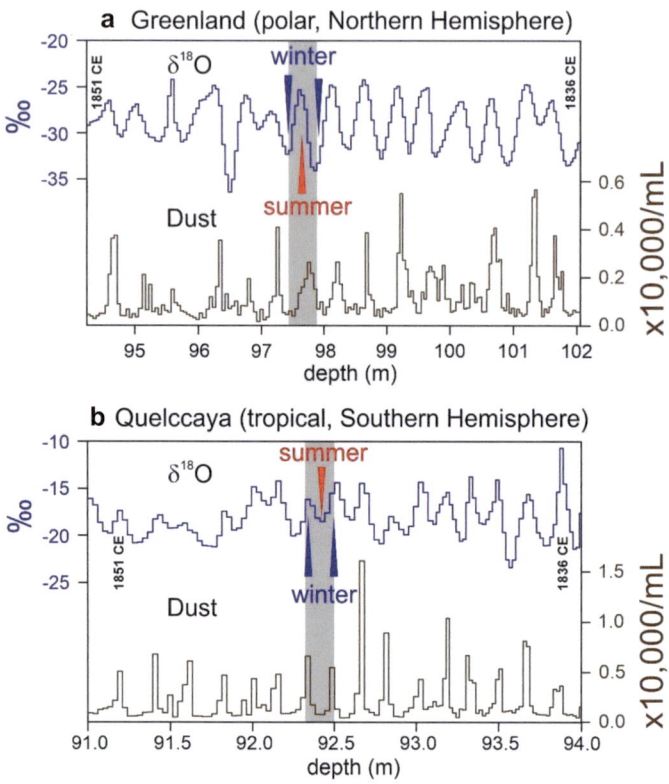

Fig. 8 Seasonal variations of ratios of $^{18}O/^{16}O$ ($\delta^{18}O$) and mineral dust in **a** a polar ice core (Core D3, Greenland) and **b** a tropical ice core (Quelccaya, southern Peruvian Andes), both dated from 1836 to 1851 CE

where lower (more negative) values of $\delta^{18}O$ indicate colder conditions (winter), while higher (less negative) values indicate warmth (summer). However, in much of the tropics, the atmospheric processes are dominated by monsoon circulation, and on a seasonal basis the values of $\delta^{18}O$ cannot be linked with any single meteorological or hydrological factor. Generally, the seasonal values of the stable isotopes in tropical snowfall in the mountains are the reverse of those in the polar regions; that is, more negative values occur in the summer and less negative in the winter (Fig. 8b). The reasons for this reversal are not completely understood, but the very high altitude at which precipitation forms in convective clouds during the monsoon season and the large amount of precipitation in the summer may play important roles.[15]

---

[15] Thompson et al. (2017a, 2017b).

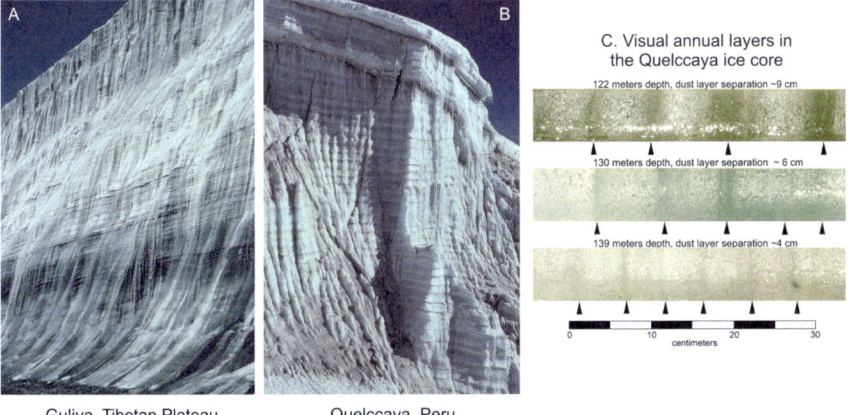

Fig. 9 The margins of (**a**) the Guliya ice cap in the western Tibetan Plateau and (**b**) the Quelccaya ice cap in the southern Peruvian Andes show distinct yearly layers composed of thicker, lighter-colored ice that formed from wet season (summer) snow and thinner darker layers that formed during dry seasons. These layers can be observed very clearly in a section of an ice core drilled on Quelccaya (**c**)

4.1.2 Impurities (e.g., Dust, Salts)

Dust is picked up and moved by wind, and the amount of dust in the air and where it is carried varies by season in many parts of the world. Larger amounts of dust are deposited on glaciers during dry seasons and are preserved as distinct annual layering. The concentration and sizes of the dust reflect atmospheric circulation (wind speed and direction) and proximity to the dust sources. Annual dust layers are particularly visible in low-latitude mountain glaciers which, unlike most polar ice sheets and glaciers, are located closer to the dust sources and receive much more snowfall during the monsoon season than during the dry season. An example is illustrated in Fig. 9, in which this annual layering is shown in the vertical edges (margins) of two lower-latitude glaciers located in and near different monsoon regions. The margin of the Guliya ice cap in the arid western Tibetan Plateau (see location in Fig. 4), located on the edge of the region influenced by the South Asian Monsoon, shows dust layers that are very dark and closely spaced (Fig. 9a) which result from the very low precipitation in this region (about 0.5 m of snow per year on average) and the exposed dust sources that surround the ice cap. On the other hand, the margin of the Quelccaya ice cap in southern Peru (see location in Fig. 4), which lies within the South American Monsoon region, shows dry season dust layers that are lighter in color and more widely spaced (Fig. 9b) due to the very high wet season snowfall (about 3 m of snow per year on average). These dust layers are also visible in three Quelccaya ice core sections (Fig. 9c). On longer time scales from decades to millennia, information about concentrations and sizes of dust particles is useful for constructing histories of changes in regional aridity and wind speed/direction, thus providing long-term proxy records of climatic and environmental conditions.

The concentrations of major ions (MSA^-, F^-, Cl^-, SO_4^{2-}, NO_3^-, Na^+, NH_4^+, K^+, Mg^{2+}, Ca^{2+}) are used to reconstruct past climatic and environmental variations. These ions originate from specific sources including sea salt, volcanic eruptions, emissions from terrestrial and marine biological sources, forest fires, and industrial emissions which contain sulfates and other compounds (Table 1). The concentrations of these chemical species are measured using ion chromatographs[16] and provide information about changing atmospheric circulation, biological activity, atmospheric chemistry, and anthropogenic activity.

Black carbon (BC) particles are produced by both natural and anthropogenic fires and by industrial activity. Records of changing BC concentrations show how industrial emissions to the atmosphere have changed before, during, and after the Industrial Revolution. Like dust and stable isotopes, BC concentrations may have a characteristic seasonality. The deposition of BC on white glacier surfaces can effectively lower the albedo (reflectivity) of the ice, resulting in heating and surface melt which may accelerate glacier thinning, although the extent to which BC is responsible for glacier retreat in the low latitudes is not known.[17]

Recent technological advances in mass spectrometry make it possible to analyze ice cores that preserve continuous, high-resolution stratigraphy of elements in trace and ultra-trace concentrations[18] that contribute to our knowledge of past climatic and environmental variability in the Earth systems. Atmospheric trace element composition is influenced by extra-terrestrial contributions (platinum group elements such as Ir, Os, Pt), aeolian dust (Al, Co, Fe, Mn, Rb, Ti, U, and V), volcanic (As, Cd, Bi) and anthropogenic emissions from mining and industrial activities (e.g., Cu, Zn, Cd, Pb). Rare earth elements (REEs) are used to help backtrack dust to its source area. An ice core record of variations in REEs can characterize dust from different sources so that scientists may determine changes in wind direction from the dust sources to the glacier or ice sheet through time.[19]

4.1.3 Gases

One of the most important forcing mechanisms of climate change is greenhouse gas (GHG) concentrations. Since the middle of the twentieth century GHGs such as

---

[16] Ion chromatographs operate by separating charged particles (anions and cations) from a liquid and measuring their concentrations.

[17] Kang et al. (2020).

[18] Trace elements are defined by chemists as having concentrations less than 100 μg per gram; ultratrace elements have concentrations of less than one microgram per gram. Rare earth elements (REE) are 17 metallic elements (scandium, yttrium and the lanthanide group). Contrary to what the name suggests, REEs are not rare. Because of their very low concentrations, in ice core research trace and rare earth elements are measured using special mass spectrometers such as the ICP-MS (inductively coupled plasma mass spectrometer).

[19] Gabrielli et al. (2010).

carbon dioxide (CO_2), methane (CH_4), and nitrous oxide (N_2O) have been measured directly in the atmosphere and the data have provided unequivocal evidence that their concentrations are continuously rising. However, to extend this modern record back in time and place the recent changes into a longer term perspective, the concentrations of these gases are measured in air extracted from bubbles which are trapped in glacier ice, providing one of the most important archives available from ice cores. Continuous, long-term records of greenhouse gases, along with other substances such as ozone-destroying chlorofluorocarbons (CFCs), have proved to be very valuable for unraveling the natural- and human-driven variability of the gases.

Because the gases are trapped in glaciers only after the ice forms and the pores within the ice close off, "gas age" is younger than "ice age" and this difference must be noted when constructing a record of gas concentrations. This time lag depends on factors that determine the thickness of the firn layer, such as atmosphere and ice temperature and the amount of snow that falls on the glacier surface each year. For example, in the interior of East Antarctica, where it is very cold and the annual snowfall is low, the porous firn layer through which air can circulate can be hundreds of meters thick and encompass thousands of years. This means that the difference between the gas age and the ice age can also be several millennia. However, at the West Antarctic Ice Sheet (WAIS) Divide drill site (in the vicinity of Byrd Station in West Antarctica, Fig. 4), where the temperatures are warmer, the annual snowfall is higher, and the firn layer is thinner, the gas-ice age difference is less (~200 years[20]) than that in the interior of East Antarctica.

4.1.4 Physical Characteristics

Dry season dust layers deposited on ice surfaces and buried between layers of cleaner, wet season snow often provide stratigraphic records that can be seen with the naked eye (Fig. 9c). However, visible stratigraphy can also be composed of ice of different densities, or layers of air bubbles or even layers of clear ice with no dust or air bubbles. The mapping of visible stratigraphy is one of the few measurements that can be made on an ice core without melting any ice, or relying on complex instruments or clean room laboratories. In regions with marked wet and dry seasons, such as in monsoon regions of Asia or South America, visible stratigraphy alone can be used to count back years in an ice core. These layers are visible not only in the ice cores but also in the glacier margins, such as the Guliya ice cap on the Tibetan Plateau and the Quelccaya ice cap in southern Peru (Fig. 9). Visible stratigraphy may sometimes be observed in polar ice cores. For example, in a deep Greenland ice core, cloudy bands of ice were used to count back ~ 50,000 years,[21] almost half the complete timescale of just over 100,000 years.

---

[20] Ahn et al. (2012).

[21] Alley et al. (1997).

4.1.5 Microbes

New areas of ice core research have developed over the past few years, and among the most intriguing is the isolation and identification of ancient microorganisms such as bacteria and viruses. Recent research has found that glaciers from the polar regions[22] to the tropics preserve microbial communities that can range in age from hundreds to hundreds of thousands years. The microbes found in ice cores typically represent those in the atmosphere at the time of their deposition, and hence may reflect climatic and environmental conditions during that time period. This new area of investigation offers the opportunity to study ancient microorganisms that survived and thrived during very different climatic regimes. For example, recent analysis of microbes in 15,000-year-old ice from the Guliya ice cap on the western Tibetan Plateau resulted in the discovery of 33 viruses, 28 of which had not been previously identified.[23] This type of research on such frozen archives has the potential to provide a better understanding of the microorganisms transported through the atmosphere today and in the past, as well as their effects on atmospheric processes and the environments where they are deposited (e.g., pathogens affecting animal or plant health). They also provide the potential to study microbial evolution through time. In addition, microbial studies in ice cores on Earth may help us understand how life may have arisen on cold planets such as Mars.

4.2 How Are Ice Cores Dated?

Establishing an accurate timescale or chronology is essential in ice core research; without a timescale an ice core cannot provide a history of past climatic and environmental changes. To be useful, the information gained from all the laboratory analyses of an ice core described above has to be transferred from depth to time. Therefore, much effort is expended initially to create the timescale. Ideally, an ice core may be dated by counting the dry season dust and chemistry layers from top to bottom, much as tree rings are counted from the outside to the inside of a tree trunk. This is demonstrated in ice core sections from the polar Greenland and tropical Quelccaya ice cores shown in Fig. 8, which show distinct wet and dry season variations in $\delta^{18}O$ and mineral dust concentrations. The visible dust layers can be observed in the Quelccaya ice core sections in Fig. 9c. The entire Quelccaya ice core $\delta^{18}O$ and dust records are shown in Fig. 10, in which the seasonal variations in the 169.6-meter-long core are visible from the top of the core to ~ 160 m depth, corresponding to the year 226 CE (common era).[24] A massive eruption of the Huaynaputina volcano in 1600 CE in southern Peru, located 300 km south of Quelccaya, deposited a thick layer of ash on the ice cap, which can be seen at about 122 m depth (Fig. 10). Since

---

[22] Knowlton et al. (2013).

[23] Zhong et al. (2021).

[24] Thompson et al. (2013b, 2013a).

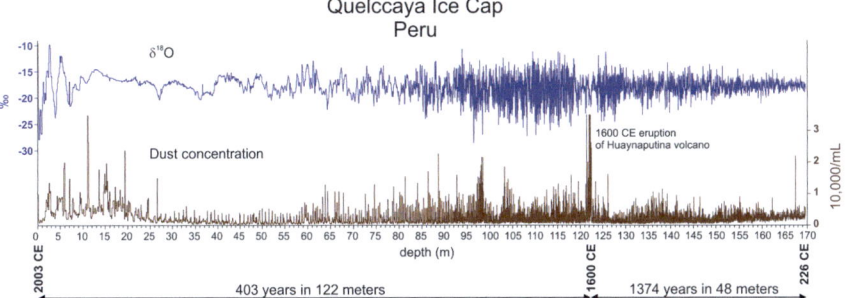

Fig. 10 Stable isotopes of oxygen ($\delta^{18}O$) and dust concentration recorded in the Quelccaya ice cap in the southern Peruvian Andes. This ice core was drilled in 2003, the date at the surface, and is dated at the bottom to 226 CE. Seasonal variations, especially in dust, are distinct and can be easily counted. Seasonal variations in $\delta^{18}O$ are smoothed in the top ~ 55 m by melting at the glacier surface and percolation of water through the firn section. The dust in the large peak at 1600 CE is volcanic ash from the eruption of the nearby Huaynaputina volcano

historical records exist from Peru that document the date of this eruption, they can help calibrate the timescale in this ice-core record.

Simple counting of dust and $\delta^{18}O$ peaks to date a climate record is not as straightforward in some ice cores as others. Although in the Quelccaya ice core seasonal variations in the $\delta^{18}O$ and dust concentrations are distinct throughout much of the core, the recent warming trend in this region has caused melting of ice at the surface and the meltwater percolating through the firn layer has smoothed the seasonal $\delta^{18}O$ signals in the top ~ 55 m (Fig. 10). Another complication involves ice compression and annual layer thinning with depth. Although thought of as a solid, ice is actually a substance that flows, and in a glacier or ice cap it flows both laterally and vertically (Fig. 11). Vertical flow results in compression with depth due to the increasing weight of successive annual layers of ice on top. This compression can be seen in the Quelccaya ice core sections in Fig. 8c in which the distance between the dust layers decreases from ~ 9 to ~ 4 cm between 122 and 139 m depth. Thus, the thinning rate, or the number of years per length of core, increases with depth until at some "flexure point" it suddenly increases greatly. At a certain depth, which depends on the average amount of snow that falls on the glacier each year (accumulation) and the thickness of the ice cap, the layers thin beyond the ability to distinguish them. For example, in the complete Quelccaya ice core record (Fig. 10) the seasonal variations in $\delta^{18}O$ and especially in dust become progressively compressed with depth in the core and at 90–95 m (corresponding to the early 1800s CE) the rate of compression increases greatly. Another way to view this compression of time with depth is illustrated by the amount of time contained within the bottom of the core compared with the top (Fig. 10). From 0 to 122 m (the depth of the 1600 CE Huaynaputina eruption), the Quelccaya ice core contains 403 years; however, from 122 to 169.6 m, the core contains 1374 years (1600 to 226 CE).

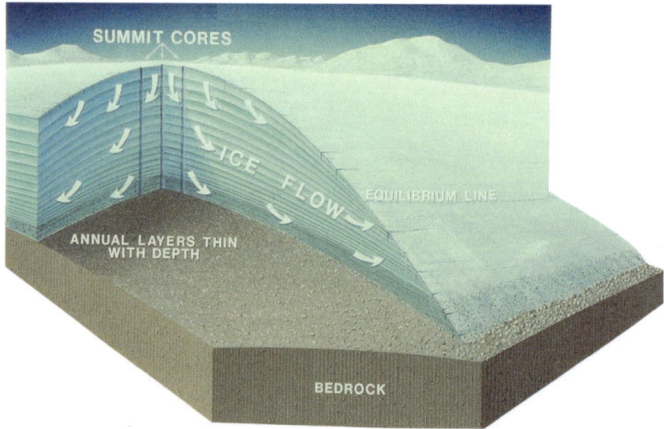

Fig. 11 A dome-shaped ice cap, showing the ideal location on the summit where ice cores should be drilled. Layers thin with depth as the weight of overlying ice presses down on them, and ice flows away from the ice divide at the summit toward the margins of the ice cap. Also shown is the equilibrium line; ice accumulation above the equilibrium line is balanced by ice ablation below the equilibrium line

Below the depth in a glacier where annual layers are no longer distinguishable, other methods must be used to establish the time scale for the ice core records. These methods include, but are not limited to: (1) matching the $\delta^{18}O$, δD, or entrapped gas (CO_2, CH_4) profiles with those from other, well-dated cores; (2) using volcanic ash layers from known ancient eruptions, if they can be found in an ice core (such as the ash layer from the 1600 CE Huaynaputina eruption in the Quelccaya ice core, Fig. 10); (3) radiocarbon dating of plant or insect fragments, if they are found (See Box 2); and (4) ice flow models that account for the rate of annual layer thinning with depth.[25]

4.3 How Are Sites Selected for the Recovery of Ice Cores?

The location that a science team selects to drill an ice core depends on the questions that are posed. For example, if the objective is to examine the history of the El Niño-Southern Oscillation (ENSO, described in Chap. 3), an ice field in the low latitudes where the effects have more impact would be selected. Because Earth is nearly spherical, 50% of its surface lies between 30°N and 30°S. This region contains immense thermal energy that is instrumental in driving the planet's weather systems such as ENSO and monsoons. On the other hand, ice fields and glaciers in the polar regions provide the best information on the variations in GHGs (such as carbon dioxide, methane, nitrous oxide) because the very cold year-round temperatures

---

[25] Bolzan (1985).

prevent melting and allow abundant gas-trapping bubbles to develop. In addition, the pristine polar atmosphere allows sampling of the natural atmospheric background concentrations of many chemical species. Once an ice field is selected, it is important to conduct preliminary field programs to collect essential in situ baseline data such as the annual accumulation rate, ice thickness, and bedrock topography to facilitate identification of an optimal drilling location.

Ice cores are drilled in areas of glaciers, ice caps,[26] and ice sheets where distortion by ice flow is minimal. The dome-shaped ice cap in Fig. 11 shows the location of ice cores drilled at the summit as close as possible to the ice divide, or the point from which the ice flows laterally outward in all directions and the vertical compression is directly downward with little lateral shear. Ice core records from different latitudes typically reflect different regional processes, therefore their integration provides more comprehensive, spatially rich histories. Long ice cores of one to three kilometers in length that are drilled through the thick ice sheets in Greenland and Antarctica provide longer histories that extend back over 100,000 years. In contrast, the ice core records from the thinner tropical mountain glaciers are shorter, typically ranging from a few hundred to ~ 20,000 years. For example, the entire climate record from the 169.6 m thick Quelccaya ice cap shown in Fig. 10 covers only about 1800 years. Because lower-latitude ice cores often come from monsoon-influenced regions with high annual precipitation, their climate records over the last few hundred years are generally more detailed than those from many polar ice cores. This makes them particularly valuable for reconstructing recent climate changes. Moreover, as many of these glaciers exist in areas that have been inhabited by human societies for many centuries, some of the ice core-derived proxy records help explain local cultural changes. The Quelccaya ice core records are discussed in greater detail in Sect. 5.2.

5 What Have We Learned From Ice Cores? Stories That Ice Cores Can Tell Us About the Past, While Raising an Alarm About the Future

American polar ice core drilling began at a secret U.S. military base called Camp Century in northwestern Greenland. The strategic value of Camp Century for U.S. national security was its proximity to the former Soviet Union. During the Cold War, the military used the ice core drilling activities at this base as a cover to hide nuclear missiles that would be transported in and out of a buried warehouse via a subsurface railway (code named Project Iceworm). However, in 1967, these plans were discarded when it was determined that the ice tunnels were becoming unstable due to ice flow.[27]

---

[26] Ice caps are smaller, flatter ice bodies than ice sheets.
[27] SciShow. (2020). *The Nuclear City Lost Under Ice, Camp Century.* SciShow. https://www.youtube.com/watch?v=0xIwzhE_pho.

In 1966, scientists succeeded in drilling the first ice core to bedrock (1390 m in length) through the Greenland ice cap and ultimately produced a climate history covering the last 100,000 years.[28] Subsequently, the drill used at Camp Century was shipped to Byrd Station, West Antarctica where in 1968 a research team drilled the first ice core to bedrock (2164 m in length) in West Antarctica.[29] Meanwhile, in 1957, the Soviet Union established a base in the coldest place on Earth, which they named Vostok Station, on the East Antarctic ice sheet. Beginning in the 1970s, they drilled through the ice to 3623 m, ultimately producing a climate record that extends back over 400,000 years.[30]

Over the last 70 years since the first deep (>300 m) ice cores were drilled in Greenland and Antarctica in the late 1950s during the International Geophysical Year, hundreds of records of climatic and environmental variations have been produced from ice cores drilled from the polar regions to the tropics (see Box 3). Deep ice core drilling through tropical and middle-latitude mountain glaciers began in 1983, much later than it began in the polar regions. Tropical mountain glaciers form under completely different climatic and environmental conditions than polar glaciers and ice sheets; thus, polar and tropical ice cores provide different information. Tropical glaciers tend to be located in regions with a monsoonal climate that is typically characterized by wet summers, dry winters, and very high annual snowfall. Tropical glaciers are much thinner (~50 to 300 m thick) than polar ice sheets (~500 to over 4000 m thick), and their climate records typically range in age from a few hundred years to hundreds of thousands of years. These differences are illustrated below by comparing two ice core records, one from the East Antarctic ice sheet and one from an ice cap in tropical South America. Note the broad range of information that may be obtained from these two extremely different environments.

---

[28] Dansgaard et al. (1969).

[29] Lonnie G. Thompson produced the very first comparisons of the dust records from the Byrd Station and the Camp Century cores in his 1977 PhD dissertation entitled: *Microparticles, Ice Sheets and Climate*.

[30] Petit et al. (1999a, 1999b).

Box 3 Ice core drilling on polar ice sheets versus high mountain glaciers and ice caps

This Box provides a brief history of the development of ice coring technologies coupled with the details of selected early projects and their accompanying discoveries that advanced our understanding of Earth's climate history. Below is a short discussion of some requirements of drilling the thick polar ice sheets versus high mountain glaciers including the successful efforts by scientists and technicians from The Ohio State University's Byrd Polar and Climate Research Center to drill lower-latitude, high-elevation mountain glaciers that resulted in the development of the first solar-powered drill.

The International Geophysical Year (IGY) occurred from 1957 to 1958 and was an international program designed to advance research in the Earth sciences. One of the areas of interest in the IGY was polar research, including ice core drilling. Although drilling efforts on glaciers had been conducted before the IGY, for the first time government-sponsored programs managed to drill cores several hundred meters long into the Greenland and Antarctic ice sheets. The first two ice cores were drilled in Greenland and Byrd Station in Antarctica. However, it wasn't until 1966 that an ice core was drilled 1390 m to the bedrock (i.e., through the entire thickness of the ice) at Camp Century in Greenland. Shortly after, a 2164 m ice core was drilled to bedrock at Byrd Station in West Antarctica.

Since these early efforts, numerous ice cores have been drilled completely through the large ice sheets on Greenland and Antarctica. The deepest core drilled to date (3623 m) was completed in 1998 at Vostok Station in East Antarctica after several years of effort.[31]

---

[31] Petit et al. (1999a, 1999b).

The logistical requirements for conducting ice core drilling projects differ greatly for polar ice sheets and for mountain glaciers and ice caps. Deep drilling projects on the polar ice sheets often take several years to complete and require many participants from different institutions that are generally supported by a government-managed base, such as McMurdo and Amundsen-Scott South Pole Stations (USA), Vostok Station (Russia), and Great Wall Station (P.R. China), to name just a few. These bases are staffed by technicians, mechanics, cooks, and medical professionals. Contractors construct and manage the drilling camps and professional ice core drillers extract and manage the core storage. Logistical support personnel and scientists, along with cargo (construction materials, heavy equipment, ice core drills, food), are ferried from the bases to the study sites using aircraft such as ski-equipped cargo planes (e.g., C130s or Twin Otters).[32] The aircraft then returns the ice to the airbase for shipment to the home institution. U.S. ice cores from Greenland are transported by aircraft and those from Antarctica typically return by ship. These deep drilling projects typically cost millions of dollars, require years of detailed planning and multiple years to complete.

Recovering ice cores from glaciers and ice caps atop high mountains requires much different logistics and equipment. Unlike the deep drilling on the Greenland and Antarctic ice sheets, drilling through a mountain ice cap or glacier is accomplished in one field season lasting one or two months. Field teams typically consist of 6 to 10 scientists who drill and package the cores, manage the camp, and organize all the logistics. The OSU team typically uses its own drills (Figs. 12 and 13) and core handling equipment, establishes its local support system, often with the cooperation and guidance of local governments and research institutions, and deals with local political issues when necessary. Conducting research in environmentally sensitive areas such as national

---

[32] To learn more about the logistical support for polar ice core research, see: (1) a video showing the Air National Guard's C130 aircraft that supports the Summit Camp in central Greenland, from Canalas, A. (2015, August 3). *New York Air Guard Unit is Lifeline for Scientists in Greenland and Antarctica.* ABC News, https://abcnews.go.com/News/video-york-air-guard-unit-lifeline-scientists-greenland/story?id=32813778; (2) a gallery of photographs that highlight the equipment and processes employed to drill an ice core to bedrock at the WAIS Divide camp on the West Antarctic ice sheet, from WAIS Divide. (accessed 2023, June 2). The National Science Foundation, http://www.waisdivide.unh.edu/gallery/; and (3) a primer on ice cores with a focus on polar ice cores, from NSF (accessed 2023, June 2). About Ice Cores. The National Science Foundation, https://icecores.org/about-ice-cores.

parks or protected areas often requires multiple permits from several different and unrelated agencies within the host country. For example, 24 permits were required to drill six ice cores from the glaciers atop Kilimanjaro in 2000. All the equipment, especially the drills, must be light and durable to withstand transport by pack animals over miles of rough terrain and ultimately by porters to altitudes typically between 18,000 and 23,000 ft. Typical drilling expeditions require about 6 tons of equipment for drilling, core processing, and camping. The remote nature of these expeditions requires the team to have special high-elevation medical kits to treat everything from toothaches to headaches, setting broken bones, and even dealing with potentially life-threatening high-altitude illnesses. Members of the field team are chosen not just for their scientific and technical expertise but also for their physical ability to work for many weeks at elevations between 14,000 and 23,000 ft in weather conditions that are often extremely cold and windy.

Fig. 12 Solar-powered drill used to recover ice cores from the summit of the Quelccaya ice cap, 1983

Fig. 13 Light-weight 200-m ice core drill system used on Quelccaya in 2003

During deep drilling programs on mountain glaciers, several tons of ice must be kept frozen in insulated boxes and then transported, first by pack animal, such as yaks in the Himalayas (Fig. 14) or burros in the Andes, then by truck, and finally by air to the researchers' institutions.[33]

Fig. 14 Yaks are used to transport frozen ice cores down the slopes of the Himalaya from the Naimona'nyi glacier to the freezer truck waiting at the end of the road below. Each insulated box contains 6 one-meter long pieces of ice core

---

[33] To learn more about the logistical support for tropical ice core research, see: (1) a video of ice core drilling on the Northwall Firn glaciers in Papua, Indonesia, from Byrd Center (2020). *Documentary on the Drilling of Ice Cores on Puncak Jaya Indonesia in 2010.* Byrd Center, The Ohio State University, https://vimeo.com/showcase/3884206/video/376564762; and (2) a video of ice core drilling in 2015 at 22,000 feet on the summit of the Guliya ice cap in northwestern Tibet, from Byrd Center (2016). *Guliya,* Byrd Center, The Ohio State University, https://vimeo.com/showcase/3884206/video/164485595.

5.1 Information from a Polar Ice Core: Dome C, Antarctica

Ice cores drilled through the large polar ice sheets in Greenland and Antarctica have the advantage of producing very long records, usually extending over tens to hundreds of thousands of years. The longest ice core records come from the interior of the East Antarctic ice sheet, where Earth's coldest temperatures are recorded and the amount of snowfall falling each year is very low. The average thickness of the East Antarctic ice sheet is 2.57 km (1.6 miles) and it is 4.77 km (2.96 miles) at its thickest. In 2004, the European Project for Ice Coring in Antarctica (EPICA) completed the recovery of a 3260 m (2.03 miles) long ice core at a place called "Dome C" (Fig. 4). The resulting ice core-derived records extend back ~ 800,000 years and cover eight complete glacial-interglacial cycles, each ~ 100,000 years long (Fig. 15). These cycles (see Chap. 9) are driven by variations in the shape of Earth's orbit around the Sun, which changes from elliptical to more circular every ~ 100,000 years, and by changes in the tilt and orientation of Earth's axis, over ~ 41,000 and ~ 26,000-year cycles, respectively. These processes control the amount of solar radiation that Earth receives any time at any location. Long Antarctic ice cores like that from Dome C provide the longest continuous ice core-derived perspectives on the changes in Earth's climate and environment, and provide very valuable tools that allow scientists to investigate the processes that drive temperature and precipitation changes over multiple glacial periods (often called ice ages) and the intervening warmer interglacial periods. For example, δD, a well-established proxy for temperature, shows that the interglacials older than 430,000 years before present (yrs BP) were longer but not as warm as the interglacials younger than 430,000 yrs BP when polar ice volume was lower and global sea level was higher. This climatic transition is called the "mid-Brunhes event."[34]

The Dome C ice core provides one of the longest records of variations in temperature, ice accumulation, and atmospheric dust concentration, while also providing a continuous record of variations in the concentrations of atmospheric gases such as CO_2 and CH_4 that act as very powerful greenhouse gases. The Dome C proxy records of temperature, accumulation, and insoluble dust concentrations, as well those from other very long Antarctic ice core records, reveal that warm interglacials periods such as the most recent one in which we live (the Holocene) are characterized by higher values of δD (warmer temperatures), higher snowfall, a less dusty atmosphere, and higher CO_2 levels. The cold glacial periods are characterized by lower δD (cooler temperatures), and lower snowfall, atmospheric dustiness, and CO_2 concentrations. The temperature difference between glacial and interglacial periods in the interior of Antarctica is about 15 °C.[35]

---

[34] The term "Brunhes" has time scale significance. It is the name for the paleomagnetic chron (interval of time) from 781,000 years ago to the present, during which Earth's magnetic field has maintained the same (normal) polarity as today.

[35] Jouzel et al. (2007).

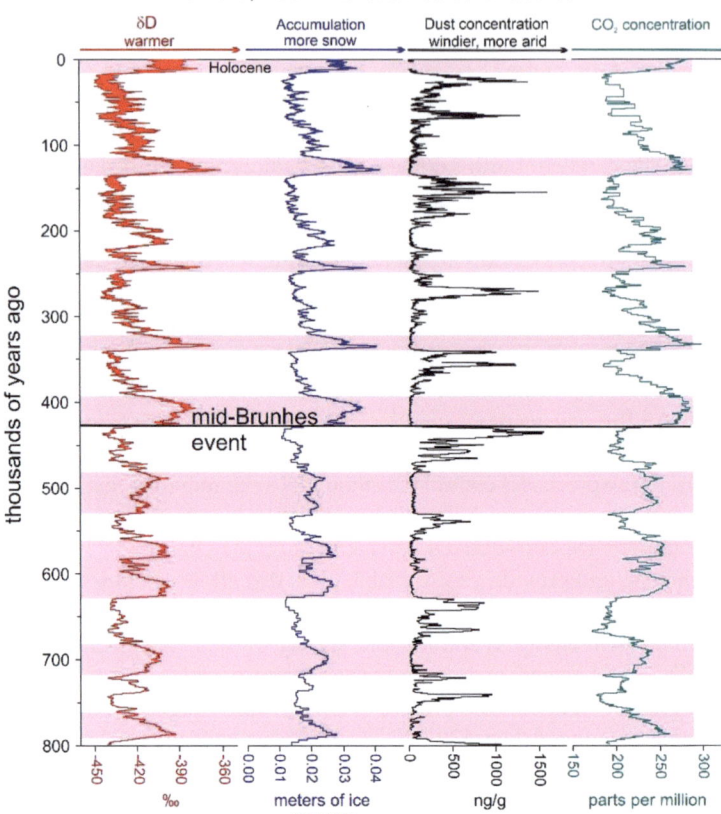

Fig. 15 Ice core records of paleoclimate from Dome C on the East Antarctic ice sheet cover the last 800,000 years (eight glacial cycles). Each warm interglacial period is marked by shaded bars. The "mid-Brunhes event", which occurred 430,000 years ago, is also marked. δD is a proxy for temperature, accumulation is a proxy for snowfall amount each year, dust concentration is an indicator of wind strength and the degree of aridity in the source area for the dust, and carbon dioxide concentration reflects the amount of that greenhouse gas in the atmosphere. Data on δD and dust are from Delmonte et al. (2008),[36] accumulation rate data are from Parrenin et al. (2007),[37] and CO_2 data are from Luthi et al. (2008)[38]

These records of temperature and greenhouse gas variations are so valuable because they tell us about natural variations in Earth's climate system long before human activity became an important factor. In Fig. 16, the 800,000-year records of temperature (derived from δD) and carbon dioxide concentrations are again compared, this time with the modern value of CO_2 included. The concentration

---

[36] Delmonte et al. (2018).

[37] Parrenin et al. (2007).

[38] Lüthi et al. (2008).

of 415 ppm as of 2020 CE is much higher than any concentration in nearly a million years and indicates that Earth's current climate is unique from a million year perspective. It is alarming to consider that according to IPCC climate model projections, if global CO_2 emissions continue to increase at the current rate, atmospheric concentrations could reach 900 to 1100 ppm by 2100 CE.[39] Unfortunately, once CO_2 is released to the atmosphere it remains there for many decades to centuries, leading to a long-term impact on Earth's climate system.

One of the misconceptions in geosciences that arises when viewing Fig. 16 is that the close relationship between the ice core-derived CO_2 and temperature (δD) provides strong evidence that CO_2 is the primary cause of the oscillating temperature pattern. On geological time scales encompassing multiple glacial-interglacial periods (Figs. 15 and 16), Earth's large temperature variations are caused primarily by the orbital cycles discussed above (and described in detail in Chap. 9). The high correlation between CO_2 and temperature in the ice core record reflects the effect of a feedback mechanism arising from the fact that CO_2 is more soluble in cold water than warm water. Thus, when Earth is colder (a glacial period), the ocean absorbs more atmospheric CO_2, and when Earth is warmer (an interglacial period), the ocean absorbs less atmospheric CO_2. The higher concentration of CO_2 remaining in the atmosphere during interglacial periods also results in additional warming via the greenhouse effect. In reality, the primary evidence that CO_2 causes warming (i.e., the greenhouse effect) comes from the empirical evidence that CO_2 absorbs infrared radiation, as discovered in 1856 by American scientist Eunice Foote and in 1869 by the British physicist John Tyndall.[40]

The take-away messages from this discussion are that orbital cycles are the fundamental driver in producing Earth's glacial-interglacial cycles, and atmospheric CO_2 concentration provides a strong positive feedback that amplifies atmospheric temperature changes associated with those cycles. However, on decade to century time scales orbital forcing is too slow to contribute to contemporary climate changes, and the warming of Earth since ~ 1850 is driven primarily by the rising atmospheric concentrations of greenhouse gases such as CO_2 and CH_4. Human activities such as the extraction and burning of fossil fuels and rampant deforestation are among the main drivers of Earth's rising temperatures that are projected to continue to rise in the coming decades.

---

[39] See Fig. 3 in Lee, J.-Y., Marotzke, J. Bala, G., Cao, L., Corti, S., Dunne, J.P., Engelbrecht, F., Fischer, E., Fyfe, J.C., Jones, C., Maycock, A., Mutemi, J., Ndiaye, O., Panickal, S., & Zhou, T., (2021). Future Global Climate: Scenario-Based Projections and Near-Term Information. In *Climate Change 2021: The Physical Science Basis. Contribution of Working Group I to the Sixth Assessment Report of the Intergovernmental Panel on Climate Change* [Masson-Delmotte, V., P. Zhai, A. Pirani, S.L. Connors, C. Péan, S. Berger, N. Caud, Y. Chen, L. Goldfarb, M.I. Gomis, M. Huang, K. Leitzell, E. Lonnoy, J.B.R. Matthews, T.K. Maycock, T. Waterfield, O. Yelekçi, R. Yu, and B. Zhou (Eds.)]. Cambridge University Press, Cambridge, pp. 553–672, https://doi.org/10.1017/9781009157896.006.

[40] Eunice Foote was the first to discover that CO_2 absorbs infrared radiation, but because her work was overlooked John Tyndall received the credit for the "rediscovery" 13 years later.

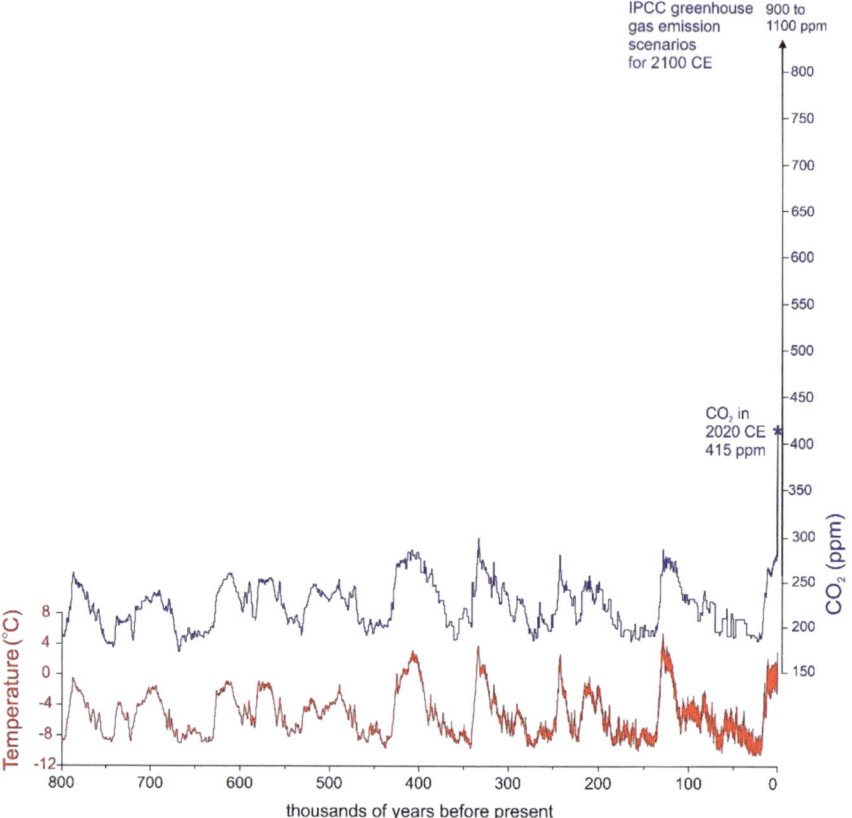

Fig. 16 The CO_2 and temperature (reconstructed from δD) records from the Dome C, Antarctica, ice core. The two curves increase and decrease together. The CO_2 concentration at present is far above the levels over the last 800,000 years and is predicted to be much higher by 2100 CE. Modified from Luthi et al. 2008.[41] Figure used with permission from AAAS. This figure is excluded from Creative Commons license

5.2 *Information from a Tropical Ice Core: Quelccaya, Peru*

The Quelccaya ice cap, Earth's largest tropical ice cap, is located at 14 °S in the southern Peruvian Andes and is perched above the western edge of the Amazon Rainforest. It receives most of its snow during the austral summer (December to February) from water vapor that originates in the tropical Atlantic Ocean and is transported westward over the rainforest, then is uplifted as it reaches the Andes Mountains and cools at high elevation. The ice cap lies in a wet tropical region windward of the mountain range, and receives abundant snowfall each year (~3 m of snow, which equals ~ 1 m of ice). The dust and many other aerosols that fall

---

[41] Lüthi et al. (2008).

on the ice cap surface originate on the Peruvian-Bolivian Altiplano, a large, high-elevation plateau that contains salt flats, numerous volcanoes, lakes, dried lake beds, and agricultural fields which have existed for centuries. Other chemical species such as nitrate and ammonium are thought to originate from plants and soils in the Amazon Rainforest.

The 1800-year ice core records from Quelccaya shown in Fig. 17 include $\delta^{18}O$ (a temperature proxy), accumulation (a snowfall proxy), nitrate, ammonium, fluoride, chloride, and mineral dust concentrations. The most prominent feature is the "Little Ice Age," which was a cool event lasting from ~ 1550 to ~ 1850 CE in the Peruvian Andes and is observed in numerous climate records throughout the world although the timing of its onset and termination varies from one region to another. This was also a time when glaciers advanced in Europe, North America, and South America, including Quelccaya. In the ice core record, it is characterized by lower $\delta^{18}O$ values and high nitrate and ammonium concentrations, while precipitation (accumulation) was very high in the beginning, and decreased in the second half. However, a climatic event in the late 1700s is marked by low precipitation and very high fluoride and chloride concentrations, which together suggest that an intense drought may have been part of this global climatic disruption.[42] The Indian summer monsoons failed at this time, causing widespread drought and death throughout South Asia.[43] The causes for this event are not fully understood, but El Niño may have played an important role. Strong El Niños are linked with arid conditions in the Peruvian highlands (where Quelccaya is located) and monsoon failure in Asia. The dry period observed in the Quelccaya ice core records may have been caused by a very strong El Niño event in 1791 CE, or a series of strong events from 1789 to 1793. Evidence for a similar drought in the late 1300s is observed in the ice core record. Although there are no written historical records from Peru before the Spanish Conquest in 1532 CE, archeological evidence from coastal Peru indicates the occurrence of El Niño conditions around this time.[44]

The current warm period began as the Little Ice Age ended in the mid-nineteenth century. Along with a warming climate, as shown by increasing $\delta^{18}O$ values, the accumulation on Quelccaya also increased. Unfortunately, the top ~ 55 m of the ice core record[45] has been altered by the recent warming and meltwater percolation from the surface through the firn (Fig. 10). This melting has concentrated aerosols such as dust, nitrate, ammonium and chloride, as well as smoothed the seasonal variations in $\delta^{18}O$.

Because many tropical glaciers are located close to centers of human activity, their records have the potential to help archeologists and historians understand what may have contributed to the rise and fall of ancient civilizations. Quelccaya is less than 100 km away from the location of the capital of the Inca Empire and the power centers of its predecessors. The information that the ice core record provides on the

---

[42] Grove (2007).

[43] Cook et al. (2010).

[44] Satterlee (1993).

[45] Fifty- five meters depth in the Quelccaya ice core record corresponds to 1948 CE.

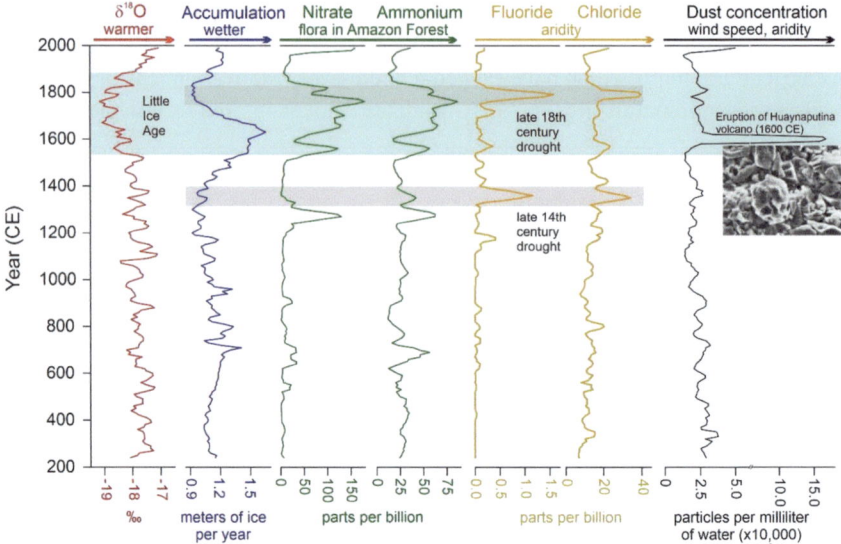

Fig. 17 Records of environmental and climatic variations, shown as 3-decade moving averages, covering 1800 years from the Quelccaya ice cap, southern Peruvian Andes. Oxygen isotope ratios ($\delta^{18}O$) are indicators of regional temperature, and accumulation of ice each year indicates precipitation. The biological activity of plants and soils in the Amazon rainforest, upwind of Quelccaya, is shown by the concentrations of nitrate and ammonium in the ice. Concentrations of fluoride and chloride are affected by widespread aridity (note that spikes in these chemical species coincide with lower accumulation), while dust concentrations indicate many atmospheric and geologic factors such as wind speed and how much local soil is exposed. The massive 1600 CE eruption of the Huaynaputina volcano, located just 300 km south of Quelccaya, is shown by a very large spike of volcanic ash. Ash particles from the eruption appear in the scanning electron microscope photograph below the dust peak. Also highlighted are the Little Ice Age (green shading) and the droughts of the late fourteenth and eighteenth centuries (gray shading)

oscillating climate between the Altiplano and the coastal lowlands over 18 centuries can potentially help explain why dominant civilizations shifted between these regions until the beginning of the Spanish Conquest in 1532 CE (Fig. 18).[46] Precipitation in the central Andes and the coastal areas of Peru is dominated by ENSO. During an El Niño, the northern coastal regions generally receive more precipitation while the Altiplano experiences more arid conditions. This precipitation oscillation between the coast and the highlands is noticeable in the Quelccaya ice accumulation record on much longer time scales. Above average values of ice accumulation occurred from ~ 600 to 1000 CE and from ~ 1450 to 1700 CE, which roughly coincide with the existence of the highland Tiwanaku/Huari and the Inca civilizations. The Tiwanaku culture was established close to the shores of Lake Titicaca about 113 miles southeast

---

[46] Thompson and Davis (2014a, 2014b).

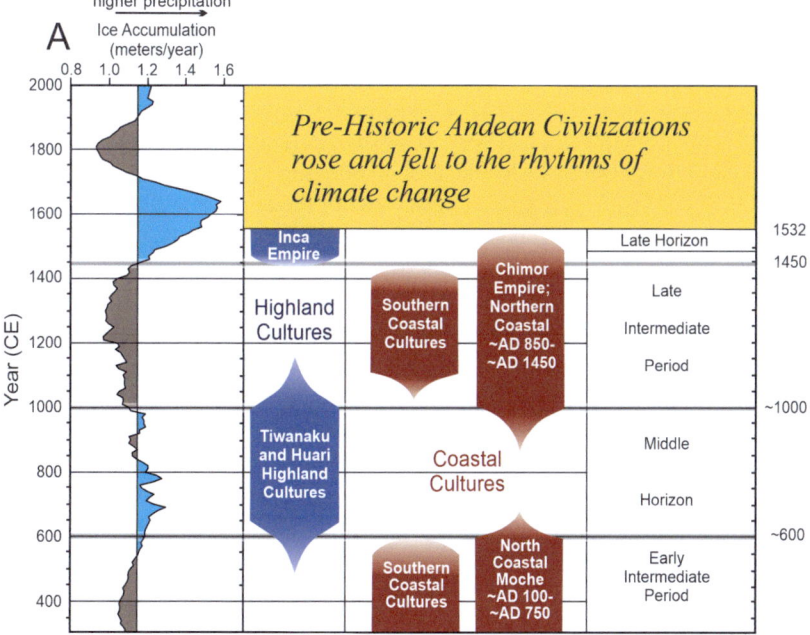
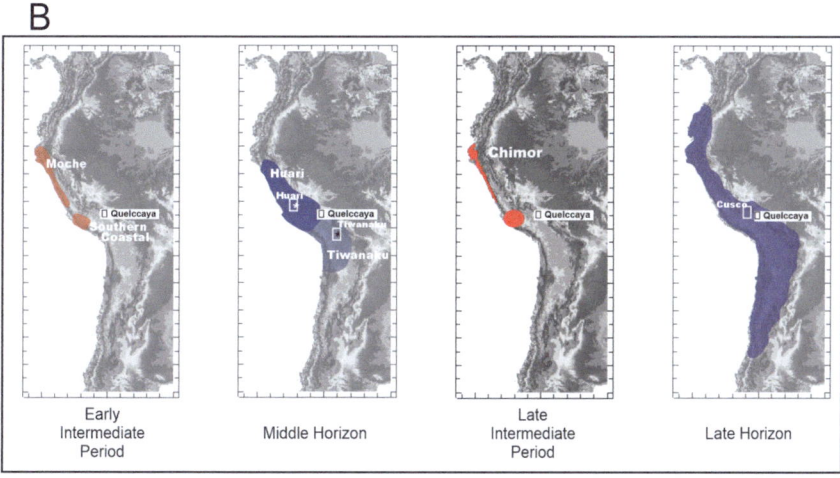

Fig. 18 a The precipitation in the southern Peruvian highlands over the last 1800 years, illustrated by the Quelccaya ice core record, is compared with the archeological evidence of the rise and fall of lowland and highland cultures in Peru, the geographical outlines of which are shown in the maps **b** During periods of favorable climate (high precipitation) on the southern Peruvian Altiplano as seen in the ice core record, highland cultures thrived while lowland cultures declined

of Quelccaya, where the farmers practiced raised field agriculture.[47] The Huari (or Wari) settled in the north, where the wetter climate may have facilitated the practice of terraced field agriculture.

The Quelccaya records show that around 1000 CE the highlands became persistently dry as the Late Intermediate Period began. Archeological evidence shows that the highland cultures began to decline around this time while the coastal cultures in Peru and Ecuador began to flourish.[48] The onset of the arid climate on the Altiplano may have played a role in the collapse of the Tiwanaku civilization, perhaps because raised field agriculture around the lakes became untenable.[49] Lake Titicaca levels declined[50] and the raised fields of the Tiwanaku were abandoned.

During the Late Horizon (mid-1400s to 1530s CE), the Inca empire rapidly expanded to cover the coastal and highland regions from central Chile to northern Ecuador. Efficient agricultural production at higher elevations alleviated the stress on food production, thus allowing the Inca to concentrate more energy and resources on expanding their geographic influence over the other populations under a centralized government located at Cusco. The Quelccaya ice accumulation record (precipitation) shows a marked increase in precipitation at the beginning of the Late Horizon (~1450 CE). The Inca Empire came to an abrupt end in the 1530s CE as a result of the incursion of Spanish forces into Peru.

The juxtaposition of the Quelccaya ice core climate record and the Peruvian archeological record suggests that before the Spanish Conquest cultural decline may have been driven, at least in part, by deteriorating environmental conditions on which agrarian societies depended.[51] However, this hypothesis is not enthusiastically embraced by all anthropologists, archeologists, and even climatologists. In fact, a convergence of events that are both natural and human-made is required to disrupt a society. Long-term climate change and short-term extreme climate events are best seen as "threat multipliers" which in some cases may serve as the tipping point leading to the decline and eventual collapse of a culture.

6 Current Loss of Glaciers

The world's cryosphere is in retreat as tropospheric temperatures warm. Today glaciers cover about 10% of Earth's surface area, compared to 30% coverage during the coldest part of the last ice age. Because this warming is amplified at higher elevations,[52] the loss of glacier ice is most drastic in mountainous areas such as the

---

[47] Kolata, A. L., (1983) The South Andes, in Jennings, J. (Ed.), *Ancient South Americans*, W. H. Freeman, 241–285.

[48] Paulsen (1976).

[49] Thompson and Kolata (2017).

[50] Abbott et al. (1997).

[51] Thompson and Davis (2014a, 2014b).

[52] Thompson et al. (2021).

Andes and the Tibetan Plateau. Rising temperatures in the polar regions are especially troubling as Earth's largest ice sheets reside there.

6.1 Ice Decline in the Polar Regions

Antarctica and Greenland contain 95% of Earth's ice.[53] Over 40 years ago the late John Mercer, a professor at The Ohio State University, predicted that the first consequence of a warming climate linked to atmospheric CO_2 increase would be the breakup of the Antarctic ice shelves in a north to south direction.[54] Average temperature on the Antarctic Peninsula has risen 2.5 °C since 1950, resulting in the breakup of the ice shelves, just as Mercer foresaw. This warming has resulted in changes to the physical and biological environment there including the distribution of penguin colonies. One of the most rapid of these deteriorations occurred in 2002, when the Larsen B ice shelf collapsed in just 31 days. The breakup of these ice shelves does not contribute to sea level rise, as they are already floating and thus displacing sea water. However, ice shelves buttress land-based glaciers so that ice shelf loss allows the velocity of those glaciers to increase and it is the more rapid flow of land-based ice into the ocean that contributes to sea level rise.[55] The West Antarctic ice sheet is also warming, particularly around its coastal embayments. There is concern that warmer water circulating under the ice shelves that are buttressing large outlet glaciers such as Thwaites and Pine Island are causing these ice shelves to become unstable. If the ice shelves disintegrate it would allow a large volume of land-based ice to flow into the Southern Ocean.[56]

The Greenland ice sheet now also experiences dramatic ice melt during the Northern Hemisphere summer. The number and the size of the melt ponds and lakes forming in the near-coastal regions of southern Greenland ice sheet has increased, and many of them now flow in deep canyons, some of which empty into crevasses. Water that flows through some of these crevasses (moulins) may reach the bottom of the ice sheet and serve as a lubricant with the potential to accelerate the flow of the basal ice[57] although this is not widely observed.[58] In the last decade, many glaciers draining Greenland and West Antarctica have accelerated their ice discharge

---

[53] This comes to ~ 14.52×10^6 km^2 of ice cover in Greenland and Antarctica.

[54] Mercer (1978).

[55] Scambos et al. (2004).

[56] Milillo et al. (2019).

[57] Das (2008).

[58] See NASA (accessed 2023, June 3). Melt Ponds. NASA. https://www.earthobservatory.nasa.gov/images/80677/greenland-melt-ponds; See also Chap. 10, Fig. 17 in Chap. 10 for a recent set of photos of Helheim Glacier in SE Greenland.

to the ocean by 20 to 100%, but this is highly variable and persists over shorter time intervals.[59]

Satellite observations of Arctic sea ice extend back to 1979. Arctic sea ice cover reaches its minimum extent each September and since 1979 has decreased at a rate of ~ 13% per decade, relative to the 1981 to 2010 average.[60] In August, 2007, the Northwest Passage first opened to ships without the assistance of an icebreaker. This presents advantages for the shipping industry but poses major problems for the indigenous people and wildlife in this region. The loss of summer sea ice also impacts the climatology in the Arctic as the highly reflective ice is replaced by the darker ocean water which absorbs more radiation and warms the water and the overlying atmosphere. This creates a positive "feedback loop" in which the warming water melts more ice which in turn increases the surface area of the ocean (read more detail about feedbacks in Chap. 2). Unless another process intervenes to break this cycle, it is possible that the Arctic sea ice may eventually reach a "tipping point" beyond which the Arctic ocean may become ice free during the boreal summer.[61]

6.2 Retreat of Mountain Glaciers

Over the last five decades, glacier retreat has been monitored using ground and aerial observations and satellite imagery. These data reveal that virtually all of Earth's high-altitude, low- and mid-latitude glaciers are losing mass, and since the beginning of the twenty-first century, the rates of loss have increased.[62] Unlike temperatures in the middle and higher latitudes, atmospheric temperatures in the tropics tend to remain fairly uniform throughout the year. Tropical mountain (or alpine) glaciers are located in the mid-troposphere (~15,000 to 20,000 ft above sea level) where the recent warming has been amplified over the last several decades. Recent El Niño events have had particularly detrimental effects on many tropical glaciers, as warming spreads almost uniformly throughout the lower tropical atmosphere during El Niños,[63] and particularly strong warm (or cold) events have immediate effects on many high-altitude tropical glaciers.[64] These glaciers respond more quickly to changing climate than the massive polar ice sheets, as their surface area to volume ratio is much greater. Glaciers outside the polar regions contain a total of 170,000 cubic kilometers of ice, enough to contribute 0.4 m to sea level rise if they all melt.

---

[59] The IPCC model projections do not incorporate these recent abrupt changes in ice flow, which have the potential to cause sea level to rise faster than predicted.

[60] National Snow and Ice Data Center (NSIDC) (accessed 2023, June 3), https://nsidc.org/cryosphere/icelights/arctic-sea-ice-101.

[61] The theory of climatic tipping points or thresholds is relatively new and is a very important line of research for paleoclimatologists and climate modelers to investigate.

[62] Zemp et al. (2019).

[63] Trenberth & Smith (2006).

[64] Thompson et al. (2017a, 2017b); Permana et al. (2019b, 2019a).

Presently, melt from mountain glaciers contributes ~ 25% to global sea level rise. Another ~ 25% is contributed by the large ice sheets [Greenland (15%) and Antarctica (10%)] while thermal expansion of the ocean water in response to the elevated global temperatures contributes the additional 50%.

It is almost certain that several tropical mountain glaciers, especially those located in the equatorial latitudes, have retreated beyond the point of recovery. Recent studies[65] conclude that much of the tropical ice may disappear by the end of this century if current rates of retreat continue or accelerate. Examples of this retreat are observed on Kilimanjaro, the Naimona'nyi glacier in the western Himalayas, the Quelccaya ice cap, Peru and near Puncak Jaya in Papua Indonesia (Fig. 19).

The populations in the South America Andes and South Asia are particularly vulnerable to glacier responses driven by global climate change. In these regions, mountain glaciers are essential as they provide freshwater (i.e., feed mountain streams) in the dry season for agriculture, hydroelectric power, commerce, and

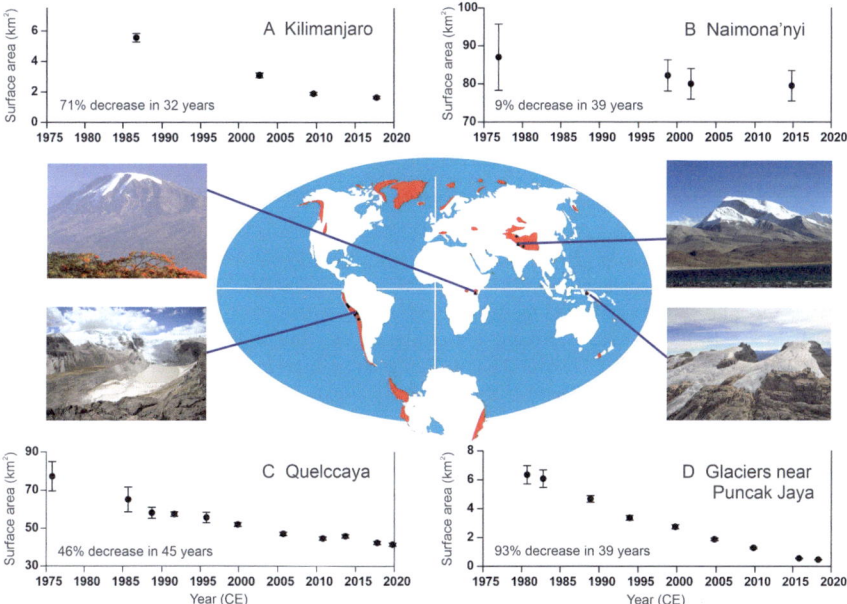

Fig. 19 The locations of 4 glaciers are shown on a global map of ice retreat (areas marked in red). The surface area changes (km^2) of four low-latitude alpine glaciers are shown, and their photographs are connected to their locations on the center map by dark blue lines. The changes in surface area from the late 20th to the early twenty-first centuries are shown for: **a** The glaciers on Kilimanjaro, East Equatorial Africa (3°S); **b** Naimona'nyi glacier, western region of the central Himalayas (30°N); **c** Quelccaya ice cap, Andes of southern Peru (14.5°S); and **d** the glaciers near Puncak Jaya, Papua, Indonesia (4°S). From Thompson et al. 2021.[66] Figure used with permission from Elsevier. This figure is excluded from Creative Commons license

---

[65] Thompson (2021).

[66] Thompson et al. (2021).

human consumption, especially during periods of drought. The current glacier retreat throughout the central and northern Andes and the Himalayan region is contributing to emerging water resource crises and environmental hazards for both urban and rural populations.

In South Asia, 72% of the streamflow in the Upper Indus River basin comes from seasonally melting snow and ice on the western Tibetan Plateau.[67] The Indus River flows through three countries (Pakistan, India, and the western Tibet region of China) where there is an increasing possibility of competition for water in the future. In regions that are already experiencing political and cultural stress, the addition of climate and water stress may further exacerbate tensions.[68] Thus, the long-term loss of glaciers which eventually reduces water supplies may serve as a 'threat multiplier' in these sensitive areas. In the past, the response of communities to natural climate disasters has included migration to more amenable environments. However, with a global population over 8 billion and increasing border protectionism by more prosperous nations, current mass migration, regardless of the cause, will result in increasing hardship and conflict. If accelerating global climate change affects rich and poor nations alike in the coming decades and centuries, the competition for arable land and water resources may greatly intensify.

References

Abbott, M. B., Binford, M. W., Brenner, M., & Kelts, K. R. (1997). A 3500 14C yr high-resolution sediment record of lake level changes in Lake Titicaca, Bolivia/ Peru. *Quaternary Research, 47*, 169–180. https://doi.org/10.1006/qres.1997.1881

Ahn, J., Brook, E. J., Mitchell, L., Rosen, J., McConnell, J.R., Taylor, K., Etheridge, D. & Rubino, M., (2012). Atmospheric CO2 over the last 1000 years: A high-resolution record from the West Antarctic Ice Sheet (WAIS) Divide ice core. *Global Biogeochemical Cycles 26*(2), https://doi.org/10.1029/2011GB004247

Allan, R. P. et al. (2021). IPCC, 2021: Summary for Policymakers. In Masson-Delmotte, V., Zhai, P., Pirani, A., et al. (Eds.), *Climate Change 2021: The Physical Science Basis. Contribution of Working Group I to the Sixth Assessment Report of the Intergovernmental Panel on Climate Change* (pp. 3–32). Cambridge University Press. https://doi.org/10.1017/9781009157896.001 https://www.ipcc.ch/report/ar6/wg1/downloads/report/IPCC_AR6_WGI_SPM.pdf. See also additional materials and information are available at https://www.ipcc.ch/report/ar6/wg1/

Alley, R. B., Shuman, C. A., Meese, D. A., Gow, A. J., Taylor, K. C., Cuffey, K. M., Fitzpatrick, J. J., Grootes, P. M., Zielinski, G. A., Ram, M., Spinelli, G., & Elder, B. (1997). Visual-stratigraphic dating of the GISP2 ice core: Basis, reproducibility, and application. *Journal of Geophysical Research Oceans, 102*, 26367–26381. https://doi.org/10.1029/96JC03837

Bolzan, J. F. (1985). Ice flow at the Dome C ice divide based on a deep temperature profile. *Journal of Geophysical Research Atmospheres, 90*, 8111–8124. https://doi.org/10.1029/JD090iD05p08111

Carey, M., Molden, O. C., Borg Rasmussen, M., Jackson, M., Nolin, A. W., & Mark, B. G. (2017). Impacts of glacier recession and declining meltwater on mountain societies. *Annals of the*

---

[67] Immerzeel et al. (2009).

[68] Schleussner et al. (2016); Wilson et al. (2017).

American Association of Geographers, 107, 350–359. https://doi.org/10.1080/24694452.2016.1243039

Cook, E. R., Anchukaitis, K. J., Buckley, B. M., D'arrigo, R. D., Jacoby, G. C., & Wright, W. E. (2010). Asian Monsoon failure and megadrought during the last millennium. *Science, 328*, 486–489. https://doi.org/10.1126/science.1185188

Dansgaard, W. (1964). Stable isotopes in precipitation. *Tellus, 16*, 436–468. https://doi.org/10.1111/j.2153-3490.1964.tb00181.x

Dansgaard, W., Johnsen, S. J., Møller, J., & Langway, C. C., Jr. (1969). One thousand centuries of climatic record from Camp Century on the Greenland ice sheet. *Science, 166*, 377–380. https://doi.org/10.1126/science.166.3903.377

Das, S. B. (2008). Fracture propagation to the base of the Greenland ice sheet during supraglacial lake drainage. *Science 320*. https://doi.org/10.1126/science.1153360

Delmonte, B., Andersson, P. S., Hansson, M., Schöberg H., Petit, J. R., Basile-Doelsch, I., & Maggi, V. (2018). Aeolian dust in East Antarctica (EPICA-Dome C and Vostok): Provenance during glacial ages over the last 800 kyr. *Geophysical Research Letters 35*(7). https://doi.org/10.1029/2008GL033382

Gabrielli, P., Wegner, A., Petit, J. R., Delmonte, B., De Deckker, P., Gaspari, V., Fischer, H., Ruth, U., Kriews, M., Boutron, C., Cescon, P., & Barbante, C. (2010). A major glacial-interglacial change in aeolian dust composition inferred from Rare Earth Elements in Antarctic ice. *Quaternary Science Reviews, 29*, 265–273. https://doi.org/10.1016/j.quascirev.2009.09.002

Grove, R. H. (2007). The Great El Niño of 1783–93 and its global consequences: Reconstructing an extreme climate event in world environmental history. *The Medieval History Journal, 10*, 75–98. https://doi.org/10.1177/097194580701000203

Immerzeel, W. W., Droogers, P., de Jong, S. M., & Bierkens, M. F. P. (2009). Large-scale monitoring of snow cover and runoff simulation in Himalayan River basins using remote sensing. *Remote Sensing of the Environment, 113*, 40–49. https://doi.org/10.1016/j.rse.2008.08.010

Johnsen, S. J., Dansgaard, W., Clausen, H. B., & Langway, C. C., Jr. (1972). Oxygen isotope profiles through the Antarctic and Greenland ice sheets. *Nature, 235*, 429–434. https://doi.org/10.1038/235429a0

Jouzel, J. et al. (2007). Orbital and millennial Antarctic climate variability over the past 800,000 years. *Science 317*(5839), 793-796. https://doi.org/10.1126/science.1141038

Kang, S., Zhang, Y., Yun Qian, Y., & Wang, H. (2020). A review of black carbon in snow and ice and its impact on the cryosphere. *Earth-Science Reviews, 210*, 103346. https://doi.org/10.1016/j.earscirev.2020.103346

Knowlton, C., Veerapaneni, R., D'Elia, T., & Rogers, S. O. (2013). Microbial analysis of ancient ice core sections from Greenland and Antarctica. *Biology, 2*, 206–232. https://doi.org/10.3390/biology2010206

Lüthi, D., Le Floch, M., Bereiter, B., Blunier, T., Barnola, J.-M., Siegenthaler, U., Raynaud, D., Jouzel, J., Fischer, H., Kawamura, K., & Stocker, T. F. (2008). High-resolution carbon dioxide concentration record 650,000–800,000 years before present. *Nature, 453*, 379–382. https://doi.org/10.1038/nature06949

McConnell, J. R. (2005). 198 Role and importance of cryospheric processes in climate system. In Anderson, M. G., (Ed.), *Encyclopedia of Hydrological Sciences*. https://doi.org/10.1002/0470848944.hsa208

Mercer, J. H. (1978). West Antarctic ice sheet and CO_2 greenhouse effect a threat of danger. *Nature, 271*, 321–325. https://doi.org/10.1038/271321a0

Milillo, P., Rignot, E., Rizzoli, P., Scheuchl, B., Mouginot, J., Bueso-Bello, J., & Prats-Iraola, P. (2019). Heterogeneous retreat and ice melt of Thwaites Glacier, West Antarctica. *Science Advances 5*. https://doi.org/10.1126/sciadv.aau3433

Parrenin, F., Barnola, J.-M., Beer, J., Blunier, T., Castellano, E., Chappellaz, J., Dreyfus, G., Fischer, H., Fujita, S., Jouzel, J., Kawamura, K., Lemieux-Dudon, B., Loulergue, L., Masson-Delmotte, V., Narcisi, B., Petit, J.-R., Raisbeck, G., Raynaud, D., Ruth, U., … Wolff, E. (2007). The EDC3 chronology for the EPICA Dome C ice core. *Climate of the past, 3*, 485–497. https://doi.org/10.5194/cp-3-485-2007

Paulsen, A. (1976). Environment and empire: Climatic factors in prehistoric Andean culture change. *World Archeology, 8*, 121–132.

Pendleton, S. L., Miller, G. H., Lifton, N., Lehman, S. J., Southon, J., Crump, S. E., & Anderson, R. S. (2019). Rapidly receding Arctic Canada glaciers revealing landscapes continuously ice-covered for more than 40,000 years. *Nature Communications, 10*, 445. https://doi.org/10.1038/s41467-019-08307-w

Permana, D. S., et al. (2019). Disappearance of the last tropical glaciers in the Western Pacific Warm Pool (Papua, Indonesia) appears imminent. *Proceedings of the National Academy of Sciences, 116*, 26382–26388.

Permana, D. S., Thompson, L. G., Mosley-Thompson, E., & Mark, B. G. (2019). Disappearance of the last tropical glaciers in the Western Pacific Warm Pool (Papua, Indonesia) appears imminent. *Proceedings of the National Academy of Sciences (PNAS), 116*(52), 26382–26388. https://doi.org/10.1073/pnas.1822037116

Petit, J. R., Jouzel, J., Raynaud, D., Barkov, N. I., Barnola, J.-M., Basile, I., Bender, M. Chappellaz, J., Davis, M., Delaygue, G., Delmotte, M., Kotlyakov, V. M., Legrand, M., Lipenkov, V. Y., Lorius, C., PÉpin, L., Ritz, C., Saltzman, E., & Stievenard, M. (1999). Climate and atmospheric history of the past 420,000 years from the Vostok ice core, Antarctica. *Nature 399*, 429–436. https://doi.org/10.1038/20859

Petit, J. R., Jouzel, J., Raynaud, D., Barkov, N. I., Barnola, J.-M., Basile, I., Bender, M., Chappellaz, J., Davis, M., Delaygue, G., Delmotte, M., Kotlyakov, V. M., Legrand, M., Lipenkov, V. Y., Lorius, C. PÉpin, L., Ritz, C., Saltzman, E., & Stievenard, M. (1999). Climate and atmospheric history of the past 420,000 years from the Vostok ice core, Antarctica. *Nature 399*, 429–436. https://doi.org/10.1038/20859

Satterlee, D. R. (1993). *Impact of a fourteenth century El Niño flood on an indigenous population near Ilo, Peru.* PhD thesis, University of Florida

Scambos, T., Bohlander, J. A., Shuman, C. A., & Skvarca, P. (2004). Glacier acceleration and thinning after ice shelf collapse in the Larsen B embayment, Antarctica. *Geophysical Research Letters* 31, L18402. https://doi.org/10.1029/2004GL020670

Schleussner, C.-F., Donges, J. F., Donner, R. V., & Schellnhuber, H. J. (2016). Armed-conflict risks enhanced by climate-related disasters in ethnically fractionalized countries. *Proceedings of the National Academy of Sciences. (PNAS)* 113, 9216–9221. https://doi.org/10.1073/pnas.1601611113

Thompson, L. G., & Davis, M. E. (2014). An 1800-year ice core history of climate and environment in the Andes of southern Peru and its relationship with Highland/Lowland Cultural Oscillations in: Meddens, F. et al., Inca Sacred Spaces: Landscape, Site and Symbol in the Andes: London, Archetype Press, pp. 261–268

Thompson, L. G. et al. (2017) Impacts of recent warming and the 2015/16 El Niño on tropical Peruvian ice fields. *Journal of Geophysical Research Atmospheres 122.* https://doi.org/10.1002/2017/JD026592

Thompson, L. G., Davis, M. E., Mosley-Thompson, E., Beaudon, E., Porter, S. E., Kutuzov, S., Lin, P.-N., Mikhalenko, V. N., & Mountain, K. R. (2017). Impacts of recent warming and the 2015/16 El Niño on tropical Peruvian ice fields. *Journal of Geophysical Research Atmospheres 122.* https://doi.org/10.1002/2017/JD026592

Thompson, L. G., & Kolata, A.L., (2017). Twelfth Century AD: Climate, environment, and the Tiwanaku State, in Weiss, H. (Ed.), *Megadrought and Collapse: From Early Agriculture to Angkor*. Oxford University Press

Thompson, L. G. (2021). The impacts of warming on rapidly retreating high-altitude, low-latitude glaciers and ice core-derived climate records. *Global and Planetary Change, 203*, 103538. and references therein.

Thompson, L. G., & Davis, M. E. (2014). An 1800-year ice core history of climate and environment in the Andes of southern Peru and its relationship with Highland/Lowland Cultural Oscillations.

In F. Meddens, C. McEwan, K. Willis, & N. Branch (Eds.), *Inca Sacred Spaces: Landscape* (pp. 261–268). Archetype Press.

Thompson, L. G., Davis, M. E., Mosley-Thompson, E., Porter, S. E., Corrales, G. V., Shuman, C. A., & Tucker, C. J. (2021). The impacts of warming on rapidly retreating high-altitude, low-latitude glaciers and ice core-derived climate records. *Global and Planetary Change, 203*, 103538. https://doi.org/10.1016/j.gloplacha.2021.103538

Thompson, L. G., Mosley-Thompson, E., Davis, M. E., Zagorodnov, V. S., Howat, I. M., Mikhalenko, V. N., & Lin, P.-N. (2013). Annually resolved ice core records of tropical climate variability over the past ~1800 years. *Science, 340*, 945–950. https://doi.org/10.1126/science.1234210

Thompson, L. G., Mosley-Thompson, E., Davis, M. E., Zagorodnov, V. S., Howat, I. M., Mikhalenko, V. N., & Lin, P.-N. (2013). Annually resolved ice core records of tropical variability over the last ~1800 years. *Science, 340*, 945–950. https://doi.org/10.1126/science.1234210

Trenberth, K. E., & Smith, L. (2006). The vertical structure of temperature in the tropics: Different flavors of El Niño. *Journal of Climate, 19*, 4956–4973.

Wilson, A. M., Gladfelter, S., Williams, M. W., Shahi, S., Baral, P., Armstrong, R., & Racoviteanu, A. (2017). High Asia: The international dynamics of climate change and water security. *Journal of Asian Studies, 76*, 457–480.

Zemp, M., Huss, M., Thibert, E., Eckert, N., McNabb, R., Huber, J., Barandun, M., Machguth, H., Nussbaumer, S. U., Gärtner-Roer, I., Thomson, L., Paul, F., Maussion, F., Kutuzov S., & Cogley, J. G. (2019). Global glacier mass changes and their contributions to sea-level rise from 1961 to 2016. *Nature 568*, 382–386. https://doi.org/10.1038/s41586-019-1071-0

Zemp, M., Huss, M., Thibert, E., Eckert, N., McNabb, R., Huber, J., Barandun, M., Machguth, H., Nussbaumer, S. U., Gärtner-Roer, I., Thomson, L., Paul, F., Maussion, F., Kutuzov, S., & Cogley, J. G. (2019). Global glacier mass changes and their contributions to sea-level rise from 1961 to 2016. *Nature, 568*, 382–386. https://doi.org/10.1038/s41586-019-1071-0

Zhong, Z.-P., Tian, F., Roux, S., Gazitúa, M. C., Solonenko, N. E., Li, Y.-F., Davis, M. E., Van Etten, J. L., Mosley-Thompson, E., Rich, V. I., Sullivan, M. B., & Thompson, L. G. (2021). Glacier ice archives nearly 15,000-year-old microbes and phages. *Microbiome, 9*, 160. https://doi.org/10.1186/s40168-021-01106-w

Lonnie G. Thompson grew up in the mountains of West Virginia and attended Marshall University where he received his B.S. degree in Geology. He began his graduate work at The Ohio State University studying coal geology, although he became interested in polar studies and glaciology after participating in an expedition to Antarctica. His PhD dissertation focused on the development and interpretation of the first records of wind-blown dust that extended back into the last ice age from ice cores from Greenland and West Antarctica. Since then he has dedicated his career to collecting and interpreting ice core records from the world's highest mountains in 16 countries in order to develop a global history of climate and environmental changes archived in glaciers. He has spent a total of over 4 years of his life at altitudes higher than 18,000 ft, and after surviving a heart transplant in 2012 he conducted research in 2015 on the 22,000 ft Guliya ice cap, Western Kunlun Mountains, a record elevation for a heart transplant recipient. He is an elected member of the National Academy of Sciences and a recipient of the United States Medal of Science. Lonnie is a Distinguished University Professor in the School of Earth Sciences and Senior Research Scientist in the Byrd and Polar Climate Research Center at The Ohio State University. Lonnie is a primary author of Chap. 4.

Ellen Mosley-Thompson grew up in West Virginia and graduated from Nitro High. She has had a life-long love of science and received her B.S. in physics and math from Marshall University and her M.A. and PhD in atmospheric science and paleoclimatology from The Ohio State University. Early in her career she participated in an ice core drilling project at South Pole Station, Antarctica where she was one of only two women. She was quickly "hooked" on the excitement of using the chemical and physical properties preserved in the ice to reconstruct Earth's climate history. She is a Distinguished University Professor, who, along with Lonnie Thompson and their research team, have acquired ice cores from both polar ice sheets and numerous high mountain glaciers to reconstruct Earth's complex climate history. Collectively, this global collection of paleoclimate histories confirms that Earth's climate has moved outside the range of natural variability experienced over at least the last 2000 years. Ellen has had the privilege of leading nine expeditions to Antarctica and six to Greenland to retrieve these valuable ice cores. Ellen teaches an undergraduate honors course on global climate and environmental change that focuses on understanding the forces driving these climate changes and facilitating collective and individual actions to change the direction in which Earth's climate system is currently heading. Ellen is a primary author of Chap. 4.

Open Access This chapter is licensed under the terms of the Creative Commons Attribution 4.0 International License (http://creativecommons.org/licenses/by/4.0/), which permits use, sharing, adaptation, distribution and reproduction in any medium or format, as long as you give appropriate credit to the original author(s) and the source, provide a link to the Creative Commons license and indicate if changes were made.

The images or other third party material in this chapter are included in the chapter's Creative Commons license, unless indicated otherwise in a credit line to the material. If material is not included in the chapter's Creative Commons license and your intended use is not permitted by statutory regulation or exceeds the permitted use, you will need to obtain permission directly from the copyright holder.

The Sedimentary Record of Past Climate Change

Adriane R. Lam, R. Mark Leckie, Steven Hovan, and Dana Royer

Guiding Question: How do we read the clues about past climates in the sedimentary record?

1 Key Take-Away Points

- Records of climate changes through time are preserved in layers of sediment and sedimentary rocks exposed in outcrops on land and in sediment cores drilled on land or recovered from lakes and the world ocean.
- By gathering and analyzing proxy data from these records, scientists can extend our understanding of climate for millions of years into the past—far deeper in time than instrumental records and ice core records.

A. R. Lam (✉)
Department of Earth Sciences, Binghamton University, Binghamton, NY, USA
e-mail: alam@binghamton.edu

R. M. Leckie
Department of Earth, Geographic, and Climate Sciences, University of Massachusetts Amherst, Amherst, MA, USA
e-mail: leckie@umass.edu

S. Hovan
Kopchick College of Natural Sciences and Mathematics, Indiana University of Pennsylvania, Indiana, PA, USA
e-mail: hovan@iup.edu

D. Royer
Department of Earth and Environmental Sciences, Wesleyan University, Middletown, CT, USA
e-mail: droyer@wesleyan.edu

- Interpreting Earth history and its paleoclimate from sediments and sedimentary rocks is a form of detective work that is applied through a number of steps, from initially choosing the study sites, to describing sediment characteristics, to analyzing and interpreting sediment components, such as fossils, used as paleoenvironmental and paleoclimatic proxies, and placing them in the context of geologic time.
- Clastic sediments and rocks provide information about where the sediments originated, the processes that worked to move the sediments, and the type of environment in which the sediments or rock were deposited. Some paleoclimatically useful clastic sediments and sedimentary rocks include paleosols, tills, rhythmically laminated sediments, wind-blown dust, and marine black shales.
- Chemical sedimentary rocks provide information about the environment in which the minerals precipitated from ion-concentrated watery solutions. Salt deposits and speleothems are two types of paleoclimatically useful chemical sedimentary rocks.
- Biogenic sediments and rocks provide important clues as to the environment in which organisms once lived, including the temperature, salinity, sunlight conditions, nutrients, and energy of the environment. Coal, limestone, calcareous ooze, and siliceous ooze are important biogenic sedimentary deposits.
- Determining the environmental conditions, ages, and rates of change in records of past climate is essential to paleoclimate reconstruction. Relative and absolute ages of sediments and sedimentary rocks are inferred through several methods involving fossils, radiometric dating, paleomagnetism, and geochemistry.
- Paleotemperatures can be reconstructed from sediments and sedimentary rocks using a variety of methods. Examples include stable oxygen isotopes, magnesium-to-calcium (Mg/Ca) ratios, and marine and terrestrial biomarkers (molecular fossils).
- Examples of the approaches used for reconstructing paleo-precipitation on land involve special horizons in paleosols that form under arid and semi-arid conditions and analysis of plant leaf wax isotopes (a type of terrestrial biomarker).
- Two examples of approaches for reconstructing atmospheric carbon dioxide levels from the sedimentary record involve fossil plant leaf stomata in terrestrial sediments and sedimentary rocks and boron isotopes in marine carbonate microfossils.
- Globally averaged change in the volume of ice on land (ice sheets and glaciers) can be inferred from marine stable oxygen-isotope records, whereas local and regional details of ice sheet history can be determined from examination of tills and other glacial deposits.
- Sea level is intrinsically linked to global temperature and ice volume changes, as well as other factors, and can be reconstructed through geologic time by mapping the positions of coastal sedimentary sequences.
- Each proxy used to reconstruct a paleoclimate variable has a specific time range over which it can be used and has its own level of temporal and spatial resolution. Therefore, different proxies are used for different time intervals and settings.

Developing a comprehensive climate history requires integrating information from many proxies.

2 Introduction

Sediment is the collective name for loose particles eroded from pre-existing rocks (e.g., mud, sand, and gravel), for material precipitated directly from concentrated solutions (e.g., evaporites or "salts"), for the shelly (i.e., biomineralized) remains of organisms, and for organic materials (i.e., non-biomineralized materials such as pollen, spores, plant debris, and the remnants of soft tissue). Billions of tons of sediment are produced every year and deposited on the continents, on the surfaces of glaciers and ice sheets, in the bottom of lakes, and in the oceans. Sedimentary particles are widespread and cover much of the surface of the Earth. In fact, if you look at your shoes right now, you are probably carrying some sediments around with you (e.g., dust or mud)! The type of sedimentary grains and the fossils and chemicals contained in them reflect the physical, biological, and chemical characteristics of the environment in which they formed, modified by processes that transported the grains and the environment in which they were deposited. For example, in arid regions small rock particles are blown around by the winds and accumulate into large sand dunes, or salt layers form as water evaporates, leaving behind the minerals that were dissolved in the water. Elsewhere, thick layers of plant vegetation accumulate in swampy regions of tropical and temperate forests, forming the precursor of coal, and tropical fish forage for food on coral reefs leaving sedimentary deposits of broken coral nearby.

As you learned in Chaps. 3 and 4, proxy data are sources of information that can be used when direct measurements of climate conditions are not available. By gathering and analyzing proxy data from sedimentary archives (i.e., data recorded in sedimentary deposits), scientists can extend our understanding of climate for millions of years into the past, far deeper in time than instrumental records or ice core records.

This chapter will explore how the sedimentary record, including its preserved fossils, is used by geoscientists to interpret the past climate (paleoclimate) of the ancient Earth. Sediments contain clues that tell us about the source of the materials, the processes and conditions involved in the formation of sediments and dissolved ions, and the processes and conditions involved in the transport and deposition (or precipitation) of sedimentary materials. Like a detective, it is the job of geoscientists to observe and interpret those clues.

2.1 Some Basics About Working with the Sedimentary Record

Over long periods of time (i.e., millions of years) and under optimum conditions at a specific location, sediments can accumulate into thick deposits, with younger sediments on top of older ones. As a result, working down through these layers of sediments is much like reading a history book of Earth and its climate; the deeper we look, the older the climate history we can read. This illustrates a fundamental geologic principle, termed **superposition**: older sediments and sedimentary rocks are deposited first, with younger layers deposited atop. In the following paragraphs, we will explore some other basic approaches that geoscientists use to access and interpret the vast layers of sedimentary "pages" through time. It is important to realize from the beginning, however, that sediment deposition—and the long-term preservation of those deposits—varies considerably through both time and space. In other words, sediments are not deposited and preserved to form a continuous "layer cake"-like record everywhere in the world. Instead, the details of sediment accumulation and its subsequent long-term preservation vary significantly from place to place. In addition, the level of detail at which any interval of Earth history is recorded by sediments also varies considerably across the globe. As a result, interpreting paleoclimate is a form of detective work that is applied through a number of steps, from initially choosing the study sites through interpreting paleoclimatic proxies and placing them in the context of geologic time.

Table 1 highlights the general flow of a sediment-based paleoclimatic study to address a research question or test a scientific hypothesis. Accessing and describing the sedimentary record are important parts of this process. These aspects are explored further in the next section assuming a best-case scenario in which the sedimentary record was deposited continuously for the time period of interest, there have been no interrupting episodes of erosion, and the sediments of interest have not been tipped on end or overturned by local or regional (tectonic) processes. The spatial and temporal complexities noted above will be mentioned, however, in situations where their effects cannot be ignored.

2.2 How Do Geoscientists Access the Sedimentary Record?

Sedimentary records generally can be accessed in two ways: (1) from rock exposures at the Earth's surface, or (2) from cores (long cylinders of sediment and sedimentary rock) extracted by drilling into the Earth. Rock exposures at the Earth's surface are called **outcrops** for short and can be the result of natural processes (e.g., erosion along the banks of a river) or of human activity (e.g., roadway construction). Outcrops provide a low-cost way to access the sedimentary record, but exposed rocks may be geographically small or of a limited time interval.

When outcrops are not available with sedimentary rocks of the appropriate age or in the appropriate location, then cores of sedimentary deposits can be collected, and then studied using similar approaches. Sediment/rock cores can be taken from the land surface, by drilling down into the subsurface sediment and rock, or from sediments at the bottom of a lake or ocean. All coring requires specialized equipment, but coring in watery environments is especially complicated because it requires a boat or a ship with onboard drilling capabilities (Fig. 1). Sediment/rock cores

Table 1 General steps of a sediment-based paleoclimatic study to address a research question or test a scientific hypothesis

Steps	Example
1. For a given research question, identify the time interval and/or the geographic region of interest, and the climatic component(s) of interest.	What were sea surface temperatures across the North Pacific Ocean between 1 and 5 million years ago? • Geographic region: North Pacific Ocean • Time interval: 1–5 million years ago
2. Identify the locations that have the most continuous and/or most detailed sedimentary record for that time interval in the area of interest.	Identify locations in the North Pacific with *thick sequences of sediments* known or estimated to span 1–5 million years ago.
3. Access the sedimentary sequences of that time interval at those locations (Sect. 2.2).	Drill cores in the seafloor from a ship (Fig. 1).
4. Describe the characteristics of the observed sediments (Sect. 3).	Primary characteristics: • The sediment types (especially composition) • The sediment ages determine if the sediment has been deposited continuously for the time period of interest, or if there have been any interrupting episodes of erosion
5. Extract and analyze the sediment components used as proxies helpful to address the research question.	Extract and analyze assemblage data and stable isotope data from fossil planktic foraminifers, a proxy for sea surface temperature.
6. Integrate the proxy results with information from the basic characteristics of the recovered sediments to interpret the climatic component of interest across the study area or through the studied interval of time.	Correlate (link together) sea surface temperature data from your coring location to data from other locations in the North Pacific for the 1 to 5 million year old time interval.

Chapter sections and figures that expand on particular steps are called out.

have the advantages that: (1) they are collected at a location chosen specifically for the scientific objectives of the study, and (2) they contain material that has not been affected by processes of weathering at the Earth's surface, so their record of Earth history and paleoclimate generally is more pristine. Using cores to access the sedimentary record is more expensive than working on outcrops, however, and choosing the proper location for coring is essential for the success of studies based on the resulting cores.

2.3 How Do Geoscientists Interpret the Broader World from an Outcrop or Core?

You are likely to have seen an outcrop, such as along a road, a riverbank, or a coastline, and that outcrop probably was a few tens of meters to a few kilometers

Fig. 1 Image of the research vessel JOIDES Resolution (JR) leaving the port of Townsville, Australia, for International Ocean Discovery Program Expedition 371. The tower-like structure in the middle of the ship is called the derrick, which supports the drill string. The front portion of the ship is where crew, staff, and scientists live and work, and the back portion of the ship includes the engine room, storage for drilling pipes, and a helicopter landing pad. The JR has the capability to drill sediment cores in water depths up to 8235 m, or just over 5 miles! Photo provided by Mark Leckie

long. In unusual circumstances, such as along the Grand Canyon of the Colorado River, an outcrop may be even longer and extend for tens of kilometers or more. However, no outcrop extends across the entire globe, across an entire continent, or even across an entire state in the United States. As a result, you may wonder how geoscientists can interpret past conditions across an area that is larger than the area of a single outcrop—and that question probably looms even larger when you realize that most sediment/rock cores are only a few cm in diameter. Several different approaches are used by geoscientists to expand their paleoclimatic and paleoenvironmental interpretations beyond the horizontal boundaries of an outcrop or core. Central to these approaches is Walther's Law.

2.3.1 Sediment Characteristics and Walther's Law

In the study of many outcrops, and of all cores, geoscientists tend to focus on the vertical stacking of sediment/rock layers. This focus is motivated by two factors: (1) the observation that sediment types often change more rapidly and more significantly in their vertical direction compared to changes observed horizontally; and (2) the application of the principle of superposition, which tells us that the vertical changes in sediment/rock type are recording a history of changes through time. This emphasis on the vertical stacking of layers is also based on another long-established principle known as **Walther's Law**. This principle is important to interpret ancient environmental conditions at a local and regional scale, outside the bounds of the outcrop or core.

Walther's Law is based on the observation that modern settings where sediment is deposited are composed of multiple adjacent subenvironments, each characterized by its own set of physical and biological processes and resulting sediments. Walther's Law states that the vertical changes in sediment types seen in an outcrop or core record the lateral migrations and stacking of those adjacent subenvironments over time. In other words, the vertical succession of sediments in an outcrop or core can be used to interpret the suite of laterally adjacent subenvironments that existed at any one time and that extended well beyond the physical boundaries of a core or outcrop.

For example, a "coastal" environment, such as along the Atlantic coast of Massachusetts, can be subdivided into three subenvironments: the beach, the sand dunes that are located landward of the beach, and the lower-lying marsh further landward behind the dunes (Fig. 2). In detail, the beach itself is affected most strongly by waves, storm waves, and tidal fluctuations of water level, which transport and deposit larger sand sizes in relatively flat and thin layers, and without terrestrial plants. Broken shells, often found in thin discontinuous layers, is a characteristic feature of ancient beach deposits. The dune environment is affected most strongly by the wind, which transports relatively fine-grained sand and deposits that sand in the form of dunes, which can be colonized by various types of terrestrial plants. Angled layers of well-sorted sand, called foresets (or cross-bedding), are a characteristic feature of ancient dune deposits. The salt marsh environment behind the dunes is a region of lower energy and terrestrial plant growth where marine and terrestrial organisms live and burrow, and where small streams cut through the muddy sediments. This region is affected most by the tidal currents, with more waters coming into the streams during high tide, and leaving the region during low tide. Organic-rich mud with intact fossils and localized channel sand deposits are characteristic features of the quieter water deposition of the salt marsh. In other words, each of these three subenvironments is distinct and characterized by a specific set of physical and biological processes.

If we now imagine that sea level rises along this coastline (that is not very difficult to imagine since it is presently happening), then the "beach" subenvironment will migrate landward over the pre-existing "dune" subenvironment, and the "dune" subenvironment will migrate landward over the location of the pre-existing "marsh" environment (Fig. 2). If sea level rises high enough, the "beach" subenvironment ultimately would migrate over the initial position of the "marsh" subenvironment.

To summarize, the vertical stacking pattern of the sediments—marsh sediments overlain by dune sediments overlain by beach sediments, as could be identified from a single core (Fig. 2)—identifies subenvironments that were once laterally adjacent and extended well beyond the physical limits of that core. In addition, this vertical sequence records the transition of a terrestrial subenvironment (i.e., the landward marsh) to a near marine subenvironment (the beach), which also indicates that this assemblage of subenvironments has shifted landward and that relative sea level has risen.

Fig. 2 Conceptual diagram of Walther's Law across beach, dune, and marsh subenvironments. Top row of images are from Plum Island, Massachusetts, showing the characteristic beach environment, the transition from the beach to dune subenvironment, and the marsh subenvironment during low tide. Bottom left figure represents the different sedimentary rock types that characterize the three subenvironments and their major characteristics (e.g., dune subenvironments are characterized by sand ripples, and marsh environments are characterized by fine-grained, darker-colored sediments). Panel on the bottom right shows how these three subenvironments and their associated sedimentary rocks would "stack" on one another through time in response to a rise in sea level. Images provided by Mark Leckie, figure by Adriane Lam

As you have seen from the example described above, identifying changes in shoreline locations are important for defining the history of global sea level change. Because global ice volume is one important control on global sea level—and global ice volume in turn is affected by global average temperature and water cycling—the study of past sea level changes is an important contributor to interpreting Earth's paleoclimate history.

A second example of the application of Walther's Law can be observed in an outcrop in southern New England that records the preserved deposits in and around an ancient lake (Fig. 3) that are now vertically stacked. In this sequence of sedimentary rocks, there are rocks that are dark gray in color, fine-grained (mud or clay size grains), and layered (or "bedded") very thinly, which contain fossils of fish. The muddy fish-fossil-bearing layers were deposited in deep lake waters, which were suitable for freshwater fish populations. The other rock layers are reddish to tan in color, coarser (sand-size grains) and contain abundant burrows and tracks from the animals that once lived, crawled, or walked across the sediment prior to it being lithified into rock. These red-colored sediments have dinosaur footprints found in the sections. The tan layers also contain ripple marks. Taken together, the assemblage of features indicate the reddish to tan layers represent a time when the lake dried up or shrank in size. This records the alternation of arid versus wet climates, which drove lake contraction and expansion, and is an example of one of many climate cycles that

Fig. 3 Preserved rock layers that were deposited in and around an ancient lake, as visible south of Hartford, Connecticut. Relative lake levels are indicated on the left of the image, which alternate from alluvial plain sediments (sediments transported by rivers) to shales deposited in the deepest section of the ancient lake. Photos provided by Mark Leckie, figure by Adriane Lam

affected the Newark-Hartford Basin during the latest Triassic and Early Jurassic time (~ 201 million years ago). Note too that each of the subenvironments (deep basin, lake edge, alluvial plain) represent what were once laterally adjacent settings and were kilometers to tens of kilometers wide at the time of deposition. Thus, Walther's Law, applied to a geographically restricted outcrop, provides paleo-information about a much larger area. Other assemblages of climatically sensitive sediments, observed in vertical stacking patterns with a variety of ages and at a variety of locations from cores and outcrops, can be used to interpret changes in shoreline position (past sea level), glacier and ice sheet extent, and other important parameters over large areas throughout Earth's history.

2.3.2 Linking Proxies to the Broader World

The discussion of proxies studied in ice cores (Chap. 4) introduced the fact that the processes affecting many of those proxies operate over areas much larger than just the location of the ice core itself. For example, the insoluble particles (i.e., dust) in an ice core may have been carried very long distances by the wind, so that studying those particles provides information about both the distant dust source and the intervening wind patterns and strength. In a similar way, many of the paleoclimatic proxies studied from sediments and sedimentary rocks (which you will learn about in later portions of this chapter) also are affected by climatically sensitive processes and conditions acting at regional to global scales. As a result, geoscientists can interpret paleoclimatic conditions across regions and/or the globe from well-designed studies of sedimentary proxies at geographically restricted outcrops and cores.

2.4 What Are Limitations of the Sedimentary Record?

Reconstructing Earth's climate history would be a much more straightforward task if a uniform sedimentary record of all geologic time was present everywhere on Earth. This is not the case, so that finding locations with sediments of the appropriate age is an essential part of any paleoclimatic study. The nature of the sedimentary record is complicated because of factors associated with: (1) conditions at the time of deposition, and (2) processes affecting sediment preservation over geologic timescales, including the availability of sediment and a place to accumulate a sedimentary deposit without subsequent erosion.

At any time in Earth's history, the Earth's surface has been a mosaic of different physical, chemical, and biological environments. The types of sediments being deposited and the rates at which those sediments are deposited vary drastically across those environments; in fact, some of those environments may be places where no sediments are being deposited or where pre-existing material is being eroded. As a result of these complexities at the time of deposition (or following it), no interval of Earth's history is recorded by sediments *everywhere* on the globe. In addition, it is important to keep in mind that sedimentary proxy data generally monitor the conditions through Earth's geologic past on much longer timescales than daily/weekly/monthly instrumental records. The sedimentary proxies usually provide records that identify changes over hundreds, thousands, or millions of years and are therefore useful to learn about long-term climate trends and patterns, rather than the day-to-day or year-to-year patterns. A few specific proxies do exist, however, that record changes over timescales of seasons, years, or decades at particular locations and during particular intervals of time. As you might imagine, these very detailed sedimentary proxies are highly valued by geoscientists.

Over the longer duration of geologic time, the second factor that affects the value of the sedimentary record as an archive of Earth's past climates is the possibility that older sediments have been removed by erosion in an area. Such erosion will leave a gap in the sedimentary record of that area; such a gap is termed an **unconformity**. As an example, consider the modern-day Himalayan Mountains, which are being uplifted and eroded at a rate of approximately 1 cm/year; this process has taken place for tens of millions of years, so that older sedimentary rocks are now exposed at the surface in the Himalayan Mountains and are being eroded. As a result, it should be clear that the Himalayan Mountains are not a good location for a study of Earth's climate ten million years ago, because it is unlikely that sedimentary rocks of that age are present there. In contrast, large amounts of this eroded sediment have been deposited in lower-lying areas adjacent to the rising Himalayan Mountains, and those adjacent areas (e.g., northern Indian Ocean) would be excellent prospects for such a study based on sedimentary paleoclimate proxies.

3 How Do We Classify Sediments and Sedimentary Rocks, and How Do Sedimentary Rocks Carry Distinctive Climatic Information?

The sediments that we see today are the product of a variety of physical, chemical, and biological processes, many of which are closely linked with climate conditions at the time the sediments were formed or with the processes involved with their transport and deposition. Therefore, it is imperative to understand how sediments are created and deposited, how sedimentary rocks form, and how geoscientists describe and interpret such sediment and rock. Figure 4 and Table 2 illustrate the important steps in the production of sediment and its conversion to sedimentary rock and highlight the fundamental difference between **clastic sediments**, which are formed by the accumulation of pieces (i.e., **clasts** such as mud, sand, and gravel) derived from pre-existing rock, and **chemical sediments**, which are formed when materials dissolve in water and are precipitated into solid forms. Some chemical sediments are formed by inorganic precipitation, meaning there are no organisms involved. Biogenic **sediments** are those chemical sediments that are formed by organisms; examples include shells and bones. Biogenic sediment grains are sometimes termed **bioclasts**, because those biologic grains can be transported away from their site of formation and deposited elsewhere, just like the non-biological clasts mentioned above.

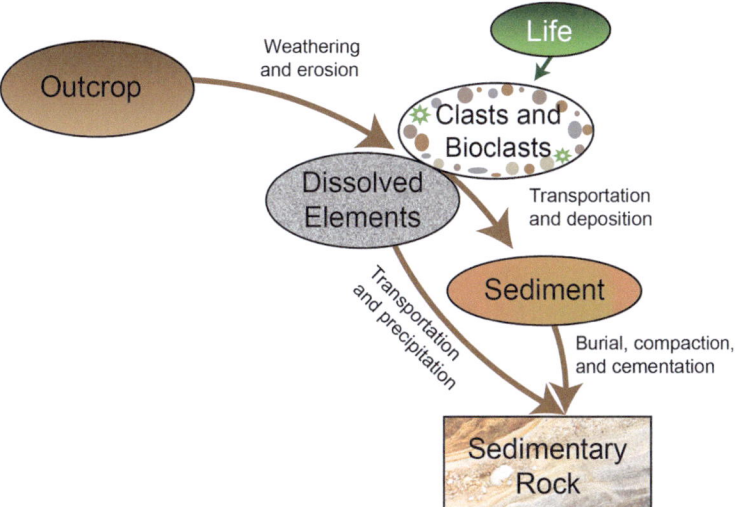

Fig. 4 Schematic diagram of the materials and processes important in the formation of sediments and sedimentary rocks. Figure by Kristen St John and Adriane Lam

Table 2 Generalized steps in the formation of sediment/sedimentary rock types

Sediment/ sedimentary rock type	Steps of formation
Clastic (detrital)	1. Pre-existing rocks are weathered and eroded into dissolved elements and/or clasts. 2. Clasts are transported by flowing water (rivers, waves, and ocean currents), air (wind), and/or ice (glaciers) and then deposited. 3. The deposited clasts (i.e., sediment) are buried, compacted, and cemented together from the precipitation of dissolved elements in groundwater.
Chemical	1. Pre-existing rocks are weathered and eroded into dissolved elements and/or clasts. 2. Dissolved elements are transported by flowing water and precipitate, forming a new mineral directly from a concentrated solution. This can happen when surface water (e.g., seawater or saline lake water) evaporates or the input of freshwater decreases. This can also happen when groundwater drips into caves, forming stalagmites and stalactites.
Biogenic	If composed of biominerals (hard parts formed biologically, such as shells, bone, and microfossil tests): 1. Pre-existing rocks are weathering and eroded into dissolved elements and/or clasts. 2. Dissolved elements are transported by flowing water to lakes or the ocean, where organisms precipitate biominerals (hard parts formed biologically, such as shells, bone, and microfossil tests). 3. After an organism dies, these hard-part remains (now called bioclasts) are either transported and deposited, or accumulate in place (where the organisms lived). Bioclasts are cemented together from the precipitation of dissolved elements.
	If composed of non-mineral organic matter (e.g., plants, humus, and soft tissue of plankton): 1. Organic matter is produced through photosynthesis on land by plants, in the ocean by phytoplankton, and through excrement of terrestrial and marine organisms. 2. After the organisms die, their organic remains settle to the bottom of poorly oxygenated waters and are compacted (no cementing agent). This can occur in swamps or if rapidly buried on the seafloor.

Modified from an unpublished table by Bill Lukens, James Madison University.

3.1 Sediment Production

Weathering is the entire suite of physical, chemical, and biological processes that break down pre-existing rocks and minerals. The products of weathering include smaller pieces of the pre-existing material (i.e., clasts), which often have been altered chemically and/or physically to some extent, and ions that have been chemically removed from clasts or other materials and are carried in solution, either in groundwater or in surface water (i.e., streams and rivers). The physical processes that produce smaller grains from pre-existing rocks are termed **physical weathering**, whereas the processes that change the chemical composition of the products and place materials into solution are termed **chemical weathering**. The details of chemical weathering, as well as its potential impacts on Earth's climate, will be explored in Chap. 7.

Important physical weathering processes include the physical actions of wind, water, or ice, and physical weathering processes dominate overall rock weathering in environments that are cold and/or dry. One example of physical weathering is called freeze–thaw and occurs when water is trapped in cracks or crevices in a large rock. That water expands and contracts as temperatures cycle below and above freezing, which expands the cracks and ultimately breaks pieces (i.e., clasts) off the original rock. The seasonal appearance of potholes in roads in cold climates develops due to similar freeze–thaw cycles of water trapped within the roadbed. Another example of physical weathering occurs beneath glaciers where the ice lies directly on the underlying bedrock. Sediment is produced and removed by the movement of the glacier across the base, but the effects of physical weathering can also be recognized by the modification of the underlying bedrock, leaving a surface that may be polished or exhibit scratches or grooves.

Sediment particles produced solely by physical weathering processes will have a mineral and chemical composition that is very similar to the composition of their parent rock. That mineral and chemical composition may include components that are removed readily by chemical weathering, so the presence of such unstable components in an ancient clastic sediment may indicate that sediment was produced, transported, and deposited under cold and/or dry climatic conditions that limited chemical weathering effects.

Chemical weathering includes a variety of processes that partially or completely remove components from the solid phases of rocks or minerals and place those components in solution. Some of these processes are inorganic, whereas others are enhanced by the presence of organisms. Warmer temperatures and moisture enhance the rate and efficiency of chemical weathering processes, so chemical weathering processes dominate overall rock weathering in environments that are warm and wet.

The susceptibility of common rocks and minerals to chemical weathering varies widely. Some materials, such as quartz (SiO_2), are very resistant to chemical weathering, whereas other materials, such as calcite ($CaCO_3$), are very susceptible and can be dissolved completely by natural chemical weathering. Rocks and minerals that contain silica (SiO_2) and aluminum (Al) are of intermediate susceptibility and can weather chemically to form one of a variety of clay minerals. As such, the presence of such clay minerals in ancient sediments can be a guide to the intensity of chemical weathering—and therefore temperature and water availability—in the past.

Chemical weathering removes components (i.e., atoms, ions, and compounds) from pre-existing rocks and minerals and places those components into solution in surface water, groundwater, and seawater. At a later time—and most likely in a different place—those dissolved components can become concentrated and **precipitate inorganically** to form new sediment. In some cases, precipitation forms completely new material in a setting at the Earth's surface. For example, the inorganic precipitation of calcium carbonate ($CaCO_3$) in shallow tropical waters supersaturated with Ca^{2+} and CO_3^{2-} ions can create new carbonate sediment (e.g., limestone), or components dissolved in groundwater can precipitate inorganically to produce speleothems (stalagmites and stalactites) in caves. In other cases, components dissolved in groundwater can precipitate as a natural cement that binds previously deposited sediment grains together, thereby converting the sediment to sedimentary rock, or can replace pre-existing material contained in the sediment. The process of ions replacing pre-existing material is the process by which petrified wood is formed, where silica dissolved in groundwater precipitates to replace woody organic material.

Sometimes chemical precipitation is mediated by an organism, so the dissolved ions derived from chemical weathering are vital for life. Many types of organisms utilize dissolved ions of Ca^{2+} and Si^{4+} to precipitate their hard parts or shells; this process is called **biomineralization**. These organisms are essential to the creation of certain types of sediment primarily through the process of building their internal or external skeletons, which become **biogenic sedimentary particles (or bioclasts)** upon the death of the organisms. For example, empty shells that wash up on a beach are now sedimentary particles.

Organisms also produce and modify sediments in their search for food (organic matter), or by physically moving rocks and particles in their environments. For example, almost all fine-grained sediment on Earth is estimated to have passed through the gut of an organism (mainly worms and other invertebrates that live in soil or sediment) at some time. In another example, parrotfish in reef environments use their hard beaks to feed on corals. The parrotfish chew up the hard skeleton of the coral, digest the soft bits of the coral animal, and excrete the pulverized coral skeleton. In one year, a single parrotfish can produce 1000 pounds of carbonate sand (which is all effectively fish poop!). The term **bioturbation** is used for the overall effects that animals have on the sediments as the animals move around, feed, and build shelters. In addition to the effects that bioturbation may have on sediment particle sizes, bioturbation can also mix sediment vertically between adjacent layers. By doing so, bioturbation "blurs" the sedimentary record, thereby reducing our ability to reconstruct Earth's climate history for short periods of time.

In addition to supplying the materials needed for biomineralization, the dissolved products of chemical weathering also provide the nutrients needed for photosynthesis and thus the production of organic matter, i.e., pollen, spores, vegetable matter, and organic soft tissue. In most cases, little of this organic matter is preserved long enough to be considered an important contributor of sediment. Under optimum conditions, however, a larger fraction of this organic matter can be preserved and become important, or even dominant, in sediment production (e.g., in order to form coal). Even when the organic component is only present in very low abundance, though, that component still may provide very important paleoclimatic proxies (as discussed later in this chapter).

3.2 Sediment Description

Geoscientists characterize sediments and sedimentary rocks by describing their composition and texture. Such observations help geoscientists infer the environmental setting and climatic conditions present at the time and place of sediment deposition. One of the most basic and important ways to describe a clastic or bioclastic sedimentary rock is by the size and shape of the sedimentary grains, called the **texture** (Fig. 5). (In contrast, inorganic chemical sediments are not composed of individual grains, so they are described as having a "crystalline texture".) **Grain size**, and thus texture, provide important clues about the energy associated with sediment transport and deposition, which can be directly related to climate in some environments. For example, grains larger than sand are rarely transported by wind, but commonly are moved by fast-moving water or by ice. Thus, larger grains present in sediments and rocks can indicate an ancient environment in which the sediment was transported and deposited under the influence of moving water or ice, and large clasts are common in a glacial deposit known as a **till**. In contrast, very small grains, such as silt and clay, can be transported 100s to 1000s of km by wind or slow-moving water before being deposited. For example, such small sediment grains are often carried by the wind across the Atlantic Ocean from northern Africa, originating from dust storms in the arid regions of the Sahara Desert. Clay and silt size sedimentary particles generally are found in deeper, lower-energy locations in lakes, salt marshes, and in the ocean, whereas beaches predominantly contain sediments of sands, pebbles, or cobbles that have been concentrated by the higher energy of the waves there.

The roundness of sediment grains is often used as a guide to how far sediments traveled before being deposited. Sediment grains with sharper edges were not transported very far from their source region (or were frozen in glacial ice) and as such did not become worn down. Very rounded grains are generally interpreted to have been smoothed during long-distance transport such as by streams and rivers, or in high energy environments such as beaches, due to each particle colliding with other particles (as in a natural rock tumbler). Alternatively, if the grains have experienced a second or third episode of erosion and transport (i.e., are recycled from older sediments), they may have become well-rounded during a prior transport episode, complicating interpretations.

Another important way to describe a clastic sedimentary rock is by the **composition** of the grains it contains. The composition of a clastic sedimentary rock reflects the composition of its parent rock and the extent of modification by chemical weathering. A sedimentary rock containing abundant components that are unstable during chemical weathering strongly suggests that its sediment was produced in an environment dominated by physical weathering (i.e., relatively cool and/or dry). On the other hand, a sedimentary rock containing only components that are resistant to chemical weathering strongly suggests that its sediment was produced in an environment dominated by chemical weathering (i.e., relatively warm and/or humid). The latter composition also can be produced by recycling sediment that was heavily weathered

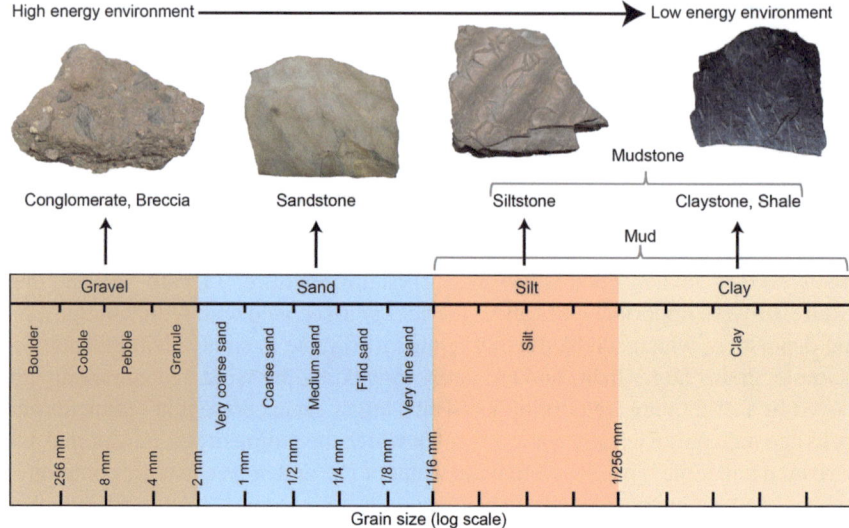

Fig. 5 Grain size chart for clastic sediments and sedimentary rocks showing grain sizes (in millimeters) along the bottom of the scale. Sediment and rock names are given within each of the major four delineations (gravel, sand, silt, and clay), along with a representative rock of each delineation. In general, larger or coarser-grained rocks and sediments are deposited in higher-energy environments (such as fast-moving rivers), whereas smaller or finer-grained rocks are deposited in lower-energy environments (such as lakes, slow-moving streams, and the deep ocean). Rock specimens, left to right: Conglomerate deposited in a braided stream on an alluvial fan. Alluvial fans are coarse sediment deposits that form at the base of a mountain. Braided streams have multiple channels that branch and merge, often related to high velocity flood events. Sandstone with rippled upper surface from fluvial (river) environment. Siltstone with ripples and mudcracks in the troughs of the ripples from the edge of a lake, and a marine black shale with graptolite fossils from a foreland basin deposit. Graptolites are extinct small colonial filter-feeding animals that floated in the ocean water. Foreland basins are elongate low spots parallel and adjacent to mountain chains, which form when rocks bend associated with tectonic plate collision

during an earlier episode of weathering, transport, and deposition, so paleoclimatic interpretations of this sediment composition must be made carefully.

3.3 How Do Sedimentary Rocks Carry Distinctive Climatic Information? Some Examples

Many types of sedimentary rocks carry a paleoclimatic signal of some sort in their bulk characteristics (e.g., their grain sizes, shapes, composition, their patterns of layering, their geographic distribution, or fossil content). Because those bulk characteristics can also be affected by controls other than paleoclimate, distinguishing the paleoclimate signal from the effects of the other controls can be challenging, and a complete discussion of these approaches for all types of sedimentary rocks is beyond the scope of this chapter. Instead, we will focus on a few specific types of sediments in which the paleoclimatic influence is most clearly recorded, in order to illustrate how paleoclimatic information can be interpreted from the bulk characteristics of sedimentary rockss. These rock types and the paleoclimate information that can be extracted from them are summarized in Table 3.

Table 3 Paleoclimatic information for selected sediments and sedimentary rocks

Sediment/ sedimentary rock type	Selected paleoclimatically-useful sediments and sedimentary rocks	Environments of formation	Paleoclimatic information provided
Clastic (detrital)	Paleosols	Terrestrial soil-forming environments	Temperature, precipitation, seasonal variations, vegetation abundance and type, groundwater abundance, and water table position
	Tills	Terrestrial to marine environments overlain by glacial ice	Glacial ice extent; locations of glacial ice
	Rhythmically-laminated sediment	Glacially influenced lakes; low-energy marine basins with low-oxygen waters	Seasonal variations in glacial discharge (ice extent and temperature) or in clastic input to the ocean (seasonal precipitation changes or seasonal wind changes)
	Wind-blown dust	Most recognizable in deep ocean settings, far from land	Weathering conditions in the source area of the dust; wind direction and wind strength
	Marine black shales	Ocean basin with low-oxygen bottom waters	Temperature; current strength in the deep ocean
Chemical	Salt deposits	Settings where evaporation exceeds water inflow; ephemeral lakes in terrestrial deserts, arid ocean coasts, possibly marine basins with restricted circulation	Temperature; arid/drought conditions (evaporation exceeds water supply)
	Speleothems	Caves	Water availability/ precipitation; temperature

(continued)

Table 3 (continued)

Sediment/ sedimentary rock type	Selected paleoclimatically-useful sediments and sedimentary rocks	Environments of formation	Paleoclimatic information provided
Biogenic	Peat → lignite → coal	Terrestrial and coastal marshes and bogs	Temperature, precipitation, and biological productivity
	Calcareous ooze → chalk → limestone	Low and mid-latitude open-ocean settings above the calcite compensation depth (CCD): ocean ridges and rises, seamounts	Carbonate productivity (sea surface temperature, and salinity); CCD level (deep water temperature and CO_2 content); global ice volume, productivity from chemical composition
	Siliceous ooze → chert	Open-ocean settings with high productivity due to upwelling (coastal regions, along the Equator, and at high latitudes); ice edge environments	Sea surface temperature and nutrient supply, productivity, sea ice extent

3.3.1 Climatically-Sensitive Clastic Sediments

Sediments dominated by clastic grains can be deposited in environments that range from the continent (also known as **terrestrial environments**) to the ocean (i.e., **marine environments**). Under appropriate conditions, sediments carrying distinctive paleoclimatic information can be deposited in specific environments in the terrestrial and marine realms. We will briefly explore five of those sediment types here (see Table 3), moving from terrestrial to marine deposits.

Paleosols: Soils form today on the continents by the weathering of the underlying bedrock, and the details of soil formation at any location depend on a variety of factors: the composition of the bedrock, the temperature, the amount of precipitation and nature of water drainage, the amount and type of vegetation, and the amount of time that weathering has been taking place. As a result, soil characteristics are directly affected by temperature and precipitation, which are components of interest in studies of climate and paleoclimate. Under favorable conditions, ancient soils have been preserved in the sedimentary record of a continental region. These ancient soils are termed paleosols, and their detailed characteristics (e.g., composition, development of soil layers or "horizons", the presence of particular chemical precipitates, and the extent and nature of any fossil plant roots; Fig. 6) provide information about the environmental conditions that existed when the original soil was forming.

Till: Till is a deposit of poorly sorted and poorly stratified (layered) sediment, generally occurring in beds a few meters to a few tens of meters thick. In other words, till is sediment that contains a wide range of grain sizes (from clay size to pebbles and cobbles; Fig. 5) and exhibits little or no internal layering. In addition, identifying a deposit as a till requires evidence that the sediment was deposited in contact with

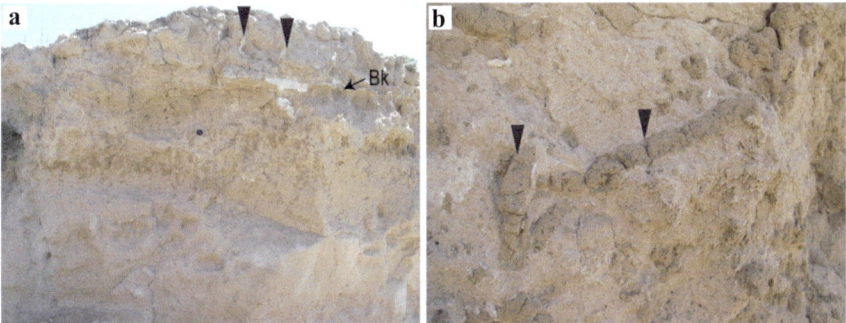

Fig. 6 Oligocene paleosol from the White River Formation in northeast Colorado. **a** An outcrop of paleosol, with black triangles pointing to structures formed in paleosols by plant roots. The black arrow with "Bk" is pointing to a Bk horizon, which is a carbonate-rich fossil soil horizon. Lens cap in the center of the image is 4 cm in width. **b** A close-up image of one of several burrows found in the paleosol, likely formed by beetle larvae. Between the arrows is approximately 5 cm. Images courtesy of Daniel I. Hembree. This figure is excluded from Creative Commons license

glacial ice, generally either underneath or along the edge of the ice. Based on this association between the presence of glacial ice and the deposition of a till, identifying tills in the sedimentary record is an important step toward mapping the times, places, and geographic extents of glaciers and ice sheets through Earth's history. The **cryosphere** is—and has been in the past—an essential component of Earth's climate system, so understanding the nature and extent of the cryosphere in the past is essential for understanding Earth's climate history.

Rhythmically laminated sediment: In contrast to tills, which generally occur as meter-thick beds of poorly sorted and poorly layered sediment, rhythmically laminated sediments consist of relatively thin layers (each less than 1 cm thick) that alternate repeatedly between two or more sediment types. Rhythmically laminated sediments can be deposited in a variety of aquatic environments, ranging from proglacial lakes (i.e., lakes located near the fronts of glaciers) to isolated depressions (i.e., basins) on the ocean floor, generally on a continental margin. These environments share three important characteristics, which are essential for the deposition and preservation of rhythmically laminated sediments: (1) the lake bed or seafloor lies below the level of wave activity, so that sediments deposited there after settling through the water column are undisturbed by physical processes; (2) the type of sediment supplied varies in a regular (and often seasonal) pattern; and (3) few or no bottom-dwelling organisms are present, so that bioturbation does not destroy the thin sediment layers through mixing.

In proglacial lakes, the amount of glacial meltwater entering the lake varies significantly with the season, thereby affecting the sediment being deposited. Most of the sediment enters the lake with the high volumes of meltwater in the spring and summer. The coarser grains settle to the lakebed relatively quickly, forming a thin layer of sand and silt. The finer-grained clays settle much more slowly and dominate the sediment that is deposited during the winter, when meltwater input is minimized by cold winter temperatures. As a result, each pair (or couplet) of a sand/silt layer plus a clay layer records one year of deposition; the name **varve** is applied when each pair of layers in rhythmically laminated sediments records one year of deposition.

In marine settings, a variety of sediment compositions can be layered together, depending on the details of the specific location. In some cases, clastic sand/silt layers are rhythmically layered with clastic clay layers, reflecting seasonal changes in the discharge from nearby rivers that empty into a marine environment. In other cases in which a marine area is influenced by rivers, clastic layers of sand/silt/clay are layered with opal or calcium carbonate produced by marine plankton. These layers reflect seasonal changes in river input (the sand/silt/clay layers) and marine biological productivity (the opal/calcium carbonate) that may be driven by seasonal changes in wind patterns and/or precipitation.

In summary, rhythmically laminated sediments are particularly useful in paleoclimatic studies for several reasons: (1) the details of the layered sediment types provide information about important climate parameters, such as temperature, precipitation, and wind patterns; and (2) these sediments provide one of the few opportunities available in the sediment record to interpret seasonal-scale changes. As a result, the uncommon occurrences of rhythmically laminated sediments attract a level of scientific interest that far exceeds their representation in the total sedimentary record.

Wind-blown clays (dust): This material, extracted from sediments in open-ocean (i.e., pelagic) environments, holds a wealth of information for geoscientists to interpret ancient atmospheric processes. Clay particles and minerals found in the middle of an ocean basin generally have reached that location via wind, and as such their presence, absence, and amount can be used to reconstruct wind patterns and strength through time. Wind-blown dust extracted from seafloor sediment cores in mid-latitude regions is often used to interpret past wind speed (from dust grain size) and shifts in the latitudinal position of wind belts (from changes in the amount and composition of dust). Because this dust originated from soils in a continental source area, the mineral and chemical composition of the dust also can be used to interpret weathering conditions in that source region at the time the dust was eroded by the wind.

Marine black shale: Black shales are sediments dominated by fine-grained clastic material, but with a "high" content of organic carbon (i.e., an organic carbon content greater than ~ 0.5%, which produces the black color). Marine black shales (Fig. 7) record deposition in ancient marine environments characterized by rapid input of organic matter, either from marine bioproductivity or washed-in from adjacent land, and bottom waters with low dissolved oxygen contents. Because of the low dissolved oxygen content, bottom-dwelling organisms were rare to absent as these sediments were deposited, and the lack of bioturbation typically preserves patterns of thin layering within the shale. The marine conditions needed to form a black shale are linked to larger-scale marine processes and paleoclimate at the time of deposition, thereby allowing geoscientists to interpret important paleoclimatic parameters from the occurrences of marine black shales. For example, oxygen-poor deep waters in

Fig. 7 Cretaceous age marine black shale overlain by interbedded gray shale and pelagic limestones near Pueblo, Colorado (~ 95–93 million years ago). A portion of this outcrop was deposited during an oceanic anoxic event (OAE). These marine deposits accumulated in a vast inland sea called the Western Interior Seaway that existed prior to the uplift of the Rocky Mountains some 40 million years later. Image provided by Mark Leckie

the oceans are most likely to be present when global carbon dioxide (CO_2) levels were high and ocean temperatures were warm, because this condition reduces the movement of oxygenated waters from the ocean surface at high latitudes to the ocean depths throughout the rest of the world. When combined with high biological productivity in the surface ocean, deposition and preservation of organic matter in marine sediments is possible. Such conditions were repeatedly present during mid-Cretaceous time when anoxic (no oxygen) to dysoxic (low oxygen) conditions existed in many parts of the world ocean, particularly in the Atlantic Ocean, and adjacent shallow-water inland seas. The most severe anoxic conditions occurred in multiple short-lived events (< 1 million years) called **oceanic anoxic events** (OAEs; Fig. 7), which are recorded as organic-rich black shales.

3.3.2 Climatically-Sensitive Chemical Sediments

Inorganic chemical sediments, which form by the precipitation of solids from components dissolved in groundwater, surface water, or seawater and without the assistance of organisms, are deposited in a much smaller number of settings than clastic sedimentary rocks. Those settings where inorganic chemical sediments do form generally have very specific climatic and/or environmental conditions. As a result, the presence of these inorganic chemical sediments provides geoscientists with well-constrained paleoclimatic and/or paleoenvironmental data. Two examples of climatically sensitive inorganic chemical sediments are summarized in Table 3 and are discussed below.

Salt deposits: Salt deposits are a type of **evaporite**, a sedimentary rock that forms from the evaporation of water. Salt deposits form when saline waters evaporate partially or completely, precipitating the mineral halite, more common known as table salt ($Na^+ + Cl^- = NaCl$), and other evaporite minerals such as gypsum ($Ca^{2+} + SO_4^{2-}$ + water = $CaSO_4 \cdot 2H_2O$). Salt deposits can form from drying of continental lakes, especially lakes that have no outflow streams. Continental lakes are very sensitive to the amounts of water supplied by precipitation and water brought into them via rivers and streams, and water is lost in these systems by evaporation. The modern Great Salt Lake in Utah is an example of such a lake. The lake's volume and extent have varied dramatically over the past century, with periods of excess evaporation marked by a reduction in the Great Salt Lake's extent and increased deposition of evaporite sediments.

Thicker and more laterally extensive salt deposits can be formed in regions with high rates of seawater evaporation. These can include coastal regions, such as regions that occur today around the Persian Gulf, and marine basins that become isolated or cut-off from the larger ocean due to sea level fall or tectonic activity. An example of the latter was the Mediterranean Sea approximately five million years ago, when sea level dropped below the level of the Mediterranean's connection to the Atlantic Ocean through the Straits of Gibraltar. High rates of evaporation in the then-isolated Mediterranean basin produced widespread and thick salt deposits, which are now buried under 100s of meters or more of younger sediment.

Additional paleoclimatic and environmental information can be gathered from salt deposits. Within these deposits, small amounts of the original lake or ocean fluids can become trapped within evaporite deposits, called fluid inclusions (Fig. 8). Geoscientists can extract this ancient water from the evaporite deposit and conduct further analyses on it, such as determining the seawater temperature at the time it was trapped through the use of Mg/Ca ratios (see Sect. 5.1.2).

Speleothems are limestone deposits (i.e., calcium carbonate, $CaCO_3$) precipitated in caves from ions carried in groundwater. The dissolved calcium and carbonate originate from weathering of carbonate rocks and are reprecipitated in the cooler temperatures of the cave. The most common speleothems used in paleoclimate research are stalactites (elongate calcium carbonate structures that hang from the ceiling of a cave) and stalagmites (features that grow from the floor of the cave). Speleothems accumulate calcium carbonate in layers slowly over time (Fig. 9). In some cases, but not all, the speleothem growth displays a banded appearance, with each pair of bands recording one year's growth.

Speleothems record valuable paleoclimatic information in several ways: (1) growth-layer thickness generally records the amount of groundwater available to the cave system, so that thicker and thinner growth layers indicate wetter and drier times, respectively; and (2) the detailed chemical and mineral compositions of a speleothem can be used to reconstruct parameters such as rainfall, temperature, and regional patterns of precipitation. And much like rhythmically laminated clastic sediments, speleothems are one of the few types of paleoclimate records that provide insights into climatic conditions and their variability at seasonal-to-annual timescales.

3.3.3 Climatically-Sensitive Biogenic Sediments

Biogenic sediments and sedimentary rocks provide important information about past environments, since the distribution of organisms' remains (fossils) in such rocks

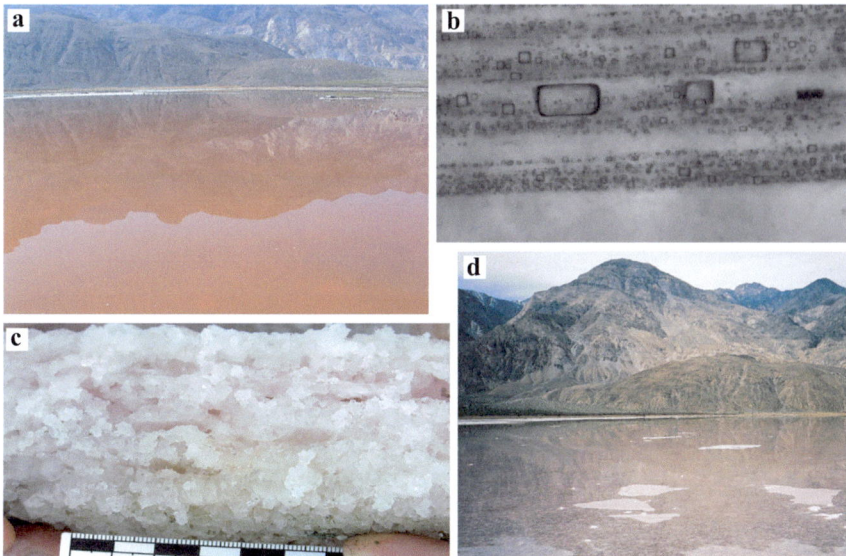

Fig. 8 a A saline lake that appears pink due to the presence of halophiles (bacteria and other microorganisms that can tolerate extreme salinity). **b** A side view through salt deposits that appear pink due to the presence of halophiles trapped within the salt. **c** Photomicrograph of fluid inclusions in salt deposits. The fluid inclusions are concentrated in horizontal bands and have a rectangular to square shape. **d** Salt-rich waters (brine) with large (up to 1 m) rafts of salt crystals forming on top of the waters. Images provided by Tim K. Lowenstein and GSA Today, used with permission of the Geological Society of America.[1] This figure is excluded from Creative Commons license

is influenced by: (1) the types of settings in which the organisms could live; and (2) environmental conditions at that time, such as temperature, salinity, and availability of sunlight and nutrients. In general, biogenic rocks and sediments most commonly are formed in shallow to deep marine, lake, and swamp settings. Biogenic sediments and sedimentary rocks are composed primarily of the remains of ancient life, including mineralized, soft tissue, and molecular remains. These remains come from a variety of organisms, some extinct and some that still exist today. Common mineralized hard parts include calcium carbonate ($CaCO_3$), silica (SiO_2), and phosphate (PO_4). Non-mineralized biogenic material, often referred to as **organic matter**, is composed of a variety of soft tissues and their component organic compounds (e.g., proteins, carbohydrates, and lipids) and rarely is preserved in large quantities. Very small amounts of specific organic molecules are preserved more commonly, though. Some specific organic molecules, called **biomarkers**, are diagnostic to particular organisms, from bacteria of the water column or soils, to single-celled photosynthetic algae, to land plants. The types of biomarkers can provide very useful information about the paleoenvironment of the sediments and changing paleoclimatic conditions. There are numerous types of biogenic sediments and sedimentary rocks, but here, we discuss

---

[1] Lowenstein et al. (2011).

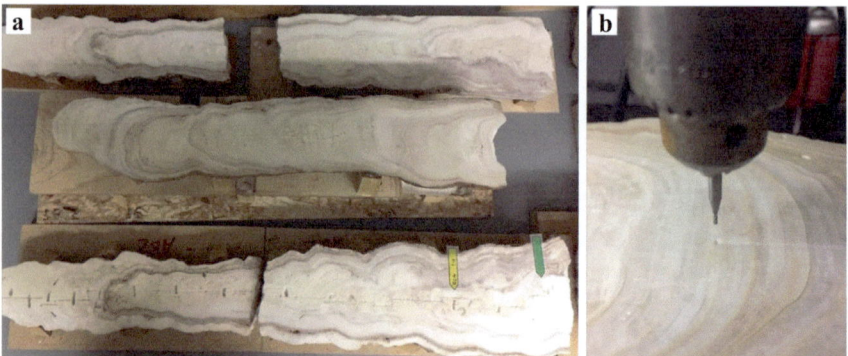

Fig. 9 Speleothems that were collected from a limestone cave, sawed in half, and polished in preparation for geochemical analyses. **a** Sliced speleothems, with the bottom speleothem already sampled for geochemical analyses. **b** A microdrill positioned over a speleothem. The microdrill will recover a small amount of $CaCO_3$ powder from the speleothem, which will be used to measure geochemical data. Note the visible "bands" or layers of the speleothem that are apparent in both images. Images provided by Adriane R. Lam

three main sets and their paleoclimatic significance: coal, calcareous biogenic, and siliceous biogenic sedimentary rocks.

Coal: Coal is a biogenic sedimentary rock that is still used today to heat homes and generate electricity in some regions of the world. Coal contains vast amounts of preserved terrestrial **organic carbon**—that is, carbon sourced from the tissue of once-living land plants. This is why fossil leaves, plant stems, and petrified logs are sometimes found in coal deposits. The most common places for organic carbon-rich sediments to accumulate and be preserved on land are in low-energy environments with abundant vegetation and a shallow-water table with low-oxygen groundwaters, such as bogs, swamps, and wetlands. Tropical and temperate rainforests are environments that have generated thick coal deposits in the geologic past. Salt marshes of large river deltas, or bordering estuaries and coastlines can also produce thin coal deposits.

When abundant plant material first accumulates and is buried, some of the plant material decomposes and the deposit is only slightly compacted. A deposit with these characteristics is called **peat**. You may have heard of peat bogs in Ireland, where the peat has been "mined" to heat homes, or you may have encountered peat as the "peat moss" used as a soil additive in gardening. When peat is further buried and subjected to additional heat and pressure, it is altered into **lignite**, a type of soft, brownish coaly material that includes traces of the original plant material. When lignite is subjected to additional heating and pressure, it is altered into coal. It takes an approximately 10 feet-thick layer of plant organic matter to create a 1 foot-thick bed of coal.

Coal is a very distinctive rock type (Fig. 10) and clearly indicates the occurrence of specific environmental conditions in the past, including conditions that: (1)

Fig. 10 a Image of an outcrop taken along Route 60, east of Charleston, West Virginia. Black arrow is pointing to a Pennsylvanian-aged (~ 323–298 million years ago) coal seam in the rocks. **b** Close-up image of the coal seam, with a rock hammer for scale. Images courtesy of Lynn Fichter. This figure is excluded from Creative Commons license

supported abundant terrestrial plant life (i.e., abundant precipitation, temperate-to-warm climate); and (2) enhanced preservation of that organic material (i.e., low-lying or coastal settings with water containing little-to-no oxygen).

During some periods of time in the geologic past, such as the Late Carboniferous (also called the Pennsylvanian Period, 323.2–298.9 million years ago), vast coal-forming swamps existed on many of the continents. As a product of such extensive swamps, much of the coal mined and used in eastern North America comes from rocks of Pennsylvanian age. The formation of such large amounts of coal, and its storage of organic carbon, led to a huge drawdown of atmospheric carbon dioxide during the Carboniferous. This removal, or **sequestering**, of carbon dioxide from the atmosphere during the Carboniferous reduced the atmospheric greenhouse effect and led to global cooling, which triggered glaciation on the supercontinent of Gondwanaland in the Southern Hemisphere.[2]

Calcareous ooze: Calcareous ooze is a biologically derived sediment that is dominated by the calcium carbonate ($CaCO_3$) shells (i.e., "tests") of microscopic, single-celled, marine plankton. The most common calcium carbonate protists[3] include coccolithophorids and other calcareous nannoplankton, and foraminifera (Fig. 11, upper row). These plankton generally live in open-ocean settings, are most abundant in the tropics and mid-latitudes, and have short life spans (weeks to months). When these organisms die (or are eaten), their tests settle through the water column. Under the proper conditions, these carbonate particles are deposited and preserved on the seafloor, sometimes in very large quantities. Along the edges of the ocean basins, the biogenic input of calcium carbonate is diluted by the rapid input of clastic sediments from land (by wind, rivers, and glaciers), so calcareous ooze is uncommon on continental margins. In the open ocean, more acidic waters that are present below

---

[2] Feulner (2017).

[3] Protists are a biological Kingdom of single-celled organisms.

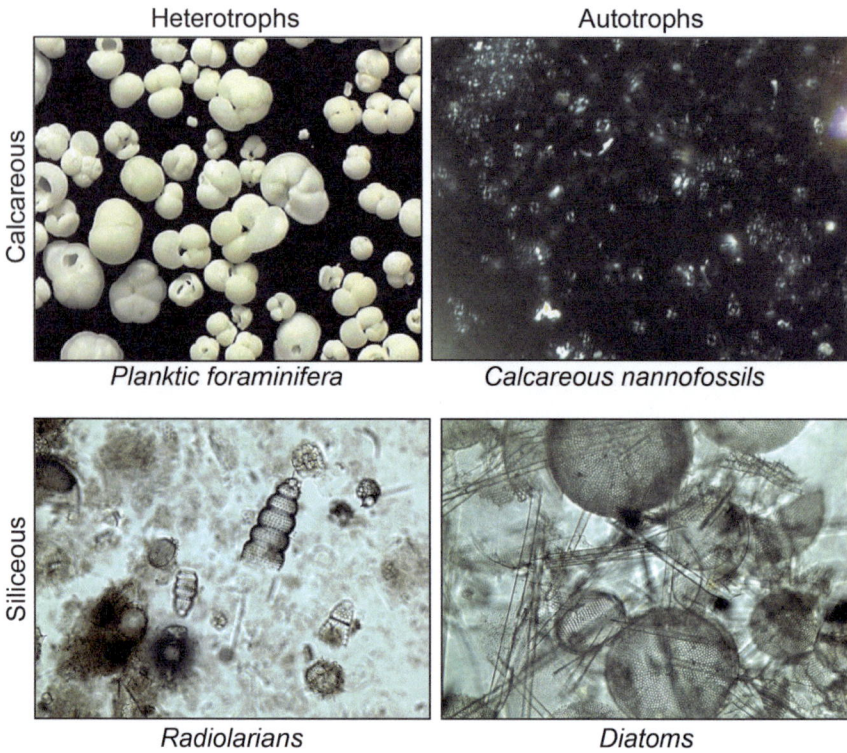

Fig. 11 Different types of single-celled, marine protists with mineralized hard parts, including heterotrophic (organisms that acquire their food) and autotrophic (organisms that create their own food) marine plankton. Upper row: Planktic foraminifera and calcareous nannofossils (coccolithophorids) are common fossils that contribute to calcareous ooze and are commonly preserved in ocean sediments above the CCD. Calcareous nannofossils are also typically indicative of low-nutrient surface waters. Lower row: Radiolarians and diatoms are common fossils that contribute to siliceous ooze. Radiolarians are associated with upwelling in tropical areas, where nutrient-rich waters are brought to the surface ocean. Diatoms are associated with high productivity in cold-water regions. Photos by Mark Leckie, from St John et al. (2021). Reused with permission from John Wiley & Sons.[4] This figure is excluded from Creative Commons license

a specific water depth remove the incoming carbonate by dissolution; this level is called the **calcite compensation depth** (or **CCD**). As a result, calcareous ooze is uncommon on the seafloor at levels deeper than the CCD at the time of deposition. Instead, calcareous ooze accumulates on bathymetric highs (think of these as mountains and large hills on the seafloor) above the CCD (Fig. 12).

Today, the CCD is located at about 5500 m water depth in the North Atlantic Ocean and at about 4500 m water depth in the North Pacific Ocean. The depth of the CCD in these ocean basins has fluctuated significantly through geologic time due to changes in sea level, deep ocean circulation, and the distribution of CO_2 between the

---

[4] St. John et al. (2021).

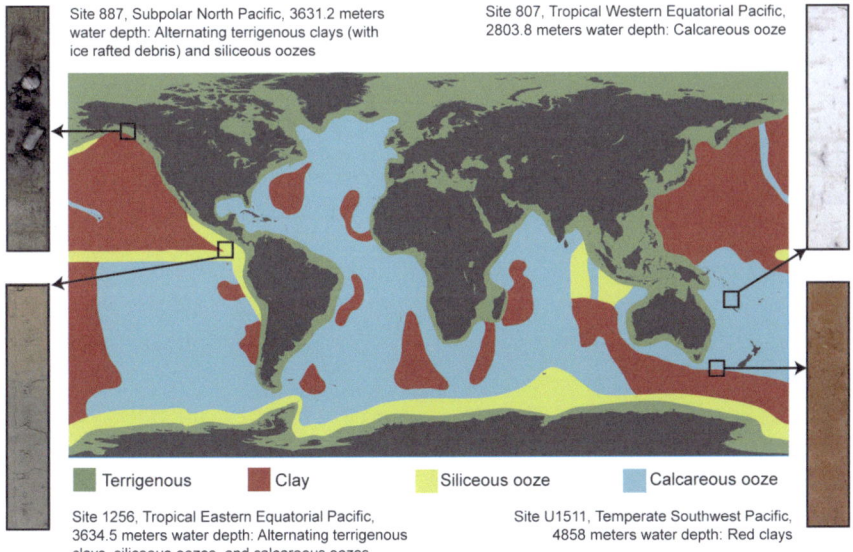

Fig. 12 Major types of seafloor sediments and their modern distribution on the ocean floor, with representative sediment core sections drilled from around the world ocean that reflect the major sediment types. Site 887 in the Gulf of Alaska shows alternating darker terrigenous sediments, including ice-rafted debris (IRD), and lighter diatom ooze; these sediments display orbitally-controlled climate cycles. Site 1256 in the eastern equatorial Pacific shows alternating gray terrigenous sediments and biosiliceous-rich lighter colored sediments, again recording climate cycles. Note that calcareous ooze from Site 807 in the tropical western equatorial Pacific Ocean was recovered from 2803.8 m water depth (1.74 miles), well above the CCD, whereas red clays from Site U1511 in the temperate southwest Pacific were recovered from 4858 m water depth (3.01 miles), well below the CCD. The deposition of sediments at Site U1511 below the CCD in the Pacific Ocean explains the lack of carbonate preserved in these sediments. Core images courtesy of the International Ocean Discovery Program[5]

ocean and the atmosphere, as will be discussed more in Chap. 8. Thus, recovery of calcareous sediments in an ocean sediment core indicates times when that site was located above the CCD and gives geoscientists an idea of how ocean chemistry has changed through time. When calcareous ooze is buried and compacted (subjected to heat and pressure), it is transformed into **chalk** (yes, just like the stuff we use to write on chalkboards! Fig. 13a). With further compaction and lithification, the chalk becomes **limestone** (Fig. 13b).

Siliceous ooze: Similar to calcareous ooze, siliceous ooze is a biogenic sediment that accumulates in the ocean and contains abundant silica (SiO_2) tests of marine plankton microfossils. Important silica-secreting protists include the diatoms and radiolarians (Fig. 11, lower row). Like the plankton that secrete calcium carbonate tests, the siliceous marine plankton also have short life spans, and their tests can accumulate

---

[5] Core images obtained from http://iodp.tamu.edu/database/coreimages.html. Public domain.

Fig. 13 a Image of the White Cliffs of Dover in England. The cliffs are made entirely of chalk, which forms from the lithification of calcareous ooze that is composed of calcareous nannofossil shells or tests. The nannofossil tests collected on the seafloor over millions of years approximately 70 million years ago during the Cretaceous Period, when sea level was much higher compared to today. This area of ancient seafloor was then tectonically uplifted above sea level. The pristine preservation of vast amounts of calcium carbonate indicates the nannofossils were deposited well above the Cretaceous calcite compensation depth (CCD). Image A courtesy of Natalie Freeman. Image A is excluded from Creative Commons license. **b** Quarry wall showing white chalk (once calcareous ooze) with numerous horizons of gray chert nodules and discontinuous beds of chert from the type Maastrichtian (uppermost Cretaceous, ~ 70 Ma) in southern Limburg, the Netherlands. Image b provided by Mark Leckie

in great quantities under the proper conditions of high production and enhanced preservation. Silica-secreting plankton thrive in regions of high productivity, where deeper, nutrient-rich waters are brought to the sunlit surface of the ocean through wind- and current-driven processes called **upwelling.** As a result, abundant siliceous plankton, and the resulting siliceous ooze, generally indicate areas of upwelling-driven high productivity (Fig. 12). The enhanced preservation of siliceous plankton shells beneath upwelling areas is due to the increased production of silica-shelled microorganisms in these nutrient-rich surface waters in combination with the very rapid burial of their shells when they die and settle to the seafloor.

A noteworthy point about diatoms, which have important paleoenvironmental and paleoclimatic implications, is that some diatom species are—and have been in the past—restricted to the surface ocean immediately beneath sea ice. Reconstructing the past history of sea ice extent is very difficult, because sea ice leaves behind little or no sedimentary record. As a result, identifying sea ice diatoms in the fossil record is a very useful contribution to understanding Earth's climate history more fully.

The lithification of siliceous oozes is one process that forms the sedimentary rock **chert** (Fig. 13b). When it forms in this way chert is indicative of an ancient open marine environment with high productivity and abundant siliceous plankton.

However, chert can also form through other processes that are not sensitive to paleoclimate, so one job of the geoscientist is to distinguish chert formed from siliceous oozes from chert formed by other processes.

4 How Do We Know the Age of Sediments?

As is true for any type of history, telling time in the sediment record is critical to reconstructing ancient climate events and the Earth system's response to those events. Therefore, much work has been done to assign ages to sediments and sedimentary rocks, and that work continues today. Geoscientists use two main approaches to assign ages to the rock and sediment record: relative age dating and absolute age dating. **Relative dating** determines the relative order of geologic events. In other words, which rock or sediment layer is older and which is younger, but without determining numerical ages. **Absolute dating** uses techniques to assign a numerical value to the age of sediments and rocks (e.g., an age of 4 million years ago versus an age of 400 million years ago, although there is an uncertainty in any numerical age).

Determining both relative and numerical ages of sediments and sedimentary rocks falls within the realm of **stratigraphy**, which is the study of the ages of those rocks and their relationships to each other and to the geologic time scale. Below are four common relative and absolute dating methods that geoscientists apply in this work: biostratigraphy, radiometric dating, magnetostratigraphy, and chemostratigraphy. Biostratigraphy was first used to date sediments several centuries ago and provides relative ages. Magnetostratigraphy was developed later, and chemostratigraphy is one of the newer relative dating techniques that has grown in use with refinement of instrumentation. In contrast, radiometric dating is the only approach that directly provides absolute ages. Absolute ages can then be integrated with (i.e., **correlated to**) sedimentary records of other types of stratigraphic data from other sedimentary rocks/sediments regionally to globally. Table 4 provides a summary of common approaches to age determination for geologic materials.

Our discussion of sediment dating will begin by considering biostratigraphy solely as a tool for relative dating, as this was the first method developed to determine the ages of rocks and sediments. We will then explore radiometric dating as the tool for absolute dating and conclude by examining how absolute dates are integrated with relative dating techniques such as biostratigraphy, magnetostratigraphy, and chemostratigraphy.

4.1 Biostratigraphy

Biostratigraphy is the use of fossils to determine the relative ages and relationships of rocks to each other and to the geologic timescale. Biostratigraphy is possible because the evolution and extinction of species through Earth history has produced

Table 4 Summary of common approaches to age determination of geologic materials

Category	Dating methods	Comments
"Absolute" ages, providing numerical dates	Radiometric dating	Based on the half-life of radioactive isotopes (e.g., ^{14}C, ^{40}K, ^{234}U, ^{238}U)
Relative ages based on pattern recognition and identification of regional or global events	Special horizons (e.g., ash layer, iridium layer) in the layers that mark one-time events (e.g., volcanic eruptions, meteorite impact)	Useful for correlating a time-specific horizon across a geographic region; some can also be radiometrically-dated
	Magnetostratigraphy	Paleomagnetic reversal pattern; "bar code" of time with global reversal events
	Biostratigraphy	Patterns of life (fossils); different forms of life lived at different times; extinction events
	Geochemical patterns	Useful for correlating patterns of change across a geographic region. For example, oxygen isotope, carbon isotope, and strontium isotope patterns in marine sediments

an ordered and recognizable pattern of fossil species appearance, occurrence and distribution, and disappearance in the rock record. The existence of this ordered and recognizable pattern is summarized as the **principle of faunal succession** and has been established through centuries of observations and correlations. As an example, we know Species A appeared in the rock record before Species B, and therefore, Species A is older. Remember, however, that these reliable patterns of first and last appearances only provide a relative age rather than a numerical age.

Only some fossil species are useful for biostratigraphy. To be useful, a species must be relatively common, distinctive (easy to identify), widely distributed, and have a short stratigraphic range. The most common fossils are those produced by relatively small and simple organisms with mineralized body parts (i.e., relatively small invertebrates with shells) and especially those that lived in the ocean. A wide geographic distribution in the rock record is controlled, at least in part, by the geographic distribution of the living organisms that produced that fossil, i.e., the **biogeography** of that species when it was alive. Very few species, today or in the past, can thrive everywhere on Earth. Instead, each species tends to be restricted to climate zones with the environmental conditions to which they are best adapted (i.e., temperature, salinity, seasonality, and resource availability). Because of this biogeographic influence on their distribution, some fossils provide both age and paleoenvironmental/paleoclimatic information.

Biostratigraphy largely relies on identifying the first occurrence (the oldest, or evolutionary first appearance) of a species and the last occurrence (the youngest appearance, or extinction) of a species. Because different types of organisms have evolved and gone extinct through time, different groups of fossils are biostratigraphically useful during different intervals of Earth history. For example, in the Mesozoic Era, while dinosaurs and their relatives dominated the land, sea, and air, the **ammonites** (cousins of octopus and squid) had rapid evolutionary rates and widespread distributions in the ocean. As a result, ammonites are ideal for Mesozoic biostratigraphy (Fig. 14). Prior to the end-Permian mass extinction (~ 252 Ma), trilobites were widespread and had high evolutionary rates making them ideal for Paleozoic biostratigraphy.

4.2 Radiometric Dating

Radiometric dating is an absolute dating technique that relies on the radioactive decay of an unstable **isotope**. Radioactive decay occurs when an unstable isotope (the "parent") spontaneously transforms into another isotope (the "daughter"). In situations not subject to other complications, geoscientists can measure how much of the parent, or original isotope, and the stable product, or daughter, are present in a sample and then calculate an age based on a known rate of decay. Every radioactive parent isotope decays at a unique rate, meaning different isotope pairs are used for dating different time intervals.

Most radiometric dating techniques are most accurately applied to igneous and metamorphic rocks, and the minerals they contain, where the calculated age begins based on a time within the cooling and crystallization history of magma-derived material (igneous rock) or recrystallization (metamorphic rock). When lava flows occur above and/or below sedimentary layers, the radiometric date(s) determined for those igneous materials can be used to assign an absolute age, or a range of absolute ages, to the associated sediments. Igneous and metamorphic rocks can be used to *bracket the absolute age of sediments* deposited on top of these rocks, or intruded by a later igneous rock. As another example, distinct layers of volcanic glass ("ash"), produced during a volcanic eruption, can be found interbedded within terrestrial sediments on land or seafloor sediments in certain parts of the ocean, such as offshore the volcanic Aleutian Islands in the Gulf of Alaska. Some volcanic ashfall deposits contain dateable minerals in addition to shards of glass. Imagine that an interval of a core of seafloor sediment from the Gulf of Alaska contains two distinct ash layers (Fig. 15a). If the grains from the lower ash layer are radiometrically-dated at 0.88 million years and the grains from the overlying ash layer are radiometrically-dated at 0.75 million years, then the sediments that lie between these two ash layers can be assigned an age range of 880,000–750,000 years old.

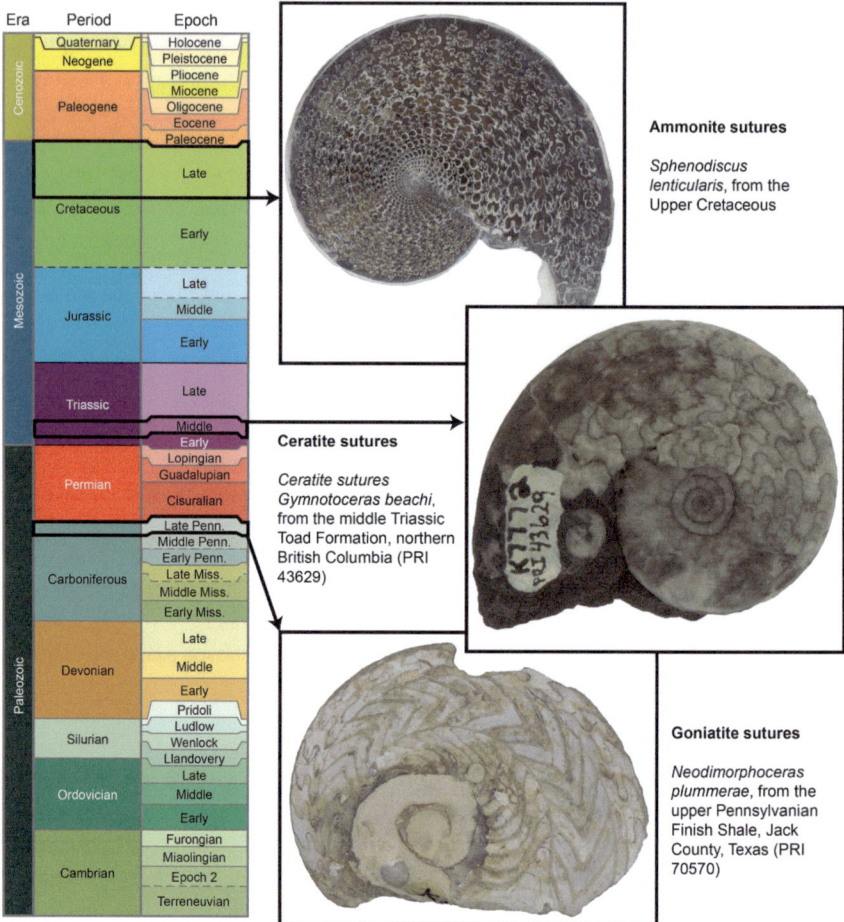

Fig. 14 Depiction of the relative geologic time scale; the Cambrian is older than the Ordovician, is older than the Silurian and Devonian. No numbers—or absolute ages—are shown, just a succession of Eras, Periods, and Epochs defined by their distinctive organisms, from older to younger. Three ammonite suture patterns are labeled where they occur in geologic time. The less complex sutures, called goniatite sutures, occur in the Paleozoic (such as the "V" like sutures on *Neodimorphoceras plummerae*), with a bit more complex ceratite sutures appearing in the early Mesozoic (wave-like sutures of *Gymnotoceras beachi*). The most complex ammonite sutures are found in the Late Cretaceous, as exhibited by the highly dendritic sutures of *Sphenodiscus lenticularis*. Photographs of ammonites provided courtesy of the Paleontological Research Institution, Ithaca, New York. This figure is excluded from Creative Commons license

The Sedimentary Record of Past Climate Change 215

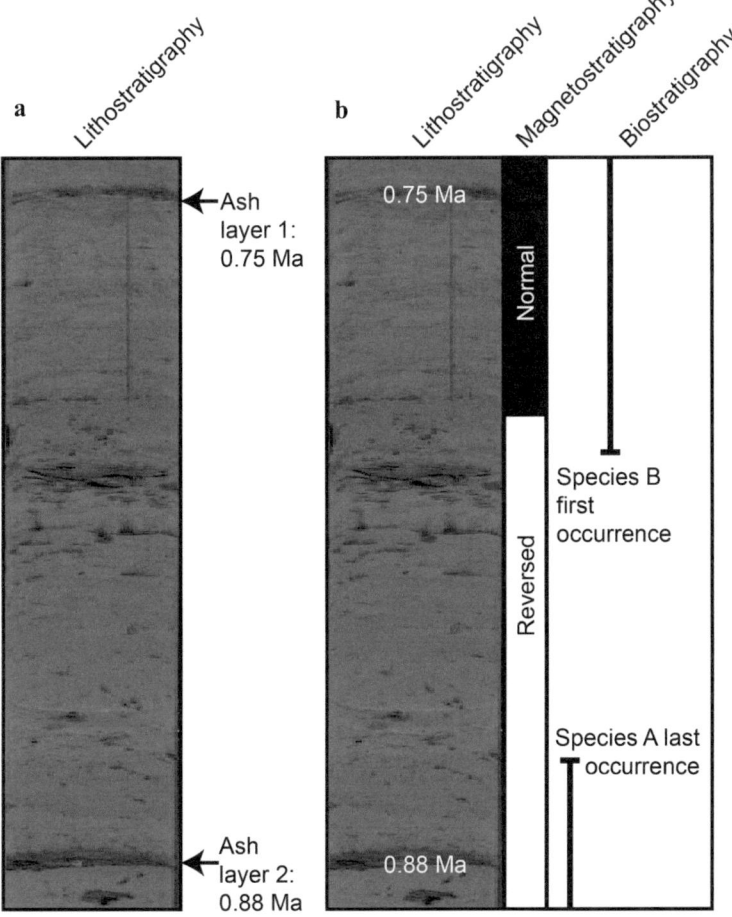

Fig. 15 a Core photos showing radiometric dating results for two volcanic ash layers in a deep-sea sediment core, with radiometric ages indicated beside each ash layer. **b** Integration of radiometric age dating with two relative age dating techniques, magnetostratigraphy and biostratigraphy, for the same core. On the magnetostratigraphy column, the black box indicates a "normal" period (North Pole in Northern Hemisphere and South Pole in the Southern Hemisphere), and the white box indicates a "reversed" period (North Pole in the Southern Hemisphere and South Pole in the Nouthern Hemisphere). The biostratigraphy column indicates the occurrences of two species of microfossils. Species A goes extinct right above the lowermost/oldest ash layer, whereas Species B evolves below the uppermost/youngest ash layer

4.3 Magnetostratigraphy

To understand how magnetostratigraphy is helpful for determining the age of sediments or sedimentary rocks, we must first have an understanding of the Earth's magnetic field. You are likely familiar with the fact that Earth has two magnetic poles, which today are located in the general vicinities of—but not at—the rotational axes of the geographic North and South Poles. These magnetic poles are part of the magnetic field, the source of which is primarily generated by fluid motions of the liquid, iron-rich outer core. Over the course of Earth's vast geologic history, the north and south magnetic poles have rapidly switched (flipped) locations! When the poles are switched, this is called **reverse polarity**; when the poles are like they are today, this is called **normal polarity**. These magnetic polarity reversals are thought to be caused by changes in the fluid convection in the outer core. Magnetic reversals are very common throughout Earth's geologic history. An individual reversal event takes place geologically very rapidly (thousands of years), and such events have occurred irregularly throughout Earth's history.

When iron-rich magnetic minerals, such as magnetite, are transported to the ocean and settle on the seafloor, those magnetic minerals will align with Earth's magnetic field. As the minerals are covered by other sediments and buried, they become "locked in place" and thus preserve a record of the magnetic polarity when the minerals were deposited. Additionally, a record of Earth's magnetic polarity is preserved in magnetic minerals that form from the cooling of magma when it is erupted at the Earth's surface as lava. Lava is erupted at many places on Earth's surface, including the mid-ocean ridges that mark the edges of tectonic plates where lava rises to the Earth's surface and creates new oceanic crust. When these igneous rocks cool, the magnetic minerals align with the Earth's polarity. Therefore, the basalts, volcanic rocks that make up oceanic crust, which are formed from cooling of lava, preserve a "barcode" of sorts that records the magnetic polarity reversals throughout Earth's history.

Radiometric dating of these volcanic rocks has established globally-applied ages for the magnetic polarity reversals. The dated history of magnetic changes developed in these volcanic rocks can be correlated to magnetic variations recorded in sediments and sedimentary rocks. This allows absolute ages to be assigned to sediments, even when those sediments cannot be dated directly by radiometric methods. Using this method to date sediments and rocks is called **magnetostratigraphy**.

4.4 Chemostratigraphy

In contrast to radiometric dating, which uses radioactive isotopes, **chemostratigraphy** uses variations in the stable isotope composition of sediments to determine their ages and age relationships. Stable isotopes do not transition spontaneously to another isotope, so the abundance of a stable isotope in a part of the sediment record does not change through time nor following its deposition. Stable isotopes of oxygen, carbon, nitrogen, hydrogen, and strontium, among others, frequently are used in chemostratigraphy. In each case, one or more physical, chemical, or biological processes act to separate a heavier stable isotope from a lighter stable isotope. This separation process is called **fractionation**, and the resulting isotopic composition is recorded in the composition of a sediment component (e.g., microfossil carbonate or organic matter). When the effects of the fractionation process influence a large area (e.g., an ocean basin or a continental region), then the resulting change in isotopic composition is recorded at the same time in sediments over that area. That distinctive change in sediment stable isotope composition can be used to link, or **correlate**, sediment records of the same age. One example of this approach is the correlation of stable oxygen-isotope records across the global ocean. In the deep sea, the stable oxygen isotopic composition of the seawater is recorded in the shells of bottom-dwelling (benthic) foraminifera. Fractionation by physical processes related to the global water cycle and to changes in the global cryosphere directly influences the oxygen-isotope composition of ocean deepwater and therefore the composition of these microfossil shells, so a global template of climate changes through time has been developed. When a new marine sediment record is recovered through coring, or a new outcrop of sedimentary rock is discovered, the new oxygen-isotope stratigraphy can be correlated to patterns observed in the previously established global deep-sea oxygen-isotope record, thereby establishing ages in the new record.

Chemostratigraphy (like biostratigraphy and magnetostratigraphy) is a relative dating technique because it does not supply absolute ages. However, absolute ages can be added if the relative dating data is integrated with radiometric measurements, either directly or indirectly as will be discussed briefly below.

4.5 Integrating Radiometric Ages into Bio-, Magneto-, and Chemostratigraphy

When materials that can be dated radiometrically co-occur with (or can otherwise be correlated to) biostratigraphic first or last occurrences, magnetostratigraphic polarity changes, or distinctive chemostratigraphic compositional changes, then the radiometrically determined absolute ages can be assigned to those biostratigraphic, magnetostratigraphic, or chemostratigraphic changes. Once such an age assignment has been confirmed by additional data, then the age assigned to that bio-, magneto-, or chemostratigraphic marker can also be used to place an age on sediments with that marker in other locations, even if those other locations do not contain materials appropriate for radiometric age determinations.

As an example, let us revisit the sediment core from the Gulf of Alaska mentioned briefly in Sect. 4.2. Re-examine Fig. 15a and consider the additional information in Fig. 15b. The presence of two volcanic ash layers in a portion of that core allowed the intervening sediments to be dated at 0.88–0.75 million years old, based on the radiometric ages of the ash layers. If micropaleontologists identify the youngest (i.e., last) occurrence of calcareous microfossil Species A just above the top of the lower ash layer, then we know that the last occurrence of Species A also was between 0.88 and 0.75 million years ago—and probably only slightly younger than 0.88 million years, if we assume that the sediments between these two ash layers accumulated at a constant rate. This age for the last occurrence of Species A can then be used to assign the same age (or age range) to the sediment intervals marked by the last occurrence of Species A at other locations.

At the same location, if the magnetostratigraphy down the core shows the first change from normal polarity to reversed polarity just below the base of the upper volcanic ash, then we also know that the first magnetic reversal occurred sometime between 0.88 and 0.75 million years—and probably only slightly older than 0.75 million years. As was true for the age assigned to the youngest occurrence of Species A, this age for the youngest reversal of Earth's magnetic field can then be used to assign the same age to the sediment intervals marked by the equivalent magnetic reversal at other locations.

As you may realize from the example above, assigning an age of 0.88–0.75 million years to either the last occurrence of Species A or to the youngest paleomagnetic reversal is not very precise. As a result, the work of biostratigraphers, magnetostratigraphers, and chemostratigraphers is never finished. These geoscientists constantly gather new data to refine the absolute age assignments for the biostratigraphic first and last occurrences of fossils, the paleomagnetic reversals, and the distinctive changes in isotopic compositions.

5 What Are Some Important Proxy Records of Ancient Climate?

A very large number of climate proxies have been developed based on components found in sediments, and skillful geoscientists are constantly searching for new proxies to better explore Earth's climate history. In this section, we will focus on a few of the proxies used to reconstruct important parts of Earth's climate system: temperature, precipitation, atmospheric CO_2 content, global and local ice volume/extent, and sea level. These climate variables and the proxies used to reconstruct them are summarized in Table 5. For some of the proxies described in this section, we also include examples of how these data aided scientific understanding of past climate conditions during different times. For example, Cenozoic long-term trends in global average temperature, regional precipitation, CO_2, and ice volume are explored in Sects. 5.1, 5.2, 5.3 and 5.4, respectively, and the Miocene Climatic Optimum and the Middle Miocene Climate Transition are explored in Sect. 5.1. These mini-case studies will be augmented with longer explorations of the application of proxies to reconstruct past climate in subsequent chapters.

This section will not consider in detail the paleoclimatic contributions arising from studies of the environmental preferences of fossils and fossil assemblages, although such studies have contributed significantly to our knowledge of Earth's climate history. Once the environmental preferences of a fossil taxon or a fossil assemblage have been identified by such studies, then the geographic movements of that environmentally sensitive taxon or assemblage can serve as the basis for a paleoclimatic interpretation. For example, the poleward expansion of a marine

The Sedimentary Record of Past Climate Change

Table 5 Selected paleoclimate variables and proxies used to reconstruct them

Selected paleoclimate variables	Selected proxies (surrogates for climate or environmental variables) in sediments/sedimentary rocks	Comments
Temperature	Stable oxygen isotopes ($\delta^{18}O$)	See Box 1. Can be measured from many types of carbonate sediment (e.g., speleothems and calcareous ooze)
	Magnesium-to-calcium ratios (Mg/Ca)	Measured from calcareous microfossils in lake and ocean sediments
	Biomarkers (molecular fossils)	Measured on specific organic compounds. Marine examples discussed here; terrestrial plant-based biomarkers also available from lake sediments and peats
Precipitation	Paleosol calcic horizon characteristics	Limited to paleosols formed under arid and semi-arid conditions
	Plant leaf wax isotopes	A different type and use of biomarkers
Carbon dioxide (CO_2)	Plant stomata	See Box 2
	Boron (B) isotopes	Measured on marine biogenic carbonate (microfossils)
Ice sheet size	Marine stable oxygen isotopes ($\delta^{18}O$)	Difficult to calculate exact ice volumes because of uncertainties in the isotopic compositions of past ice sheets and oceans. Provides a globally integrated value; does not provide regional details
	Tills and other glacial deposits	Used to assess local/regional details of ice sheet history. Detailed age dating often difficult or impossible. Record is often fragmentary due to erosion by later ice advances
Sea level	Marine stable oxygen isotope record	Significant uncertainties caused by unknown isotopic compositions of past ice sheets and oceans
	Coastal sedimentary sequences	Important to separate global and local changes. Can provide reasonable estimates of the true magnitude of global sea level changes

planktic species that has been interpreted to require subtropical temperatures can be used to infer poleward warming. Similarly, the migration of a marine plankton species that required lower salinity surface waters into a region previously inhabited by a higher-salinity species can be used to infer an increase in precipitation or a reduction in evaporation (to lower the sea surface salinity).

5.1 Sediment Proxies to Reconstruct Temperature

In the context of climate and paleoclimate, "temperature" can have many meanings: the temperature of the atmosphere, the surface ocean, or the deep ocean; the yearly average temperature; or the summer or winter average temperature, to name a few examples. Because the components of Earth's climate system are interconnected, and because some aspect of "temperature" is often the first characteristic that comes to mind when "climate change" is considered, it is important to reconstruct as many of these temperature parameters as possible. As a result, many sedimentary proxies have been developed to reconstruct a wide range of Earth's "temperatures". Here we will focus on three proxies: the stable oxygen isotopic composition of carbonates, the magnesium-to-calcium (Mg/Ca) ratio of biogenic carbonates, and molecular recorders (i.e., "molecular fossils" or biomarkers) in certain organic compounds.

5.1.1 Stable Oxygen Isotopic Composition of Carbonates

You were introduced to the stable isotopes of oxygen in Chap. 4, including the use of the stable oxygen isotopic composition of the water molecules in ice cores as a proxy for the air temperature at the site where the water vapor condensed and was deposited. The isotopic composition of the water varies as temperatures change because of differences in the likelihood of evaporation and condensation for heavier versus lighter water molecules at warmer temperatures versus cooler temperatures. This difference in behavior during evaporation and precipitation, due to differences in isotopic weights (mass), is an example of **fractionation**. Within sediments and sedimentary rocks, we rarely can sample the water that was present at the time of their deposition. We are fortunate, however, that the stable oxygen isotopic composition of that water influenced the stable oxygen isotopic composition of several forms of calcium carbonate that were precipitated from that water, including the shells of fossils and microfossils. As a result, the stable oxygen isotopic composition of the calcium carbonate provides information about the isotopic composition of the water from which the carbonate formed. An overview of stable oxygen isotopes in land and marine carbonates is described in Box 1.

> **Box 1 Generalizations and complexities of temperature reconstruction from stable oxygen isotopes**
>
> The generalized relationships between the stable oxygen isotopic composition measured from common paleoclimate materials (ice and carbonate) and temperature are summarized in Fig. 16. All stable oxygen-isotope compositions are measured as ratios of $^{18}O/^{16}O$ and are denoted as $\delta^{18}O$ after they

are calculated with respect to the isotopic ratios of internationally distributed laboratory standards.[6]

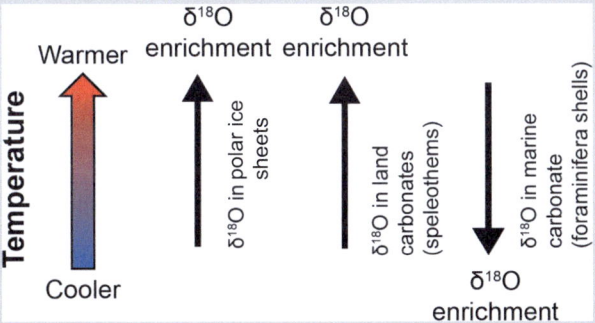

Fig. 16 Generalized relationship between temperature and stable oxygen isotopes of three common materials used for paleotemperature reconstructions: glacial ice in polar regions (see Chap. 4 for discussion of oxygen isotopes in polar vs tropical ice), speleothems (i.e., land-based inorganic carbonate), and foraminifera microfossil shells (marine-based biogenic carbonate)

The relationship between the oxygen isotopic composition of carbonate and its interpreted temperature depends on whether the carbonate formed on land, so that it was formed from water that had undergone evaporation and precipitation, or the carbonate formed in the ocean from the water that was left behind during evaporation. Speleothems are one example of carbonates that form on land. In the ideal case of speleothem formation, as is generally true for polar ice cores, the carbonate is expected to become more enriched in the heavy stable isotope (^{18}O) as the surface air temperature at the site warms, due to increased precipitation of water molecules containing ^{18}O. However, many speleothem records do not exhibit this straightforward relationship because other physical, chemical, and biological processes take place in the soil and rocks overlying a cave that influence the isotopic composition of the water as it migrates from the ground surface into the cave system. In addition, details of the chemical reaction of carbonate precipitation within the cave reduce the abundance of ^{18}O within the carbonate as the air temperature of the cave increases, thereby working against the isotopic change driven by warming at

---

[6] One of the most common laboratory standards used for analysis of stable isotopes of water (e.g., rainwater or melted glacial ice) is the Vienna Standard Mean Ocean Water (VSMOW), which is composed of distilled deep offshore ocean water. A common laboratory standard used for analysis of stable isotopes of carbonate (e.g., speleothems and marine microfossil shells) is Vienna Pee Dee Belemnite (VPDB). See Copelen, T. (1996). USGS. https://wwwrcamnl.wr.usgs.gov/isoig/res/guide.html.

the ground surface. As a result, the stable oxygen isotopic composition of a speleothem can provide a record of temperature at the site, but interpreting the causes and meaning of that temperature record can be challenging.[7]

The marine carbonates whose stable oxygen isotopic compositions are considered here are the products of marine plankton, especially calcareous nannoplankton and planktic foraminifera, as well as seafloor dwelling benthic foraminifera. For more than 70 years, geoscientists have known that biogenically-produced carbonate is enriched slightly in ^{18}O relative to the abundance of ^{18}O in the seawater from which an organism precipitates its shell, and that the amount of ^{18}O enrichment increases as the water temperature becomes cooler. In a simple system, where water temperature is the only variable that changes, this relationship means that one could measure the oxygen isotopic composition of biogenic carbonates down a sediment core and use those compositions to reconstruct a history of water temperature changes at that location. As you probably realize, such simple systems are uncommon in nature, and indeed a number of other factors have been shown to also affect the stable oxygen isotopic composition of seawater, and therefore marine biogenic carbonate. These other factors include: variations in seawater salinity, variations due to ecological and physiological differences among plankton species, and variations in the isotopic composition of the ocean water itself. With careful selection of the materials used and an expanded understanding of where and how marine species lived, the effects of the first three factors can be minimized or corrected for. However, the effects of changes in the isotopic composition of the ocean overwhelm the changes due to water temperature shifts during times in Earth history when global ice volume is undergoing major changes (e.g., changes from an ice-free Earth to one with ice at the poles, and vice versa). This is because it is difficult to quantify temperature versus ice volume effect using stable isotopes alone. Therefore, other paleotemperature proxies are used in parallel with stable isotopes in order to isolate the temperature effect from the ice volume effect. Such complementary paleotemperature proxies include Mg/Ca ratios of the biogenic carbonate shells and the detailed chemistry of organic compounds (molecular fossils or biomarkers) preserved in the sediments. These additional paleotemperature proxies are described in Sects. 5.1.2 and 5.1.3, respectively.

Figure 17 illustrates the results of a widely-cited study of paleotemperatures during the past 66 million years, based on the stable oxygen isotopic composition of marine carbonate. The curve records the average isotopic composition of benthic (bottom-dwelling) foraminifers from 14 deepwater, open-ocean sites throughout the world ocean, and the proxy-interpreted mean difference in deepwater temperature of the oceans at that time compared to the average temperature of the deep ocean

---

[7] For a detailed review of the use of oxygen isotopes in cave formations see Lachniet (2009).

today. The temperature scale shown on Fig. 17 clearly indicates that the deep ocean was warmer—and often was significantly warmer—than today for much of the last 66 million years. This record also shows a long-term trend of cooling that took place from ~ 50 Ma to 34 Ma. In other words, a long period of deepwater cooling (likely indicative of cooling of the surface ocean and of the overlying atmosphere) preceded the development of large ice sheets on Antarctica at 34 Ma (see bar graph in Fig. 17). The increased thickness of the line in the younger part of the curve is the result of large changes in the isotopic composition of the oceans as global ice volume fluctuated in both Antarctica and the Northern Hemisphere.

5.1.2 Elemental Ratios: Mg/Ca

Another way in which geoscientists can estimate ocean temperatures in the geologic past is by measuring the ratio of magnesium (Mg) to calcium (Ca; Mg/Ca) in the calcite tests of planktic or benthic foraminifera. When a foraminifera secretes the carbonate of its test, it incorporates some Mg from the surrounding seawater into the calcite crystal structure of the test. The amount of Mg that is incorporated is controlled by the temperature of the seawater, with more Mg incorporated as the water temperature increases. In general, the incorporation of Mg is affected by different environmental variables than oxygen-isotope compositions.

An example of the application of Mg/Ca ratios for temperature reconstruction is shown in Fig. 18, which records two major climate events that occurred during the Miocene. The older climatic event was the Miocene Climate Optimum (MCO; also shown in Fig. 17) which occurred from approximately 16.9 to 14.7 million years ago, and the second was the Middle Miocene Climate Transition (MMCT; also shown in Fig. 17) which occurred from 14.7 to 13.8 million years ago. The MCO occurred when atmospheric CO_2 levels increased and the Earth became about 3 °C warmer compared to today. The MCO may have been facilitated, if not triggered, by rapid flood basalt volcanism in the Pacific Northwest of the United States (See Columbia River Flood Basalt in Fig. 17) that released large volumes of CO_2 into the atmosphere. The Yellowstone hotspot plume also erupted close to this time (~ 16.7–15.9 million years ago) based on radiometric dating of volcanic ash beds between basalt lava flows. Later, the Earth began to cool, possibly due to changes in the amount of incoming solar radiation (insolation). In response, Antarctic ice sheets rapidly grew in volume, culminating in the MMCT. The MMCT also coincides with cooling of surface and deep ocean waters. This event was a major climate step that brought the Earth into its more modern-day climate configuration, although both the East and West Antarctic Ice Sheets were dynamic and responsive to warmer-than-present events throughout the Neogene and Quaternary, such as the MCO, mid-Pliocene warmth, and super-interglacials of the Pleistocene.

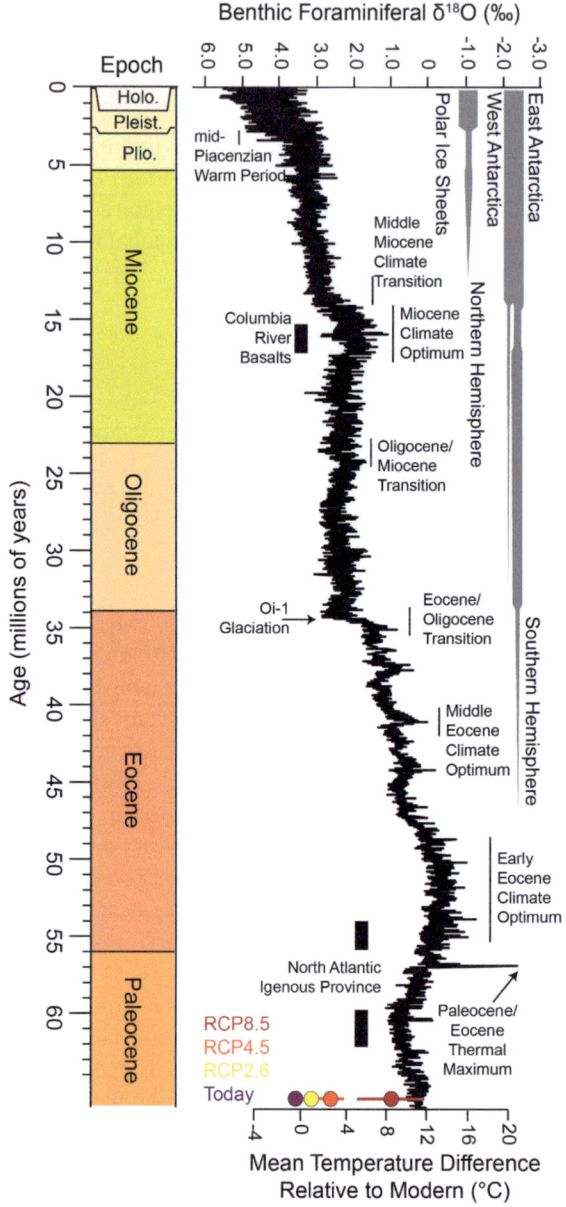

Fig. 17 Cenozoic global reference benthic foraminiferal oxygen isotope dataset from ocean drilling core sites spanning the past 65 million years. Data are mostly generated by using benthic foraminifera tests of two genera extracted from carbonate-rich deep-sea sediments. Genus-specific corrections were applied, as the two species of foraminifera live slightly differently on the ocean floor. Oxygen-isotope data were used to infer average deepwater temperatures, which are shown relative to modern deepwater temperature on the bottom of the figure. Several paleoclimate events (e.g., Paleocene/Eocene Thermal Maximum) and intervals (e.g., Miocene Climate Optimum) are labeled. Future projections for global average temperature in the year 2300 based on different climate model scenarios (high, medium, and low greenhouse gas emissions) are indicated on the bottom temperature scale (dark red, orange, and yellow dots, respectively). Gray vertical bars on the right of the temperature curve mark rough estimates of ice volume in each hemisphere. Times of large-scale volcanism are indicated by black vertical bars (e.g., Columbia River Flood Basalts; and the North Atlantic Igneous Province which is described in Chap. 8). Data replotted with permission from AAAS from Westerhold et al. (2020) and references therein.[8] This figure is excluded from Creative Commons license

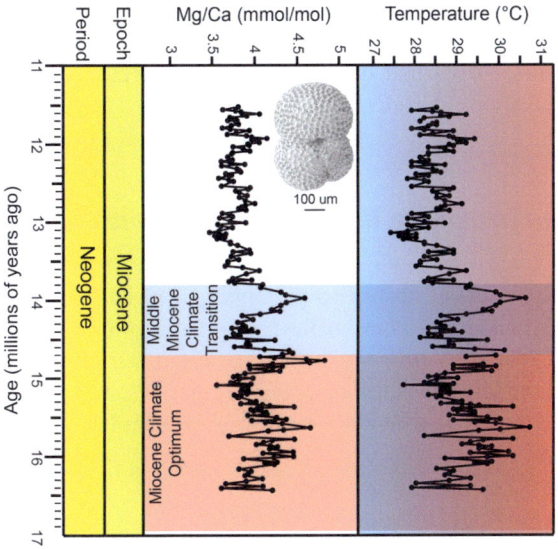

Fig. 18 Left panels are geologic time in millions of years ago, with Epoch and Stages for the Neogene Period. Middle panel is the ratio of magnesium-to-calcium as measured from the surface-dwelling planktic foraminifera species *Trilobatus trilobus*, pictured (scale bar is 100 μm). Right panel is the Mg/Ca ratio converted to temperature in degrees Celsius. Notice that the older part of the Miocene was much warmer than the younger part of the Miocene. Data from Sosdian et al. (2020),[9] foraminifera SEM provided by Adriane Lam

5.1.3 Molecular Recorders

Fossils with mineralized skeletons can carry excellent chemical indicators of Earth's climate and environments through time, such as the stable isotopes and Mg/Ca

---

[8] Westerhold et al. (2020).
[9] Sosdian et al. (2020).

ratios discussed above. In other cases, chemical evidence of ancient climates can be extracted from **molecular fossils**, which are specific organic molecules produced by ancient organisms and preserved in the sediment record. Often such molecular fossils are called **biomarkers**. One advantage of using molecules extracted from the sediments is that all organisms produce soft tissue, including the vast majority of taxa who do not now—or have not in the past—produced mineralized skeletons. In fact, the majority of molecular fossils come from three domains of life: Archaea, Bacteria, and Eukarya, from organisms that rarely produce mineralized skeletons.

Biomarkers and other molecular fossils can be used as proxies for a range of environmental conditions, but a particular focus has been on proxies for sea surface temperature. The most widely used temperature biomarkers are U^k_{37} (derived from photosynthetic phytoplankton, coccolithophorids (Fig. 11)), and TEX_{86} (derived from the cell wall lipids of Archaea), which are based on complex organic molecules whose compositions and/or structures change slightly as water temperature changes. Laboratory and field studies of the modern producers of these biomarkers have yielded calibration curves that can be used to interpret paleotemperatures from biomarker compositions in ancient sediments.

5.2 Sediment Proxies to Reconstruct Precipitation

The number of proxies to reconstruct precipitation is smaller than the number of proxies to reconstruct temperature, but the precipitation proxies that have been developed are very useful. Two such proxies will be discussed here, which are both for continental environments: the characteristics of carbonate-rich (i.e., calcic) horizons in paleosols, and the stable isotopic composition of plant leaf waxes.

5.2.1 Carbonate-Rich Horizons in Paleosols

In Sect. 3.3.1 of this chapter, you were introduced to fossil soils, or paleosols, and the fact that the characteristics of a paleosol can provide valuable information about the environmental conditions under which that soil formed. While this paleoenvironmental information often is qualitative (e.g., soil type X formed under warm and humid conditions, while soil type Y formed under cold and dry conditions), some specific paleosol characteristics—if they are present—can provide more quantitative paleoclimatic information. One example of the latter is the presence and nature of a carbonate-rich horizon within the soil; such a horizon is termed a calcic horizon (or the "Bk horizon", in a terminology often used by those who study soils and fossil soils; a Bk horizon is marked in Fig. 6). When very well-developed, this calcic horizon is often called a caliche or hardpan. Today, calcic horizons are best developed in soils that form under arid to semi-arid conditions, especially when mean annual precipitation is less than ~ 90 cm. Calcic horizons are uncommon when mean annual precipitation exceeds 100 cm.

A calcic soil horizon forms through the interaction of chemical and biological processes in the soil and groundwater, which act to dissolve carbonate from the overlying or underlying material, transport the dissolved carbonate to the level of the calcic horizon, and reprecipitate the carbonate at that level. In modern soils, the depth to the calcic horizon varies with the mean annual precipitation, from as shallow as ~ 10 cm within the soil when precipitation averages ~ 20 cm/yr to a depth of ~ 150 cm when precipitation averages ~ 80 cm/yr. This relationship of calcic horizon depth versus mean annual precipitation can then be applied to a paleosol that contains a calcic horizon, thereby providing a quantitative estimate of past precipitation at that location. Some research has also suggested that the thickness of the calcic horizon can be used to reconstruct the annual range, or seasonality, of precipitation, with the calcic horizon thickening as the seasonal difference in precipitation increases.[10]

5.2.2 Stable Isotopic Composition of Leaf Waxes

In Sect. 5.1.3, you were introduced to the concept of molecular fossils, with the specific examples of several biomarkers used to reconstruct sea surface temperatures of the past. Molecular fossils of land-based organisms also can be found in ancient organic matter. These include a variety of compounds produced by terrestrial plants. Among these terrestrial plant biomarkers, the compounds known as leaf waxes (i.e., the long-chain *n*-alkyl lipids, in the terminology of organic chemistry) have become the focus of an increasing amount of research in the last several decades. The stable hydrogen isotopic composition of leaf waxes (i.e., deuterium isotopes, δD) provides important insights into the hydrologic cycle of the past. When sampled at the appropriate sites and/or combined with other types of data, the leaf wax composition can be used to interpret the history of moisture sources and/or the amounts of precipitation at a location or region.[11]

5.3 Important Proxy Data to Reconstruct Atmospheric CO_2

Carbon dioxide levels have fluctuated greatly through Earth's geologic past, which has greatly influenced climate and life. By quantifying past levels of atmospheric CO_2, geoscientists can better understand how the various components of Earth's climate system interacted to influence those atmospheric CO_2 concentrations, and how those climate components responded to the changes in atmospheric CO_2. There are several proxies for reconstructing ancient CO_2 levels. The two highlighted here are structural characteristics of ancient leaves (i.e., stomata) and the boron stable isotopic composition of marine microfossil carbonate. A combined record of these and other CO_2 proxy data for the Cenozoic is shown in Fig. 19. With some exceptions,

---

[10] Retallack (2005).
[11] Inglis et al. (2022).

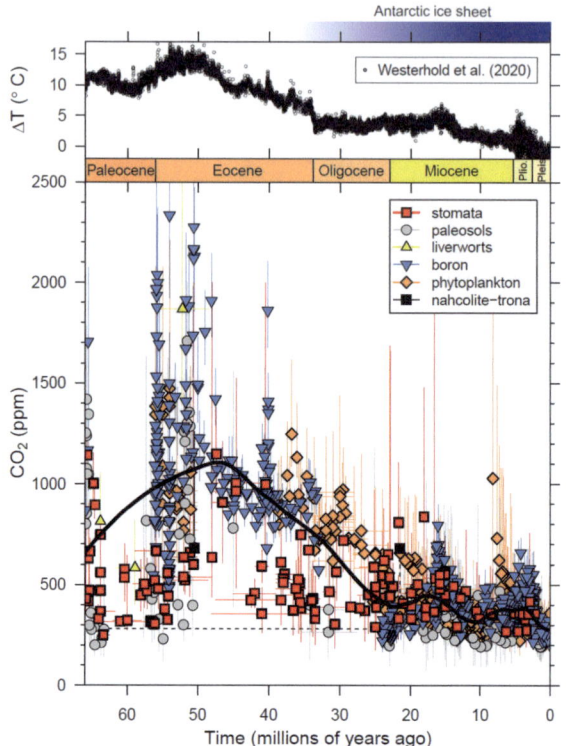

Fig. 19 Relationship between deep-sea temperatures and atmospheric CO_2 changes during the Cenozoic (65 million years ago to the present). Deep-sea temperatures (upper panel; from Westerhold et al., 2020)[12] generally track the estimates of atmospheric CO_2 (lower panel) reconstructed from terrestrial and marine proxies. CO_2 compilation updated from Foster et al. (2017).[13] CO_2 uncertainties (error bars) are shown with 95% confidence. The horizontal gray-dashed line marks the pre-industrial CO_2 level of 280 ppm. The top blue bar indicates approximate timing of ice sheet development on Antarctica. Same as Fig. 7.11. Figure courtesy of Dana Royer, Wesleyan University

there is a strong consensus among different proxies in the CO_2 reconstructions. For example, Fig. 19 shows globally high CO_2 levels in the Eocene and decreasing levels in the Oligocene, corresponding to the time when Antarctic ice sheets formed and the Himalayan Mountains began to rise.

Stomatal density of leaves

Certain structural characteristics of terrestrial plant leaves have been observed to vary today with the CO_2 content of the atmosphere in which the plants grow. This motivated geoscientists to explore these characteristics as possible proxies for reconstructing past atmospheric CO_2 concentrations. The history and status of one of these proxies, **stomatal density** of plant leaves, is explored in detail in Box 2. The approach to proxy development, verification, and application described in Box 2 is an excellent example of the approach that must be used to develop any reliable paleoclimatic proxy.

---

[12] Westerhold et al. (2020).

[13] Foster et al. (2017). Data sources used in the update are listed at the end of Chap. 7.

Box 2 Reconstructing atmospheric CO_2 levels with fossil plants[14]

By Dana L. Royer, Department of Earth & Environmental Sciences, Wesleyan University.

Plants are sensitive indicators of their environment because individual plants do not migrate. Leaves, with their high surface-area-to-volume ratio, are especially well suited to provide information about their environment. Leaves are also engines of food production, converting CO_2 into sugar via photosynthesis.

Beginning in the 1980s, plant scientists observed an inverse relationship between the density of **stomatal pores** on leaf surfaces—which are the conduits for taking in CO_2 for photosynthesis—and the amount of CO_2 in the atmosphere. This makes sense because fewer pores are needed for CO_2 uptake when there is a lot of CO_2 in the atmosphere, and vice versa when atmospheric CO_2 is low. Furthermore, plants lose water through their stomata (this process is called transpiration), and in most environmental settings, conserving water makes a plant more competitive. There is a strong evolutionary pressure for a plant to optimize the tradeoffs between CO_2 uptake and water loss. The inverse relationship between stomatal density and atmospheric CO_2 concentration is one outcome of this optimization.

Stomatal pores are microscopic, typically tens of microns in length and width (Fig. 20). Cuticle is a waxy substance that covers a leaf. Owing to its chemistry, cuticle is very resistant to decomposition and therefore is common to find in fossil leaf-bearing sediments. Crucially, cuticle is imprinted with the outermost layer of leaf cells (analogous to a wax seal), including the stomatal complexes, which are composed of the pore and two bounding guard cells. This means that stomatal density can be measured wherever cuticle fragments are recovered in the sedimentary record, all the way back to the origins of land plants over 400 million years ago. So far, over 300 estimates of atmospheric CO_2 for the Phanerozoic have been made with the stomatal proxy. This CO_2 proxy, along with a proxy based on fossil soils, are presently the most reliable paleo-CO_2 proxies for the pre-Cenozoic. The measurements of fossil stomata have helped shape our understanding of ancient CO_2 levels, for example, the presence of very high CO_2 during the early Paleozoic, followed by a decline to near-present-day levels during the massive glaciation in the Carboniferous and early Permian, some 300 million years ago.

---

[14] References for Box 2: Franks et al. (2014) and McElwain and Steinthorsdottir (2017).

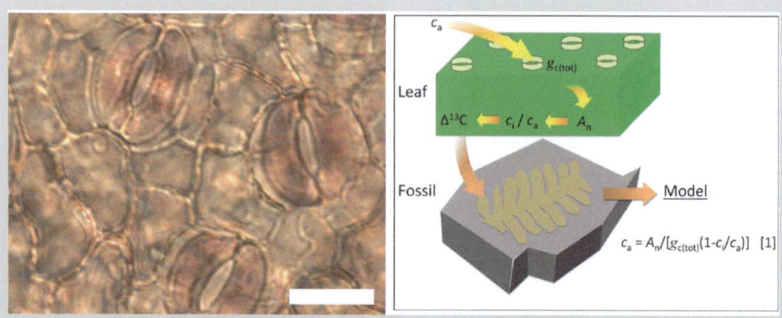

Fig. 20 (Left) Cuticle from a present-day sycamore leaf. Note the stomatal pores and kidney-bean-shaped guard cells. Scale bar = 25 μm. (Right) Conceptual framework for leaf gas-exchange CO_2 proxies. $\Delta^{13}C$ is the carbon isotopic fractionation between atmospheric CO_2 and the leaf. The right panel comes from Franks et al. (2014).[15] Figure by Dana Royer

The stomatal proxy has some limitations. First, it is empirical in nature. The response of stomatal density to CO_2 is measured in a living species, which is then applied to fossils of the same species. In contrast, a proxy based on a mechanistic understanding of the proxy system should be more reliable, especially for deep-time applications, because relative to an empirical relationship the chemistry and physics that control a proxy system are less likely to change over geologic time. Second, the empirical stomatal relationships should only be applied to fossils of the same (or a closely related) species. This means that the CO_2 estimates for the Paleozoic and Mesozoic are associated with more uncertainty because the fossil species are less related to the calibrated living species. Also, creating present-day calibrations is time-consuming—less than ten species have been calibrated thus far. These two limitations have constrained the proxy's applicability to the geologic record.

Starting in the 2000s, several mechanistic proxies were developed that depend mostly on the principles of leaf gas exchange. Plant physiologists have a well-vetted conceptual model for photosynthesis (A_n, the carbon assimilation rate) that depends on the capacity for leaf gas exchange ($g_{c(tot)}$, the total leaf conductance to CO_2) and the gradient between atmospheric CO_2 (c_a) and intercellular CO_2 (c_i).

$$A_n = g_{c(tot)} \times (c_a - c_i)$$

---

[15] Franks et al. (2014).

Solving for atmospheric CO_2, the required inputs are the total leaf conductance to CO_2, c_i/c_a ratio, and leaf assimilation rate (Fig. 20). These inputs can be estimated, respectively, from measurements of fossil stomatal size and density, measurements of fossil leaf $\delta^{13}C$ along with inferences for the associated air $\delta^{13}C$, and inferences of assimilation rate from a nearest living relative. These leaf gas-exchange proxies show promise for producing more accurate paleo-CO_2 estimates, especially for the Paleozoic and Mesozoic, and for opening up much of the fossil cuticle record for paleo-CO_2 inference, because laborious calibrations in living plants are not needed.

5.3.1 Boron Stable Isotope Composition of Marine Biogenic Carbonate

Another common method used to reconstruct ancient levels of atmospheric CO_2 is the boron (B) stable isotopic composition of marine plankton carbonate. In seawater, boron forms two compounds, and the relative abundances of these two compounds vary with the seawater pH. The ratios of the two stable isotopes of B are different in these two compounds, and biogenic carbonate records the B stable isotope ratio of the water in which the carbonate is formed. Therefore, by measuring the boron isotopic composition of ancient microfossil carbonate, such as planktic foraminifera tests, geoscientists can reconstruct seawater pH through time. The concentration of CO_2 dissolved in seawater is a major control on seawater pH, so the seawater pH can be used to back-calculate the seawater concentration of CO_2 and subsequently the atmospheric concentration of CO_2, since atmospheric CO_2 and CO_2 dissolved in the surface ocean are related. Boron-based CO_2 reconstructions are included in Fig. 19.

5.4 *Important Proxy Data to Reconstruct Ice Sheet Size*

Reconstructing the size of past ice sheets is complicated by several factors. For example, an advancing ice sheet often is very effective at eroding the deposits of previous ice advances and retreats, so that the records of older ice behavior are highly fragmentary and poorly preserved. Additionally, much of the record of older ice sheet behavior during the Cenozoic (the last ~ 65 myr of Earth history) lies beneath the existing ice sheets in Greenland and Antarctica. As a result, proxy data are essential for inferring past ice volumes. Two proxies are particularly important: (1) the stable oxygen-isotope record derived from marine biogenic carbonate, and (2) the ages and spatial distribution of glacially influenced sediments, such as tills.

5.4.1 Stable Oxygen Isotopic Composition of Marine Biogenic Carbonate

Section 5.1.1 and Box 1 introduced the fundamentals of stable oxygen isotopes contained in carbonates and highlighted the use of carbonate stable isotopic compositions to interpret past temperatures in caves (from speleothems) and in the ocean (from surface-dwelling and bottom-dwelling foraminifera). The interpretation of Fig. 17, however, also emphasized the fact that oxygen-isotope fluctuations in the carbonate record younger than 34 Ma have been controlled by changes in the isotopic composition of the seawater in which the carbonate was formed, in addition to changes in ocean temperature. What has caused these large changes in the oxygen isotopic composition of seawater during this time?

As mentioned briefly in Sect. 5.1.1, several lines of evidence indicate that large ice sheets first formed in Antarctica ~ 34 Ma. Large changes in the oxygen isotopic composition of the ocean at and since that time have been a result of repeated transfers of large volumes of water between the ocean and the land as the large ice sheets formed, expanded, and contracted. These isotopic changes have been driven as follows:

- During ice sheet growth, large quantities of water are evaporated from the ocean surface, moved by the atmosphere to land, and precipitated as snow at higher latitudes;
- The evaporation process preferentially removes water molecules containing ^{16}O from the ocean, because those water molecules weigh less and evaporate more readily; as a result, the remaining ocean water is enriched in water molecules containing ^{18}O (this is the process of fractionation of isotopes due to the hydrologic cycle). In other words, the process of building an ice sheet leaves the ocean isotopically "heavier", while the ice in the ice sheet is, relatively speaking, isotopically "lighter";
- Marine microfossil carbonate produced during a time of increased global ice volume, therefore, is forced to be isotopically heavier because it is formed from seawater containing relatively more ^{18}O.

The opposite effect is produced when global ice volume decreases (i.e., when large ice sheets shrink), because the water returned to the ocean from the ice sheets is isotopically "light". This returned water decreases the relative abundance of ^{18}O in seawater, and carbonate formed during these times of reduced ice volume records the isotopically "light" composition of seawater at that time.

Because of these relationships, the benthic foraminiferal oxygen-isotope record provides a qualitative indication of the magnitude of global ice volumes and their changes through time. Making quantitative estimates of global ice volume from this method is complicated, but promising efforts are currently underway. The trend toward isotopically heavier values (i.e., more positive values of the $\delta^{18}O$ parameter) in Fig. 17 from the Middle Miocene toward the present indicates a progressive increasing trend in global ice volume, while the increased width during the late

Pliocene and the Pleistocene records larger ice volume fluctuations (larger glacial-interglacial cycles) than occurred during the Miocene. These distinctive characteristics of the global ice volume record during the late Pliocene and the Pleistocene have been attributed to the onset of large ice sheets in the Northern Hemisphere and to their repeated growth and retreat behavior during this time.

5.4.2 Tills and Other Glacial Deposits

Using the oxygen-isotope record of benthic foraminifera to define past global ice volumes—even if only qualitatively—has several advantages: (1) the record is distant from the direct erosional effects of glacial advances, so the record is more likely to be complete, and (2) the record provides a globally averaged view that is relatively easy to access. The globally averaged aspect of this record can also be a disadvantage, however, if one is interested in knowing the regional geographic details of the ice sheets and their histories. Instead, the ages, geographic distributions, and depositional characteristics of sediments associated with glacial activity must be mapped to construct those more localized histories of specific ice sheets. This often includes mapping some of the climatically sensitive sediment types described in Sect. 3.3.1, such as glacial tills and rhythmically laminated sediments deposited in proglacial lakes, as well as other ice-associated deposits. Such efforts often are complicated by the scarcity of material that can be used for detailed age dating and by the fragmentary nature of a glacial record affected by later ice advances. When studying glacial records older than a few million years old, it is also important to recognize that plate tectonic information must be used to position the study area at its proper location (especially the proper latitude) at the time of interest. Research on the presence and distribution of glacial deposits informed the summary interpretation of the onset and presence of ice sheets in the Northern Hemisphere and Antarctica shown in Fig. 17.

5.5 *Important Proxy Data to Reconstruct Sea Level*

Global (i.e., eustatic) sea level is a very dynamic surface, and it has been throughout our planet's history. Global sea level is controlled over time spans of tens to hundreds of thousands of years by the growth and retreat of land-based ice (glaciers and ice sheets), as well as the thermal contraction and expansion of seawater itself as it cools and warms. Over periods of millions to tens of millions of years global sea level is also affected by tectonic influences on the size and shape of seafloor features, which can displace the overlying ocean water (described further in Chap. 6). Two major approaches have been used to reconstruct the history of global sea level: extracting sea level from biogeochemical records and tracing the geographic locations of coastal environments through time (e.g., Fig. 2).

5.5.1 Extracting Sea Level Information from the Benthic Foraminiferal Geochemical Record

Because sea level is closely correlated to ice volume and temperature, geoscientists work to infer sea level changes through time from benthic foraminiferal oxygen-isotope and Mg/Ca analyses. This paired biogeochemical approach is promising, but challenging. Complications in correctly interpreting global ice volume and temperature changes from benthic foraminifera also affect efforts to accurately determine the magnitudes of sea level changes. Comparing sea level reconstructions from the geochemistry of benthic foraminifera with sea level reconstructions from coastal sedimentary sequences (described below) is one way to crosscheck interpretations (Fig. 21).

5.5.2 Extracting Sea Level Information from Coastal Sedimentary Sequences

A second widely used approach to reconstructing past global sea level is based on identifying, mapping, and dating coastal sedimentary sequences and their lateral migrations, which is an application of Walther's Law (Sect. 2.3.1). Data for this approach can come from outcrops (Fig. 22), from cores, and from information collected indirectly from the subsurface, called **seismic reflection data** (based on soundwaves being directed into the ground and reflected back from sedimentary layers in the subsurface. Think of this as a geologist's mega-version of a medical ultrasound system).

Each study using this approach begins by working in a specific area, and the resulting history of shoreline migrations has been caused by a combination of global sea level changes and local geologic effects, such as local uplift or subsidence. A global sea level record is then created by combining results from a large number of widely distributed local studies—preferably with each coming from a location where the local geologic effects are relatively small and predictable—and looking for times of similar sea level changes across all the records. Since it is unlikely that all locations would experience the same local changes at the same time, those changes that occur at the same times across the larger study area are then considered to reflect global changes in sea level.

6 Evidence of Climate Change from the Sedimentary Record in Chaps. 6–9

This chapter provided you with a foundation to appreciate the materials and processes by which geoscientists reconstruct past climate change from the sedimentary record. Included within the presentation of information were selected examples from specific

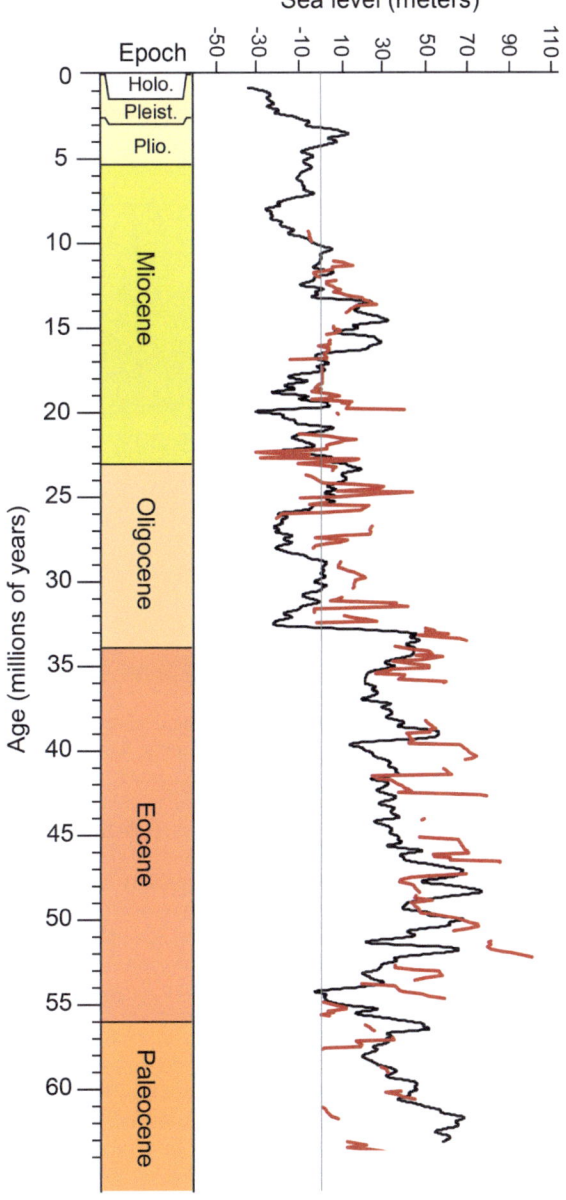

Fig. 21 Comparison of smoothed sea level estimates in meters relative to modern sea level from $\delta^{18}O$ and Mg/Ca (black line) and coastal sedimentary sequences from onshore New Jersey and Delaware Coastal Plain records (red). Light gray vertical line denotes the "0" line. Uncertainty of sea level estimates vary among published studies but vary from ± 10 to ± 20 m. Modified with permission from AAAS. *Source* Miller et al. (2020).[16] This figure is excluded from Creative Commons license

time intervals (e.g., trends of global temperature reconstructions for the Cenozoic Era from marine carbonate oxygen-isotope records; more focused reconstructions of temperatures during the Miocene Climate Optimum and the Middle Miocene Climate

---

[16] Miller et al. (2020).

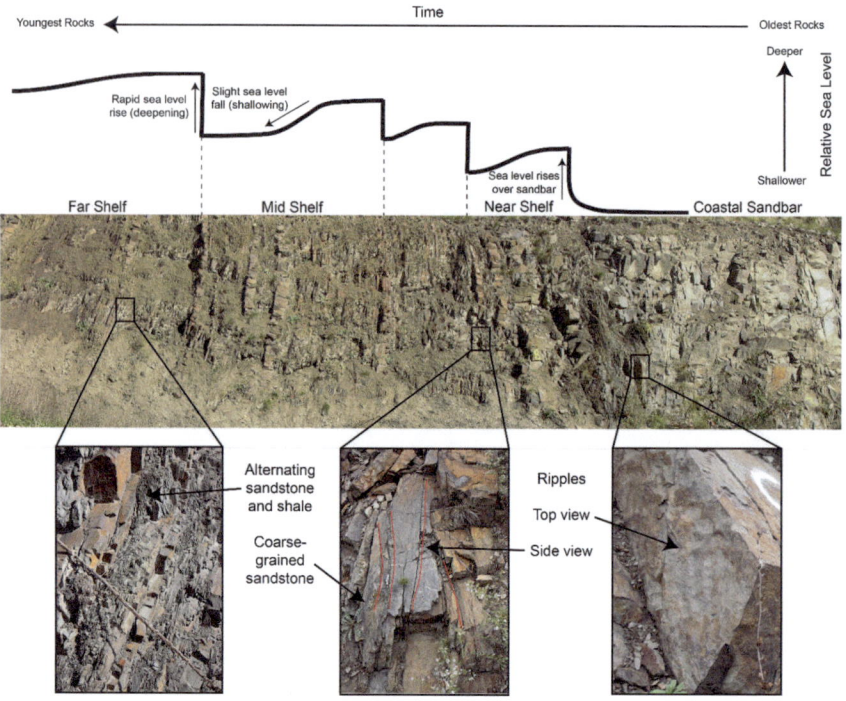

Fig. 22 Section of alternating sandstone and shale that records sea level changes through time, exposed at Briery Gap Road, Pendleton County, West Virginia. The oldest rocks are located to the right of the image, and the younger rocks are to the left. These rock layers, which began as sediments, started as flat-lying, but due to tectonic mountain-building activity, they were tilted ~ 90° (nearly vertical). An interpretation of relative sea level (shallower to deeper) is indicated above the outcrop image. Paleo-environmental interpretations are also included on the sea level curve and range from nearshore (coastal sandbar) to offshore (far shelf). The three images below the outcrop image are close-ups of the Briery Gap section that show alternating layers of coarser-grained sandstone and finer-grained, darker shale (left), a side view of ripples in coarser-grained sandstone (middle), and a top view of ripples (right). Using sequence stratigraphic principles, it is clear that the coarser-grained sandstones on the right of the outcrop image were deposited in shallower, higher-energy water depths, whereas the darker and finer-grained shales were deposited in deeper, more distal water depths. Preserved features in the rocks, such as ripples, indicate that the sediments were deposited in an environment that was subjected to wave action. The sea level changes depicted in this image occurred over a much shorter amount of time compared to the sea level curve in Fig. 21; this image likely depicts sea level changes that occurred over millions of years, whereas Fig. 21 depicts sea level changes over tens of millions of years. Images and interpretations of past sea level and paleo-environments courtesy of Lynn Fichter. This figure is excluded from Creative Commons license

Transition based on carbonate Mg/Ca ratios). The following four chapters explore additional paleoclimatic trends, patterns, and extreme events throughout geologic history. Most of the explorations are based on observations and proxies contained in the sedimentary record, many of which you have been introduced to in this chapter.

References

Feulner, G. (2017). Formation of most of our coal brought Earth close to global glaciation. *Proceedings of the National Academy of Sciences,* 11333–11337.
Foster, G. L., Royer, D. L., & Lunt, D. J. (2017). Future climate forcing potentially without precedent in the last 420 million years. *Nature Communications, 8*, 14845. https://doi.org/10.1038/ncomms14845
Franks, P. J., Royer, D. L., Beerling, D. J., Van de Water, P. K., Cantrill, D. J., Barbour, M. M., & Berry, J. A. (2014). New constraints on atmospheric CO_2 concentration for the Phanerozoic. *Geophysical Research Letters, 41*, 4685–4694.
Inglis, G. N., Bhattacharya, T., Hemingway, J. D., Hollingsworth, E. H., Feakins, S. J., & Tierney, J. E. (2022). Biomarker approaches for reconstructing terrestrial environmental change. *Annual Review of Earth and Planetary Sciences, 50*, 369–394. https://doi.org/10.1146/annurev-earth-032320-095943
Lachniet, M. S. (2009). Climatic and environmental controls on speleothem oxygen-isotope values. *Quaternary Science Reviews, 28*, 412–432. https://doi.org/10.1016/j.quascirev.2008.10.021
Lowenstein, T. K., Schubert, B. A., & Timofeeff, M. N. (2011). Microbial communities in fluid inclusions and long-term survival in halite. *GSA Today, 21*(1), 4–9.
McElwain, J. C., & Steinthorsdottir, M. (2017). Paleoecology, ploidy, paleoatmospheric composition, and developmental biology: A review of the multiple uses of fossil stomata. *Plant Physiology, 174*, 650–664.
Miller, K. G., Browning, J. V., Schmelz, W. J., Kopp, R. E., Mountain, G. S., & Wright, J. D. (2020). Cenozoic sea level and cryospheric evolution from deep-sea geochemical and continental margin records. *Science Advances, 6*, eaaz1346. https://doi.org/10.1126/sciadv.aaz1346
Retallack, G. J. (2005). Pedogenic carbonate proxies for amount and seasonality of precipitation in paleosols. *Geology, 33*, 333–336.
Sosdian, S. M., Babila, T. L., Greenop, R., Foster, G. L., & Lear, C. H. (2020). Ocean carbon storage across the middle Miocene: A new interpretation for the Monterey Event. *Nature Communications, 11*, 134. https://doi.org/10.1038/s41467-019-13792-0
St. John, K., Leckie, R. M., Pound, K., Jones, M., & Krissek, L. (2021). *Reconstructing Earth's climate history: Inquiry-based exercises for the class and lab* (2nd ed., 560 p.). Wiley-Blackwell. ISBN: 978-1-119-54411-1
Westerhold, T., Marwan, N., Drury, A. J., Liebrand, D., Agnini, C., Anagnostou, E., Barnet, J. S. K., Bohaty, S. M., De Vleeschouwer, D., Florindo, F., Frederichs, T., Hodell, D. A., Holbourn, A. E., Kroon, D., Lauretano, V., Littler, K., Lourens, L. J., Lyle, M., Pälike, H., … Zachos, J. C. (2020). An astronomically dated record of Earth's climate and its predictability over the last 66 million years. *Science, 369*, 1383–1387. https://doi.org/10.1126/science.aba6853

Adriane R. Lam A person who hated high school and education—that was Adriane R. Lam before she attended Reynolds Community College, where she earned her AS and found her love of rocks and super old dead things. She completed her BS in geology at James Madison University, MS in invertebrate paleontology at Ohio University, and PhD in paleoceanography and micropaleontology at University of Massachusetts Amherst before landing a postdoc at Binghamton University, where she is now an Assistant Professor in the Department of Earth Sciences. Adriane is the co-founder and co-President of Time Scavengers, a non-profit organization, which supports the next generation of Earth stewards. Adriane's research interests include science communication, reconstructing ocean surface currents across major climate events, inferring how biota respond to Earth systems changes, and biostratigraphy, using fossils to put time into the sedimentary record. When she's not in the lab, Adriane enjoys spending time with her friends, cats and dog; hiking, reading, and pretending she knows a lot about gardening. Adriane is a primary author of Chap. 5.

R. Mark Leckie has been interested in rocks and fossils since he was knee-high to a grasshopper, but he wasn't aware that 'geology' was a major until applying to college. He became a micropaleontologist and stratigrapher with research interests in the Cretaceous Western Interior Seaway of North America, paleoceanography of the Cretaceous and Cenozoic deep sea, and Antarctic ocean-climate change. He has been active in scientific ocean drilling throughout his career. Mark earned a BS and MS at Northern Illinois University and a PhD at the University of Colorado. He is a Professor at the University of Massachusetts Amherst where he teaches Oceanography, Earth History, Field Methods, and Paleoceanography. Mark is a primary author of Chap. 5.

Steven Hovan became an oceanographer by chance. Growing up in the Motor City (Detroit, MI), he set off to college to become an automotive engineer. Or so he thought until an introductory oceanography class piqued his interest enough to join a 4-week research expedition in the Pacific Ocean. Since then, he has never looked back. He switched his major and earned a PhD in Marine Geology and Geophysics from the University of Michigan and has enjoyed a career as an ocean scientist focused on teaching, research and leadership and currently serves as the Dean of Natural Sciences and Mathematics at Indiana University of Pennsylvania. His research interests surround the paleoclimatic record of terrigenous inputs to the deep sea, particularly those involving dust transport to map global winds patterns throughout time and understand how they relate to changes in the global climate system. In all aspects of his work, providing undergraduate students with the opportunity to become involved in genuine research activities is a primary goal. Steve is a primary author of Chap. 5.

Dana Royer In college, Dana Royer was drawn to big-picture ideas like global biogeochemical cycling. This led to an interest in plants, both living and fossil, because plants play such an important role in these cycles. Dana uses fossil plants as tools to unlock information about the climate and ecology of ancient terrestrial ecosystems. He is particularly interested in ways to reconstruct CO_2 and temperature from information preserved in fossil leaves, and in turn how CO_2 and temperature are related to one another on geologic time scales. Dana is a Professor at Wesleyan University. Dana is the author of Box 2 in Chap. 5.

Open Access This chapter is licensed under the terms of the Creative Commons Attribution 4.0 International License (http://creativecommons.org/licenses/by/4.0/), which permits use, sharing, adaptation, distribution and reproduction in any medium or format, as long as you give appropriate credit to the original author(s) and the source, provide a link to the Creative Commons license and indicate if changes were made.

The images or other third party material in this chapter are included in the chapter's Creative Commons license, unless indicated otherwise in a credit line to the material. If material is not included in the chapter's Creative Commons license and your intended use is not permitted by statutory regulation or exceeds the permitted use, you will need to obtain permission directly from the copyright holder.

Plate Tectonics and Long-Term Climate Change

Mary H. Schultz

Guiding Question: How does plate tectonics help drive long-term climate change?

1 Key Take-Away Points

- Modern plate tectonic theory states that Earth's rigid outer layer, called the lithosphere, is broken into about a dozen rigid tectonic plates that move relative to each other on top of a softer layer called the asthenosphere—plates are formed, destroyed, and shift position over time.
- There are three main plate tectonic boundary types: divergent, convergent, and transform. Importantly, the major transfers of heat, carbon, and water from the lithosphere to the ocean and atmosphere occur associated with volcanism at divergent and convergent boundaries.
- Scientists understood that tectonic plates moved based on compiling a wealth of observational data, which involved sonar mapping of the seafloor, heat flow, and paleomagnetic measurements, and other clues from the rock record.
- At a fundamental level, tectonic plates move because of two forces: heat energy transfer and gravity.
- Because tectonic recycling of Earth materials includes carbon-rich matter, plate tectonics has an important role in the geologic storage, movement, and transfer of carbon, thus impacting long-term climate.
- There are two primary climate states that our planet has experienced over long-timescales: Greenhouse (also called Hothouse) climate states are defined by higher-than-average sea levels, high levels of atmospheric CO_2 and warm temperatures. Icehouse climate states are marked by lower sea levels, relatively cool

M. H. Schultz (✉)
Department of Physical Sciences, San Jacinto College, Houston, TX, USA
e-mail: Mary.Schultz@sjcd.edu

temperatures, the presence of continental ice sheets, and low levels of CO_2. Though our climate is currently in an Icehouse state (and our species has only existed during an Icehouse state), the majority of our planet's climate experience has been in a Greenhouse state.
- Volcanism associated with divergent and convergent plate boundaries and with intraplate hotspots (i.e., mantle plumes) plays a key role in the long-term carbon cycle due to the release of carbon dioxide. Scientists can point to paleoclimatic trends and recent climatic events that can be directly connected to volcanism.
- A primary example of when Earth experienced a Greenhouse climate state was the mid-Cretaceous Period, ~ 90 million years ago. Scientists know this time-period experienced a Greenhouse climate state because of the widespread distribution of climatically distinctive types of sedimentary rock, preserved fossils, and marine algal biomarkers.
- In the early to mid-Cenozoic Era, Earth's climate transitioned to an Icehouse state, where it remains today. The tectonically-driven openings and closings of oceanic gateways, which changed ocean circulation, were likely contributors to global cooling and the development of widespread ice cover of the highest latitudes. This cooling transition was augmented by tectonic uplift in central Asia, which may have increased freshwater input to the Arctic Ocean, and in turn the expansion of the North Hemisphere glaciation.
- Global sea level was higher in the Cretaceous than today. Though widespread glaciation and sea water thermal contraction accounts for most of the decrease in sea level over the past 50 million years, plate tectonics also impacted sea level through a variety of factors.

2 Introduction

It's very likely that you've heard about plate tectonics at some point during your primary and secondary school education. You also likely associate the subject with the "Ring of Fire" and geologic hazards such as volcanoes and earthquakes. While this is certainly a common and useful association, this chapter is going to explore the connections among plate tectonic processes and long-term climate change, including shifts between the two major long-term climate states of the Earth system: Greenhouse and Icehouse states. In the first half of the chapter, we will establish the foundation of plate tectonic theory: its early origins, essential components of the theory, and evidence that supports it. In the second half of the chapter, we will explore how plate tectonics influences climate. You will discover the important roles that plate tectonics has in the long-term carbon cycle via volcanism, and in the redistribution of heat and moisture via both the opening and closing of oceanic gateways and the uplift of continental plateaus and mountain ranges—all of which impact climate over million-year timescales.

3 Introduction to Plate Tectonics

3.1 Emergence of an Idea: What Was the Origin of Plate Tectonic Theory?

Plate tectonic theory had its origins in fundamental observations about the Earth. In the early twentieth century, German meteorologist and explorer Alfred Wegener observed that the eastern coast of South America and the western coast of Africa resembled pieces of a jigsaw puzzle that could fit together. Though not the first to make this perplexing observation, Wegener was the first to devote the time to scrupulously research the fossil record, geological features, and evidence of past glaciers on the two separate continents. The evidence that Wegener compiled had him convinced that Earth's continents were once joined together in a supercontinent he called *Pangaea*, translating to "all Earth" (Fig. 1). He then suggested that Pangaea began to break apart around 250 million years ago, with the continental pieces eventually moving through a stationary seafloor, like icebreakers plowing through ice, into their present-day configuration. Wegener published this hypothesis of continental drift in a book titled *Entstehung der Kontinente und Ozeane* (*The Origin of Continents and Oceans*) in 1915. Wegener's proposed ideas drew strong skepticism from the scientific community as he could not provide reasonable explanations as to *why* or *how* the continents moved. Indeed, one component of his hypothesis that remains incorrect is his proposal that the continents moved by themselves. As we'll soon see, the ocean floor is anything but stationary. It would be decades before advanced technologies led to improved understanding of the solid Earth and for parts of Wegener's hypothesis of **continental drift** to be proven correct. Wegener's key contribution—the idea that the continents moved—is the foundation of the modern theory of plate tectonics (from Greek *tekton,* "builder"): Earth's surface is broken up into rigid plates of continental and oceanic rock that move relative to one another through time. Sadly, Wegener passed away returning from an expedition to Greenland and would never know the significant impact of his contributions. Let's explore the modern understanding of plate tectonics in the following sections.

3.2 What Are Tectonic Plates Made of, and How Do They Behave?

Through careful observation and use of the scientific method, geologists established that Earth consists of three main compositional layers: the crust, mantle, and the core (Fig. 2). How do we know this? After all, the deepest humans have ever drilled into the Earth is through a borehole about 12 km (7 miles) deep—barely making it into the mid-crust! That being said, some clues about Earth's interior can be found by direct geological evidence. For example, features called **tectonic windows** allow scientists

Fig. 1 Paleogeographic reconstruction of what the supercontinent Pangaea likely resembled at about 200 million years ago. German meteorologist Alfred Wegener first proposed the existence of Pangaea in the early twentieth century. From Scotese (2016)[1]

to access rocks that would otherwise be hidden deep under the surface, either because they are in a location where the crust is anomalously thin (and thus can be accessed via coring),[2] or they've been uplifted/emplaced by faulting. In addition, geologists use indirect methods like seismology, the study of earthquakes, to explore the interior of our planet. When stress is released on a fault, seismic waves radiate out in all directions and travel through the Earth. The velocity and paths of the seismic waves will change depending on the material they travel through. The combination of direct geological evidence from coring or observations at or near the surface, and indirect seismic wave data, provides a comprehensive understanding of the structure and composition of Earth's interior.

The **crust** is the Earth's strong, rocky outer shell that is primarily composed of silicate minerals (minerals made of silicon and oxygen compounds, along with a wide range of elements). Additionally, carbonate minerals such as calcite ($CaCO_3$) that comprise limestone, a carbonate rock, can be found in the upper portion of Earth's crust, although silicate minerals and rocks are far more abundant. The formation and weathering of both silicate and carbonate minerals and rocks are important in Earth's long-term carbon cycle, which will be introduced here and addressed in greater detail

---

[1] Scotese, C.R., 2016. PALEOMAP PaleoAtlas for GPlates and the PaleoData Plotter Program, PALEOMAP Project, http://www.earthbyte.org/paleomap-paleoatlas-for-gplates/. Christopher Scotese, PaleoAtlas v. 3 licensed under the Creative Commons Attribution 4.0 International License.

[2] In 2023, long sections of largely intact mantle were reached by drilling into a tectonic window below the seafloor located near the Lost City Hydrothermal Field along the Atlantis Massif in the mid-Atlantic Ocean (IODP Expedition 399, Site U1601). Research continues in order to determine if those rocks were recovered from within the mantle itself. See: Voosen (2023).

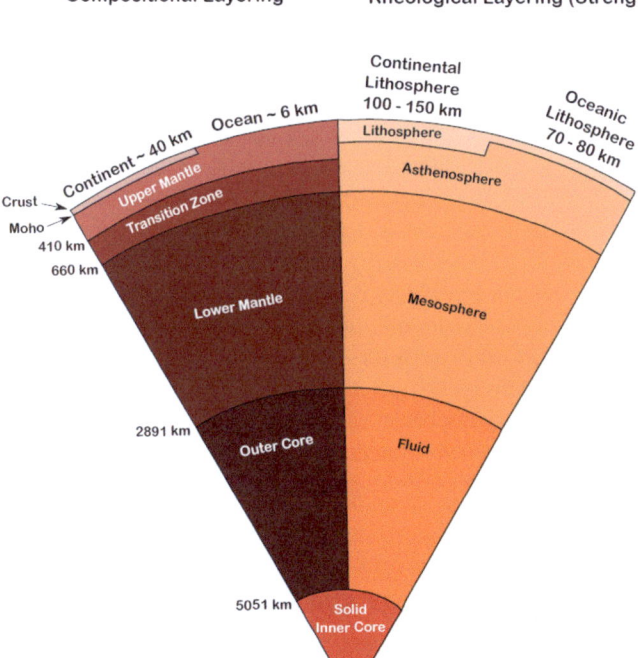

Fig. 2 Categorization of Earth's layers. The left-hand side of this figure displays the compositional divisions: the crust, upper mantle, transition zone, lower mantle, outer core, and solid inner core. The lighter colors on the right indicate the physical layering of the planet, which are defined by strength: the lithosphere, asthenosphere, mesosphere, fluid outer core, and solid inner core. Redrawn after Kearney et al. (2009).[3] This figure is excluded from Creative Commons license

in Chap. 7. Underlying the crust is a thick **mantle**. Although the mantle is also made of silicate minerals, these minerals incorporate more of the heavier elements such as magnesium (Mg) and iron (Fe) in their structures. A dense and hot iron-rich **core** makes up the center of our planet.

It is logical that geoscientists know more about the crust than either of the other two layers, as the crust is the layer most accessible to us; however, that knowledge was based solely on the continental crust until relatively recently. It wasn't until the middle of the twentieth century that samples were recovered from below the seabed to confirm what rocks and minerals form the ocean floor that underlies 70% of Earth's surface. Observations and data from extensive exploration, including drilling into the subsurface, and seismic surveys that reach into the interior, reveal that oceanic crust is quite different compared to continental crust in several ways. Foremost, **oceanic crust** is more dense compared to continental crust. Important silicate minerals of the oceanic crust include olivine [$(Mg,Fe)_2SiO_4$] and pyroxene

---

[3] Kearney et al. (2009).

[e.g., (Ca,Na)(Mg,Fe)Si$_2$O$_6$], which contain the elements magnesium and iron in addition to silicon and oxygen. Prominent minerals in the **continental crust** include quartz (SiO$_2$) and feldspar (e.g., KAlSi$_3$O$_8$) that contain lighter elements such as aluminum, potassium, and silicon. The continental crust is thus more buoyant with an average density of around 2.7 g/cm^3, while oceanic crust has an average density of about 3.3 g/cm^3. Another difference can be found in the thicknesses of the two crustal types. Continental crust thickness ranges in total from about 30 to 70 km due to its lower density. Oceanic crust is much thinner in comparison with an average thickness of about 7 km (Fig. 2). Finally, continental crust is much older on average than oceanic crust. While the oldest oceanic crust present in the oceans today is approximately 180 million years old, the oldest continental crust is over 4 billion years old. We'll find out why this is the case later in Sect. 3.5.

The differences in density, composition, and thickness between continental and oceanic crust are certainly important in the story of plate tectonics. However, the dominant influence on how the continents and ocean basins move lies in the physical, rather than chemical, properties of the crust as well the upper mantle.

3.3 What Are the Lithosphere and the Asthenosphere?

In contrast to the compositional subdivisions of crust/mantle/core that we just defined, an alternate subdivision used in plate tectonics identifies two important layers on the basis of their physical properties (or strength; Fig. 2). The materials of Earth's outermost layer are rigid and often break. We call this layer the **lithosphere**, and it constitutes Earth's crust (continental and oceanic) as well as the uppermost mantle.

Underneath the lithosphere is a layer of the upper mantle called the **asthenosphere**. Whereas the lithosphere is rigid, the asthenosphere acts more like a viscous (or sticky) fluid and flows over long periods of time. A Snickers candy bar might offer a useful analogy. If you break a Snickers bar in half, the peanut layer will break immediately, however, the caramel will spread apart and be the last layer to separate.

The lithosphere-asthenosphere boundary occurs at a temperature of about 1300 °C, and its exact depth depends upon the geothermal gradient or the rate at which temperature increases with depth in the Earth. A general estimate of an average geothermal gradient is about 30 °C per kilometer. At mid-ocean ridges where magma cools to form new oceanic crust, and the geothermal gradient is much higher, the lithosphere-asthenosphere boundary is essentially at 0 km depth. In the deeper ocean basins, the boundary is likely around 60–70 km depth, and it may deepen to 100–250 km beneath the continental lithosphere because the geothermal gradient is lower than average through continental crust. The location of the base of the asthenosphere in Earth's interior is more difficult to estimate, but it may reach a depth of about 700 km according to evidence from seismic waves.

3.4 What Tools and Evidence Did Scientists Use to Propose Plate Tectonic Theory?

Wegener had no way of providing thorough evidence for his hypothesis of continental drift when he first published it in 1915. It took decades of advancing instrumentation and methodologies in multiple fields of the physical sciences to assemble the expansive puzzle of plate tectonic theory. We've seen, for example, that our understanding of the interior of the Earth requires the study of seismic waves, an indirect method not widely applied until well after Wegener's time. The composition and topography of the ocean floor wasn't known well until the widespread use of sonar and other sophisticated instruments during World War II in the 1940s. The study of volcanoes and heat flow data also greatly contributed to the theory. Even today, the puzzle remains unfinished and there are many questions left to be explored.

The observations and data scientists have gathered over decades point to the following statement as a summary of plate tectonic theory: the lithosphere is broken into about a dozen rigid tectonic plates (Fig. 3) that move relative to each other on top of a viscous asthenosphere. These plates are formed, destroyed, and shift position over time. They move across Earth's surface at rates[4] generally ranging from less than one to up to 10 cm per year (or about the rate at which fingernails grow). We'll learn much more about how we know the past and present rates of plate movements in the next two sections.

3.5 What Happens at the Boundaries of Tectonic Plates?

On January 7, 2020, a magnitude 6.4 earthquake rattled the island of Puerto Rico, an American territory that was still recovering from the destruction of Hurricane Maria in 2017. Puerto Rico sits at the boundary of two tectonic plates, the North American Plate, which is moving southwest at about 2.2 cm/yr and converging with the Caribbean Plate. Areas that lie on or near the boundaries of tectonic plates are likely to experience large earthquakes like the magnitude 6.4 event in Puerto Rico or the devastating 2015 magnitude 7.8 earthquake that destroyed many villages and killed nearly 10,000 people in Nepal, southern Tibet (China), and northern India. Volcanism may also result at certain types of plate boundaries and is readily observed at plate boundaries surrounding the Pacific Ocean, famously known as "The Ring of Fire."

There are three major plate boundary types: divergent, convergent, and transform (Fig. 4).

---

[4] Note that there are ridge segments with super-fast spreading rates of 20–30 cm/yr, but they are uncommon, and it is also unknown for how long these very fast rates can be maintained. See for example: Zhang et al. (2018).

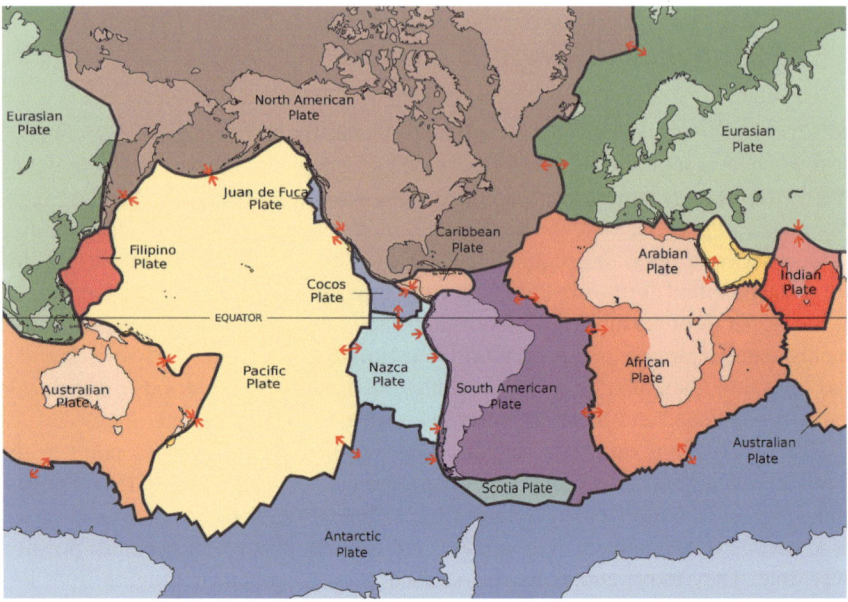

Fig. 3 Earth's primary tectonic plates and their relative motions. From the USGS and Wikimedia Commons[5]

3.5.1 Divergent Boundaries

Divergent plate boundaries or margins are regions where the lithosphere is moving apart (Fig. 4a). It is important to point out that two diverging plates don't create a giant crack or gap in the Earth, but rather new lithosphere is formed at the boundary as the plates slowly move in opposite directions. Iceland is one of the few places on Earth where divergent motion takes place at the surface instead of below sea level. Analyses of seismic wave velocities indicate that we can see the exposed rift valley separating the North American plate from the Eurasian plate on Iceland's surface because of a plume of heated magma from the mantle that sits beneath Iceland, augmenting the volume of magma that rises to the surface that typically occurs at divergent plate margins. See Box 1 for an explanation of **mantle plumes** and **volcanic hotspots**.

It wasn't until about 40 years after Wegener presented his hypothesis of continental drift that advancing technologies developed during World War II allowed scientists to collect the data needed to pick up where Wegener left off. Geoscientist and U.S. Navy officer Harry Hess proposed the concept of **seafloor spreading,** after mapping the North Pacific Ocean's more clearly defined features called **mid-ocean ridges** that stand higher in elevation than the rest of the ocean basin (see Sect. 3.6). Hess hypothesized that new oceanic lithosphere is created at these ridges by the cooling of

---

[5] Original figure by the USGS (n.d.). Tectonic Plates of the Earth. https://www.usgs.gov/media/images/tectonic-plates-earth. Downloaded from Wikimedia Commons File:Plates tect2 en.svg. (August 29, 2022) https://commons.wikimedia.org/wiki/File:Plates_tect2_en.svg. Public domain.

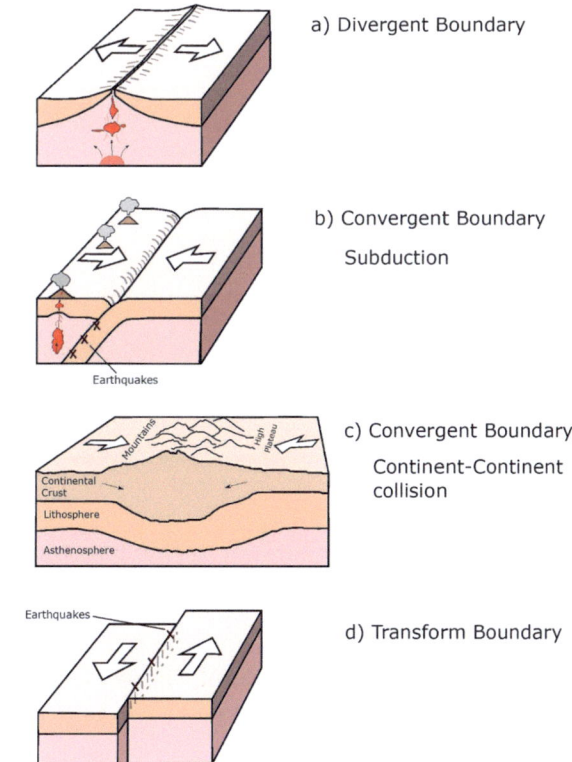

Fig. 4 Plate tectonic boundary types. **a** A divergent boundary where two plates are pulling apart from each other and new oceanic lithosphere is formed. This is what occurs at mid-ocean ridges. **b** A simple depiction of what occurs at a subduction zone as an older, denser oceanic plate subducts beneath a more buoyant oceanic or continental plate. **c** Also shows convergence, but with two continental lithospheric plates. Since continental lithosphere is more buoyant than oceanic lithosphere, subduction will not occur. What occurs instead is the formation of mountain belts (not volcanically active) and plateaus. **d** Depiction of a transform boundary, where lithosphere is neither created nor destroyed, but plates move laterally in opposite parallel directions

molten material from the asthenospheric mantle. The molten material emerges at the Earth's surface from **fissures**, long narrow openings or cracks along the mid-ocean ridges. New oceanic lithosphere, composed of the volcanic rock basalt, is formed as the lava cools, crystallizing solid minerals that collectively form rock. As the new lithosphere is formed at mid-ocean ridges, the older lithosphere moves away from the ridge crest in opposite directions. As a result, the oldest lithosphere is always farthest away from the mid-ocean ridge where it formed (Fig. 5). Some of the oldest oceanic lithosphere found on the seafloor today, approximately 180–200 million years old, is located in the western Pacific Ocean, many thousands of kilometers away from the spreading center. For decades, the western Pacific's lithosphere was considered the oldest on Earth, but scientists have very recently discovered[6] patches of lithosphere in the Mediterranean Sea that are estimated to be as old as 340 million years!

Geologist Marie Tharp created the first scientific map showing the bathymetry of the ocean floor in intricate detail. The crest of the 60,000 km-long global mid-ocean ridge system is the most prominent example of a divergent boundary. The rate at which the oceanic lithosphere diverges, called the seafloor spreading rate, is thought

---

[6] Granot (2016).

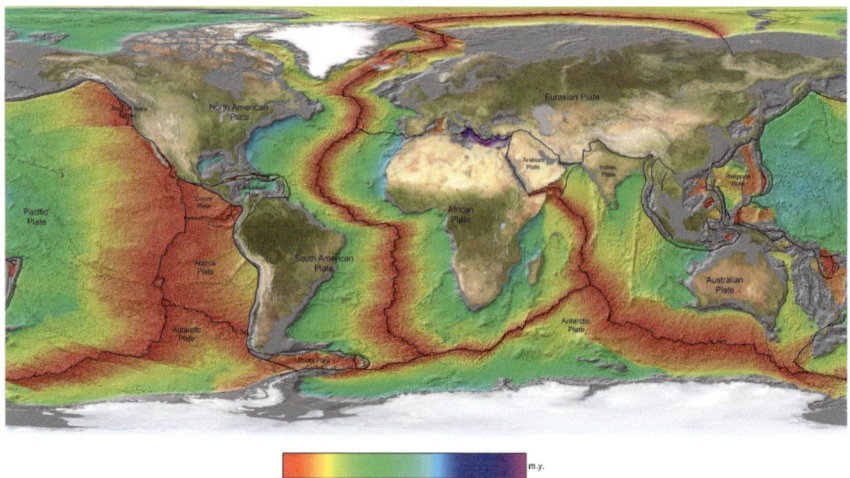

Fig. 5 Map showing the age of the oceanic lithosphere. Notice how mid-ocean ridges are well-defined as new oceanic lithosphere is being created. The oldest oceanic lithosphere can be found away from mid-ocean ridges. Images credit: Mr. Elliot Lim, CIRES & NOAA/NCEI; data from Müller et al. (2008)[7]

to exert a primary control on global climate over the past several hundred million years through the amount and composition of gases released during mid-ocean ridge volcanism, as we'll see later.

Divergence occurs in the continental lithosphere as well in a process called **rifting**. The breakup of supercontinent Pangaea around 200 million years ago began with the formation of numerous rift valleys, and the opening of some of these rifts eventually led to the creation of the Atlantic Ocean. The valleys of the East African Rift System and the Red Sea to the north are the most pronounced examples of present-day continental divergence.

3.5.2 Convergent Boundaries

Plates come together or collide at **convergent** plate boundaries. There are several types of convergent boundaries because tectonic plates can have different lithosphere at their leading edges (i.e., the edges that are converging). Thus, there exists: (1) oceanic-oceanic convergence, (2) oceanic-continental convergence, and (3) continental-continental convergence.

When oceanic lithosphere converges with other oceanic lithosphere or with continental lithosphere, the older, colder, and denser oceanic plate will sink beneath the less dense, younger plate in a process called subduction (Fig. 4b). Subduction is

---

[7] Images credit: Mr. Elliot Lim, CIRES & NOAA/NCEI, https://www.ngdc.noaa.gov/mgg/image/crustalimages.html; Data from Müller et al. (2008).

the counterbalance to seafloor spreading. While new lithosphere is created at mid-ocean ridges, old oceanic lithosphere is being consumed at subduction zones. By the time oceanic lithosphere reaches a subduction zone, it has cooled substantially as it moved far away from its spreading center. This decrease in temperature also results in thermal contraction, and thus an increase in rock density. In addition, the down-going plate, called the **slab**, may have a significant cover of sediments, which can reach 4 km in thickness. Importantly, these sediments often contain carbonate material and can release carbon dioxide (CO_2) when heated during the subduction process. As the down-going oceanic plate descends into the asthenosphere beneath the overriding plate, a potentially extensive and narrow depression, called a **trench,** forms on the seafloor. Some of the deepest places on Earth are found at oceanic trenches. The Mariana trench in the western Pacific Ocean reaches a depth of over 10,900 m (about 35,700 ft), which is about 1800 m (6000 ft) deeper than Mount Everest is tall! Not all trenches are that deep, however, and for some subduction zones, a trench may not be present at all. Shallower trenches occur in places where the subducting slab is relatively young, warm, and buoyant. Additionally, large quantities of land-derived sediment can fill the depression of the trench, as is the case on the Cascadia margin off the coasts of Oregon and Washington state.

Heat flow, carbon (especially CO_2), and water play essential roles throughout the subduction process. The cold slab and accompanying subducted sediments begin to heat as they subside into the asthenospheric mantle. Hydrous (water-containing) minerals begin to release their water and carbon content, and when the down-going plate reaches depths of 65–130 km, water and other volatiles generate partial melting of the nearby mantle. This process results in volcanism at the surface of the overriding plate about 150–200 km away from the trench axis.

The chain of volcanoes that develops as a result of oceanic-continental convergence is called a **continental arc**. The Cascade Arc of the Pacific Northwest is the result of the Juan de Fuca plate subducting beneath the North American plate. The famous eruption of Mt. St. Helens in Washington state in 1980 killed approximately 57 people and caused about $1.1 billion in damage. Volcanologists are keeping a close eye on nearby Mount Rainier as it has the potential to be even more devastating than Mt. St. Helens, given that it is adjacent to more densely populated areas including Seattle and Tacoma.

Similarly, an **island arc** may develop when older oceanic lithosphere subducts beneath other, younger oceanic lithosphere, such as the Mariana Islands in the western Pacific. Another example of an island arc is the Tonga Islands in the southwest Pacific, where the western edge of the Pacific tectonic plate subducts below the Indo-Australian tectonic plate. Like most arc volcanism associated with subduction zones, the Tonga arc system includes both emergent island volcanoes and submarine volcanoes (underwater volcanoes). On January 15, 2022, one of the shallowly submerged volcanoes in this arc system—the Hunga-Tonga-Hunga-Ha'apai volcano—captured the world's attention as it produced the most powerful eruption on Earth in at least

30 years. The eruption produced huge plumes of ash that reached 30 km into the atmosphere and thick ash layers blanketed parts of the Tonga islands affecting drinking water supplies and resulting in crop damage. The eruption also triggered a series of earthquakes, some of which generated a tsunami, destroying homes in the Tonga islands.[8]

Importantly, subduction and resulting arc volcanism are essential in the global recycling of climatically-important components including carbon and water. As we saw earlier in this section, an extensive swath of 180–200-million-year-old oceanic lithosphere can be found in the western Pacific, some of the oldest on Earth. The average lifespan of oceanic lithosphere is about this old, or perhaps a bit younger. Thus, all oceanic lithosphere, and the carbon and water that is contained therein, is fully recycled every 100–200 million years.

3.5.3 Convergent Boundaries (Continent–Continent)

The third type of convergent boundary has some important differences compared to the two previously described. When two plates of continental lithosphere converge, they collide and crumple as continental lithosphere is simply too buoyant to be subducted (Fig. 4c). Since subduction can't occur in this scenario, both trenches and volcanic arcs will be absent. The collision of continental lithospheric plates can, however, still produce large earthquakes. Continent–continent collisions built some of the most impressive mountain belts on Earth including the Himalayas, the Alps, and the much-older Appalachians. Later, and in the next chapter, we'll see how tectonic uplift can affect long-term global climate.

3.5.4 Transform Boundaries

Finally, plates may also maneuver past one another at **transform boundaries**, where lithosphere is neither created nor destroyed (Fig. 4d). Note that with the addition or destruction of lithosphere, there are no volcanoes or fissure eruptions associated with transform margins. Numerous transform boundaries exist at mid-ocean ridges; these boundaries accommodate the details of divergent plate movement by forming at an angle perpendicular to the mid-ocean ridge. These transform boundaries offset the ridge, breaking it into segments along its length. The transform faults may develop to accommodate differences in mid-ocean ridge magma supply, rates of spreading, direction of spreading, and the details of spreading-caused plate rotation related to motion across a (mostly-)spherical Earth. However, one famous example of a transform boundary is not on the seafloor but is in North America: the San Andreas Fault System that spans about 1200 km (800 miles) down the length of California. Here, the Pacific Plate shifts northwestward at about 5 cm/yr relative to the North American plate. Other major transform boundaries include the Alpine Fault in New

---

[8] Witze (2022).

Zealand, the Dead Sea Transform in Israel and Jordan, and the East Anatolian Fault in northern Turkey. Though the highest magnitude earthquakes typically occur at convergent boundaries, the various stresses acting at transform boundaries can cause substantial earthquakes as well.

3.6 How Do We Know Plates Moved?

3.6.1 Paleomagnetic Evidence and Seafloor Spreading

How did geoscientists come to understand that part of Wegener's hypothesis was correct all along? As we saw in Sect. 3.5, the need for mapping the ocean floor during World War II resulted in some of the greatest discoveries surrounding plate tectonics. Harry Hess proposed the theory of seafloor spreading after sonar mapping revealed the presence of mid-ocean ridges and contemporaneously, paleomagnetists (scientists that study the Earth's magnetic field and its history) sought to learn about the ocean floor's geophysical characteristics. What could the magnetic nature of minerals in ocean floor rock reveal?

As you learned in Chap. 5, Earth's magnetic field is hypothesized to exist because of convection in the Earth's molten outer core as Earth itself rotates on its axis. Like a bar magnet, our magnetic field has repelling and attracting ends (Fig. 6). Modern magnetic field lines emerge from the Southern Hemisphere's magnetic pole at an angle perpendicular to the Earth surface ($-90°$ dip) and encase the Earth as they travel north. The field lines are parallel to the Earth's surface (i.e., $0°$ dip) close to the equator before reentering at the Northern Hemisphere's magnetic pole ($+90°$ dip). Note that magnetic poles are not the same as geographic North and South Poles (the poles of Earth's rotational axis), and therefore compasses need a correction factor in order to point today toward the geographic North Pole when they are responding to a force oriented toward the Northern Hemisphere's magnetic pole. Our current magnetic field configuration, or **polarity**, is coined *normal*, whereas *reverse polarity* describes a magnetic field with the opposite configuration (i.e., the "attracting end" of the imaginary bar magnet; $+90°$ dip) is located near the south geographic pole.

Key evidence surrounding the past behavior of the magnetic field is found in the seafloor. As the lava crystalizing into volcanic rocks at the crests of mid-ocean ridges cools below a specific temperature called the Curie temperature, the magnetic minerals in these rocks align with the orientation of the magnetic field at the time of cooling (i.e., at the time of crystallization). Geophysicists in the 1950s observed that the magnetic orientations of minerals in older oceanic crust contrasted at different distances away from the ridge crest, alternating between normal and reversed polarity in a pattern that is symmetric on the two sides of the ridge. This means that there have been changes in the polarity of our magnetic field since at least 180 million years ago (the age of the oldest in situ oceanic crust), and theoretically earlier in the geologic past as well. In fact, every polarity change of the magnetic field since 180 million years ago could be documented by additional mapping of the seafloor

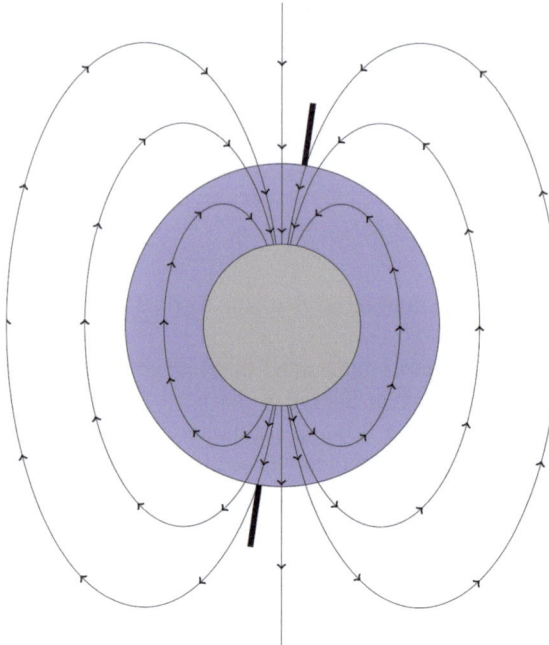

Fig. 6 Simplified representation of Earth's magnetic field. This representation does not take into account the altering influence of the solar wind on the magnetic field. The magnetic field emanates from the outer core (gray). The orientation of the field lines with respect to Earth's surface varies from parallel near the equator to perpendicular near the geographic rotational poles. Paleogeographers use this feature to estimate paleolatitudes of rocks, on the basis of the magnetic field orientation they acquired at the time of their formation. Heavy black line segments represent Earth's rotational poles (geographic North and South Poles). Illustration by William St. John

in other locations, as well as in measurements made in thick sequences of terrestrial basalts. The complete set of ocean crust magnetic data revealed a stripe-like pattern with the stripes, or **magnetic anomalies**, running nearly parallel to the mid-ocean ridge axis. Furthermore, the striking symmetry of the stripes on either side of the ridge suggested the existence of seafloor spreading. Dating revealed that the ages of the magnetic anomalies on either side of the ridge matched and displayed a clear pattern of increasing age away from the ridges.[9] In addition, these magnetic anomaly sets were found to be age-equivalent across the various ocean basins (Fig. 5).

---

[9] Radiometric dating of seafloor basalts was initially challenging because alteration of freshly erupted basalt by seawater caused significant errors in the early techniques of bulk-rock ^{40}K- ^{40}Ar dating. The first testing of ages predicted from the magnetic anomaly pattern came by sampling and biostratigraphically dating the oldest ocean sediments, immediately overlying basaltic "basement" rock below the seafloor. (See: Maxwell & Von Herzon, 1970.) Radiometric dates determined by subsequently developed, more sophisticated techniques are not affected by basalt-seawater interactions, and confirm this pattern of seafloor age increasing away from a mid ocean ridge crest.

Age dating methods have been essential in transitioning hypotheses about plate tectonics to theories. We know from age dating methods that since the formation of the oldest in situ oceanic crust some 180 million years ago, our magnetic field has experienced a rapid[10] change in polarity over 150 times at irregular intervals—there is no regular timing or periodicity in these polarity shifts. Furthermore, dating of terrestrial basalt sequences has extended the record of documented global magnetic polarity changes back to ~ 250 million years ago.[11] Age dating methods have also been essential in determining the directions and rates (i.e., velocities, or distance/time) of seafloor spreading (and thus of plate motion) for the last 180 million years. Later in this chapter, you will see how changing rates of seafloor spreading directly impact volcanic CO_2 emissions, and thus influence long-term global climate.

3.6.2 Evidence from Ancient Continental Rocks

Paleomagnetists suggest that a supercontinent called Rodinia existed about 1 billion to 800 million years ago, long before the accretion of Pangaea. How do we know the configuration of the continents that far back in Earth's history if oceanic crust older than 180 million years old has been consumed by subduction? The geologic clues lie on Earth's continents instead of on the ocean floor.

Paleomagnetists help reconstruct Earth's paleogeography, or past arrangement of the continents, by examining the preserved magnetic orientations in rocks. Earth's magnetic poles don't stay in a fixed position but tend to drift, called apparent polar wander. In fact, the Northern Hemisphere magnetic pole has meandered nearly 50 km per year since the turn of the twentieth century!

To begin to untangle the position of the continents hundreds of millions of years ago, geophysicists studied rocks of different ages on continents that are now thousands of kilometers apart: North America and Asia (Eurasia). For each study site, they were able to reconstruct the location of the Northern Hemisphere magnetic pole at the time of its rock's formation using the orientations of the rock's magnetic minerals. To determine how the position of the paleomagnetic pole changed through time, they then compared magnetic mineral orientations in rocks of the *same age*. For example, the orientations in rocks 200 million years old but from vastly different continents revealed that they "pointed" to two *different* magnetic poles in the Northern Hemisphere: the magnetic pole identified from 200-million-year-old rocks in North America was different from the magnetic pole identified from 200-million-year-old rocks in Eurasia. Of course, these geophysicists knew there had to be more to the

---

[10] When polarity reversals occur they are geologically rapid (estimated to be hundreds to thousands of years).

[11] Earth's magnetic polarity changes result from changes in the circulation of liquid iron in Earth's outer core—the source of the magnetic field itself. The random occurrence in timing of magnetic polarity changes suggests spontaneous or random circulation changes occur within the outer core, which is influenced by conditions in Earth's inner core and by Earth's rotation. Spontaneous reversals of Earth's magnetic field have been produced in computer model simulations. See: Glatzmaier (1999).

story as there could only be a single magnetic pole in the Northern Hemisphere at any given time, whether that's 200 million years ago or ten years ago.[12] Bringing Wegener's idea to the forefront, these geophysicists asked themselves, "Where would North America and Eurasia be located if we repositioned them so that their respective magnetic poles were in the same position?" When they did this repositioning, they found that North America and Eurasia were located next to one another (Fig. 1), suggesting that continents (and the plates they are a part of) have moved over time.

Paleomagnetism isn't the only method geologists use to figure out the ancient positions of the continents. As you read in Chap. 5, the presence of climatically sensitive sedimentary deposits also provides some indication of the paleoposition of the continents as well as the paleoclimate. For example, glacial deposits indicate paleotemperatures cold enough for ice to have existed at that time, while also indicating that the region where the glacial deposits are found today was located at mid- to high latitudes (i.e., latitudes where glaciers are most likely to form) at that time in the past. Eolian deposits are records of wind-carried sediments that accumulate in dry, desert-like climates. Other climatically sensitive sedimentary deposits include red beds, which are red soils that form in areas where there is a sharp contrast in seasonal moisture; and evaporites, which are salts that precipitate out of water in lakes and marginal seas in places where evaporation outpaces precipitation. Paleogeographic inferences can sometimes be made from such sedimentary records because global atmospheric circulation patterns have some latitudinal relationships; for example, where upper level air descends toward the Earth's surface there tend to be dry climates. This occurs at approximately 30° latitude and at the poles. Where surface winds converge and rise up there tends to be wet climates. These occur at the equator and near 60° latitude.

3.6.3 Evidence from GPS

Though paleomagnetism can tell us about the positions and movements of the continents in the past, geoscientists use the Global Positioning System (GPS) to monitor the current velocities of the plates (Fig. 7). The GPS receivers are anchored firmly into the bedrock of different tectonic plates and satellite tracking measures their movement. Each measurement is based on receiving information simultaneously from at least three satellites and has an accuracy of less than 1 mm. GPS measurements of plate movements match closely with predicted plate motions (compare Figs. 7 and 3), which were based on plate geometries, orientation of the mid-ocean ridge crest, and long-term spreading rates calculated from seafloor magnetic anomaly patterns.

---

[12] However, model simulations suggest that just prior to when a magnetic polarity change occurs, the magnetic field may weaken and possibly contain multiple poles for a short period of time. See: Glatzmaiers and Roberts (1995).

Plate Tectonics and Long-Term Climate Change 255

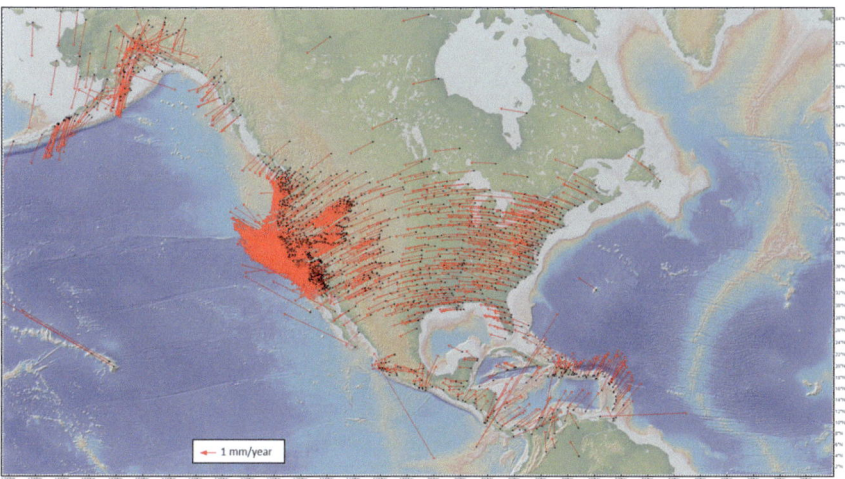

Fig. 7 Map showing the current-day plate motion (red arrow vectors) of over 2000 locations in North America, Central America and the Caribbean. Vector arrow indicates direction of plate motion; vector length indicates velocity (mm/year). Compare the plate movement to the plate boundaries in Fig. 3. Notice the density of GPS stations in the western US. These measurements involve satellite radio positioning that uses the Global Positioning System (GPS)[13]

3.7 Why Do Plates Move?

As you read about the nuts and bolts of plate motion and the various processes that occur at the margins of plates described thus far, the question may have occurred to you: what is the driving mechanism behind all of these tectonic processes? In other words, *why* do plates move? This is a fundamental theoretical question of planetary scale! At a first order level, the primary reason for plate motion relates to two forces: the transfer of heat energy and gravity. Lithospheric plates are part of Earth's dynamic thermal and gravity-driven convective system.[14] Hot (and therefore lower density) magma rises up to the surface from the asthenospheric portion of the mantle, adding new rock to the margin of lithospheric plates at ocean ridges. In contrast, at some distance away from the ridges, the lithosphere is older and colder, and therefore more dense—thus these slabs subduct (sink) into the underlying asthenosphere. Lithospheric plates move across Earth's surface driven by these density differences and aided by asthenosphere convection below it. The gravitational force at the spreading ridges acts to "push" the plate along. Meanwhile, the gravitational force in subduction zones acts to "pull" the plate along. This general theoretical model is depicted in Fig. 8.

---

[13] Map created using GeoMapApp 3.6.15 with overlying EarthScope PBO Plate Velocities-IGS08 (February 2017). CC BY 4.0.

[14] See introductory videos on plate tectonic mechanisms hosted by IRIS, a consortium of over 100 US universities dedicated to the operation of science facilities for the acquisition, management, and distribution of seismological data. https://www.iris.edu/hq/inclass/animation/what_are_the_forces_that_drive_plate_tectonics (accessed June 2023).

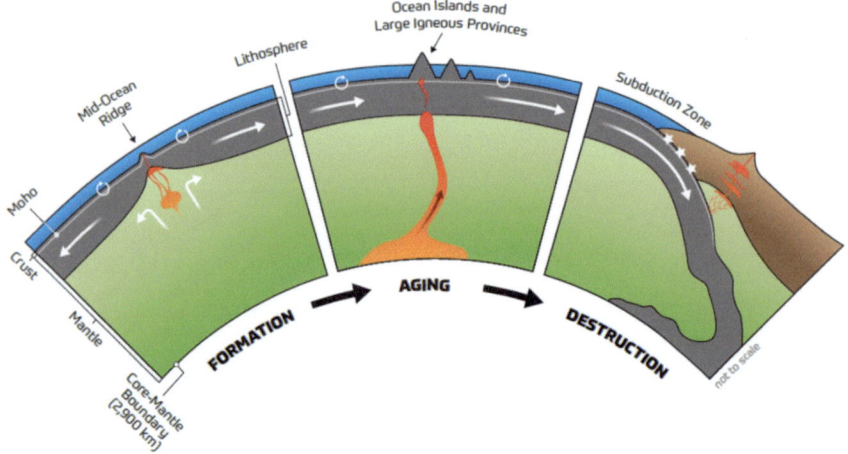

Fig. 8 Representation of the plate tectonics model, including: (left) formation of new oceanic lithosphere at the mid-ocean ridge, (middle) movement away from the mid-ocean ridge, and (right) destruction at a subduction zone. The oceanic plate is dark gray, the continental lithosphere is brown. An upwelling mantle plume (orange) results in intraplate hotspot volcanism (see discussion of mantle plumes in Box 1). The white arrows show plate motion, white circular arrows depict hydrothermal circulation, and white stars are large earthquakes. Illustration by Geo Prose in Koppers and Coggon (2020)[15]

Note that this convective system results in a recycling of Earth materials: the formation of new rocks at the ridges and the destruction of older rocks in subduction zones. Because the tectonic recycling of Earth materials includes carbon-rich matter, plate tectonics has an important role in the geologic storage, movement, and transfer of carbon—thus impacting long-term climate (more on this in Sect. 4).

The plate tectonic process can also be viewed as the control on a metaphorical "life cycle" or evolution of an ocean basin in which an ocean basin is born, grows (when the rate of seafloor spreading exceeds the rate of subduction for that basin), and eventually declines and "dies" (via subduction when the rate of subduction exceeds the rate of seafloor spreading for that basin).[16] The stages of ocean basin evolution are summarized in Table 1, including modern-day examples of each stage. You will see in Sect. 4 that the tectonic reconfiguration of the ocean basins also has consequences for Earth's climate through time.

While the rates of seafloor spreading and subduction may change over time for an individual ocean basin, thereby driving that basin's "life cycle," you should realize

---

[15] Koppers and Coggon (2020) (p. 28); illustration by Geo Prose, inspired by Fig. 1 in Crameri et al. (2019).

[16] The life cycle of an ocean basin is also known as the Wilson Cycle, named after a twentieth century geophysicist, J. Tuzo Wilson (1908–1993), for his contributions to plate tectonics and hotspot research. When emphasis is placed on the associated changes in the positions of landmasses—in other words, the assembly and breakup of a supercontinent, rather than on the changes to an ocean basin—the cycle is instead referred to as the Supercontinent Cycle.

Table 1 Stages in the generalized evolution of an ocean basin, driven by plate tectonic activity

	Embryonic	Juvenile	Mature	Declining	Terminal	Suturing
Description	Rift valleys form as continents begin to split due to divergent plate motion under a continent.	A narrow sea forms as plate divergence continues and seafloor spreading begins.	A broad ocean basin exists from the ongoing seafloor spreading originating at a mid-ocean ridge.	Trenches and volcanic arcs develop as seafloor is subducted due to convergent plate motion some distance from the ridge.	Narrow ocean basin as ongoing subduction at a convergent margin has destroyed most of the seafloor.	Ocean basin is entirely consumed as continents along its margins fully collide forming large mountains.
Modern-day example	East African Rift Valley	Red Sea	Atlantic Ocean	Pacific Ocean	Mediterranean Sea	Himalayas

that the global rate of lithosphere addition by seafloor spreading is always matched by the global rate of lithosphere loss by subduction. If the global addition rate of new lithosphere exceeded the global destruction rate of old lithosphere, then the earth would expand; conversely, a global destruction rate of old lithosphere faster than the global addition rate of new lithosphere would cause the earth to contract. There is no geologic evidence that earth has expanded or contracted in this way through its history, implying that the global rates of lithosphere production and lithosphere destruction have remained balanced.

> **Box 1 What are mantle plumes, hotspots, and large igneous provinces (LIPs)?**
>
> In Sect. 3.5 on tectonic plate boundaries, you learned that divergence between the North American and Eurasian plates can be observed on the surface of Iceland, and that the reason we can see this phenomenon above sea level is due to a plume of magma from the mantle. Iceland is just one example of many so-called "hotspots" in the world that are formed by mantle plumes. The distribution of mantle plumes is intriguing. While the Icelandic mantle plume (and its resulting hotspot) occurs at a divergent plate margin, other mantle plumes have been identified where three plate boundaries meet (a "triple-junction"), and many mantle plumes exist far from plate boundaries—at the interior of tectonic plates. The potential relationship (if any) between large-scale tectonics and mantle plumes is an area of active research.

Buoyant magma is theorized to rise from a source deep and relatively stationary in the mantle. Once the plume reaches the lower lithosphere, it partially melts, and at times that melt rises to the surface and produces volcanic activity called a hotspot. Because each mantle plume is located at a relatively constant position within the Earth, when a tectonic plate moves over a plume, a chain of volcanoes can form on the plate's surface (see Fig. 8, center panel). For example, the Hawaiian Islands and their associated submerged seamount chain that stretches for about 2450 km into the northwestern Pacific Ocean is an example of a hotspot track. Seafloor spreading of the Pacific plate causes each new volcano to move to the northwest after it is formed. With this motion, the oldest part of the Hawaiian island hotspot track is in the northwestern Pacific. As the plate moves in that direction, the melt from the relatively stationary mantle plume erupts at locations to the southeast of the previous volcanic activity. Indeed, the island of Hawaii (also known as the Big Island) is the youngest and most active volcano in the chain and is also the furthest to the southeast (Fig. 9).

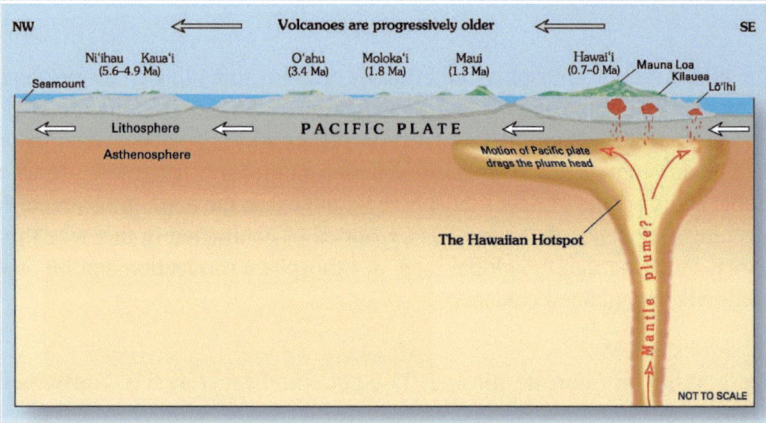

Fig. 9 Cross-section depiction of the Hawaiian island chain and the inferred mantle plume that has fed the Hawaiian hotspot on the overriding Pacific Plate. The geologic ages of the oldest volcano on each island are progressively older to the northwest, consistent with the hotspot model for the origin of the Hawaiian Ridge-Emperor Seamount Chain. Illustration by Joel E. Robinson, USGS[17]

---

[17] Illustration by Joel E. Robinson (2012). USGS. From File:Hawaii hotspot cross-sectional diagram.jpg, Wikimedia Commons, https://commons.wikimedia.org/wiki/File:Hawaii_hotspot_cross-sectional_diagram.jpg. Public Domain.

Additionally, at certain times in the geologic past extremely active mantle plumes—sometimes called "superplumes"—have produced enormous quantities of magma during relatively short periods of time. These eruptions, on a scale not seen on Earth at the present time, produced geologic features called **large igneous provinces (LIPs)**. These provinces include massive emplacements of volcanic rocks in the form of continental flood basalts and giant oceanic plateaus. The size, volume, and production rate of igneous rock in LIPs are impressive! For example, at the "small" end of the spectrum, the Columbia River Flood Basalts in the northwestern US span an area of ~ 210,000 km^2 and a volume of ~ 175,000 km^3; this means that the average thickness of the Columbia River Basalts is just over a half mile. They were emplaced as ~ 350 individual lava flows, some exceeding 1000 km^3, emerged from dikes connected to a subsurface mantle plume. Most of this material was emplaced during just a 1.1-million-year time period in the Miocene,[18] meaning that the average location affected by the Columbia River Basalts was buried by ~ 2.5 feet of new basalt every 1000 years. The largest LIP is the submarine Ontong Java Plateau (OJP) in the southwest Pacific Ocean. You will learn more about its emplacement and its impact on climate during the Cretaceous Period in Sect. 4.

4 Plate Tectonic Processes and Long-Term Climate Change

In Sect. 3, you learned the fundamentals of plate tectonics. The reason for reviewing plate tectonics in a textbook on climate change may not have been obvious at first—and if you learned about plate tectonics sometime during primary or secondary school its connections to climate change were likely not raised at all. However, the connections are significant: plate tectonics, as well as hotspot volcanism (Box 1), have played major roles in the evolution of Earth's climate over time scales of millions to billions of years. We will explore those connections explicitly; focusing on cases that illustrate the potential effects of changes in the carbon budget, the opening and closing of ocean gateways, and mountain building. Additionally, the climatic impact of explosive volcanism over much shorter time scales (just years) is explored in Box 2, and the direct long-term impact of tectonics on sea level is addressed in Box 3. As a preface to the case studies, some fundamentals need to be established to augment the background you have on plate tectonics itself: (a) recognition that Earth has two main climatic states, (b) familiarization with the role of volcanism in

---

[18] For more information on the Columbia River flood basalts see the USGS website (accessed 2023, June 14): https://www.usgs.gov/observatories/cvo/columbia-river-basalt-group-stretches-oregon-idaho

Earth's long-term carbon cycle, and (c) familiarization with the concept of oceanic gateways.

4.1 What Are Greenhouse and Icehouse Climate States?

Since at least the Neoproterozoic Era (starting 1 billion years ago), Earth's climate has fluctuated between two climatic states: Greenhouse (also called Hothouse) and Icehouse states. Greenhouse conditions are warm, with high atmospheric CO_2 levels, no long-lived large ice sheets, and high sea levels. In contrast, Icehouse conditions are cooler, with low atmospheric CO_2 levels, long-lived large ice sheets, and low sea levels. Earth is currently in an Icehouse state, as it has large ice sheets in Greenland and Antarctica.[19] Such Icehouse states, however, account for far less of Earth's history than Greenhouse states (Fig. 10). The transitions between Greenhouse and Icehouse conditions in Earth's history have sometimes been gradual and other times abrupt, but in all cases—*except* for the climate state transition that we appear to be in the midst of now (see Chap. 1, Fig. 2)—geologic processes were the driving influences.

4.2 What Is the Role of Volcanism in the Long-Term Carbon Cycle?

As you read in Chap. 2, the carbon cycle is central to the story of Earth's climate evolution. The global carbon cycle (Fig. 11) can be subdivided into a short-term cycle and a long-term cycle. The short-term carbon cycle operates on time scales of seconds up to hundreds of years. Photosynthesis and respiration/decomposition are key processes in the short-term carbon cycle that convert carbon compounds into new forms. In contrast, the long-term carbon cycle operates on time scales of thousands to millions of years. Volcanism, carbon burial (rock formation) and rock weathering are important processes in the long-term carbon cycle that convert carbon compounds into new forms, and exchange carbon between rocks and the Earth's other systems, including the ocean, the biosphere, the soil, and the atmosphere. Here, we will briefly focus on the roles of volcanism, and by association tectonics and/or mantle plumes (hotspots, Box 1), in the long-term carbon cycle because these are significant factors in the regulation of Earth's climate over geologic time. In Chap. 7, you'll explore the long-term carbon cycle in additional detail and will find out how the formation and weathering of rocks has a direct impact on atmospheric carbon concentrations, and thus global temperature as well.

---

[19] Glacial and interglacial periods (described in Chap. 4 and addressed more in Chap. 9) are relatively colder and warmer times within an Icehouse state. Even in an interglacial period, ice sheets still exist (as they do today). Greenhouse states, however, have no long-lived ice sheets of any size.

Fig. 10 Timeline of Greenhouse (red-pink shading) and Icehouse states over the last 850 million years. Well-defined Icehouse states, periods for which there is evidence of continental ice sheets at one or both poles, are marked by dark blue bars. Light blue shading represents cool climatic conditions, but currently lack evidence of the continental ice sheets characteristic of well-defined Icehouse states. Figure used with permission from the National Research Council (NRC). Source NRC (2011).[20] This figure is excluded from Creative Commons license

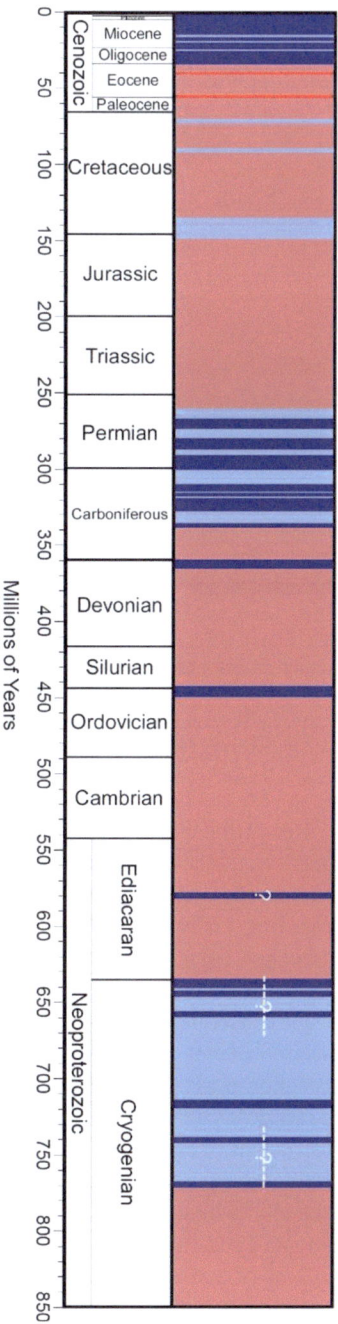

---

[20] National Research Council (2011). Understanding Earth's Deep Past: Lessons for Our Climate Future. The National Academies Press. https://doi.org/10.17226/13111. Used with permission.

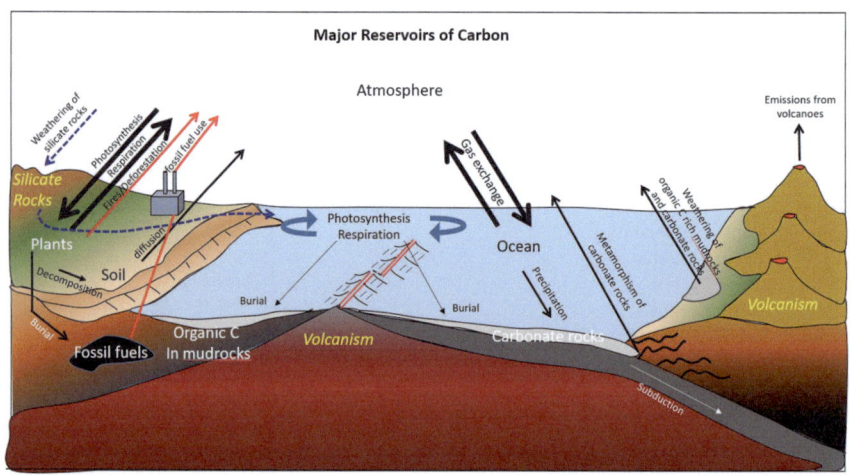

Fig. 11 Earth's carbon cycle. Major reservoirs of carbon include the lithosphere (carbonate rocks and organic carbon in rocks), ocean, soil, atmosphere, and terrestrial plants. Processes that exchange carbon between reservoirs are noted with arrows, with the size of the arrow representing relative rates of exchange; thicker arrows indicate faster rates of exchange (i.e., their importance on short time scales), and thinner arrows indicate slower rates of exchange (i.e., their importance on long-timescales). Black and blue arrows are natural processes. Red arrows are anthropogenic processes. Natural, short-term carbon cycle processes include biological processes of photosynthesis and respiration/decomposition, and physical processes of gas exchange between the atmosphere and ocean. Long-term carbon cycle processes include volcanism, carbon burial (rock formation) and rock weathering (and rock metamorphism). Figure by Kristen St John and Lee Kump

Recall that the volcanism occurs in association with both divergent and convergent plate margins: fissure eruptions take place at the divergent rift zones, and volcanic arcs exist on the overriding plate adjacent to subduction zones. In addition, you learned from Box 1 (Figs. 8 and 9) that plumes of hot magma from the mantle can also result in hotspot volcanism. Such hotspots can be found all around the globe, resulting in additional volcanism on both continental and oceanic portions of plates. One of the major consequences of volcanism is the release of various gases, including copious amounts of water vapor, as well as sulfur dioxide, hydrogen sulfide, and the important greenhouse gas, carbon dioxide. According to the United States Geological Survey, the world's volcanoes, both on land and below sea level, generate about 260 million metric tons of carbon dioxide annually.[21] While that's less than one percent of global greenhouse gas emissions coming from human industrial and automotive activity annually (which is around 36 billion tons![22]), the long-term effect of volcanic carbon dioxide emissions played a crucial role in our planet's climate regulation over the billions of years of Earth's history before the human Industrial Revolution. Both

---

[21] USGS (accessed 2023, June 14). Volcanoes Can Affect Climate. https://www.usgs.gov/programs/VHP/volcanoes-can-affect-climate.

[22] Liu et al. (2023).

Plate Tectonics and Long-Term Climate Change 263

trends of past climate change and specific climatic events have been connected to volcanism.

> **Box 2 Explosive volcanic eruptions and short-term cooling**
>
> Even though we often associate volcanic eruptions with lava, ash, and pyroclastic flows that can reach temperatures of well over 1000 °C, some of these calamitous events actually cause cooling at the Earth's surface. Such volcanic activity was discussed in Chap. 3 as a climate forcing mechanism important and well documented in human-recorded, instrumental, and proxy records of the Common Era (i.e., the last 2000 years; see Fig. 1a in Chap. 3).
>
> As discussed in this chapter volcanoes act as vital players in the long-term global carbon cycle via the outgassing of CO_2, however highly explosive volcanoes can have a very different and significant short-term impact on climate through their output of sulfur dioxide from occasional very large explosive eruptions. If an eruption is large enough, it can inject sulfur dioxide and ash into the stratosphere. The sulfur dioxide combines with oxygen and water vapor producing sulfuric acid in the stratosphere, which then condenses into fine sulfate aerosols:
>
> $$SO_2 + O_2 + H_2O \rightarrow H_2SO_4$$
>
> These sulfate aerosols increase reflection of radiation from the Sun, and that radiation is sent back to space, cooling the troposphere and the Earth's surface. Volcanic aerosol clouds can last in the stratosphere for a few years, dissipating by chemical reactions and the settling of aerosols to the Earth's surface (note here that the emplacement of magma and warming effects of carbon dioxide take place over longer time scales). A closer look at this short-lived but significant cooling effect is highlighted below using two major, and famous, volcanic eruptions.
>
> - **Mount Tambora, Indonesia, 1815**: The 1815 eruption of Mount Tambora in the Indonesian volcanic arc was the largest volcanic eruption in human recorded history. The main blast took place on the evening of April 10, 1815, and the resulting debris, pyroclastic flows, and tsunamis immediately killed at least 10,000 islanders, and likely many more. Mount Tambora released an estimated 60 million tons of sulfur into the atmosphere, and the aerosol cloud produced by the massive eruption blocked radiation from the Sun reducing the average global temperature by 3 °C. The initial impact of the temperature decrease was felt in areas surrounding the eruption, with an estimated 80,000 losing their lives due to crop failure and famine. Cold temperatures were felt as far away as eastern North America and western Europe. In fact, the summer of 1816 is known as "the year without a summer," as snow and ice lingered in these regions throughout

June, July, and August. Interestingly, the dark and cold conditions in the summer of 1816 are said to have inspired Mary Shelley to write her famous book *Frankenstein*. Additionally, some have argued that these conditions also inspired the invention of what would eventually become the bicycle, as there was a lack of oats to feed horses and other livestock.
- **Mount Pinatubo, Philippines, 1991**: Mount Pinatubo erupted on June 15, 1991 and was one of the largest and most consequential eruptions of the twentieth century (Fig. 12). The volcano emitted the largest sulfur dioxide cloud (15–20 million tons) ever recorded by satellites. The 1991 eruption of Mt. Pinatubo was sufficient to cool the entire planet by about 0.4 °C for ~ 15 months.[23]

Fig. 12 Mt. Pinatubo in the Philippines began to display eruptive events in April of 1991 and erupted spectacularly on June 12. Three days later, the volcano exploded and sent ash ~ 35 km into the air. From the USGS[24]

---

[23] While the 2022 Tonga eruption was larger than the 1991 Pinatubo eruption, it did not emit enough sulfur dioxide to impact global climate.

[24] Photo by Dave Harlow, USGS. https://www.usgs.gov/media/images/june-12-1991-eruption-column-mount-pinatubo-taken-th.

The cooling effect on the planet of explosive volcanic eruptions is of great interest not just to paleoclimate researchers, but also to scientists and engineers interested in climate intervention (i.e., geoengineering) options for addressing modern global warming. Stratospheric aerosol injection (SAI) is a topic discussed in detail in Chap. 11. SAI is designed to replicate the volcanic cooling effect by artificially introducing sulfur dioxide into the stratosphere, and maintaining it by replenishing the aerosol layer at regular intervals.

4.3 What Is an Oceanic Gateway?

An **oceanic gateway** is a passage between ocean basins, which opens or closes due to tectonic processes. The opening of an ocean gateway is associated with tectonic plate divergence and the early stages of an ocean basin's development as continental rifting occurs and landmasses separate. For example, the Drake and Tasman Passages that exist between Antarctica and South America, and Antarctica and Australia, respectively, are modern oceanic gateways that allow the Antarctic Circumpolar Current to flow unimpeded around Antarctica. In contrast, the closing of oceanic gateways is associated with tectonic plate convergence. For example, the Isthmus of Panama that exists today connects South and North American landmasses. However, when it formed it severed the east to west flow of tropical surface ocean currents from the Atlantic to the Pacific Ocean. As ocean basins and the gateways that connect them evolve over long-timescales, ocean circulation does as well. Changes in ocean circulation have been central to transitions in global climate as these circulation changes have affected the direction and rates of water flow and heat transfer across latitudes.

4.4 Case Studies of Long-Term Climate Change

In the preceding sections, you were introduced to tectonic-related geologic processes that operate on long-timescales and influence the presence of Greenhouse versus Icehouse states. These include volcanic impact on the carbon cycle and tectonic influences on ocean circulation. With these processes in mind, we'll now explore case studies of particular times and places where one or both of these geologic phenomena contributed to either maintaining a particular climatic state or transitioning the planet from one climatic state to another. First, we'll look at the Cretaceous Period as a classic example of the Earth system in a Greenhouse state. This will be followed by case studies that examine tectonic factors over the past ~ 50 million years that influenced Earth's climate transition to the Icehouse state that has persisted to present day. In Chap. 7, you will revisit some of the same case studies to explore the additional influence that carbon burial (rock formation) and chemical weathering of continental rocks had on long-term climate.

4.4.1 Case Study 1: Greenhouse Conditions of the Mid-Cretaceous Period

Let's start by looking at a paleogeographic reconstruction of the Cretaceous ~ 90 million years ago (Fig. 13). Such maps are made based on evidence from the time-specific rock record regarding tectonic plate configurations, geography, and environmental and climatic conditions. We see that at this time, the supercontinent Pangaea was in the process of breaking up. Although the opening of the Atlantic Ocean between North American and Africa was well on its way, the southern Atlantic Ocean was still in the earliest stages of development, and the northernmost North Atlantic (between Greenland and northern Europe) had not yet rifted (recall stages in the life cycle of an ocean basin, Table 1). The Indian subcontinent had rifted from the southeastern coast of Africa, but Australia was still connected to Antarctica. If we look to the north, we observe that the paleopositions (90 million years ago) of both Greenland and Canada were similar to today. We also see how pervasive shallow seas were around the globe in the middle of the Cretaceous. In North America, a vast inland sea called the Cretaceous Western Interior Seaway dominated what is now the Great Plains. Shallow seas were also prominent in what is currently northern Europe, and even extended into what is now one of the driest regions on Earth: the Sahara Desert. Note too, the absence of ice sheets on the Cretaceous map, even given the polar position of Antarctica.

Geologic mapping of Cretaceous rocks reveals that this was a time of extensive limestone and chalk formation (see Chap. 5, Fig. 13a). In fact, the name "Cretaceous" is derived from the Latin word *creta*—chalk. Limestone formation most often occurs in warm and shallow marine waters. Most of the limestone-forming environments on Earth today, including the Caribbean Sea, Indian Ocean, and the tropical Pacific, can be found between 30° N and 30° S. Widespread limestone formation, including units deposited in the Cretaceous Western Interior Seaway, which stretched across a wide range of latitudes during this period, thus suggests overall warmer global temperatures.

Other pieces of direct evidence pointing to warmer temperatures during the middle of the Cretaceous come from fossils preserved in sedimentary rocks. For example, Cretaceous rock units from northern Greenland reveal fossils of breadfruit trees, which are a type of fig tree. Modern-day breadfruit trees can be found on the tropical islands of French Polynesia in the Pacific Ocean. Additionally, fossils of the *champsosaur*, an alligator-like creature from the Late Cretaceous, were found in the Canadian Arctic. In the Southern Hemisphere, Cretaceous plant fossils have been found on the Antarctic Peninsula and other sedimentary rocks dated to the Cretaceous point to evidence of a temperate rainforest community at a paleoposition of 82° S latitude. Collectively, these fossil data imply that the polar latitudes were at the very

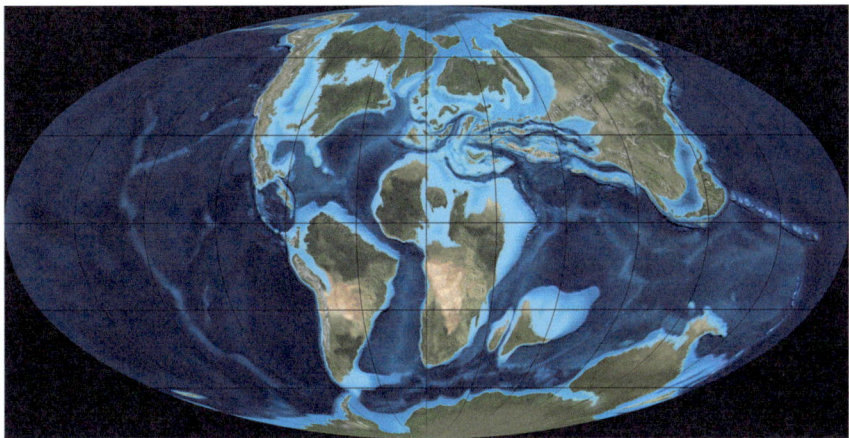

Fig. 13 Paleogeographic map of the Late Cretaceous (~90 million years ago). From Scotese (2016).[25]

least above freezing. Additionally, compilations of oxygen isotope measurements from surface-dwelling foraminifera from the Falkland Plateau near the southern tip of South America suggest sea surface temperatures approximately 90 million years ago were at least 30 °C (86°F).[26]

Other forms of evidence of the Cretaceous warmth come from marine algal biomarkers. As you read in Chap. 5, biomarkers are molecular fossils and organic material preserved in the sedimentary record that serve as proxies for environmental conditions, such as temperature and CO_2, at the time they were deposited. Another line of evidence comes from global climate models that take into account paleogeography and proxy-data for paleotemperatures to predict paleo-CO_2 levels. Results from such models predict high levels of atmospheric carbon dioxide in the middle Cretaceous, much higher than today's levels of about 420 parts per million (ppm). In fact, most climate model estimates for atmospheric carbon dioxide during the mid-Cretaceous range between 600 and 2400 ppm.

In sum, a wide range of geologic evidence and model results points to warm temperatures with very high CO_2 levels during the middle Cretaceous. Now we can examine the question: why? What tectonic-related climate factors were especially influential at this time to maintain a Greenhouse climate state? The answer lies in the fact that the Cretaceous was an exceptional time for volcanic activity *from both tectonic and hotspot* (mantle plume) processes. Recall, the breakup of Pangaea was underway during the Cretaceous. As continental lithosphere thinned and rifted throughout this period, new basins formed and grew into oceans, drastically increasing the amount of volcanic activity at mid-ocean ridges. Globally, this

---

[25] Scotese, C.R., 2016. PALEOMAP PaleoAtlas for GPlates and the PaleoData Plotter Program, PALEOMAP Project, http://www.earthbyte.org/paleomap-paleoatlas-for-gplates/. Christopher Scotese, PaleoAtlas v. 3 licensed under the Creative Commons Attribution 4.0 International License

[26] Bice et al. (2003).

increase must have been balanced by increased subduction around the perimeter of the paleo-Pacific, with increased volcanism at volcanic arcs. Some atmospheric-ocean model results specifically point to the formation and opening of the North Atlantic Ocean and the outgassing of CO_2 at its mid-ocean ridge as a main contributing factor of the middle Cretaceous Greenhouse. Increased outgassing of CO_2 from newly formed mid-ocean ridges in general has been cited by scientists as one of the main contributors of the warm Cretaceous climate.

Augmenting the volcanic CO_2 emissions from the newly formed mid-ocean ridges was additional CO_2 from the Large Igneous Province (LIP)-style hotspot volcanism in the middle Cretaceous. Recall from Box 1 that what defines LIPs are their extremely large emplacements of igneous material over a relatively short time. Some LIPs cover areas of more than 1 million km^2. The Cretaceous Period was a time of intense LIP activity, particularly in the western Pacific and the Caribbean. The largest LIP is the submarine Ontong Java Plateau (OJP) in the southwest Pacific Ocean; it is ~ 2 million km^2 (the size of western Europe) and the volume of igneous rock forming the plateau is ~ 60 million km^3! The main emplacement event at the OJP occurred about 120 million years ago, with a second, smaller event occurring around 90 million years ago at the height of the Cretaceous Greenhouse. The amount of magma production was immense—likely exceeding the entire production from all of the mid-ocean ridges at the time.[27] Additionally, on the other side of the globe, volcanic eruptions were building the plateau of the Caribbean large igneous province, with most of its eruptive episodes occurring between 95 and 88 million years ago.

Although the exact volume of gas produced by these LIP events is difficult to know, the sheer size and estimated volume of magmatic material associated with LIPs suggest that copious outgassing occurred during these events, transferring geologically-stored greenhouse gases from the lithosphere to the atmosphere, thus warming the planet. While there are still many intriguing questions regarding the origin and formation of LIPs, scientists have found that they have greatly impacted the entire Earth system, from climate to the biosphere, and from ocean chemistry to sea levels.

4.4.2 Transition to an Icehouse in the Mid to Late Cenozoic Era

With the exception of short-lived cool episodes in the mid to late Cretaceous, the Earth maintained warm Greenhouse state conditions well into the Cenozoic Era (Fig. 10). Geologic evidence indicates that the transition to an Icehouse began in the mid to late Cenozoic Era, with the onset of widespread glaciation of Antarctica occurring ~ 34 million years ago and the development of large-scale ice sheets in the Northern Hemisphere occurring only 2.75 million years ago. Several hypotheses and discussions about why the Greenhouse to Icehouse transition took place will be featured in this chapter, as well as in Chap. 7. Here, we'll examine the possible influences that tectonically-driven changes to land and ocean configuration had on

---

[27] Fitton et al. (2004).

the climatically-important global distributions of heat and moisture. This will involve an examination of the opening and closing of key oceanic gateways in the Cenozoic Era. In Chap. 7, you will examine the prominent roles of other factors, including changes in the effects of the weathering of continental rocks following the formation of new mountain ranges.

Case Study 2: Earth System Connections Among the Opening of the Tasman and Drake Gateways, Ocean Circulation, and the Onset of Antarctic Glaciation

Two regions of our planet that experienced dramatic change over the course of the Cenozoic Era are Antarctica and the Arctic. Recall that during the Cretaceous there were no ice sheets in either of the polar regions. Today, of course, Antarctica and Greenland host the planet's largest continental ice sheets. Let's examine the possible roles that the tectonically-driven opening of ocean gateways played in the onset of Antarctic glaciation, and we'll turn our attention to the northernmost latitudes in the next section.

As we read earlier in this chapter, the southern half of the supercontinent Pangaea (called Gondwana) consisted of the following modern landmasses: Africa, India, Australia, South America, and Antarctica. Early in the Cenozoic, while Africa and India had rifted and shifted away from Antarctica, both Australia and the southern tip of South America were still connected to our southernmost continent. Any west-to-east oceanic flow adjacent to Antarctica was restricted to geographically separated portions of the southernmost Atlantic and southernmost Pacific, because today's interconnected Southern Ocean did not yet exist. As a result, gyres of oceanic surface circulation transported warm waters from lower latitudes relatively close to Antarctica; an ancient Southern Hemisphere analog of the warm waters presently transported northward by the modern-day Gulf Stream and North Atlantic Currents, which help to moderate climates in the British Isles and northern Europe, and ultimately bring heat to the Arctic.

The Drake Passage developed when the southern tip of South America fully detached from Antarctica. When Australia eventually rifted and separated from Antarctica, the oceanic gateway called the Tasman Passage opened. The opening of these two oceanic gateways over millions of years eventually resulted in unrestricted flow around the Antarctic continent, and the creation of the Antarctic Circumpolar Current (ACC). Researchers in the 1970s proposed that the surface of Antarctica was topped by vast continental ice sheets only *after* Australia and South America separated from it.[28] They hypothesized a cause-and-effect connection: the tectonically-controlled opening of the oceanic gateways allowed the formation of the ACC which blocked the transport of heat to Antarctica by southward-flowing ocean currents. No longer were the warm western boundary ocean currents that flow from low to high latitudes on the western side of ocean gyres able to bathe the Antarctic continent

---

[28] Kennett and Brunner (1973).

with warm water. In essence, Antarctica, already positioned in a polar setting, was no longer in a refrigerator, but was put into an insulated freezer. The presentation of this hypothesis kick-started paleoclimate research in and around Antarctica to understand the timing of the onset of glaciation, the timing of the opening of the Drake and Tasman Passages, and the relative importance (among other possible drivers) of the formation of the ACC in cooling Antarctica.

Similar to the development of plate tectonic theory, climate scientists also use a variety of methods, both direct and indirect, to gather different types of data to prove or disprove hypotheses and make additional discoveries. As the opening and closing of oceanic gateways takes place over million-year timescales, we will be examining evidence and data primarily collected from the sedimentary record, using approaches described in Chap. 5.

Let's explore the question of when glaciation first appeared in Antarctica. Since ice sheets blanket Antarctica today, scientists must look for clues regarding million-year-old periods of glaciation away from the main continental landmass. Of particular interest in this case is off-shore sedimentary strata from the continental slope and shelf that have been dated to around 40 to 35 million years using biostratigraphy and radiometric dating. The presence of glacially influenced sediments from these locations indicate that the earliest Antarctic ice sheets began to develop about 35 million years ago.

With data pointing to widespread expansion of Antarctic ice sheets at about 35 million years ago, we can now explore the *why* of the problem: why did widespread ice sheets appear in Antarctica at that time and not before?

The Drake Passage is the relatively narrow ocean gateway that currently connects the South Pacific and South Atlantic Ocean basins (Fig. 14). Datasets derived from a variety of techniques point to the Drake Passage opening between about 40 and 20 million years ago. This date range roughly aligns with the timing of the development of Antarctic glacially influenced sediments, suggesting that the establishment of the ACC did indeed have a significant impact in cooling the continent. The tectonics of the Drake Passage region are exceedingly complex, and research is ongoing to refine when exactly the crucial opening of this gateway occurred.

To the east, the Tasman Passage formed as a result of Australia and later, Tasmania, separating from Antarctica. Many research studies investigating this topic have used an array of methods on sediment core samples, including micropaleontologic, sedimentologic, geochemical, and paleomagnetic techniques. These studies indicate that the opening of the Tasman Passage began around 37 million years ago, with the presence of shallow waters and a deltaic environment in place about 35 million years ago, and a fully opened ocean gateway between Australia and Antarctica a few million years later, by about 33 to 30 million years ago.

In summary, the openings of the Drake and Tasman Passages are consistent in timing with the age of the first glacially influenced sediments from Antarctica. Though research is ongoing, the tectonically-driven opening of these two oceanic gateways established the ACC and transformed global ocean circulation and climate.

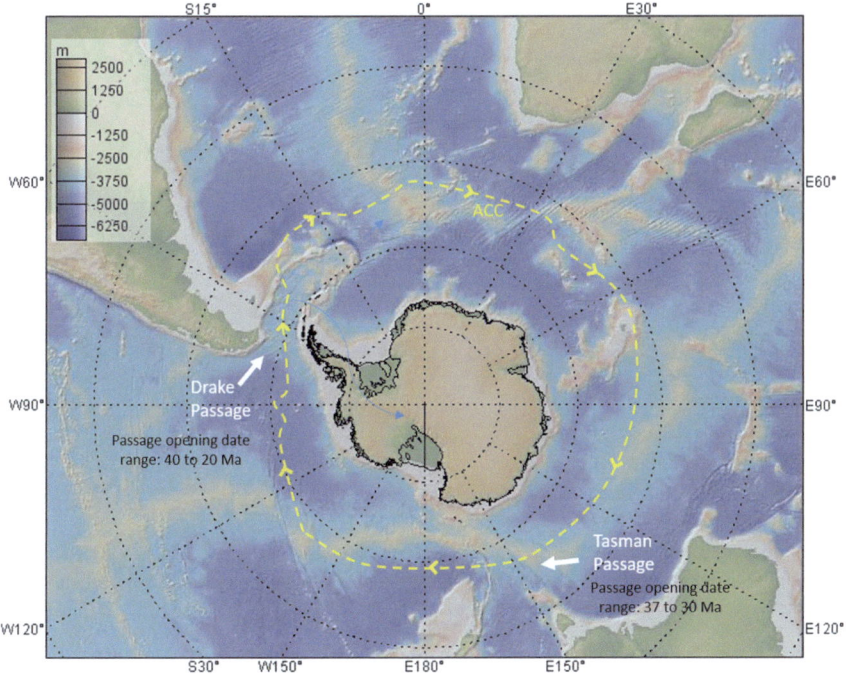

Fig. 14 Map showing the generalized position of the Antarctic Circumpolar Current (ACC), which developed after the opening of the Drake and Tasman Passages[29]

Case Study 3: Earth System Connections Between Neogene Tectonics, Arctic River Discharge, and Northern Hemisphere Glaciation

Though large ice sheets were established in Antarctica at the Eocene–Oligocene transition, 34 million years ago, it was a very different story for high latitudes in the Northern Hemisphere. Widespread cooling and associated glaciation in the Arctic region wouldn't take place until over 30 million years *later*. It is hypothesized that two far-distant tectonic-driven processes—in central America and in the interior Asian continent—may have played contributing roles in creating suitable conditions for ice formation in the Arctic. Let's consider how.

The Isthmus of Panama in Central America is famous today for the Panama Canal, a great engineering marvel that, once completed in 1914, allowed ships to transit from the Pacific Ocean to the Atlantic Ocean (and vice versa) without having to traverse all the way around the southern tip of South America. However, this narrow strip of land wasn't always there. Evidence from sediment cores and other data indicate that about 5 million years ago, North America and South America were not connected as they are today. At that time, tropical waters of the Pacific and the Atlantic mixed

---

[29] Map created using GeoMapApp 3.6.15. CC BY 4.0. Generalized position of AAC based on AAC polar front data from Orsi and Harris (2019).

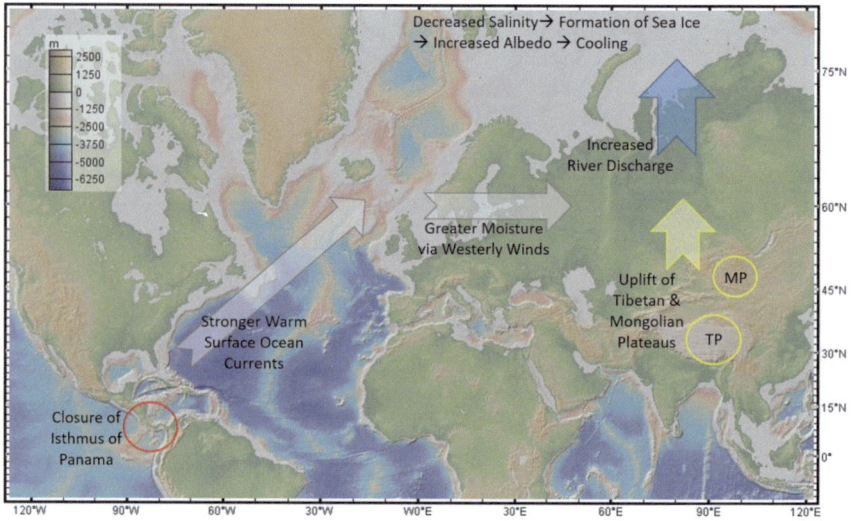

Fig. 15 Tectonic-initiated influences on ocean circulation and precipitation that may have influenced high latitude cooling, leading to Northern Hemisphere glaciation. After Driscoll and Haug (1998) and Wang (2004), as depicted in Stein et al., (2021; Fig. 6)[30]

and were allowed to flow freely between these continents. However, over the next ~ 2 million years, active plate tectonics of the region drastically altered the geology as well as global oceanic and atmospheric circulation.

Five million years ago, the Central American Seaway (where the Isthmus of Panama is located today) was peppered with underwater volcanoes as the Pacific plate subducted beneath the Caribbean plate. Evidence from sediment cores indicates that sediment from the coasts of North and South America made its way in between these underwater volcanoes; eventually, there was enough sediment to completely fill in the Central American Seaway. The Isthmus of Panama was created once these sediments, the associated volcanics, and fragments of other microplates accumulated and were uplifted by tectonic processes at this convergent margin (Fig. 15). Though uncertainties remain about the exact timing, most evidence from paleogeographic studies, sediment core samples, and clues from biodiversity point to the existence of a shallow passage 4 to 3 million years ago, and a complete isthmus closure 3 to 2.7 million years ago.

Geoscientists hypothesized that the closure of the Isthmus of Panama significantly impacted the Cenozoic Earth system. Once the Panama isthmus formed, ocean currents that once flowed between the Atlantic and the Pacific were completely cut off. Salinity levels rose sharply in the Atlantic and Caribbean as the tropical climate accelerated evaporation, whereas the Pacific Ocean gained more freshwater from precipitation supplied by the easterly trade winds. The Gulf Stream, a western boundary current that flows from the tropics to the North Atlantic, intensified once

---

[30] Stein et al. (2021); Driscoll and Haug (1998); and Wang (2004).

the isthmus formed, bringing more heat and moisture to northwestern Europe during the mid-Pliocene.[31] At this point, you might be thinking: wait—if the timing of the isthmus closure coincided with *warming* of the Northern Hemisphere, how did ice cover develop?

The rest of the story and a possible solution can be found on the other side of the globe, where the ongoing collision of India with Eurasia resulted in the uplift of the Tibetan Plateau as well as the Mongolian Plateau to the northeast. Paleoclimate researchers[32] suggested that the moisture sourced from the warmer and stronger Gulf Stream following the closure of the Isthmus of Panama was then transported to Eurasia by westerly winds (Fig. 15). This in turn produced precipitation that fell in the form of either snow or rain. Crucial to our story, the uplift of plateaus in central Asia and the associated mountain belts (including the Himalayan Mountains) determined the flow direction of all rivers on the Asian landmass—including Siberian rivers. It has been hypothesized[33] that Siberian rivers began flowing northward in the late Pliocene to early Pleistocene, causing an increase in freshwater delivered to the region to flow toward the Arctic Ocean. The Arctic region thus experienced an increase in freshwater input, both directly from the atmosphere as well as from the northward flow of Siberian rivers.

With added freshwater, the Arctic region was primed for sea ice formation (lowering the salinity means it is easier for water to freeze). Sea ice growth results in higher albedo, increasing the reflection of sunlight and heat back to space, while also preventing heat stored in the ocean from escaping to the atmosphere; both of these factors contributed to cooling the northernmost latitudes. Thus, with these two major factors—moisture and cooling—in place, the stage was set for glacial ice growth in the Northern Hemisphere. As proposed, the reorganization of rivers in Siberia, driven by tectonic uplift in Asia, may have been a key mechanism for the increased river input to the Arctic Ocean and in turn the expansion of the North Hemisphere glaciation.

From these examples, it is evident that the tectonic evolution of the solid Earth is crucial for establishing boundary conditions that affect oceanic and atmospheric circulation and atmospheric composition, and thus impacts the long-term evolution of Earth's climate. Other important processes also influenced long-term climate, especially those involved in the long-term carbon cycle (Fig. 11), such as silicate rock weathering, organic carbon burial, and feedbacks within the Earth systems. You will learn about these geologic processes and their climate-regulating roles in Chap. 7.

---

[31] In addition to initiating major changes in oceanic and atmospheric circulation, the closure of the Isthmus of Panama had a considerable effect on our planet's biodiversity. Mammals such as the opossum, the porcupine, and the armadillo that exist in North America today all have ancestral roots in South America. Other animals including dogs, cats, llamas, raccoons, and horses migrated south across the isthmus. Paleontologists call this event the Great American Interchange.

[32] Driscoll and Haug (1998).

[33] Wang (2004).

Box 3 Plate tectonics' influence on sea level

As you learned in previous chapters, changes in global sea level are primarily controlled by an increase or decrease in the amount of land-based ice (i.e., glaciers and ice sheets), and secondarily controlled by the thermal expansion or contraction of seawater as it warm and cools. However, over periods of millions to hundreds of millions of years, changes in plate tectonics can be another influence on global sea level.[34] We can examine these tectonic factors in the context of global sea level fall of ~ 170 m since the Cretaceous highstand (see map Fig. 13) that occurred 100 to 80 million years ago:

- **Tectonic Factor #1—Changing Sea Floor Spreading Rates**: Inspections of magnetic anomalies on the ocean floor indicate that seafloor spreading rates have decreased over the past 80 million years. One team of researchers used tectonic plate reconstructions and models to calculate seafloor spreading rates since the Early Cretaceous. They found that seafloor spreading rates in the Pacific were higher during the Early Cretaceous, as high as 9.2 cm/yr, and had slowed to 6 cm/yr at the dawn of the Cenozoic Era (a rate similar to today's).[35] Though the mid-Atlantic ridge is considered a slow spreading ridge with much lower rates of seafloor spreading than the Pacific, magnetic anomalies signal a similar slowing at both divergent plate margins since the Cretaceous. A slower rate of seafloor spreading means a smaller *area* of newly generated oceanic crust (see Sect. 3.5.1, Fig. 5). As the ocean crust moves away from the ridge and ages, it cools and becomes more dense, lowering the level of the ocean floor. In other words, a slower-spreading ridge will have a smaller volume than a fast-spreading ridge, thereby displacing less water upward and onto the continents. In conclusion, the slowing of seafloor spreading rates over the last 80 million years has been a contributing factor to sea level lowering.

---

[34] Miller et al. (2005).
[35] Seton et al. (2009).

- **Tectonic Factor #2—Changing Length of Mid-Ocean Ridges**: Another possible reason for sea level fall over the past 80+ million years has been a reduction in the total length of mid-ocean ridges around the globe. If you examine a map of present-day plate tectonics (e.g., Figs. 3 and 5), you'll find that the primary ridge of the Pacific is on the eastern side of the ocean. This ridge is very appropriately named the East Pacific Rise. If you trace the ridge north from the southeast Pacific, you'll eventually run into the southwestern coast of North America. Here, the mid-ocean ridge converts into the continental San Andreas transform fault system, where the Pacific plate slides to the northwest relative to the North American plate. The San Andreas fault system runs about 1200 km through California before the mid-ocean ridge reappears off the coasts of northern California (north of Cape Mendocino), Oregon and Washington state. Of course, it wasn't always this way. Throughout the Cretaceous, subduction was ongoing along the entire length of the North and South American coasts. The plate that once was subducted beneath North America is called the Farallon plate. Geologists hypothesize that Farallon plate subduction caused the uplift of the Rocky Mountains 80 to 50 million years ago. The majority of the Farallon plate has now been subducted; the parts of the Farallon plate that remain at the earth's surface are now known as the Juan de Fuca, Gorda,[36] and Cocos plates (Fig. 3). Geologic and seismologic evidence suggest that the subducted parts of the Farallon plate remain in the upper mantle beneath North America. For the sections of the Farallon plate that were fully subducted, the corresponding portions of the East Pacific Rise also were destroyed as subduction rates exceeded regional seafloor spreading rates. Moreover, the Farallon plate wasn't the only ancient plate to meet its end via subduction in the time since 80 million years ago. Two other ancient tectonic plates that once dominated Panthalassa (the proto-Pacific Ocean during the time of Pangaea), the Phoenix plate and the Izanagi plate, met a fate similar to that of the Farallon plate; mid-ocean ridges associated with these plates were also subducted in the process. The loss of hot, and thus less dense, oceanic material at mid-ocean ridges left more area for cooler crust and deeper ocean basins as we find in the western Pacific today, lowering sea level overall.

---

[36] The Gorda plate is located off the coast of northern California, just south of the Juan de Fuca plate. It is too small to be shown in Fig. 3.

- **Tectonic Factor #3—Changing Amount of Continental Material**: In Sect. 4.4, we discussed the collision of India with Eurasia in the early Cenozoic and the rise of the Himalayan Mountains and Tibetan plateau. The Alps of Europe and the Zagros Mountains of Iran were also formed around this time due to continent–continent collisions. When ongoing subduction eventually causes continents to collide, the continent to ocean ratio decreases, meaning there is more room available for oceans. Sea levels fall as a result.
- **Tectonic Factor #4—Changing Igneous Activity**: Also in Sect. 4.4, we considered how the emplacement of Large Igneous Provinces during the middle to Late Cretaceous might have contributed to the Cretaceous Greenhouse. They also contributed to sea level rise in the middle to Late Cretaceous as they added a significant volume of oceanic crust. Once these extensive igneous bodies ceased activity, however, they began to cool. Just as we observe with any aging oceanic crust, cooling results in an increase in density and these bodies subsided over time, producing a greater volume in the ocean basin and thereby lowering sea levels.

Thus, the controls on sea level are a combination of primary climate-related factors: the amount of ice on land and the thermal property of the water, but also secondary influences on sea level, several of which involve plate tectonics over million-year time scales.

References

Bice, K. L., Huber, B. T., & Norris, R. D. (2003). Extreme polar warmth during the Cretaceous greenhouse? Paradox of the late Turonian $\delta^{18}O$ record at Deep Sea Drilling Project Site 511. *Paleoceanography, 18*(2), 1031. https://doi.org/10.1029/2002PA000848

Crameri, F., Conrad, C. P., Montesi, L., & Lithgow-Bertelloni, C. R. (2019). The dynamic life of an oceanic plate. *Tectonophysics, 760*, 107–135. https://doi.org/10.1016/j.tecto.2018.03.016

Driscoll, N. W., & Haug, G. H. (1998). A short circuit in thermohaline circulation: A cause for northern hemisphere glaciation? *Science, 282*(5388), 436–438. https://doi.org/10.1126/science.282.5388.436

Fitton, J. G., Mahoney, J. J., Wallace, P. J., & Saunders, A. D. (Eds.). (2004). Origin and evolution of the Ontong Java Plateau. *Geological Society London Special Publications, 229*, 1–8. 0305-8719.

Glatzmaier, G. A. (1999, October 21). What causes the periodic reversals of the earth's magnetic field? Have there been any successful attempts to model the phenomenon? *Scientific American.* https://www.scientificamerican.com/article/what-causes-the-periodic/

Glatzmaiers, G., & Roberts, P. (1995). A three-dimensional self-consistent computer simulation of a geomagnetic field reversal. *Nature, 377*, 203–209. https://doi.org/10.1038/377203a0

Granot, R. (2016). Palaeozoic oceanic crust preserved beneath the eastern Mediterranean. *Nature Geoscience, 9*, 701–705. https://doi.org/10.1038/ngeo2784

Kearney, P., Klepeis, K. A., & Vine, F. J. (2009). *Global tectonics* (3rd ed., 496 p.). Wiley-Blackwell. ISBN: 978-1-405-10777-8

Kennett, J. P., & Brunner, C. A. (1973). Antarctic Late Cenozoic Glaciation: Evidence for initiation of ice rafting and inferred increased bottom-water activity. *GSA Bulletin, 84*(6), 2043–2052. https://doi.org/10.1130/0016-7606(1973)84%3c2043:ALCGEF%3e2.0.CO;2

Koppers, A. A. P., & Coggon, R. (Eds.). (2020). *Exploring Earth by scientific ocean drilling: 2050 science framework* (124 pp.). https://doi.org/10.6075/J0W66J9H

Liu, Z., Deng, Z., Davis, S., & Ciais, P. (2023). Monitoring global carbon emissions in 2022. *Nature Reviews Earth & Environment, 4*, 205–206. https://doi.org/10.1038/s43017-023-00406-z

Maxwell, A. E., & Von Herzon, R. P. (1970). *Initial reports of the Deep Sea Drilling Project 3*. U.S. Government Printing Office. http://www.deepseadrilling.org/03/dsdp_toc.htm

Miller, K. G., Kominz, M. A., Browning, J. V., Wright, J. D., Mountain, G. S., Katz, M. E., Sugarman, P. J., Cramer, B. S., Christie-Blick, N., & Pekar, S. F. (2005). The Phanerozoic record of global sea-level change. *Science, 310*, 1293–1298. https://doi.org/10.1126/science.1116412

Müller, R. D., Sdrolias, M., Gaina, C., & Roest, W. R. (2008). Age, spreading rates and spreading symmetry of the world's ocean crust. *Geochemistry. Geophysics, Geosystems 9*, Q04006. https://doi.org/10.1029/2007GC001743

Orsi, A. H., & Harris, U. (2019). Fronts of the Antarctic Circumpolar Current—GIS data, Ver. 1. Australian Antarctic Data Centre. https://data.aad.gov.au/metadata/records/antarctic_circumpolar_current_fronts. Accessed June 14, 2023.

Seton, M., Gaina, C., Muller, R. D., & Heine, C. (2009). Mid-Cretaceous seafloor spreading pulse: Fact or fiction? *Geology, 37*(8), 687–690. https://doi.org/10.1130/G25624A.1

Stein, R., St. John, K., & Everest, J. (2021). Expedition 377 Scientific Prospectus: Arctic Ocean Paleoceanography (ArcOP). International Ocean Discovery Program. https://doi.org/10.14379/iodp.sp.377.2021

Voosen, P. (2023, May 25). At long last, ocean drillers exhume a bounty of rocks from Earth's mantle. *Science, 380*(6648), 876–877. https://doi.org/10.1126/science.adi9181

Wang, P. (2004). Cenozoic deformation and the history of sea-land interactions in Asia. In P. Clift, W. Kuhnt, P. Wang, & D. Hayes (Eds.), *Continent-Ocean interactions within East Asian marginal seas*. Geophysical Monograph 149. https://doi.org/10.1029/149GM01

Witze, A. (2022). Why the Tongan eruption will go down in the history of volcanology. *Nature, 602*, 376–378. https://doi.org/10.1038/d41586-022-00394-y

Zhang, T., Gordon, R. G., & Wang, C. (2018). Oblique seafloor spreading across intermediate and superfast spreading centers. *Earth and Planetary Science Letters, 495*, 146–156. https://doi.org/10.1016/j.epsl.2018.05.001

Mary H. Schultz After becoming captivated by plate tectonics as an undergraduate student at Bryn Mawr College, Mary Schultz went on to earn her doctoral degree in Geological Sciences at Arizona State University where she explored the relationship between climate and tectonics in the evolution of the central Himalayan Mountains. In graduate school, Mary became interested in how earth processes and the environment impact communities in the United States and around the world. This led her to Washington, DC, where she served as a Congressional Fellow, working in the United States Senate on policy issues ranging from climate change to the space industry. She then turned to one of her primary passions, teaching, and was a Visiting Assistant Professor of Geology at James Madison University, a position she held for two years. Mary's fellowship experience and interest in translating complex scientific ideas to broader audiences led her to a position as a scientific research writer at Texas Children's Hospital in Houston. Mary remained in Houston and is now a Professor of Geology at San Jacinto College. Mary is the primary author of Chap. 6.

Open Access This chapter is licensed under the terms of the Creative Commons Attribution 4.0 International License (http://creativecommons.org/licenses/by/4.0/), which permits use, sharing, adaptation, distribution and reproduction in any medium or format, as long as you give appropriate credit to the original author(s) and the source, provide a link to the Creative Commons license and indicate if changes were made.

The images or other third party material in this chapter are included in the chapter's Creative Commons license, unless indicated otherwise in a credit line to the material. If material is not included in the chapter's Creative Commons license and your intended use is not permitted by statutory regulation or exceeds the permitted use, you will need to obtain permission directly from the copyright holder.

The Roles of Rock Formation and Weathering in Long-Term Climate Change

Lee Kump and James Kasting

Guiding Question: How does rock formation and weathering help drive long-term climate change?

1 Key Take-Away Points

- The weathering of silicate rocks at Earth's surface includes chemical processes that remove carbon dioxide (CO_2) from the atmosphere and convert it into limestone.
- Because the rate of chemical weathering is dependent on climate, and climate on atmospheric CO_2 level, there is a natural feedback between climate and weathering that controls climate on geologic time scales.
- Under the right conditions, organic carbon (produced through photosynthesis) can be buried in sediments and incorporated into sedimentary rocks. This burial removes CO_2 from the atmosphere for long time periods.
- Thus there is a balance over long geologic time scales between the input of CO_2 to the atmosphere from volcanoes and its removal through silicate rock chemical weathering and through photosynthesis followed by organic carbon burial.
- For a planet to support life, it must have a solid or liquid surface, have an abundant, available source of energy, have a source of essential nutrients, and have liquid water. On Earth the latter requirement is facilitated through the carbonate–silicate feedback loop.

L. Kump (✉)
College of Earth and Mineral Sciences, Pennsylvania State University, University Park, PA, USA
e-mail: lrk4@psu.edu

J. Kasting
Department of Geosciences and Meteorology, Pennsylvania State University, University Park, PA, USA
e-mail: jfk4@psu.edu

K. St. John and L. Krissek (eds.), *Climate Change*,
https://doi.org/10.1007/978-3-031-82869-0_7

- This feedback mechanism stabilized Earth's climate over the last 4 billion years during a time in which the Sun has increased its luminosity by 40%. Without climate regulation, Earth could have fallen into a frozen state early in its history or been driven to uninhabitable warm temperatures as solar luminosity increased.
- Living organisms can also affect the rate of silicate rock weathering, through their production of organic acids and CO_2 in soils that accelerate mineral dissolution. In doing so, they create a cooler climate than what would otherwise exist in their absence.
- Uplift of mountain belts also enhances chemical weathering by bringing fresh rocks to the surface. The uplift of the Himalayas caused a gradual cooling of Earth over the last 40 million years, driving the world toward a glacial "icehouse" state.
- Unlike the weathering of silicate rocks which consumes CO_2, the weathering of organic-carbon-rich rock (e.g., coal) releases CO_2 to the atmosphere. Therefore, fossil fuel burning is an enhancement of the natural process of organic carbon-rich rock weathering. This enhancement, though, is hundreds of times faster than the natural rate, and is thus driving rapid increases in atmospheric CO_2 level that cannot be countered by increased silicate rock weathering.

2 Introduction: How Has Earth Remained Habitable over the Eons?

You may have a thermostat where you live. If you do, you set the thermostat at your preferred temperature, and the furnace or air conditioner works to keep the home at that temperature. Now imagine instead that you had a new kind of temperature controller in which you programmed your broad preferences: the lowest temperature you can stand, your optimal temperature, and the hottest temperature you're willing to tolerate. For example, in the summer that might be a 68 °F (20 °C) minimum, an optimal temperature of 72 °F (22 °C), and a maximum temperature of 80 °F (27 °C). The controller would also determine the air conditioner's efficiency curve: how much energy it would need to expend to keep the room at a given temperature, given the outside temperature, the rate of heat transfer through your walls and roof, even the cost of energy. The intersection of these two curves would give the thermostat the temperature that satisfies both sets of requirements best. Why hasn't someone invented that?

Actually, Earth has. The long-term climate state of our planet has been regulated over its history by a system that involves volcanoes, atmospheric carbon dioxide (CO_2) levels, and the weathering of rocks on land and on the seafloor (as well as other long-term geologic processes). As you learned in Chap. 6, volcanoes emit carbon dioxide, and as we shall see, weathering of many common rocks consumes carbon dioxide. Interactions between these rates of CO_2 supply and CO_2 consumption set the atmospheric carbon dioxide level, and through the greenhouse effect, the average temperature of the planet on time scales of 100,000 years or longer. If an extended interval of greater volcanic activity on these time scales occurs, atmospheric CO_2 levels rise, the temperature increases, and warmer temperatures stimulate rock weathering, increasing weathering rates until they rise to meet the increased rate of volcanic CO_2 outgassing.

To understand how this climate regulation system works to maintain planetary habitability, we need to understand what determines the rate of CO_2 consumption by rock weathering. And to do that we have to learn some of the basics of geology: what are the different kinds of rocks and which are important to climate regulation, what the rock cycle is, what is weathering, and what controls it. Then we can take the cover off our global climate controller and see how it works. In doing so, we'll learn that it's not just about temperature; the supply of water and its retention in soils is important to rock weathering as well, as is biological activity. For these reasons, subtropical deserts on Earth are hot but dry and thus have very low rates of silicate weathering. We'll also learn that the weathering of some rocks actually releases CO_2 to the atmosphere. Ultimately the sustainability of Earth's climate system over the eons depends on the **rock cycle** and its fundamental driver, plate tectonics (see Chap. 6), because the surface emplacement of cooled (solidified) magma from Earth's interior and the tectonic uplift of mountains are needed to provide fresh rock to keep the weathering process going.

3   What Is the Rock Cycle?

Earth is a living, vibrant planet because it is so efficient at recycling. For example, about half the rain falling in the Amazon rainforest is pumped into the atmosphere from the soil by the rainforest trees themselves, a process called *evapotranspiration.* Soil water not pumped by trees enters the great Amazon River, where it returns to the ocean and can evaporate again. The other essential building blocks of life, carbon and nutrients, are similarly recycled. The most persistent, slowest, but ultimately most essential recycling is of the Earth itself (Fig. 1): the solid outer crust is formed through igneous processes of volcanic eruption and subsurface magma crystallization (*the latter forming intrusive igneous rocks*); vertical motions of the crust (**uplift**) driven by plate tectonics bring crust-building materials from Earth's interior to the surface, where they are exposed to rainfall, cycles of cold and warm temperatures (the seasons), and biological activity, all of which decomposes (weathers) the rock. Wind, ice, and rivers erode the weathered material (solids from soils, as well as dissolved materials) and transport it to the ocean, where sediments including the solid materials eroded from the continents and newly crystallized minerals, mostly the skeletons of organisms, are deposited on the seafloor. Downwarping of the crust (**subsidence**, the opposite of uplift), because of sediment loading and tectonic compression, and deposition of newer sediments buries seafloor sediments to depths where they are compacted and cemented together at high temperatures and pressures. Subduction at convergent margins is another way sediments (and sedimentary rocks) can be carried into the subsurface, where the heat and pressure can transform these materials into

The Rock Cycle

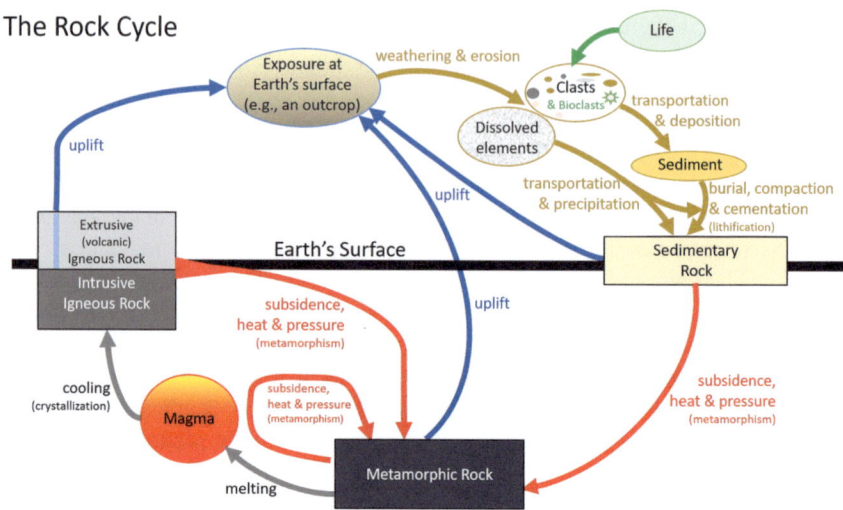

Fig. 1 Rock cycle. Figure by Kristen St. John, modified from a diagram by Steven Earle (2015)[1]

metamorphic rocks. In some cases the buried and/or subducted material is incorporated into the mantle, closing the cycle with the igneous processes of melting, eruption, and intrusion.

If plate tectonics stopped, there would be no subduction-zone volcanoes, uplift would cease, mountains would erode but not be renewed, and re-formation of the crust would only be by "hotspot" volcanoes emanating from mantle plumes (see Chap. 6). The planet would all but "die." This was the likely fate of Mars, a planet too small, whose interior cooled too quickly, and where plate tectonics, if it ever got started, wound down early in the planet's history.

In this chapter we're focusing on one segment of the rock cycle: weathering and the consequent formation of carbon-containing sediments and sedimentary rocks. As mentioned in the introduction, weathering is so important because it is the process that ensures that inputs of CO_2 to the atmosphere from volcanoes and the weathering of organic-carbon-bearing rocks are balanced by outputs from the atmosphere, in this case from the consumption of atmospheric CO_2 during silicate rock weathering and the formation of organic carbon-containing sediments. Moreover, that balance is achieved through feedbacks between weathering and atmospheric CO_2 levels. Without this CO_2 regulation mechanism, Earth may have become uninhabitable: without mechanisms to remove CO_2 from the atmosphere, volcanic CO_2 would simply accumulate, gradually increasing Earth's surface temperature to levels intolerable to life. The dry but very hot planet of Venus has no silicate weathering because there is no water available to drive the chemical reactions that degrade the minerals.

---

[1] Modified from Earle, S. (2015). Physical Geology. BCcampus Open Education. Figure 3.2. https://opentextbc.ca/geology/chapter/3-1-the-rock-cycle/. Creative Commons Attribution 4.0 International License

Venus has thus experienced a "runaway greenhouse effect" leading to a very dense CO_2 atmosphere and extremely hot temperatures.

3.1 How Is Carbon Dioxide Taken up During Rock Weathering?

Rocks are composed of minerals. Key here are the *silicate minerals*: minerals that contain the elements silicon and oxygen. Silicates are the most abundant minerals in Earth's crust, and are represented primarily by the *feldspars*, minerals that contain sodium, potassium, calcium, and magnesium along with aluminum, in addition to the silicon and oxygen that also compose another abundant mineral, quartz. Igneous, metamorphic, and sedimentary rocks that contain silicate minerals can informally be called *silicate rocks*. To keep the chemistry simple, let's consider a very simple silicate, a calcium-silicate mineral *wollastonite* ($CaSiO_3$). Wollastonite is found in metamorphic rocks. Having formed at high temperatures and pressures deep underground, it is unstable when exposed to the low temperatures and pressures at Earth's surface and is subject to weathering: its mechanical disintegration and chemical decomposition.

As you learned in Chap. 5, **physical weathering** (also called mechanical weathering) results from a variety of processes. In temperate climates, the seasonal cycle of freezing and thawing breaks apart rocks and minerals along cracks: as water freezes it expands, and this exerts great pressure, extending cracks and ultimately breaking the rock or mineral apart. The roots of plants can also work their way into these cracks, and their tips also exert pressures sufficient to break apart rocks. At high latitudes and high altitudes where glaciers form, their movement across the land surface grinds up rock. The finer material produced from physical weathering can be carried away (eroded) by streams or the wind, or accumulate in soils.

Physical weathering then sets the stage for **chemical weathering** (a process also briefly described in Chap. 5), the decomposition of minerals by chemical reactions, often driven by rain water that percolates into soil and dissolves the mineral. Gases that are dissolved in rainwater, including carbon dioxide and oxygen, participate in those chemical reactions as acids and oxidants (see below). The small particles produced by physical weathering are more reactive than the solid rocks from which they are derived and tend to dissolve in water. That dissolution can be complete (*congruent*) or partial (*incongruent*). Congruent weathering results in dissolved materials that can be carried in solution by rivers to the ocean where they become the "salts" that make the ocean salty. Incongruent weathering typically occurs when components of the original minerals contain aluminum, a highly insoluble element that typically is retained in soils in the form of *aluminosilicate* clay minerals (Fig. 2).

Fig. 2 Weathering of granite. The materials shown are sections of core drilled from the Occoquan granite, Fairfax County, Virginia, USA. The lower cylinder is fresh granite from 20 m below the surface and shows unaltered feldspar, quartz and mica grains. The loose material above it is from closer to the surface; in this material, the feldspars have been largely converted to clay minerals; quartz and mica remain. Core materials courtesy of the United States Geological Survey

The dissolution process is aided by the presence of natural acids. The most common of these is **carbonic acid**, the acid that results when carbon dioxide is dissolved in water. Rainwater is naturally acidic (Fig. 3) because CO_2 from the atmosphere dissolves in it. The carbonic acid content of soil water is further enhanced through the incorporation of CO_2 produced by root and microbe respiration (similar to our respiration of oxygen to form CO_2). Plants also exude other organic acids that facilitate the dissolution of silicate minerals. Importantly, dissolution of silicate minerals by carbonic acid in soil waters neutralizes the acid. In other words, it converts the CO_2 to a non-acidic carbon compound, bicarbonate (HCO_3^-), a negatively charged carbon species in water.

The overall process of silicate mineral dissolution by CO_2 in soil water can be described by a simple chemical formula, using the mineral wollastonite as the silicate example:

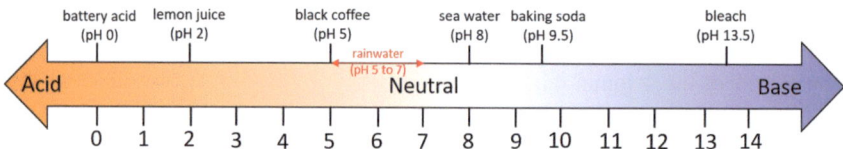

Fig. 3 pH scale. Rainwater has a pH ranging from 5 to 7

$$2(CO_2 + H_2O) + CaSiO_3 \rightarrow Ca^{2+} + 2HCO_3^- + SiO_2 + H_2O \qquad (1)$$

2 carbonic acid (carbon dioxide dissolved in water)
+ example silicate mineral → calcium ion
+ 2 bicarbonate ions + silica (all in solution in water) + water

Note that the CO_2 in soil water either got there through direct absorption of atmospheric CO_2 into rainwater, or indirectly from the atmosphere via intermediate processes of photosynthesis and respiration, as described above. So, ultimately, chemical weathering of silicate minerals removes CO_2 from the atmosphere and converts it into HCO_3^- which is carried by rivers and groundwater to the oceans.

3.2 What Is the Fate of the Neutralized Bicarbonate in the Ocean?

Bicarbonate will reside in the oceans for tens of thousands of years, and ultimately will be removed with the precipitation of carbonate minerals, most likely through the process of *biomineralization*, the formation of minerals by organisms forming shells or skeletons. *Carbonate biomineralizers* include corals, bivalves such as clams and oysters, and microscopic foraminifera and coccolithophorids. The latter microscopic organisms (see Chap. 5) are responsible for the calcareous ooze deposits found in some ocean basins, and the chalk forming the White Cliffs of Dover (see Chap. 5, Fig. 13a). The carbonate minerals are dominated by the mineral *calcite*, the main component of *limestone*. The process of carbonate mineral formation can also be expressed chemically:

$$Ca^{2+} + 2HCO_3^- \rightarrow CaCO_3 + (CO_2 + H_2O) \qquad (2)$$

calcium ion + 2 bicarbonate ions (both dissolved in water)
→ calcite + carbonic acid

Although CO_2 is actually released during this process, this is only a minor setback in the uptake of atmospheric CO_2 through weathering since two CO_2 molecules were consumed during the weathering process and only one was released during calcite precipitation; a net of one CO_2 molecule was removed from the atmosphere, as we'll see below.

Limestone rocks exposed at the Earth's surface also undergo chemical weathering; the reaction is the reverse of Eq. 2. The reaction is congruent, and the calcium and bicarbonate ions are carried to the sea by rivers and precipitate once again as $CaCO_3$ (Eq. 2). Thus, weathering of limestones leads to no net consumption of atmospheric CO_2. This is why we focus on silicate weathering when considering long-term controls on atmospheric CO_2 and climate.

The silica resulting from chemical weathering of silicate minerals on land and delivered to the ocean in dissolved form is used by *siliceous biomineralizing organisms* who produce siliceous skeletons from the silica in seawater. Notable examples of silica biomineralizers are the sponges and the microscopic diatoms and radiolarians (see Chap. 5, Fig. 11). Although the skeletons they produce are formed of material with a disordered crystal structure (opal), with time at the seafloor, upon death and sedimentation, crystalline quartz is formed. The chemistry of this process is simple:

$$SiO_2 \rightarrow SiO_2 \rightarrow SiO_2 \tag{3}$$

(silica dissolved in water) $\rightarrow$ (opal) $\rightarrow$ (quartz)

The sedimentary rock produced is called *chert* (see Chap. 5, Fig. 13b). So the products of silicate mineral weathering on land provide the basic ingredients for the organisms making shells in the ocean and supplying biomineralized components of sediments on the seafloor.

Note that if we combine Eqs. 1–3 we get the overall reaction representing weathering and biomineralization, a series of processes that converts silicate rocks and atmospheric CO_2 to limestone and chert:

$$CO_2 + CaSiO_3 \rightarrow CaCO_3 + SiO_2 \tag{4}$$

carbon dioxide + example silicate mineral $\rightarrow$ calcite
+ quartz (limestone + chert)

The limestone on the seafloor serves as a stable repository for the carbon that, before its participation in the weathering process, was in the atmosphere as a CO_2 molecule, contributing to the greenhouse effect. Moreover, carbon locked in limestone remains sequestered for tens to hundreds of millions of years, even after exposure on the continents through the rock cycle. Thus, the weathering of silicate rocks and the subsequent formation of carbonate rocks tends to cool the planet. Below we will see what balances this tendency toward cooling, and how weathering itself provides the most important climate-stabilizing mechanism on Earth.

3.3 Is Limestone Formation the Only Way Carbon Is Removed from the Ocean and Atmosphere?

Some carbonate rocks also contain significant amounts of magnesium in the form of the mineral *dolomite*. Dolomite rock formation thus represents another final product of the silicate weathering process, sequestering atmospheric CO_2. Sedimentary rocks also contain ancient organic matter in mudrocks and fossil fuels (see Chap. 5, and Box 1 on the Global Carbon Cycle), which we can express simply as C_{org}. C_{org} represents the carbon initially removed from the atmosphere through the process of photosynthesis that is not returned as CO_2 to the atmosphere during respiration and decomposition and becomes buried in sedimentary rocks. When exposed to the atmosphere during weathering, C_{org} reacts with oxygen (O_2) to produce CO_2, a process that is essentially the reverse of photosynthesis.

$$C_{org} + O_2 \rightarrow CO_2 \quad (5)$$

Ancient organic C + oxygen → carbon dioxide

This means that weathering both consumes atmospheric CO_2 (when silicate rocks are weathered) and releases CO_2 to the atmosphere (when C_{org} is weathered). Weathering also releases nutrients like phosphorus though, and this release of rock-bound P to soils, rivers and the ocean during weathering and erosion stimulates photosynthesis and the production of organic matter, closing the loop of the organic carbon cycle. Thus, CO_2 is released to the atmosphere by ancient C_{org} weathering and removed from the atmosphere by photosynthesis, deposition, and burial of C_{org}. Variations in the global rate of organic carbon burial have occurred over Earth history, and these are likely to have significantly impacted atmospheric CO_2 levels and climate. Nevertheless, much of the research addressing long-term climate regulation has been on feedback mechanisms involving silicate weathering and climate, since silicate rocks are vastly more abundant than rocks rich in organic carbon. As a result, our discussion will focus on silicate weathering, rather than on mechanisms that control C_{org} burial and how those might regulate long-term climate.

Box 1 The global carbon cycle

In Chap. 6, you were introduced to the importance of volcanism in the global carbon cycle. From the carbon perspective, weathering is part of the global carbon cycle (Fig. 4). Large but nearly balanced fluxes of carbon are associated with photosynthesis and respiration, and with gas exchange between the atmosphere and ocean. Much smaller fluxes of carbon are associated with the geological processes of weathering, volcanism, burial and metamorphism, as indicated by the width of the arrows (not to scale). Fossil fuel use and deforestation are smaller than the largest of the natural fluxes but much larger than the geological fluxes. Interesting, although the fluxes involved in geological processes are much smaller, their reservoirs of carbon in limestones, mudrocks, and fossil fuels are much larger than those associated with the "fast" cycles: plants, algae in the ocean (which is so small that it is not even shown), and the atmosphere. What this means is that a CO_2 molecule emitted from a volcano might spend 10 years in the atmosphere before being taken up by a plant or the ocean. If taken up by a plant, it will on average be returned to the atmosphere within a decade. A fraction of plant material becomes incorporated into the soil where it decays, and it might spend a few decades in the soil before being converted from organic material to CO_2 by bacteria and fungi. The CO_2 released in the soil might react with a silicate mineral and be converted to bicarbonate and carried by rivers to the ocean, or it may be released to the atmosphere. A small fraction of organic material escapes decomposition and is buried in terrestrial and marine sediments, and an even smaller fraction accumulates in sufficiently high concentrations that it can serve as a fossil-fuel resource (oil, natural gas, or coal). These geological reservoirs of organic carbon are so large that carbon atoms spend tens of millions of years on average before weathering or metamorphism returns them to the atmosphere again as CO_2. Subduction carries sedimentary rocks on the seafloor down into deep-sea trenches and then more deeply into the mantle. Ultimately this flux of carbon out of the Earth's surface environment balances the release of carbon from Earth's interior during volcanism.

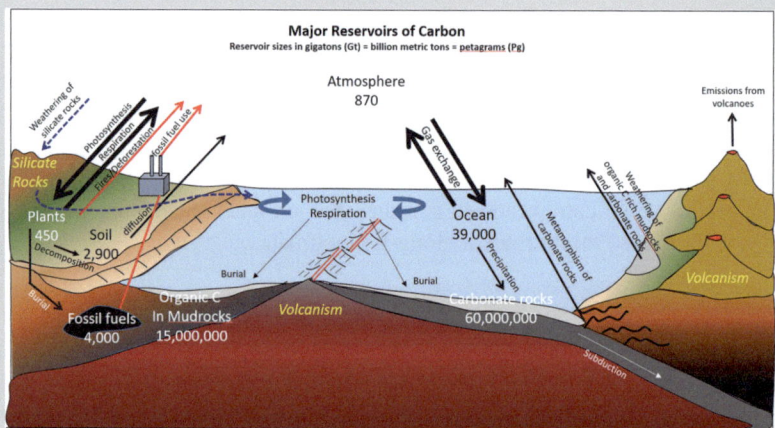

Fig. 4 Representation of the global carbon cycle. It includes the same information as Fig. 10, Chap. 6, with the addition of reservoir sizes. The sizes of reservoirs are in gigatons (1 gigaton = billion metric tons). For example, the atmosphere reservoir is 870 gigatons. The arrows in the figure mark the flow of carbon (also called carbon flux). It shows the routes carbon can take among and between the different reservoirs. Black and blue arrows are used for natural processes, whereas red arrows are used for processes related to human activities. Figure by Kristen St John and Lee Kump

4 How Does Weathering Fit into Global Climate Regulation?

Given the vigor of the carbon cycle (see Box 1), and the resulting short residence time (Table 1) of carbon in critical reservoirs like the atmosphere, even small imbalances in inputs and outputs could be catastrophic: for example, a 10% excess of photosynthesis over respiration could deplete the atmosphere of its CO_2 in a matter of a hundred years or so if uncompensated for from other fluxes of material; within a few decades, the atmospheric CO_2 content would fall below that which Earth had at the height of the last glaciation 18,000 years ago.

In order to consider how weathering fits in global climate regulation, it's important first to recognize that three climate-related factors affect the *rate* of rock weathering: temperature, rainfall, and biological activity (notably plant root and microbial respiration and organic acid production). Temperature affects rock weathering as it does for all chemical reactions: as is common in chemistry, the rate of dissolution of minerals increases as the temperature increases. Think of how much faster sugar dissolves in hot water compared to cold water, or how long it takes to make "sun tea" compared to that for tea steeped with boiling water. Moreover, temperature affects biological activity, and this can affect weathering rates: the plants and microbes that aid in the weathering process have minimum, optimum, and maximum temperatures. And all of these processes depend on the availability of water: higher rainfall means that more plants and bacteria can grow, and more minerals can dissolve into that water. Saving the biological effects for later, we can represent the interactions with weathering and rainfall in a systems diagram (Fig. 5).

The diagram uses arrows to show connections between components of the system, and the + or − indicates whether the connection is positive or negative, in other words, whether a change in the component at the arrow's tail causes a change in the same direction (+) or in the opposite direction (−) in the component in the arrow's tip. For example, note that there's an arrow from global average temperature to silicate weathering rate, and this has a + sign beside it: an increase in global average temperature causes an increase in silicate weathering rate. So too does an increase

Table 1 Residence time of carbon in various reservoirs

Carbon reservoir	Residence time[a] (years)
Atmosphere	10
Living biomass	10
Soils	50
Ocean	500
Marine sediments	5000
Organic carbon in sedimentary rocks	200,000,000
Limestone	200,000,000

[a] Residence time is the time it takes to completely replace the carbon in a reservoir through balanced inputs and outputs

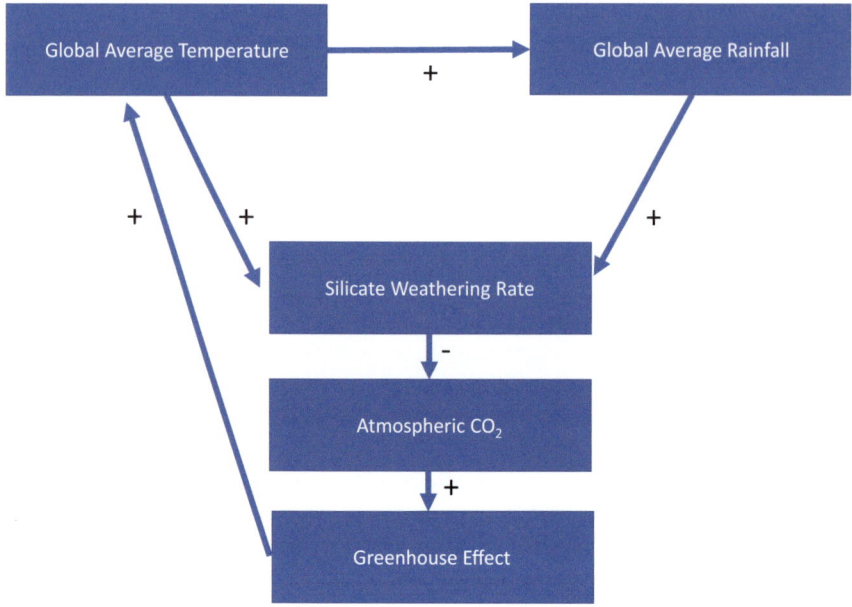

Fig. 5 Feedbacks of the silicate-weathering climate regulation mechanism

in rainfall. Similarly, a decrease in global average temperature causes a decrease in silicate weathering rate: the change is in the same direction. Moreover, as we learned in Chap. 2, an increase in global temperature causes an overall increase in global rainfall.[2] We also learned in that chapter about the greenhouse effect: that an increase in the atmospheric concentration of certain gases like CO_2 causes an increase in global temperature. These are also positive connections. Note though that the arrow connecting silicate weathering rate to atmospheric CO_2 has a negative sign next to it; it is a negative connection: an increase in silicate weathering rates causes atmospheric CO_2 concentrations to fall. The change is in the opposite direction. Similarly, a decrease in silicate weathering rate will lead to an increase in atmospheric CO_2 (presuming that other factors such as volcanism remain unchanged).

Note that the set of arrows in the systems diagram above creates a **feedback loop** (also known simply as a **feedback**). There are two loops in the diagram: one that routes from temperature to silicate weathering rate to atmospheric CO_2 to the greenhouse effect and back to temperature. The other follows the same general path but extends to rainfall. Feedbacks represent a round-trip flow of information, where cause

---

[2] Although rainfall is regionally varying, dry areas tend to get drier and wet areas tend to get wetter as global temperatures rise.

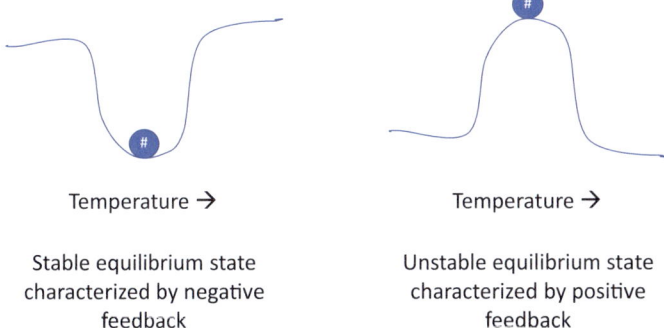

Temperature →   Temperature →

Stable equilibrium state   Unstable equilibrium state
characterized by negative   characterized by positive
feedback   feedback

Fig. 6 Stable and unstable equilibrium states. Perturbing the temperature from a stable equilibrium state is like rolling a ball up a hill from a valley: it returns to the original state. From an unstable equilibrium state, a disturbance takes the system to an entirely new state (as a ball disturbed from a precarious balance at the top of a hill)

and effect, a feature of the arrows connecting two system components, become impossible to separate! Feedbacks can either be positive or negative. **Positive feedbacks** amplify disturbances: initial disturbances to one system component, for example, a decrease in temperature, causes further decreases in temperature. **Negative feedbacks** diminish disturbances: a decrease in temperature ultimately leads to a restorative increase in temperature. So a system initially at equilibrium (relatively steady conditions) tends to return to that equilibrium after a disturbance if it is regulated by a negative feedback; the equilibrium state is stable. In contrast, if the system is characterized by a positive feedback, then when disturbed from equilibrium the system will evolve further from that equilibrium state. The equilibrium state is unstable. Think of these two states as balls sitting in a valley (stable, negative feedback) or a hilltop (unstable, positive feedback) (Fig. 6).

To determine whether the two feedbacks in Fig. 5 are positive or negative, let's start our thought experiment by disturbing temperature from its equilibrium state with an imagined instantaneous warming. The increase in temperature will cause an increase in rainfall, and both the increase in rainfall and the increase in temperature will cause an increase in silicate weathering rate. The increase in silicate weathering rate will cause a reduction in atmospheric CO_2, which will reduce the greenhouse effect, which will lead to a reduction in temperature. Given that this last effect counters the initial disturbance, the feedback loop is negative, and the equilibrium state is stable. Similarly, if the planet were to cool, rainfall would diminish, and both of these changes would reduce silicate weathering rates. With lower rates of CO_2 consumption by weathering and a continued input of CO_2 from volcanoes, atmospheric CO_2 levels would increase, restoring global temperature to its original state.

Think of what this means: Earth regulates its own temperature, within limits and at rates set by the residence times of the carbon reservoirs involved (Table 1; and see below regarding limits) through the interaction of multiple components of the Earth system: the atmosphere, oceans and land surface, rocks, and water, together with the carbon dioxide in the air, soil, and dissolved in water, all work in combination, without foresight or planning, to stabilize the climate of the planet over long time scales. Planetary regulation is an emergent feature of the multiple interactions among these components. It has the appearance of intelligence; Earth does just the right thing to counter all sorts of disturbances. While foresight and planning are capabilities of humans, planetary temperature regulation is, instead, only accidental, yet essential to sustaining habitability over the history of the planet. Temperature regulation apparently failed on Venus and Mars, if it ever existed. For Venus, the sun likely eventually became too hot for these feedbacks to keep the planet habitable; for Mars, tectonics seems to have failed to continue the rock cycle and provide sufficient volcanic CO_2 to sustain a sufficient greenhouse effect. Earth's position in the solar system and its continued tectonics have been "just right" for sustaining life on our planet (Box 2).

Box 2 What makes Earth habitable?

By James Kasting, Department of Geosciences and Meteorology, Pennsylvania State University

Our planet Earth is the only planet within our Solar System that is habitable at its surface. This, of course, is a good thing for us, as we would not be here if that were not true. One may ask, however, what are the general properties that define a habitable planet, and could such a planet exist elsewhere, perhaps orbiting a nearby star?

To begin, we should acknowledge that there is no general agreement about a definition of planetary habitability, or even a definition of life itself. Some authors have argued that we should not even try to define life until we have some other examples besides life here on Earth. We are not purists, though, so we'll look for definitions that many astrobiologists would find acceptable. We like the so-called NASA definition of life, which is very broad: Life is a self-sustaining chemical system capable of Darwinian evolution. One might point out that sterile animals like mules cannot reproduce, yet we consider them to be alive. This just illustrates the fact that this problem is currently not well-constrained.

What makes a planet habitable, then? We'll list four factors that become increasingly Earth-centric as we go down the list:

- *A habitable planet must have either a solid or a liquid surface.* The reason is that life as we know it requires a stable pressure-temperature (P–T) regime. Here, we are being less optimistic than the late Carl Sagan. Sagan's book/TV series *Cosmos* contained a description of hypothetical "floaters" living in Jupiter's atmosphere. They adjust to the right P–T level by changing their internal pressure, much as some fish can do in the oceans by using air bladders. But Sagan's hypothetical floaters and terrestrial fish with air bladders are both highly evolved organisms. If the first organisms were single-celled, as most biologists assume, then this type of P–T regulation would not have been available. So, we consider this requirement to be universal.

- *A habitable planet must contain free, readily available energy that organisms can use to sustain metabolism.* If life is a chemical system, then the type of free energy we are talking about is Gibbs free energy, which is conserved in chemical reactions under constant pressure and temperature conditions. Much of the free energy that powers life here on Earth's surface comes ultimately from abundant sunlight. Photosynthetic autotrophs (e.g., plants, phytoplankton) use the energy from sunlight to convert CO_2 to organic carbon. They derive their metabolic energy by reacting some of this organic matter with oxidants such as O_2 or sulfate, or through fermentation. Some chemotrophic organisms can use other sources of free energy. For example, methanogens derive energy by combining CO_2 and H_2 to form CH_4 and H_2O; no sunlight is required. But all organisms require an energy source, so this is also considered a universal requirement for life.

- *A habitable planet must provide carbon and other bioessential elements, e.g., CHNOPS.* This requirement is admittedly more Earth-centric; nevertheless, we consider it likely to apply to life in general. Carbon (C) forms far more compounds than any other element, and terrestrial life takes advantage of many of them. Hydrogen (H), Nitrogen (N), and Oxygen (O) are also ubiquitous in biological organic compounds. Phosphorus (P), of course, is an essential element in both DNA and RNA, as it forms the tri-phosphate linkages that connect nucleotides. Sulfur (S) bonds with many of these other atoms and can access multiple redox states. Importantly, as they are found relatively early in the periodic table, all these elements are relatively abundant in the cosmos, so it is not surprising that terrestrial life takes advantage of them.

- *A habitable planet must provide access to liquid water.* This requirement is even more Earth-centric, but it is sufficiently respected that the mantra of NASA's astrobiology exploration program is: Follow the water! Some terrestrial organisms, e.g., ones that form spores, can get by without water for long time periods; however, all organisms require water to metabolize and reproduce. Water derives much of its utility to life by being a highly polar molecule (possessing regions of positive and negative charge), capable of dissolving biomolecules, most of which are polar. Polar water molecules also help stabilize the helical structure of DNA. Water is also thought to be ubiquitous in the universe, as the elements of which it is composed, H and O, are the first and third most abundant elements, respectively.

> A key issue for planetary habitability is where the liquid water is present. On Earth, liquid water is found at the surface, and this enables a vigorous biosphere powered by photosynthesis. Complex life, taken here to mean animal life (including humans), almost certainly requires surface liquid water. That said, single-celled organisms are found in many subsurface locations on Earth[3] and could conceivably exist in subsurface regions of Mars, Europa, and Enceladus. So, these worlds, too, may be "habitable", depending on what one means by the word.
>
> Finally, to connect more directly with the rest of this chapter, Earth's habitability for surface organisms like us depends on processes that help to maintain surface liquid water. Earth's climate system contains a strong negative feedback between atmospheric CO_2 and climate that plays an important role in doing this. For example, see the case study on the faint young sun problem at the end of the chapter. A planet that can actively recycle carbon, as Earth does, is much more likely to remain habitable over long time periods than one that is tectonically inactive.

5 What Role Does Life Play in Planetary Climate Regulation?

This ability to regulate global temperature within habitable limits for life is likely to have its own limitations, and those limitations arise because of the role of life itself in planetary regulation. Recall that so far we've been ignoring the effects of life on the silicate weathering process. We can insert biological activity as a separate system component into the diagram (see Fig. 7), because both temperature and rainfall (water availability) affect life. The effect of changes in temperature and rainfall on biological activity depends on the state of the system: whether the temperature or rainfall amounts are presently below, at, or above the optimum temperatures and moisture states for the particular plants and bacteria that facilitate the weathering process. That's why we showed these connections as ± in Fig. 7. And to be clear, decades of scientific research have not led to a definitive answer as to the magnitude of this biological effect on global weathering rates and thus climate regulation, much less what its minimum, optimum, and maximum temperatures are. The scientific consensus is that the current global climate state is below the optimum for silicate weathering: a somewhat warmer and wetter world should stimulate the biota and cause an increase in silicate weathering rate, averaged at the global scale. The optimum global temperature for biotic enhancement of weathering is perhaps 10 °C warmer than today (i.e., a global average temperature of 25 °C). Above that

---

[3] For example see this review of subseafloor microbial life: D'Hondt et al. (2019).

temperature, on a global average, many organisms, especially those living in low latitudes, will experience thermal stress. If this is correct, then above a global average temperature of 25 °C, the sign of the connection between temperature (and perhaps rainfall) will switch from positive to negative, creating the possibility that the overall feedback loop will become positive instead of negative. There is high level of uncertainty though in these temperature estimates; there is a vast number of organisms involved in terrestrial ecosystems, each of which has its own physiological response to temperature and water availability. Another part of the problem is the heterogeneity of climates, soil types, bedrock types, and topographies across the globe that makes it challenging to establish what the overall global connections are between climate (temperature and rainfall) and the biological enhancement of silicate weathering rate.

We refer to the *possibility* that the feedback may become positive instead of negative, because temperature and rainfall have direct effects on silicate weathering rate as well as indirect effects, via biological activity. Let's compare three situations: a situation below, just above, and well above the optimum temperature for biological enhancement of weathering (Fig. 8).

When the climate is cool, we expect a warming to definitely lead to increased silicate weathering because both biological activity and weathering itself increase with temperature. Above the optimum temperature, further warming will actually reduce silicate weathering below the maximum rate because of a weakened biological enhancement. At higher temperatures though, above the tolerance limits for

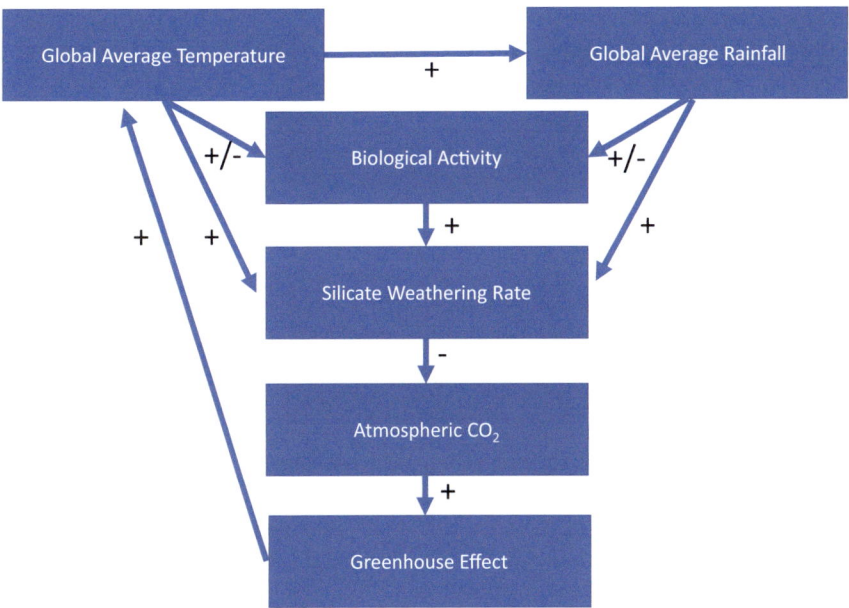

Fig. 7 Feedback loops in the silicate-weathering climate regulation mechanism involving biological activity

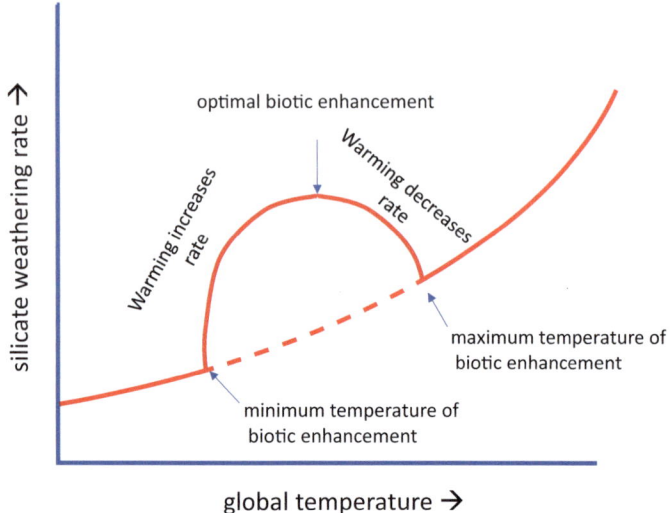

Fig. 8 Effect of biological enhancement on silicate weathering rate. The solid line is the effect of temperature on the silicate weathering rate without biotic enhancement; the dashed line shows the effect of temperature on the silicate weathering rate with biological enhancement

life, weathering rates again increase with temperature. Thus the *overall sign* of the feedback involving temperature and weathering (in Fig. 8) might go from negative at suboptimal temperatures for the biota, to positive just above the optimal temperature, to back to negative at even higher temperatures.

We can also explore the silicate weathering climate feedback from the perspective of carbon balance: the equilibrium climate states discussed above will be those in which the rate of CO_2 input to the ocean and atmosphere is equal to the rate at which silicate weathering, at the global scale, converts CO_2 to limestone (or dolomite).

The conceptual diagram below (Fig. 9a) depicts this balance. In this figure we have added a line representing the rate of volcanic input of CO_2 to the atmosphere and ocean (in green, representing today's rate[4]) to the curve representing silicate weathering rates at different global average temperatures. Where the red and green lines cross are the temperatures where the rates of CO_2 input and output are equal; these are called "steady states" because when input and output are equal, the amount of CO_2 in the atmosphere remains steady even as CO_2 is moving into and out of the atmosphere. These are also called equilibrium climate states. There are three equilibrium states (ES 1–3): one at a cool temperature (ES 1) like today's global average of around 15 °C (59 °F), another at an intermediate temperature (ES 2), above the optimum temperature for biotic enhancement, and a third (ES 3) above the upper limit for biological enhancement.

---

[4] Using today's rate is for comparison purposes only, and allows for a simplified model. We know from Chap. 6 that the rate of volcanism has changed over geologic time.

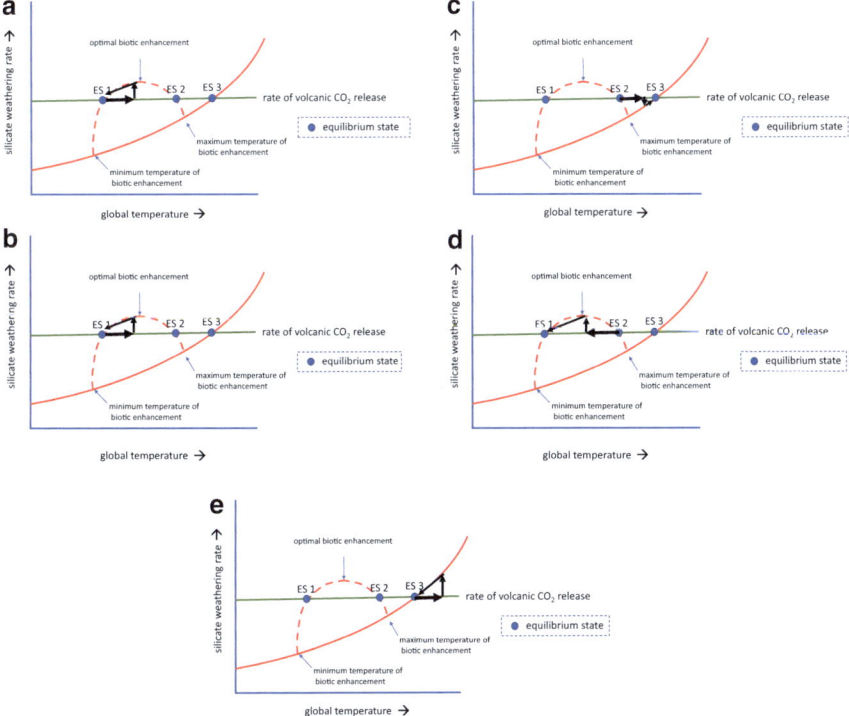

Fig. 9 **a** Balance between volcanic CO_2 release (which is reasonably assumed to be independent of global temperature) and silicate weathering rate (which is dependent on temperature, modified by biotic enhancement) occurs at the places where the two curves cross. These are steady states representing stable (ES 1 and ES 3) and unstable (ES 2) equilibrium states. **b** Response to a sudden warming from the equilibrium state ES 1 (black arrows). **c** Response to a sudden warming from the equilibrium state ES 2 (black arrows). **d** Response to a sudden cooling from the equilibrium state ES 2 (black arrows). **e** Response to a sudden warming from the equilibrium state ES 3 (black arrows)

Note that the stability of each equilibrium state can be inferred from this diagram. For example, if the temperature of the planet was originally at stable equilibrium state ES 1 and then were to increase for some reason (the horizontal black arrow in Fig. 9b), weathering rate would increase (vertical black arrow) and exceed the rate of volcanism: atmospheric CO_2 levels would decrease to counter the warming (black arrow returning to ES 1), demonstrating this is a stable equilibrium state.

The opposite would happen if temperature were to fall. Note also that this steady state, and the equilibrium temperature it creates, is below the optimal temperature for biotic enhancement. Equilibrium states are not necessarily (in fact rarely) optimum states of systems.

ES 2 is at an intermediate temperature and represents an unstable equilibrium state (Fig. 9c). Warming leads to reduced silicate weathering because of the reduction in the biotic enhancement, which would allow atmospheric CO_2 to accumulate from volcanic input, leading to further warming. Ultimately this would drive the system to ES 3, and in the process, severely reduce biological activity and the weathering enhancement it would otherwise produce.

In contrast, a cooling that took the system below the temperature at unstable ES 2 would lead to an increase in biotic enhancement and thus an increase in silicate weathering, drawing down atmospheric CO_2 (because silicate weathering rates exceeded volcanic input rates), cooling the planet below the optimal for biotic enhancement, ultimately reducing the biotic enhancement of silicate weathering and shifting the state of the system to ES 1 (Fig. 9d).

ES 3 is the equilibrium state of the system above the upper limit for biological enhancement of weathering. It represents a stable equilibrium state: if the temperature of the planet was originally at ES 3 and then were to increase for some reason (the horizontal black arrow in Fig. 9e), weathering rate would increase (vertical black arrow) and exceed the rate of CO_2 input by volcanism: atmospheric CO_2 levels would decrease to counter the warming (black arrow returning to ES 3).

This section, addressing the biotic enhancement of weathering, is admittedly complex and has introduced considerations that, frankly, most models of the global climate system have yet to incorporate. Thus, it may be useful to review the important points made:

- Interactions between silicate weathering rate, biological enhancement of silicate weathering, and temperature set one or more climatic equilibrium states for any particular rate of CO_2 release by volcanoes.
- Purely physical and chemical processes have many linear relationships. But in biology physiological rate laws are more parabola-shaped, which are important to feedback dynamics. They can give rise to alternative stable states.
- Some of these equilibrium states are stable, meaning that a disturbance of the climate system will be moderated back to the original equilibrium state.
- Some of these equilibrium states are unstable, meaning that a disturbance of the climate system will not be moderated back to the original equilibrium state.
- If the climate system is disturbed from an unstable equilibrium state, however, the interactions between silicate weathering, biological enhancement of weathering, and temperature will drive the climate system toward one of the stable equilibrium states.
- Because these equilibrium climate states exist, temperature conditions on Earth have self-regulated within a relatively narrow range throughout Earth history.

6 Case Studies of Past Climate Change

As you've learned in the sections above (and in Chap. 6), there are several geological and biological processes and system feedbacks important in regulating our planet's long-term climate. Now let's apply these concepts, using a case study approach, to better understand climate over two timespans of Earth's history. The first case study will take the long view—a *really* long view—by considering Earth's climate over its entire history, with special emphasis on conditions across billions and hundreds of millions of years (e.g., the Phanerozoic Eon). The second case study will re-examine a timeframe you were introduced to in Chap. 6, the mid-to late Cenozoic Era transition from a Greenhouse to an Icehouse state.

6.1 Case Study: Climate Record over Earth's Long-Term History

With a fossil record that extends back 3.5 billion years, we have strong evidence that Earth has remained habitable to an increasing variety of life forms for most of its 4.5 billion year history. This is despite some considerable challenges to habitability, including a sun that has increased in luminosity by 33%, massive volcanic eruptions and huge asteroid impacts, repositioning of continents and encroaching and retreating seas.

According to the theory of solar evolution, stars like our sun, perhaps counterintuitively, become brighter as they age. The energy of our sun comes from the fusion of four hydrogen (H) atoms to form one helium (He) atom, the mass of which is just slightly less than the sum of the four original hydrogen atoms. According to Einstein's famous formula, we can calculate the energy (E) that excess mass (m) is converted into by using $E = mc^2$, where c is the speed of light. The energy heats the star, creating very hot surface temperatures. The star radiates electromagnetic energy into space, a very small fraction of which impinges on Earth, a flux we call **insolation** (*in*coming *sol*ar radi*ation*). With time, the sun's core contracts, pressure increases, and so too does its temperature, as more hydrogen converts to helium. As a result, the star radiates more energy to space as it ages. The famous astronomer Carl Sagan recognized this creates a problem in thinking about ancient climates: if everything else in the climate system was the same as today, but the sun was 75% as bright as it is today, Earth surface temperatures would have been well below freezing (Fig. 10). There would have been no ocean, which we know is an important part of Earth's climate system as we understand it. However, this model is in contrast to available evidence in the rock record that Earth's long-term average global temperature remained (with some exceptions) within a fairly limited—and habitable—range for most of geologic history. This has become known as the **Faint Young Sun Paradox** (Fig. 10).

If the Sun was dimmer in the past, sending less sunlight to Earth and thus tending to create a cooler planet, then the negative feedback illustrated in Fig. 5 implies that atmospheric CO_2 should have accumulated in response until the planet warmed sufficiently such that silicate weathering rates came to balance volcanic emission rates. Earth's interior was hotter in its past, and thus rates of volcanism were likely higher in the past as well. This suggests that in fact early Earth temperatures might have been as high as or higher than today, because of very high atmospheric CO_2 levels,

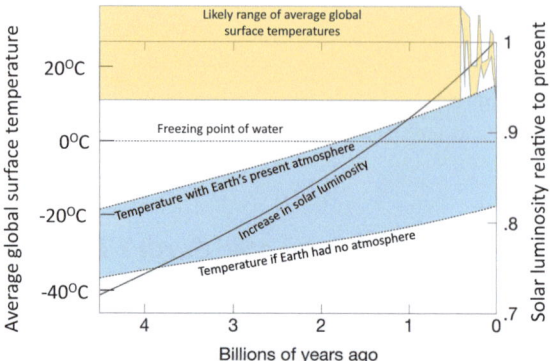

Fig. 10 Increase in solar luminosity (solid line) relative to its present value (right vertical axis) over Earth history and the global average surface temperatures (in degrees C; left vertical axis) that would have resulted had the Earth had no atmospheric greenhouse gases (*lower dashed line*) or if it always had its current composition (*upper dashed line*); this range in temperatures is shaded blue. The yellow swath represents the range of Earth's long-term global average temperatures based on evidence in the geologic record, including data from Fig. 11. Figure modified from Kump et al. (2010)[5]

despite a faint sun. In any event, the silicate weathering feedback likely stabilized climate in a fairly clement range, avoiding fates suffered by Venus (too hot) and Mars (too cold) (see Box 2).

That's not to say that life on our planet didn't come close to complete annihilation at times in the past. Indeed, life may have had false starts early in Earth history (prior to 3.8 billion years ago), during what's known as the "period of heavy bombardment" by planetesimals and large asteroids, including one 4.5 billion years ago that led to the formation of the moon. An impact with extraterrestrial objects of this size likely vaporized any primordial ocean and melted Earth's crust into a "magma ocean", clearly an environment unsuitable for sustaining life.

Another challenge came 700 million years ago, and then again at 650 million years ago when, for reasons not completely understood, Earth's climate descended into a **Snowball Earth** state with sea ice covering most of the oceans and thick ice sheets covering all continents. These conditions lasted for millions of years. Surface temperatures, even in the tropics, have been estimated to have been below − 20 °C except where open ocean conditions might have prevailed (a variant of the Snowball Earth state referred to as a "Slushball"). These relatively warm (near 0 °C) surface waters may have served as refugia for life that otherwise could have been obliterated had the ocean become completely covered in sea ice. Note that complex animal life had yet to evolve on Earth during these Snowball events; rapid evolutionary diversification followed the events, as Earth returned to more hospitable conditions. Why Earth's climate was driven into this state is unknown, but we might speculate that

---

[5] Kemp et al. (2010).

either volcanism waned, or factors accelerating weathering increased, until atmospheric CO_2 levels fell to such a critically low level that sea ice and continental ice spread across the globe. This failure of the silicate weathering feedback was not irreversible: volcanoes seem to have pierced the ice sheets on land to release CO_2 to the atmosphere, or submarine volcanic CO_2 escaped through cracks in the sea ice to the atmosphere, which accumulated CO_2 under very low weathering rates on the frozen planet until those CO_2 contents became high enough to warm the planet above the freezing point, melting the ice sheets and sea ice.

A third example of the challenges life faced is the greatest of all extinction events, at the end of the Permian Period 252 million years ago. The great **end-Permian Extinction** was accompanied by a dramatic warming that persisted for at least a few million years into the ensuing Triassic Period.

Given the challenges of ascertaining Earth's temperature in deep time from an incomplete geological record and based on indirect ways of determining temperature, geologists are only now beginning to piece together a relatively continuous temperature history for the planet. The record for the past 500 million years shown in Fig. 11 is the product of a recent interdisciplinary collaboration directed by the Smithsonian Institution. It's based on a combination of geological information (such as the paleo-temperature proxies described in Chap. 5) collected over the last century, combined with computer modeling of Earth's climate system (such as described in Chap. 10) and how the climate system has evolved in response to changes in solar luminosity, atmospheric composition, and changes in the land surface topography and geography resulting from plate tectonics, and evolution of terrestrial vegetation. The diagram displays repeated variations driven by these factors, but no overall trend toward warmer or colder climates. These changes are quite large: the warm intervals of the Mesozoic and Paleozoic represent Greenhouse times (red in Fig. 11) when there were no ice sheets at high latitudes and tropical warmth extended toward the poles, allowing organisms whose descendants today can only be found in equatorial regions to thrive in the Arctic and Antarctic. In contrast, the Icehouse times of the Paleozoic and Cenozoic (blue in Fig. 11) witnessed massive ice sheets that intermittently extended across the middle latitudes, restricting forest ecosystems to low latitudes. Despite these large fluctuations, life on planet Earth has been sustained; it would appear that climate regulation, likely involving the weathering feedbacks discussed in this chapter, has been effective in preventing extreme icehouses and protracted, uninhabitable greenhouse climates.

6.2 Case Study: Himalayan Uplift and Cenozoic Climate Cooling

A particularly important period of Earth history is the time marked by progressive cooling, apparent in Fig. 11, beginning about 40 million years ago and culminating in the relatively cool climate Earth has been exhibiting for the last 34 million years.

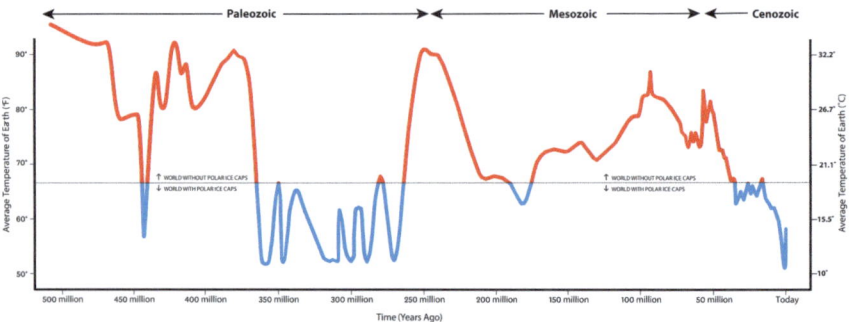

Fig. 11 Estimates of how global average temperature has varied through the Phanerozoic (Anagnostou et al., 2020; Badger et al., 2013a, 2013b; Bartoli et al., 2011; Beerling et al., 2009; Breecker & Retallack, 2014; Brown et al., 2022; Burgh et al., 1993; Cotton & Sheldon, 2012; Da et al., 2015; Doria et al., 2011; Ekart et al., 1999; Erdei et al., 2012; Fletcher et al., 2008; Foster et al., 2012; Franks et al., 2014; Greenop et al., 2014, 2019; Greenwood et al., 2003; Grein et al., 2011, 2013; Guillermic et al., 2022; Gutjahr et al., 2017; Harper et al., 2020; Henehan et al., 2019, 2020; Huang et al., 2013; Hyland & Sheldon, 2013; Hyland et al., 2013; Ji et al., 2018; Koch et al., 1992; Kohn et al., 2015; Kowalczyk et al., 2018; Kürschner et al., 1996, 2001, 2008; Liang et al., 2022a, 2022b; Londoño et al., 2018; Martínez-Botí et al., 2015; Maxbauer et al., 2014; McElwain, 1998; Milligan et al., 2019, 2022; Moraweck et al., 2019; Nordt et al., 2003; Pearson et al., 2009; Reichgelt et al., 2020; Retallack, 2009a, 2009b; Roth-Nebelsick et al., 2012, 2014; Royer, 2003; Royer et al., 2001; Seki et al., 2010; Smith et al., 2010; Sosdian et al., 2018; Srivastava et al., 2013; Stap et al., 2016; Steinthorsdottir et al., 2016a, 2016b, 2019a, 2019b, 2021; Stults et al., 2011; Sun et al., 2012, 2017; Super et al., 2018; Tanner et al., 2020; Tesfamichael et al., 2017; Wang et al., 2015, 2020; Witkowski et al., 2018; Zhang et al., 2013, 2018) Modified from an image produced by the Smithsonian Institution. © Smithsonian Institution. This figure is excluded from Creative Commons license

This time began with the growth of large ice sheets on Antarctica and then, for the last 2+ million years, has been characterized by the waxing and waning of ice sheets across North America and Fennoscandia (see Chap. 9). What explains this persistent cooling? The tectonically driven opening and closing of oceanic gateways was a likely contributor to global cooling and the development of widespread ice cover of the highest latitudes. This was augmented by tectonic uplift in central Asia, which may have increased freshwater input to the Arctic Ocean and in turn the expansion of the North Hemisphere glaciation (explored in Chap. 6). Perhaps most important, however, was a decrease in the greenhouse effect related to rock weathering. We will explore this essential piece of the puzzle here.

Similar to the efforts to establish changes in temperature through time, geologists have also developed a number of ways to estimate the CO_2 content of ancient atmospheres. Note that there is no direct way to make these measurements, with the exception of the ice-core record, from which bubbles of ancient air can be extracted and analyzed (see Chap. 4). Instead, scientists have used a combination of modern investigations (e.g., in sealed greenhouses with atmospheres adjusted to higher CO_2 levels) and theory to establish proxies for ancient atmospheric CO_2 levels. These include the density of pores (stomata) on fossil leaves (when CO_2 levels are high,

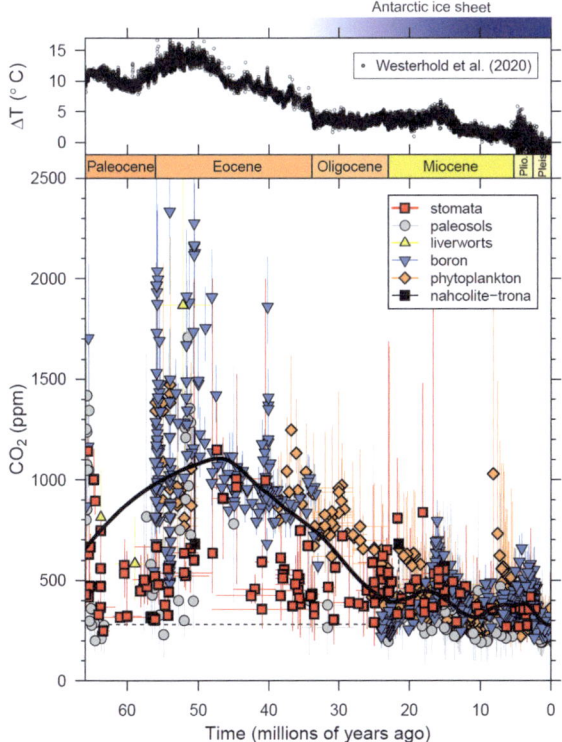

Fig. 12 Relationship between deep-sea temperatures and atmospheric CO_2 changes during the Cenozoic (65 million years ago to the present). Deep-sea temperatures (upper panel; from Westerhold et al., 2020)[6] generally track the estimates of atmospheric CO_2 (lower panel) reconstructed from terrestrial and marine proxies. CO_2 compilation updated from Foster et al. (2017).[7] CO_2 uncertainties (error bars) are shown with 95% confidence. The horizontal gray dashed line marks the pre-industrial CO_2 level of 280 ppm. The top blue bar indicates approximate timing of ice-sheet development on Antarctica. Same as Fig. 18, Chap. 5. Figure courtesy of Dana Royer, Wesleyan University

plants need fewer and smaller stomata to bring in CO_2 into their cells for photosynthesis; see Chap. 5) and various proxies based on the way in which the isotopes of carbon and boron separate during various chemical reactions based on the amount of CO_2 in the atmosphere, soil, and surface ocean (see Chaps. 5 and 8). There's considerable uncertainty in these estimates, but by and large they are consistent with the broad feature of progressive cooling (in this case of the polar regions, as reflected in the temperature of the deep sea) over the past 40 million years (Fig. 12).

So, if the cooling was the result of a gradual drawdown of atmospheric CO_2 levels (in addition to tectonic-drive changes in ocean circulation and heat transport), what

---

[6] Westerhold et al. (2020). Data sources used in the update are listed at the end of the chapter.
[7] Foster et al. (2017).

caused the CO_2 drawdown? Our earlier discussion suggests two possibilities: either the rate of volcanism has declined over the last 40 million years, and/or a factor other than temperature itself enhanced the rate of silicate weathering. Quantifying how volcanic CO_2 emissions have varied over geologic time has proved challenging. Evidence such as preserved ash layers or lava flows give us a sense of the magnitude and frequency of volcanic eruptions in the past; as discussed in Chap. 6, the abundance of such deposits from the Cretaceous Period (145–66 million years ago), for example, establishes this time as one of globally significant, elevated volcanic activity, and with that came increased CO_2 emissions. Although there is evidence to support continued volcanism in the early part of the Cenozoic Era, there is no convincing evidence for a decline in volcanic activity beginning 40 million years ago and continuing to the present. Without evidence of decreased volcanism then we must focus on the effects of silicate weathering, so we must look for factors other than temperature itself that might have enhanced silicate weathering.

A leading contributor to enhanced silicate weathering has been increased mountain-building since the Cretaceous, most notably associated with the collision between India and Asia over the last 40 million years. Plate tectonic reconstructions of this time interval show a fast-moving Indian subcontinent racing (in geologic terms; around 5–10 cm per year) toward Asia, and upon collision, uplifting the Himalayas and Tibetan Plateau at around a centimeter per year (Fig. 13). In addition to the change in river drainage (i.e., freshwater input) to the Arctic as described in Chap. 6, this uplift of the collision zone increased the exposure, weathering and erosion of silicate minerals previously deeply buried on both continents. In other words, it made the region more "weatherable." The scale of this event was sufficient in magnitude to increase the overall weatherability of the planet, and thus the global silicate weathering rate. The result was a drawdown in atmospheric CO_2, since weathering rates came to exceed the rate of volcanic CO_2 input.

This imbalance of input and output could not have been sustained for long. For example, with an estimated steady state weathering (and volcanism) rate of 0.1 Pg of C per year and a combined atmosphere + ocean reservoir of C of 40,000 Pg (Box 1, Fig. 4), the entire ocean plus atmosphere reservoir of C could be depleted in just 4 million years with only a 10% imbalance between volcanism and weathering. Long before that, the atmospheric CO_2 level would fall to levels that would initiate a "snowball Earth" state. What prevented this from happening?

The answer to this question again involves the sensitivity of global silicate weathering rates to global average temperature. As the uplift of the Himalayas and the Tibetan and Mongolian Plateaus increased silicate weathering rates in that region of the world, the resulting drawdown of atmospheric CO_2 and associated cooling and aridification reduced silicate weathering rates elsewhere. This cause and effect re-established steady state with the original globally averaged rate of silicate weathering (with regionally higher rates in the Himalayas and reduced rates elsewhere) balancing volcanism, but with lower atmospheric CO_2 and cooler global climate. As the rate of uplift continued to increase, rates of central Asian mountain and plateau weathering continued to increase, the planet further cooled, and weathering rates

Fig. 13 Collision of India with Asia during the Cenozoic. **a** 60 million years ago; **b** 40 million years ago; **c** 20 million years ago; and **d** modern map. Maps courtesy of Chris Scotese[8]

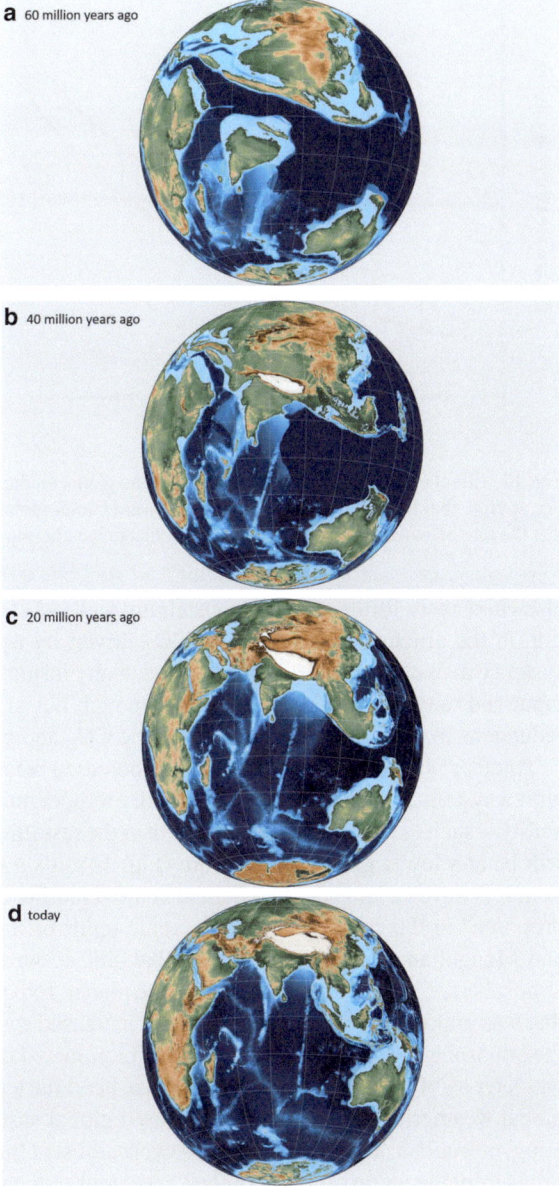

---

[8] Scotese (2021).

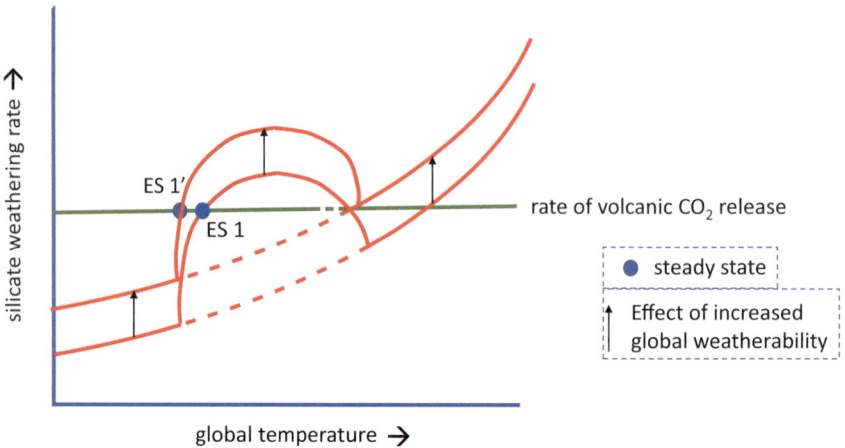

Fig. 14 Effect of increased "weatherability" of the planet on the stable equilibrium state ES 1 from Fig. 9. Note that ES 1' has shifted to the left, toward cooler global temperatures. Note also, though, that the rate of volcanic CO_2 release and the silicate weathering rate are the same as in Fig. 9

elsewhere were further reduced, sustaining a global steady state. This steady reduction in the amount of atmospheric CO_2 driven by uplift is an example of how a system can evolve at "quasi" steady state: at any instant in time, the balance between input and output is very close, but just enough out of balance to drive progressive reductions in that steady state (atmospheric CO_2 and global temperature).

Another way to think of the effect of increased weatherability is this: if you start with a system at steady state at a particular temperature and then increase weatherability—such as by large-scale uplift—then the resulting steady state weathering rate will be at a lower global temperature (Fig. 14). It's a subtle, but important distinction. It resolves a longstanding debate in the scientific community as well. Scientists proposed[9] in the early 1990s that tectonic uplift of the Himalayas and the Tibetan and Mongolian plateaus increased global silicate weathering rates, drawing down atmospheric CO_2 content, and cooling the planet. Experts on the carbon cycle[10] said that was impossible: had weathering rates increased even by small amounts globally, the atmospheric CO_2 content would have plummeted catastrophically. This paradox was later resolved with the arguments made here: the tectonic activity didn't increase global weathering *rates*, rather it increased global susceptibility of rocks to weathering, or *weatherability*.[11] The carbon cycle adjusted through a series of quasi-steady states from the warm climates of the Cretaceous to the relatively cool climates of the modern. Paradox resolved!

Since that time, other phenomena have been invoked to drive weatherability-related climate change. A good example is the shifting geographic position through

---

[9] Raymo and Ruddiman (1992).

[10] Caldeira et al. (1993).

[11] Kump and Arthur (1997).

global plate motions of mountainous island "arcs," covered in highly weatherable igneous rocks (such as the Japanese archipelago or the Aleutian Islands). The movement of such arcs from drier to more rainy tropical latitudes increased the regional weathering rates and the global weatherability, causing other cooling episodes in Earth history. These and other drivers (as discussed in Chap. 6), including intensified volcanic activity leading to warming and subdued volcanism leading to cooling, opening and closing of ocean gateways, changes in freshwater input to sensitive (polar) oceans, and changes in other greenhouse gases such as methane, have been invoked to explain "greenhouse" and "icehouse" episodes of Earth history. Collectively these drivers explain why long-term climate change has happened in the past.

The long-term habitability of our planet is testament to the effectiveness of feedbacks, such as the climate regulation by silicate-weathering discussed here, in stabilizing Earth's climate in the face of forces that might otherwise drive it toward the fate of our neighboring planets, Venus and Mars: too hot, and too cold. Earth's position in the solar system, together with active climate regulation, has made Earth "just right" for the origin and persistence of life. Earth history shows, however, that a massive injection of carbon dioxide into the atmosphere, whether it be by volcanoes or by the "artificial weathering" of ancient organic carbon by fossil-fuel burning, can shift the climate state from icehouse to greenhouse and take tens of thousands of years to recover. From a long-term, Earth systems perspective, this is a demonstration of resilience of the climate system. For humanity, however, such a shift could be devastating.

7 Conclusion

In this chapter, we've come to understand how the weathering process is not only an important component of the global carbon cycle, but an integral part of a planetary climate regulation mechanism that has stabilized Earth's climate against potentially much larger swings that otherwise would have occurred. Over Earth's history, global weathering rates have responded to plate tectonic activity, with periods of active mountain building enhancing global weatherability and a drawdown of atmospheric CO_2. Weathering has also responded to, rather than driven, climate change: periods of increased volcanic activity and emission of CO_2 to the atmosphere have been times of global warming that led to higher weathering rates. In so responding, silicate weathering rose to balance the higher rates of volcanic CO_2 emission, stabilizing the carbon cycle and the climate upon which it depends. This robust climate system thus has not maintained constant global temperature; rather it has allowed for modest swings in temperature leading to ice ages and "greenhouse" climate intervals during which polar ice sheets were absent. As emissions of CO_2 from fossil fuel use cause our atmospheric CO_2 content and greenhouse effect to increase, we expect that global weathering rates will rise, countering the anthropogenic disruption to the atmosphere and climate. Unfortunately, the effect is negligible on human timescales given the slow rates of weathering and the tremendous rate of fossil-fuel emissions, which are nearly three orders of magnitude (1000×) the rate of weathering or volcanism. Natural silicate weathering will ultimately restore the atmosphere to a post-Anthropocene state, but the recovery will take tens to hundreds of millennia. Humans are experimenting now with engineered processes (see Chap. 11) that simulate and accelerate weathering (and CO_2 removal) in a "geoengineering" approach called CO_2 "mineralization." Promising experiments are being conducted in Iceland, for example, involving CO_2 injection into the fresh basalt subsurface of this volcanic island; weathering reactions seem to quantitatively remove the CO_2 and convert it to carbonate minerals. Scaling this approach up to significantly offset current rates of fossil-fuel burning will be a challenge, though. Reducing CO_2 emissions remains the primary goal in addressing climate change.

References

Anagnostou, E., John, E. H., Babila, T. L., Sexton, P. F., Ridgwell, A., Lunt, D. J., Pearson, P. N., Chalk, T. B., Pancost, R. D., & Foster, G. L. (2020). Proxy evidence for state-dependence of climate sensitivity in the Eocene greenhouse. *Nature Communications, 11*, 4436. https://doi.org/10.1038/s41467-020-17887-x

Badger, M. P. S., Lear, C. H., Pancost, R. D., Foster, G. L., Bailey, T. R., Leng, M. J., & Abels, H. A. (2013). CO_2 drawdown following the middle Miocene expansion of the Antarctic Ice Sheet. *Paleoceanography, 28*, 42–53. https://doi.org/10.1002/palo.20015

Badger, M. P. S., Schmidt, D. N., Mackensen, A., & Pancost, R. D. (2013). High-resolution alkenone palaeobarometry indicates relatively stable pCO$_2$ during the Pliocene (3.3–2.8 Ma). *Philosophical Transactions of the Royal Society A, 371*, 20130094. https://doi.org/10.1098/rsta.2013.0094

Bartoli, G., Hönisch, B., & Zeebe, R. E. (2011). Atmospheric CO_2 decline during the Pliocene intensification of Northern Hemisphere glaciations. *Paleoceanography, 26*, PA4213. https://doi.org/10.1029/2010pa002055

Beerling, D. J., Fox, A., & Anderson, C. W. (2009). Quantitative uncertainty analyses of ancient atmospheric CO_2 estimates from fossil leaves. *American Journal of Science, 309*, 775–787. https://doi.org/10.2475/09.2009.01

Breecker, D. O., & Retallack, G. J. (2014). Refining the pedogenic carbonate atmospheric CO_2 proxy and application to Miocene CO_2. *Palaeogeography Palaeoclimatology Palaeoecology, 406*, 1–8. https://doi.org/10.1016/j.palaeo.2014.04.012

Brown, R. M., Chalk, T. B., Crocker, A. J., Wilson, P. A., & Foster, G. L. (2022). Late Miocene cooling coupled to carbon dioxide with Pleistocene-like climate sensitivity. *Nature Geoscience, 15*, 664–670. https://doi.org/10.1038/s41561-022-00982-7

Caldeira, K., Arthur, M. A., Berner, R. A., & Lasaga, A. C. (1993). Cooling in the late Cenozoic. *Nature, 361*, 123–124. https://doi.org/10.1038/361123b0

Cotton, J. M., & Sheldon, N. D. (2012). New constraints on using paleosols to reconstruct atmospheric pCO$_2$. *Geological Society of America Bulletin, 124*, 1411–1423. https://doi.org/10.1130/b30607.1

D'Hondt, S., Inagaki, F., Orcutt, B. N., & Hinrichs, K.-U. (2019). IODP advances in understanding of subseafloor life. *Oceanography, 32*(1), 198–207. https://doi.org/10.5670/oceanog.2019.146

Da, J., Zhang, Y. G., Wang, H., Balsam, W., & Ji, J. (2015). An Early Pleistocene atmospheric CO_2 record based on pedogenic carbonate from the Chinese loess deposits. *Earth and Planetary Science Letters, 426*, 69–75. https://doi.org/10.1016/j.epsl.2015.05.053

Doria, G., Royer, D. L., Wolfe, A. P., Fox, A., Westgate, J. A., & Beerling, D. J. (2011). Declining atmospheric CO_2 during the late Middle Eocene climate transition. *American Journal of Science, 311*, 63–75. https://doi.org/10.2475/01.2011.03

Ekart, D. D., Cerling, T. E., Montañez, I. P., & Tabor, N. J. (1999). A 400 million year carbon isotope record of pedogenic carbonate: Implications for paleoatmospheric carbon dioxide. *American Journal of Science, 299*, 805–827. https://doi.org/10.2475/ajs.299.10.805

Erdei, B., Utescher, T., Hably, L., Tamás, J., Roth-Nebelsick, A., & Grein, M. (2012). Early Oligocene continental climate of the Palaeogene Basin (Hungary and Slovenia) and the surrounding area. *Turkish Journal of Earth Sciences, 21*, 153–186. https://doi.org/10.3906/yer-1005-29

Fletcher, B. J., Brentnall, S. J., Anderson, C. W., Berner, R. A., & Beerling, D. J. (2008). Atmospheric carbon dioxide linked with Mesozoic and early Cenozoic climate change. *Nature Geoscience, 1*, 43–48. https://doi.org/10.1038/ngeo.2007.29

Foster, G. L., Lear, C. H., & Rae, J. W. B. (2012). The evolution of pCO_2, ice volume and climate during the middle Miocene. *Earth and Planetary Science Letters, 341–344*, 243–254. https://doi.org/10.1016/j.epsl.2012.06.007

Foster, G. L., Royer, D. L., & Lunt, D. J. (2017). Future climate forcing potentially without precedent in the last 420 million years. *Nature Communications, 8*, 14845. https://doi.org/10.1038/ncomms14845

Franks, P. J., Royer, D. L., Beerling, D. J., Van de Water, P. K., Cantrill, D. J., Barbour, M. M., & Berry, J. A. (2014). New constraints on atmospheric CO_2 concentration for the Phanerozoic. *Geophysical Research Letters, 41*, 4685–4694. https://doi.org/10.1002/2014gl060457

Greenop, R., Foster, G. L., Wilson, P. A., & Lear, C. H. (2014). Middle Miocene climate instability associated with high-amplitude CO_2 variability. *Paleoceanography, 29*, 845–853. https://doi.org/10.1002/2014pa002653

Greenop, R., Sosdian, S. M., Henehan, M. J., Wilson, P. A., Lear, C. H., & Foster, G. L. (2019). Orbital forcing, ice volume, and CO_2 across the Oligocene-Miocene transition. *Paleoceanography and Paleoclimatology, 34*, 316–328. https://doi.org/10.1029/2018pa003420

Greenwood, D. R., Scarr, M. J., & Christophel, D. C. (2003). Leaf stomatal frequency in the Australian tropical rainforest tree *Neolitsea dealbata* (Lauraceae) as a proxy measure of atmospheric pCO_2. *Palaeogeography Palaeoclimatology Palaeoecology, 196*, 375–393. https://doi.org/10.1016/S0031-0182(03)00465-6

Grein, M., Konrad, W., Wilde, V., Utescher, T., & Roth-Nebelsick, A. (2011). Reconstruction of atmospheric CO_2 during the early Middle Eocene by application of a gas exchange model to fossil plants from the Messel Formation, Germany. *Palaeogeography Palaeoclimatology Palaeoecology, 309*, 383–391. https://doi.org/10.1016/j.palaeo.2011.07.008

Grein, M., Oehm, C., Konrad, W., Utescher, T., Kunzmann, L., & Roth-Nebelsick, A. (2013). Atmospheric CO_2 from the late Oligocene to early Miocene based on photosynthesis data and fossil leaf characteristics. *Palaeogeography Palaeoclimatology Palaeoecology, 374*, 41–51. https://doi.org/10.1016/j.palaeo.2012.12.025

Guillermic, M., Misra, S., Eagle, R., & Tripati, A. (2022). Atmospheric CO_2 estimates for the Miocene to Pleistocene based on foraminiferal $\delta^{11}B$ at Ocean Drilling Program Sites 806 and 807 in the Western Equatorial Pacific. *Climate of the past, 18*, 183–207. https://doi.org/10.5194/cp-18-183-2022

Gutjahr, M., Ridgwell, A., Sexton, P. F., Anagnostou, E., Pearson, P. N., Pälike, H., Norris, R. D., Thomas, E., & Foster, G. L. (2017). Very large release of mostly volcanic carbon during the Palaeocene-Eocene Thermal Maximum. *Nature, 548*, 573–577. https://doi.org/10.1038/nature23646

Harper, D. T., Hönisch, B., Zeebe, R. E., Shaffer, G., Haynes, L. L., Thomas, E., & Zachos, J. C. (2020). The magnitude of surface ocean acidification and carbon release during Eocene thermal maximum 2 (ETM-2) and the Paleocene-Eocene thermal maximum (PETM). *Paleoceanography and Paleoclimatology, 35*, e2019PA003699. https://doi.org/10.1029/2019pa003699

Henehan, M. J., Ridgwell, A., Thomas, E., Zhang, S., Alegret, L., Schmidt, D. N., Rae, J. W. B., Witts, J. D., Landman, N. H., Greene, S. E., Huber, B. T., Super, J. R., Planavsky, N. J., & Hull, P. M. (2019). Rapid ocean acidification and protracted Earth system recovery followed the end-Cretaceous Chicxulub impact. *Proceedings of the National Academy of Sciences, USA*, 201905989. https://doi.org/10.1073/pnas.1905989116

Henehan, M. J., Edgar, K. M., Foster, G. L., Penman, D. E., Hull, P. M., Greenop, R., Anagnostou, E., & Pearson, P. N. (2020). Revisiting the Middle Eocene climatic optimum "carbon cycle conundrum" with new estimates of atmospheric pCO_2 from boron isotopes. *Paleoceanography and Paleoclimatology, 35*, e2019PA003713. https://doi.org/10.1029/2019pa003713

Huang, C., Retallack, G. J., Wang, C., & Huang, Q. (2013). Paleoatmospheric pCO_2 fluctuations across the Cretaceous-Tertiary boundary recorded from paleosol carbonates in NE China. *Palaeogeography Palaeoclimatology Palaeoecology, 385*, 95–105. https://doi.org/10.1016/j.palaeo.2013.01.005

Hyland, E. G., & Sheldon, N. D. (2013). Coupled CO_2-climate response during the Early Eocene Climatic Optimum. *Palaeogeography Palaeoclimatology Palaeoecology, 369*, 125–135. https://doi.org/10.1016/j.palaeo.2012.10.011

Hyland, E., Sheldon, N. D., & Fan, M. (2013). Terrestrial paleoenvironmental reconstructions indicate transient peak warming during the early Eocene climatic optimum. *Geological Society of America Bulletin, 125*, 1338–1348. https://doi.org/10.1130/b30761.1

Ji, S., Nie, J., Lechler, A., Huntington, K. W., Heitmann, E. O., & Breecker, D. O. (2018). A symmetrical CO_2 peak and asymmetrical climate change during the middle Miocene. *Earth and Planetary Science Letters, 499*, 134–144. https://doi.org/10.1016/j.epsl.2018.07.011

Kemp, L. R., Kastings, J. F., & Crane, R. G. (2010). *The Earth system* (420 p.). Pearson. ISBN-10: 0321597796.

Koch, P. L., Zachos, J. C., & Gingerich, P. D. (1992). Correlation between isotope records in marine and continental carbon reservoirs near the Palaeocene/Eocene boundary. *Nature, 358*, 319–322. https://doi.org/10.1038/358319a0

Kohn, M. J., Strömberg, C. A. E., Madden, R. H., Dunn, R. E., Evans, S., Palacios, A., & Carlini, A. A. (2015). Quasi-static Eocene-Oligocene climate in Patagonia promotes slow faunal evolution and mid-Cenozoic global cooling. *Palaeogeography, Palaeoclimatology, Palaeoecology, 435*, 24–37. https://doi.org/10.1016/j.palaeo.2015.05.028

Kowalczyk, J. B., Royer, D. L., Miller, I. M., Anderson, C. W., Beerling, D. J., Franks, P. J., Grein, M., Konrad, W., Roth-Nebelsick, A., Bowring, S. A., Johnson, K. R., & Ramezani, J. (2018). Multiple proxy estimates of atmospheric CO_2 from an early Paleocene rainforest. *Paleoceanography and Paleoclimatology, 33*, 1427–1438. https://doi.org/10.1029/2018PA003356

Kump, L. R., & Arthur, M. A. (1997). Global chemical erosion during the Cenozoic: Weatherability balances the budgets. In W. R. Ruddiman (Ed.), *Tectonic uplift and climate change* (pp. 399–426). Springer. https://doi.org/10.1007/978-1-4615-5935-1_18

Kürschner, W. M., van der Burgh, J., Visscher, H., & Dilcher, D. L. (1996). Oak leaves as biosensors of late Neogene and early Pleistocene paleoatmospheric CO_2 concentrations. *Marine Micropaleontology, 27*, 299–312. https://doi.org/10.1016/0377-8398(95)00067-4

Kürschner, W. M., Kvacek, Z., & Dilcher, D. L. (2008). The impact of Miocene atmospheric carbon dioxide fluctuations on climate and the evolution of terrestrial ecosystems. *Proceedings of the National Academy of Sciences USA, 105*, 449–453. https://doi.org/10.1073/pnas.0708588105

Kürschner, W. M., Wagner, F., Dilcher, D. L., & Visscher, H. (2001). Using fossil leaves for the reconstruction of Cenozoic paleoatmospheric CO_2 concentrations. In L. C. Gerhard, W. E. Harrison, & B. M. Hanson (Eds.), *Geological perspectives of global climate change: APPG studies in geology* (Vol. 47, pp. 169–189). The American Association of Petroleum Geologists.

Liang, J., Leng, Q., Höfig, D. F., Niu, G., Wang, L., Royer, D. L., Burke, K., Xiao, L., Zhang, Y., & Yang, H. (2022). Constraining conifer physiological parameters in leaf gas-exchange models for ancient CO_2 reconstruction. *Global and Planetary Change, 209*, 103737. https://doi.org/10.1016/j.gloplacha.2022.103737

Liang, J., Leng, Q., Xiao, L., Höfig, D. F., Royer, D. L., Zhang, Y. G., & Yang, H. (2022). Early Miocene redwood fossils from Inner Mongolia: CO_2 reconstructions and paleoclimate effects of a low Mongolian plateau. *Review of Palaeobotany and Palynology, 305*, 104743. https://doi.org/10.1016/j.revpalbo.2022.104743

Londoño, L., Royer, D. L., Jaramillo, C., Escobar, J., Foster, D. A., Cárdenas-Rozo, A. L., & Wood, A. (2018). Early Miocene CO_2 estimates from a Neotropical fossil assemblage exceed 400 ppm. *American Journal of Botany, 105*, 1929–1937. https://doi.org/10.1002/ajb2.1187

Martínez-Botí, M. A., Foster, G. L., Chalk, T. B., Rohling, E. J., Sexton, P. F., Lunt, D. J., Pancost, R. D., Badger, M. P. S., & Schmidt, D. N. (2015). Plio-Pleistocene climate sensitivity evaluated using high-resolution CO_2 records. *Nature, 518*, 49–54. https://doi.org/10.1038/nature14145

Maxbauer, D. P., Royer, D. L., & LePage, B. A. (2014). High Arctic forests during the middle Eocene supported by moderate levels of atmospheric CO_2. *Geology, 42*, 1027–1030. https://doi.org/10.1130/g36014.1

McElwain, J. C. (1998). Do fossil plants signal palaeoatmospheric CO_2 concentration in the geological past? *Philosophical Transactions of the Royal Society London B, 353*, 83–96. https://doi.org/10.1098/rstb.1998.0193

Milligan, J. N., Royer, D. L., Franks, P. J., Upchurch, G. R., & McKee, M. L. (2019). No evidence for a large atmospheric CO_2 spike across the Cretaceous-Paleogene boundary. *Geophysical Research Letters, 46*, 3462–3472. https://doi.org/10.1029/2018GL081215

Milligan, J. N., Flynn, A. G., Kowalczyk, J. B., Barclay, R. S., Geng, J., Royer, D. L., & Peppe, D. J. (2022). Moderate to elevated atmospheric CO_2 during the early Paleocene recorded by *Platanites* leaves of the San Juan Basin, New Mexico. *Paleoceanography and Paleoclimatology, 37*, e2021PA004408. https://doi.org/10.1029/2021PA004408

Moraweck, K., Grein, M., Konrad, W., Kvaček, J., Kova-Eder, J., Neinhuis, C., Traiser, C., & Kunzmann, L. (2019). Leaf traits of long-ranging Paleogene species and their relationship with depositional facies, climate and atmospheric CO_2 level. *Palaeontographica Abteilung B, 298*, 93–172. https://doi.org/10.1127/palb/2019/0062

Nordt, L., Atchley, S., & Dworkin, S. (2003). Terrestrial evidence for two greenhouse events in the latest Cretaceous. *GSA Today, 13*(12), 4–9.

Pearson, P. N., Foster, G. L., & Wade, B. S. (2009). Atmospheric carbon dioxide through the Eocene-Oligocene climate transition. *Nature, 461*, 1110–1113. https://doi.org/10.1038/nature08447

Raymo, M. E., & Ruddiman, R. F. (1992). Tectonic forcing of late Cenozoic climate. *Nature, 359*, 117–122. https://doi.org/10.1038/359117a0

Reichgelt, T., D'Andrea, W. J., Valdivia-McCarthy, A. C., Fox, B. R. S., Bannister, J. M., Conran, J. G., Lee, W. G., & Lee, D. E. (2020). Elevated CO_2, increased leaf-level productivity, and water-use efficiency during the early Miocene. *Climate of the past, 16*, 1509–1521. https://doi.org/10.5194/cp-16-1509-2020

Retallack, G. J. (2009). Greenhouse crises of the past 300 million years. *Geological Society of America Bulletin, 121*, 1441–1455. https://doi.org/10.1130/b26341.1

Retallack, G. J. (2009). Refining a pedogenic-carbonate CO_2 paleobarometer to quantify a middle Miocene greenhouse spike. *Palaeogeography Palaeoclimatology Palaeoecology, 281*, 57–65. https://doi.org/10.1016/j.palaeo.2009.07.011

Roth-Nebelsick, A., Grein, M., Utescher, T., & Konrad, W. (2012). Stomatal pore length change in leaves of *Eotrigonobalanus furcinervis* (Fagaceae) from the Late Eocene to the Latest Oligocene and its impact on gas exchange and CO_2 reconstruction. *Review of Palaeobotany and Palynology, 174*, 106–112. https://doi.org/10.1016/j.revpalbo.2012.01.001

Roth-Nebelsick, A., Oehm, C., Grein, M., Utescher, T., Kunzmann, L., Friedrich, J.-P., & Konrad, W. (2014). Stomatal density and index data of *Platanus neptuni* leaf fossils and their evaluation as a CO_2 proxy for the Oligocene. *Review of Palaeobotany and Palynology, 206*, 1–9. https://doi.org/10.1016/j.revpalbo.2014.03.001

Royer, D. L., Wing, S. L., Beerling, D. J., Jolley, D. W., Koch, P. L., Hickey, L. J., & Berner, R. A. (2001). Paleobotanical evidence for near present-day levels of atmospheric CO_2 during part of the Tertiary. *Science, 292*, 2310–2313. https://doi.org/10.1126/science.292.5525.2310

Royer, D. L. (2003). Estimating latest Cretaceous and Tertiary atmospheric CO_2 concentration from stomatal indices. In S. L. Wing, P. D. Gingerich, B. Schmitz, & E. Thomas (Eds.), *Causes and consequences of globally warm climates in the early Paleogene* (pp. 79–93). Geological Society of America Special Paper 369.

Scotese, C. R. (2021). An Atlas of Phanerozoic paleogeographic maps: The seas come in and the seas go out. *Annual Review of Earth and Planetary Sciences, 49*(1), 679–728. https://doi.org/10.1146/annurev-earth-081320-064052

Seki, O., Foster, G. L., Schmidt, D. N., Mackensen, A., Kawamura, K., & Pancost, R. D. (2010). Alkenone and boron-based Pliocene pCO_2 records. *Earth and Planetary Science Letters, 292*, 201–211. https://doi.org/10.1016/j.epsl.2010.01.037

Smith, R. Y., Greenwood, D. R., & Basinger, J. F. (2010). Estimating paleoatmospheric pCO_2 during the Early Eocene Climatic Optimum from stomatal frequency of *Ginkgo*, Okanagan Highlands, British Columbia, Canada. *Palaeogeography Palaeoclimatology Palaeoecology, 293*, 120–131. https://doi.org/10.1016/j.palaeo.2010.05.006

Sosdian, S. M., Greenop, R., Hain, M. P., Foster, G. L., Pearson, P. N., & Lear, C. H. (2018). Constraining the evolution of Neogene ocean carbonate chemistry using the boron isotope pH proxy. *Earth and Planetary Science Letters, 498*, 362–376. https://doi.org/10.1016/j.epsl.2018.06.017

Srivastava, P., Patel, S., Singh, N., Jamir, T., Kumar, N., Aruche, M., & Patel, R. C. (2013). Early Oligocene paleosols of the Dagshai Formation, India: A record of the oldest tropical weathering in the Himalayan foreland. *Sedimentary Geology, 294*, 142–156. https://doi.org/10.1016/j.sedgeo.2013.05.011

Stap, L. B., de Boer, B., Ziegler, M., Bintanja, R., Lourens, L. J., & van de Wal, R. S. W. (2016). CO_2 over the past 5 million years: Continuous simulation and new δ^{11}B-based proxy data. *Earth and Planetary Science Letters, 439*, 1–10. https://doi.org/10.1016/j.epsl.2016.01.022

Steinthorsdottir, M., Porter, A. S., Holohan, A., Kunzmann, L., Collinson, M., & McElwain, J. C. (2016). Fossil plant stomata indicate decreasing atmospheric CO_2 prior to the Eocene-Oligocene boundary. *Climate of the past, 12*, 439–454. https://doi.org/10.5194/cp-12-439-2016

Steinthorsdottir, M., Vajda, V., & Pole, M. (2016). Global trends of pCO_2 across the Cretaceous-Paleogene boundary supported by the first Southern Hemisphere stomatal proxy-based pCO_2 reconstruction. *Palaeogeography, Palaeoclimatology, Palaeoecology, 464*, 143–152. https://doi.org/10.1016/j.palaeo.2016.04.033

Steinthorsdottir, M., Vajda, V., & Pole, M. (2019). Significant transient pCO_2 perturbation at the New Zealand Oligocene-Miocene transition recorded by fossil plant stomata. *Palaeogeography, Palaeoclimatology, Palaeoecology, 515*, 152–161. https://doi.org/10.1016/j.palaeo.2018.01.039

Steinthorsdottir, M., Vajda, V., Pole, M., & Holdgate, G. (2019). Moderate levels of Eocene pCO_2 indicated by Southern Hemisphere fossil plant stomata. *Geology, 47*, 914–918. https://doi.org/10.1130/g46274.1

Steinthorsdottir, M., Jardine, P. E., & Rember, W. C. (2021). Near-future pCO_2 during the hot mid Miocene climatic optimum. *Paleoceanography and Paleoclimatology, 36*, e2020PA003900. https://doi.org/10.1029/2020pa003900

Stults, D. Z., Wagner-Cremer, F., & Axsmith, B. J. (2011). Atmospheric paleo-CO_2 estimates based on *Taxodium distichum* (Cupressaceae) fossils from the Miocene and Pliocene of eastern North America. *Palaeogeography Palaeoclimatology Palaeoecology, 309*, 327–332. https://doi.org/10.1016/j.palaeo.2011.06.017

Sun, B.-N., Ding, S.-T., Wu, J.-Y., Dong, C., Xie, S., & Lin, Z.-C. (2012). Carbon isotope and stomatal data of late Pliocene Betulaceae leaves from SW China: Implications for palaeoatmospheric CO_2-levels. *Turkish Journal of Earth Sciences, 21*, 237–250. https://doi.org/10.3906/yer-1003-42

Sun, B.-N., Wang, Q.-J., Konrad, W., Ma, F.-J., Dong, J.-L., & Wang, Z.-X. (2017). Reconstruction of atmospheric CO_2 during the Oligocene based on leaf fossils from the Ningming Formation in Guangxi, China. *Palaeogeography, Palaeoclimatology, Palaeoecology, 467*, 5–15. https://doi.org/10.1016/j.palaeo.2016.09.015

Super, J. R., Thomas, E., Pagani, M., Huber, M., O'Brien, C., & Hull, P. M. (2018). North Atlantic temperature and $p$$CO_2$ coupling in the early-middle Miocene. *Geology, 46*, 519–522. https://doi.org/10.1130/G40228.1

Tanner, T., Hernández-Almeida, I., Drury, A. J., Guitián, J., & Stoll, H. (2020). Decreasing atmospheric CO_2 during the late Miocene cooling. *Paleoceanography and Paleoclimatology, 35*, e2020PA003925. https://doi.org/10.1029/2020PA003925

Tesfamichael, T., Jacobs, B., Tabor, N., Michel, L., Currano, E., Feseha, M., Barclay, R., Kappelman, J., & Schmitz, M. (2017). Settling the issue of "decoupling" between atmospheric carbon dioxide and global temperature: $[CO_2]_{atm}$ reconstructions across the warming Paleogene-Neogene divide. *Geology, 45*, 999–1002. https://doi.org/10.1130/G39048.1

van der Burgh, J., Visscher, H., Dilcher, D. L., & Kürschner, W. M. (1993). Paleoatmospheric signatures in Neogene fossil leaves. *Science, 260*, 1788–1790. https://doi.org/10.1126/science.260.5115.1788

Wang, Y., Momohara, A., Wang, L., Lebreton-Anberrée, J., & Zhou, Z. (2015). Evolutionary history of atmospheric CO_2 during the late Cenozoic from fossilized *Metasequoia* needles. *PLoS ONE, 10*(7), e0130941. https://doi.org/10.1371/journal.pone.0130941

Wang, Y., Wang, L., Momohara, A., Leng, Q., & Huang, Y.-J. (2020). The Paleogene atmospheric CO_2 concentrations reconstructed using stomatal analysis of fossil *Metasequoia* needles. *Palaeoworld, 29*, 744–751. https://doi.org/10.1016/j.palwor.2020.03.002

Westerhold, T., Marwan, N., Drury, A. J., Liebrand, D., Agnini, C., Anagnostou, E., Barnet, J. S. K., Bohaty, S. M., De Vleeschouwer, D., Florindo, F., Frederichs, T., Hodell, D. A., Holbourn, A. E., Kroon, D., Lauretano, V., Littler, K., Lourens, L. J., Lyle, M., Pälike, H., … Zachos, J. C. (2020). An astronomically dated record of Earth's climate and its predictability over the last 66 million years. *Science, 369*, 1383–1387. https://doi.org/10.1126/science.aba6853

Witkowski, C. R., Weijers, J. W. H., Blais, B., Schouten, S., & Damsté, J. S. S. (2018). Molecular fossils from phytoplankton reveal secular P_{CO_2} trend over the Phanerozoic. *Science Advances, 4*, eaat4556. https://doi.org/10.1126/sciadv.aat4556

Zhang, Y. G., Pagani, M., Liu, Z., Bohaty, S. M., & DeConto, R. (2013). A 40-million-year history of atmospheric CO_2. *Philosophical Transactions of the Royal Society A, 371*, 20130096. https://doi.org/10.1098/rsta.2013.0096

Zhang, L., Wang, C., Wignall, P. B., Kluge, T., Wan, X., Wang, Q., & Gao, Y. (2018). Deccan volcanism caused coupled $p$$CO_2$ and terrestrial temperature rises, and pre-impact extinctions in northern China. *Geology, 46*, 271–274. https://doi.org/10.1130/G39992.1

Lee Kump has always struggled to define his scientific discipline in conventional terms, but was attracted to Earth system science in its fledgling years and has maintained an interdisciplinary approach to the study of Earth ever since. He has spent his entire career at Penn State as a geosciences faculty member and affiliate of the Earth System Science Center (now Earth and Environmental Systems Institute), department head, and now dean of the College of Earth and Mineral Sciences. Through his research he and his students strive to unlock the mysteries of Earth's climate and biotic evolution, with a focus on the causes and consequences of the establishment of an oxygen-rich atmosphere, and climate and ocean change during past episodes of global warming and mass extinction. Lee is the primary author of Chap. 7.

James Kasting is a retired colleague of Lee Kump's at Penn State. Like Lee, he was an early affiliate of the Earth System Science Center. He worked at NASA Ames for seven years before that. James' research interests are in planetary science and astrobiology. He has published numerous papers about ancient atmospheric photochemistry and climate, with an emphasis on the Archean

Earth. He is known by astronomers for his work on habitable zones around stars. James is the author of Box 2 in Chap. 7.

Open Access This chapter is licensed under the terms of the Creative Commons Attribution 4.0 International License (http://creativecommons.org/licenses/by/4.0/), which permits use, sharing, adaptation, distribution and reproduction in any medium or format, as long as you give appropriate credit to the original author(s) and the source, provide a link to the Creative Commons license and indicate if changes were made.

The images or other third party material in this chapter are included in the chapter's Creative Commons license, unless indicated otherwise in a credit line to the material. If material is not included in the chapter's Creative Commons license and your intended use is not permitted by statutory regulation or exceeds the permitted use, you will need to obtain permission directly from the copyright holder.

Abrupt Climate Change: The PETM

Debbie Thomas and Kristen St. John

Guiding Questions: What were the causes and consequences of an abrupt global warming event of the past? How does this inform our understanding of climate change today?

1 Key Take-Away Points

- The accumulation of more, and more types, of data does not always lead directly to more agreement or consistent interpretations of the geologic record. In fact, such advances often reveal unknowns, or questions that were not known earlier in the investigation.
- The Earth systems are complex and investigating climate change from the geologic record requires reconciling different types and scales of data.
- One of the best analogs for modern climate change occurred approximately 56 million years ago, an event known as the Paleocene-Eocene Thermal Maximum (PETM).
- Climate change can impact different groups of organisms in unequal, and potentially contrasting, manners—in the case of the PETM, some groups of organisms suffered extinction while others evolved new species.

D. Thomas (✉)
College of Marine Sciences and Maritime Studies, Texas A&M University, College Station, TX, USA
e-mail: dthomas@tamug.edu

K. St. John
Department of Geology and Environmental Science, James Madison University, Harrisonburg, VA, USA
e-mail: stjohnke@jmu.edu

- The same components of the carbon cycle that control long-term climate (i.e., volcanism and weathering) also have the potential to drive rapid and transient climate change under unique circumstances.
- The prominent change in global carbon isotope values is evidence for a massive and rapid release of geologically stored carbon into the ocean–atmosphere–biosphere system.
- Evidence for global warming throughout the ocean–atmosphere–biosphere system coincides with the carbon isotope change, indicating greenhouse gas-induced warming.
- Ocean acidification is a predictable consequence of excess CO_2 release into the ocean–atmosphere–biosphere system, and is occurring in modern times.
- PETM recovery occurred over ~ 100,000 years, through a combination of CO_2 drawdown out of the ocean–atmosphere–biosphere system related to rock weathering and burial in the sediments.
- Modern rates of CO_2 release to the ocean–atmosphere–biosphere system are nearly 10 times faster than during the PETM, and could have more dire consequences given the potential to melt glacial ice.

2 Introduction

So far through this book, you have explored the major controls on climate, both short and long term. You also now are familiar with the strategies and techniques we can employ to construct records of past climate from the materials preserved in the geologic record.

In this chapter, we explore the geologic record for evidence of some event or episode resulting from natural processes that might resemble the current perturbation. If we describe the current climate perturbation in geologic terms, it would be characterized as the rapid transfer of carbon from long-term storage in the earth's crust (i.e., buried fossil fuels) into the actively cycling ocean–atmosphere–biosphere system. Two of the most predictable and diagnostic signatures of a rapid carbon release into the ocean–atmosphere–biosphere system are (1) an increase in the greenhouse gas concentration of the atmosphere and (2) a decrease in the pH of the oceans, and we have observed both of these signatures directly in the context of the current carbon release. Was there a time in the geologic past characterized by a similarly rapid transfer of carbon from long-term storage in the crust or mantle, into the ocean–atmosphere–biosphere system? What aspects of the geologic record could be explored for evidence of these predictable signatures? How could we learn from the geologic record about the impact of the change in carbon cycling on the rest of the earth system?

As you reflect on what you have learned through the first seven chapters of this book, you can begin to appreciate the advantages of a paleoclimatic perspective in understanding the causes and consequences of rapid climate change. Perhaps the biggest advantage is that we can "see" the entire event from start to finish, along with all the context and associated environmental change that accompanied the warming. That is, we have the ability to study the conditions at the onset of the event, we can observe how the event began, and also investigate the recovery back to some long-term "equilibrium state." We also have the potential to tease apart how perturbations in the carbon cycle may have impacted the climate and biota, and vice versa.

This perspective is considerably different from our current vantage point. Recall from Chap. 3 that at this moment in geologic time, we only have a few decades of earth systems direct observations of the current "event." Our current perspective also is biased by the fact that the response of the climate system naturally lags the climate forcing, so we don't see a full and immediate cause and effect. The current perturbation to the energy cycle, i.e. the increase in the concentration of atmospheric greenhouse gases and the consequent increase in atmospheric retention of outgoing long-wave radiation, doesn't translate immediately as a global increase in surface temperatures because of the thermal inertia that resides largely in the oceans and ice sheets. In other words, the warming our planet has recently experienced isn't yet the full effect of the change in radiative forcing. And its impact isn't uniformly distributed across the planet, nor is the change a simple and steady trend through time.

In fact, there was a time in the geologic past characterized by the geologically rapid transfer of carbon from long-term storage in the crust or mantle, into the ocean–atmosphere–biosphere system! At the close of the Paleocene Epoch ~ 56 million years ago, the Earth experienced one of its most dramatic episodes of short-term global warming, the **Paleocene-Eocene Thermal Maximum**, or **PETM**, event. In this chapter, we begin by exploring the conditions that existed on Earth during the Paleocene Epoch, prior to the onset of the event. Then we explore the onset of the rapid warming and the known associated changes in the energy, carbon, and water cycles, with an eye to what ultimately caused the warming. Finally, we seek to understand what conditions or feedbacks within the cycles of energy, carbon and water contributed to the "recovery" of the PETM back to the general equilibrium state in the Eocene Epoch. The multiple and interconnected dimensions of the earth system involved in the PETM provide an excellent opportunity to apply the concepts developed throughout this book.

3 Setting the Stage—What Conditions Existed at the Onset of the PETM?

Let's revisit a figure introduced in Chap. 1. Because deep-ocean temperatures change very slowly and reflect the average conditions of a substantial part of the planet, climate scientists use the record of ocean temperatures from the deep ocean as a gauge of overall climate state. The dark blue and purple portions of the temperature change record shown in Fig. 1 are based on the oxygen isotope composition of seafloor-dwelling organisms called benthic foraminifera.

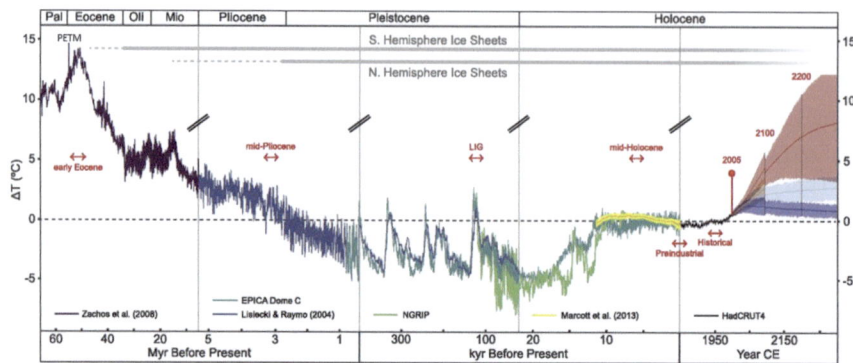

Fig. 1 Temperature trends for the past 65 M.y. and projected to the year 2250 CE. Six potential geologic analogs for future climates are noted with red labeled arrows (e.g., early Eocene). Major patterns include a long-term cooling trend, periodic fluctuations driven by changes in the Earth's orbit, and recent and projected warming trends. Temperature anomalies (ΔT) are relative to 1961–1990 global means and are composited from proxy-based reconstructions, modern observations (instrumental measurements), and future temperature projections for four emissions pathways. The presence of ice sheets in the Southern and Northern Hemisphere are shown as gray horizontal lines. Geologic time epochs are noted along the top, with Paleocene abbreviated as Pal; Oligocene as Oli; and Miocene as Mio. The **Paleocene-Eocene Thermal Maximum** event is abbreviated as PETM. Figure modified (PETM label added) and used with permission from PNAS, Burke et al., (2018).[1] This figure is excluded from Creative Commons license

Recall from Chap. 5 that the oxygen isotopic composition of zooplankton calcium carbonate shells (foraminifera) provides a record of the water temperature in which the shell grew. For the portion of the compilation spanning the older geologic history of global temperatures (i.e., the blue and purple parts of the curve), it is important to note that the typical duration of a specific "event" or spike that can be identified is on the order of several thousands of years. This time resolution is substantially different from our ability to detect specific decade- or century-scale events in ice core records (the green and teal portions of the curve).

Global conditions during the late Paleocene through early Eocene (approximately 66 to 50 million years ago) represent the warmest climate state of the entire Cenozoic. In Chap. 1, we identified the very warm Early Eocene as an analog for our future climate state, and we can include the preceding Paleocene Epoch as part of that analog state as well. The overall warmth of the time period indicated by the deep-sea data in Fig. 1 is corroborated by data from terrestrial environments at all latitudes. For example, fossil flora indicate substantially warmer temperatures in the high latitudes.[2] The fossil record also indicates that fauna associated with warm conditions, such as crocodilians and turtles, as well as large, flightless birds, inhabited the regions above the Arctic Circle.[3]

---

[1] Burke et al. (2018).

[2] Irving and Wynne (1991).

[3] Stidham and Eberle (2016) and McKenna (1980).

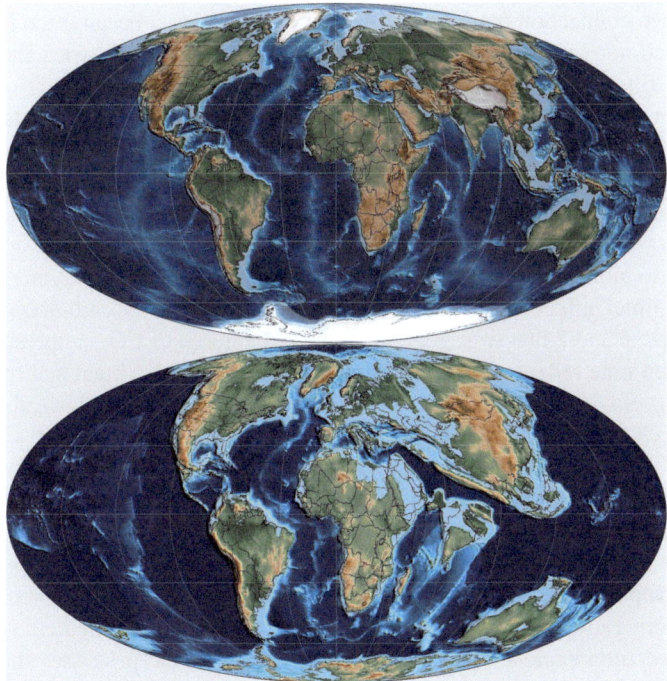

Fig. 2 Upper: modern geography and physiography. Lower: paleogeographic reconstruction for the PETM (55.8 Ma). Maps courtesy of Chris Scotese[4]

The combination of high-latitude marine and terrestrial evidence for overall warmth is consistent with geologic evidence for very little ice at the poles. In fact, paleoclimate proxy data and fossil evidence suggest a climate state characterized by little to no permanent glacial ice in either hemisphere (e.g., the "Greenhouse" world compared to the more recent "Icehouse" world characterized by major continental ice at both poles). The lack of significant continental-scale glaciers such as those presently found on Greenland or Antarctica would have led to vastly higher global sea levels, depicted in Fig. 2 by the flooding of many modern-day continental surfaces.

Why was the earth so much warmer during the Paleocene and Eocene? As a starting point, we need to examine the geography of the planet ~ 56 million years ago (Fig. 2). "Rewinding" plate tectonic motions by 56 million years reveals several major differences in the distribution of ocean basins and land masses from today. The very northern portion of the Atlantic Ocean near what is modern-day Greenland was significantly more restricted during the late Paleocene and early Eocene. Toward the other pole, the ocean passages between South America and Antarctica (the modern Drake Passage) as well as Australia and Antarctica were much more restricted. There

---

[4] Scotese (2021).

was an open connection between North and South America because the Isthmus of Panama had not yet formed. There also was an ocean between Africa and the land masses of Europe and Asia (the modern Mediterranean Sea is far more restricted). The Indian subcontinent had not yet collided with Asia to form the Tibetan Plateau and Himalayas. You also will note that substantial continental areas were covered by shallow seas.

Most, if not all, of these major plate tectonic differences played a significant role in the overall climate differences that existed during the Paleocene and Eocene compared to more recent climate. We will focus here on the plate tectonic processes that had a direct influence on climate through their impact on the carbon cycle. Recall from Chaps. 6 and 7 that over geologic time, the amount of CO_2 in the atmosphere at any given time reflects a general balance between volcanic input and removal by weathering. So, if the rate of volcanic activity increased relative to weathering, the amount of CO_2 in the atmosphere would have increased. One of the major "checks" on volcanic inputs of CO_2 into the ocean–atmosphere–biosphere system is weathering of exposed rock and soil that removes the CO_2 during the chemical reaction that also breaks down the rock. For example, although not captured explicitly in Fig. 2, one of the most prominent tectonic features of the Paleocene and Eocene was the volcanic eruption of Iceland and associated regions bordering eastern Greenland and northwestern Europe, known as the North Atlantic Igneous Province. Long-term and passive (non-explosive) volcanism very similar to the processes that created the Hawaiian Islands (as opposed to the violently explosive eruptions that typically result in short-term cooling, such as Krakatoa) resulted in the emplacement of nearly 2 million cubic kilometers of volcanic rock[5] roughly equivalent to the volume of the modern-day Gulf of Mexico! These eruptions introduced approximately 10,000,000 metric tons of mantle-derived CO_2 into the ocean–atmosphere–biosphere system.[6] In comparison, over 400,000,000 metric tons of carbon have been added to the ocean–atmosphere–biosphere system through fossil fuel combustion since 1850.[7]

The North Atlantic Igneous Province is the most prominent early Cenozoic example of voluminous volcanic inputs that served to keep earth's climate significantly warmer for much of the past 200 million years (its importance and timing is shown in Fig. 17 of Chap. 5) . As you might imagine, such voluminous volcanism supported much higher atmospheric CO_2 concentrations than later in the Cenozoic. Figure 3 presents the global compilation of deep-sea oxygen isotope values from benthic foraminifera (upper panel, similar to the data in Fig. 1, updated with new data) along with the corresponding carbon isotope values obtained from those same analyses (middle panel). The lower panel of Fig. 3 depicts the range of compiled atmospheric CO_2 concentration estimates reconstructed from a variety of techniques. Note the very prominent trend toward negative values (indicated by an upward trend in the figure) in the carbon isotope data from ~ 60 to 50 million years ago (interval shaded in yellow in Fig. 3). The carbon stable isotope composition recorded by the calcium carbonate shells of benthic foraminifera reflects a mixture of all the different

---

[5] Eldholm and Thomas (1993).

[6] Same as above.

[7] Friedlingstein et al. (2019).

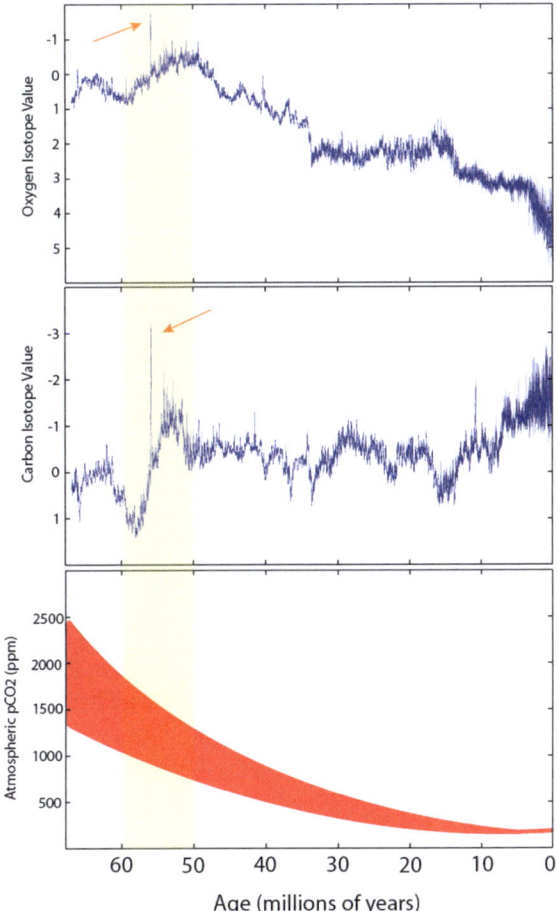

Fig. 3 Records of the deep-sea oxygen ($\delta^{18}O$, top) and carbon ($\delta^{13}C$, middle) isotope composition from benthic foraminfera (Data from: Westerhold et al., 2020),[8] shown with the range of reconstructed atmospheric carbon dioxide (Data from: Zachos et al., 2008; Zhang et al., 2013; Anagnoustou et al., 2016)[9]

sources of carbon that has become dissolved in the seawater. Such sources vary from the CO_2 that have been mixed in from the atmosphere, to the dissolved carbon that was released during the decomposition of organic matter within the oceans, to the dissolved carbon delivered to the oceans via rivers. A thorough characterization of carbon isotope variations requires a textbook unto itself, but Box 1 and the following discussion provide you with the fundamental background on stable carbon isotopes so you can apply it to understanding the PETM.

---

[8] Westerhold et al. (2020).
[9] Zachos et al. (2008); Zhang et al. (2013); and Anagnostou et al. (2016).

Box 1 Stable carbon isotopes as a proxy for changes in the global carbon cycle

By Kristen St. John, Department of Geology and Environmental Science, James Madison University.

Carbon isotopes are present in all carbon-containing materials in the Earth system. They are present in all organic matter—including the tissues (e.g., fats, muscles, flesh) of all living organisms on land and in the ocean, as well as in the geologically-stored (buried) organic remains of past life (e.g., sedimentary organic carbon: peat, coal, oil and methane gas). They are present in inorganic carbon-bearing materials too, including the shells of carbonate ($CaCO_3$) secreting organisms (such as foraminifera, which you learned about in Chap. 5), and in CO_2 gas present in the atmosphere, dissolved in ocean water, and dissolved in magma.

The most abundant carbon isotope is carbon-12 (^{12}C); it has 6 protons and 6 neutrons in the nucleus, and comprises nearly 99% of all carbon in the Earth system. In much smaller abundances are "heavier" isotopes of carbon: ^{13}C (6 protons and 7 neutrons in the nucleus) and ^{14}C (6 protons and 8 neutrons in the nucleus), comprising only ~ 1% and 0.0000000001% of the total carbon in the Earth system, respectively. ^{14}C is unstable, meaning that it is radioactive, and therefore changes in the nucleus over time can be used to determine the age of relatively young (< 60,000 years old) carbon-containing materials. In contrast, ^{12}C and ^{13}C are stable (nonradioactive) but are used in another very special way: the ratio of these isotopes is used as a proxy for changes in the global carbon cycle. Understanding how this proxy works is key to understanding the PETM.

The relative abundance of stable isotopes of carbon in a given sample are determined by mass spectrometry, by which ions of different masses are separated by passage through a magnetic beam and measured. The ratio of $^{13}C/^{12}C$ is denoted as $\delta^{13}C$ after values are calculated with respect to the isotopic ratio of a laboratory standard. Notice that all of the figures in this chapter that display stable carbon isotope data use the $\delta^{13}C$ notation. The unit for $\delta^{13}C$ is parts per thousand (i.e., per mil, with the unit symbol "‰").

The $\delta^{13}C$ values of different carbon reservoirs in the global carbon cycle are not uniform in composition due to physical (thermodynamic) and biological **fractionation** processes that partition lighter and heavier isotopes. One example of fractionation will be given here: it is biologically easier for an organism to use ^{12}C (rather than ^{13}C) during any metabolic reaction involving a carbon-based molecule. As a result, the $\delta^{13}C$ values of organic carbon (e.g., plants on land, algae in the ocean, and sedimentary organic carbon) are more negative (i.e., have lower, or isotopically 'lighter', values) than $\delta^{13}C$ of inorganic carbon in the atmosphere and ocean. A typical range of $\delta^{13}C$ in the terrestrial and marine biosphere is $-$ 10 to $-$ 25‰. Similar values are seen in sedimentary (i.e., buried) organic carbon deposits in the form of coal and petroleum. Buried methane deposits have even more negative $\delta^{13}C$ values, ranging between -40 and -70‰, due to additional fractionation effects as anaerobic bacteria break down more complex buried organic carbon molecules into this simpler form.

In contrast, $\delta^{13}C$ of pre-industrial atmospheric CO_2 was -6.5‰. $\delta^{13}C$ of dissolved inorganic carbon in the pre-industrial ocean was between $+2$ and 0‰, with the surface waters typically being more positive compared to deep waters because of the downward transport of isotopically light ^{12}C by sinking and decomposing marine biota.[10] Because volcanic CO_2 is sourced from the subduction of both inorganic and organic carbon, its $\delta^{13}C$ values can vary widely, but a value of -6‰ is generally representative.

A key takeaway from the discussion above is that there is spatial heterogeneity of $\delta^{13}C$ in the global carbon cycle; with different reservoirs having different $\delta^{13}C$ compositions due to fractionation effects. Importantly, reservoir changes to $\delta^{13}C$ also occur through time. As carbon is redistributed among the different reservoirs via processes of the short- (e.g., photosynthesis and decomposition, atmosphere–ocean gas exchange) and long-term carbon cycles (e.g., volcanism, burial of organic carbon, weathering of sedimentary organic carbon) (topics you learned about in Chaps. 6 and 7), the isotopic compositions of the reservoirs change too. Some of these relationships are illustrated in Fig. 4; these relationships are useful to keep in mind as you explore the causes and effects of the PETM described in this Chapter.

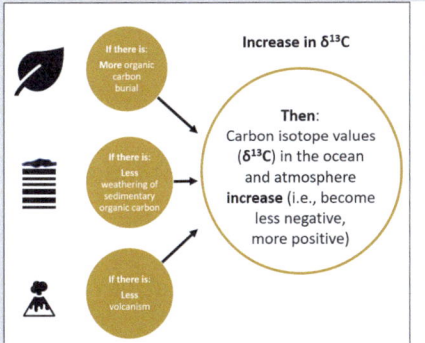 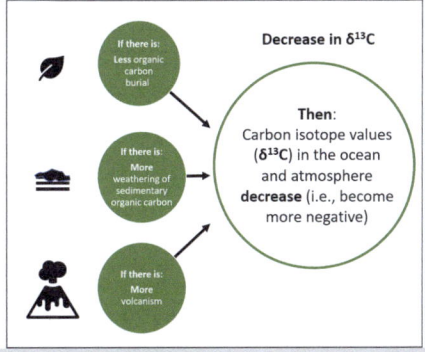

Fig. 4 Simplified representations of how changes in the carbon cycle impact stable carbon isotopes ($\delta^{13}C$) in the ocean and atmosphere. These changes are recorded in the carbonate shells of marine organisms, such as foraminifera

For the purposes of this discussion, we begin with a basic example in order to assess a potential relationship between the trends in the concentration of CO_2 in the

---

[10] Kwon et al. (2022) and Saltzman and Thomas (2012).

atmosphere and the trend toward more negative carbon isotope values recorded by the benthic foraminiferal shells. A typical carbon isotope value for carbon dioxide released from volcanic activity is $-6‰$ (parts per thousand, or per mil). Volcanism from the North Atlantic Igneous Province described above contributed an estimated 10,000,000 metric tons of CO_2 to the atmosphere and oceans. With the relatively negative isotopic composition of $-6‰$, such a huge mass of new CO_2 added through volcanic activity during the Paleocene and Eocene is entirely consistent with the prominent global trend in deep-water carbon isotope values as shown in Fig. 3 (red arrow middle panel). Of course, such an increase in atmospheric CO_2 also is consistent with the decrease in the corresponding oxygen isotope values (i.e., increase in temperature) observed in the top panel.

So, one way to understand the long-term, overall warmth of the late Paleocene and early Eocene is to consider that the overall balance between CO_2 supplied to the ocean–atmosphere–biosphere system and its removal through weathering (as discussed in Chap. 7) was set at a higher value than modern. Over the course of the past 66 million years, the balance between supply of CO_2 from volcanism and its removal via weathering evolved. For example, radiometric dating of the volcanic deposits indicates that the major phase of North Atlantic Igneous Province volcanism ended by approximately 54 million years ago, although Iceland, perched atop the Mid-Atlantic Ridge, remains volcanically active today. The dramatic decrease in this major volcanic input of CO_2 to the atmosphere was followed closely in time (geologically speaking) by the initiation of one of the most prominent weathering sinks of carbon the planet has ever known—uplift of the Tibetan Plateau through the plate tectonic collision of the Indian subcontinent with Asia approximately 50 million years ago. The combination of these two major changes in input and removal was a significant factor in the overall global cooling observed in the geologic record since 50 million years ago (Fig. 1).

4 History of PETM Research as a Lens into Paleoclimate Reconstruction

Now that we have had a chance to explore the longer-term climate of the Paleocene and Eocene, we can examine the boundary between the Paleocene and Eocene epochs, which in Fig. 3 is characterized by a sharp spike in both the carbon and oxygen isotope records (indicated with the orange arrow in the top two panels). But before we try to understand these spikes, we need to walk through a little history of how the PETM was first recognized and then how the records of these spikes were developed. Such a walk-through illustrates the nature of scientific investigation as described in Chap. 1, in which exploration and discovery, testing ideas, and community analysis and feedback advance scientific understanding with benefits for society. Table 1 provides an overview of the history of the investigation of the PETM and we will refer back to the "steps" in this history throughout the following discussion.

Table 1 Timeline of investigations and findings key to the understanding of the PETM

Timeframe	Type of Investigation	Step #	Key finding	Figure
1970s–1980s	Paleontology of benthic foraminifera	1	Major extinction event of seafloor-dwelling protists	Figure 5
1990–1991	Oxygen and carbon isotopes of foraminifera	2	Spike in O and C isotope values coincident with benthic extinction	Figures 6 and 7
1992	Oxygen and carbon isotopes in terrestrial deposits	3	Recognition that same isotope spike recorded on land	Figure 8
Mid 1990s	Oxygen and carbon isotopes of foraminifera	4	Confirmation of global occurrence of isotope spikes	Figures 9
Mid 1990s–2000s	Calcium carbonate content of marine sediments	5	Recognition of deep ocean acidification during PETM	Figures 12 and 13
2010s	Boron proxy for pH	6	Recognition of surface ocean acidification during PETM	Figure 14
Mid 1990s–2010s	Carbon isotopes of bulk sediment	7	Attempts to better define rates of onset and total duration; and to identify source of light carbon released into ocean/atmosphere.	Figure 14

4.1 Biotic Change at the Paleocene/Eocene Boundary Reconstructed from Fossils ("Step" 1)

Typically, the first indications of environmental change noted in the geologic record are discovered by paleontologists studying evolutionary trends evident from different groups of fossils. Changes in the abundance and types of organisms that existed in a given region at a given time are clues that something may have altered the broader environmental conditions. In fact, such fossil changes were the very foundation upon which the geologic time scale was developed, as the boundaries between major divisions of geologic time were associated with prominent changes in the fossil record. It is no surprise that the first evidence of environmental change recognized from the time of the Paleocene/Eocene boundary was recorded by paleontologists. Different teams of paleontologists in different parts of the world noted major changes in the fossils preserved in the marine sedimentary record as well those preserved in terrestrial sediments. Paleontologists who specialize in the fossil record of mammals have been studying the interval of time surrounding the Paleocene/Eocene boundary for decades, because it marks a major burst in the evolution and inter-continental

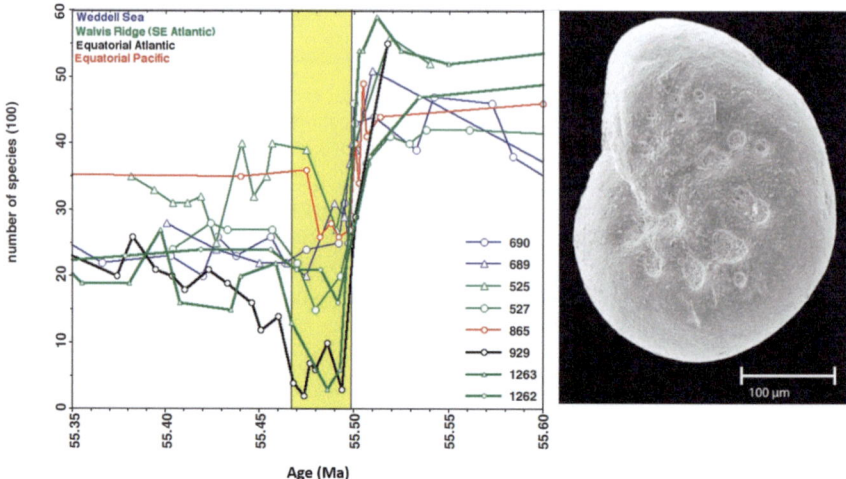

Fig. 5 Left: benthic foraminifera data from eight drill cores (the numbers in the plot legend refer to the location of the drill core (e.g., Site 690 was drilled in the Weddell Sea off Antarctica, and Site 929 was drilled from a location in the equatorial Atlantic Ocean) in four regions indicating the rapid decrease in the total species. The number of species on the y-axis refers to the number identified from examination of 100 individuals. Right: scanning electron microscope images of *Gavelinella beccariiformis*, one of the benthic foraminifera species that went extinct at the PETM. The white line indicates a scale of 100 µm. Graph from Thomas (2007)[12] used with permission from the Geological Society of America. This figure (left) is excluded from Creative Commons license. Image (right) courtesy of Serena Dameron, University of Massachusetts—Amherst

dispersal of mammals. Sediment sections preserved in nearly every continent indicate that numerous new species of mammals originated at this time, including the primates.[11]

Global studies of the marine environment during the late Paleocene—early Eocene time interval were made possible largely through coordinated international efforts to recover sedimentary cores from the global ocean basins. Paleontologists studying sediments recovered by the Deep Sea Drilling Project (precursor to the successor phases of scientific ocean drilling known as the Ocean Drilling Program, the International Ocean Drilling Program, and the International Ocean Discovery Program) discovered an extinction event at the Paleocene-Eocene boundary within the benthic foraminifera (the same organisms whose shells are analyzed to obtain the carbon and oxygen isotope data). Figure 5 indicates the dramatic decrease in the number of all species of benthic foraminifera observed in the sediment samples at the five locations. The photograph in Fig. 5 shows a specimen of *Gavelinella beccariiformis*, one of the species that went extinct at the event. This benthic foraminiferal extinction event is now known to be the most prominent extinction of seafloor-dwelling protists since a major event approximately 90 million years.

---

[11] Gingerich (2003).

[12] Thomas (2007). Permission granted from GSA.

4.2 Building on the Fossil Clues–Stable Isotope Data from Sea and Land ("Steps" 2 and 3)

It is important to point out that the incredible detail evident in the global compilation of oxygen and carbon isotope data in Fig. 3 is the product of decades of scientific ocean drilling core recovery and analysis. This level of detail was not available at the time that paleontologists were unraveling the intriguing evolutionary history of the mammals on land, and the benthic foraminifera on the seafloor. For perspective, Fig. 6 shows the state-of-the-art compilation of oxygen isotope data that existed by 1987. During the 1970s and 1980s as the inventory of deep-sea drill cores began to grow through the expeditions of the Deep-Sea Drilling Project and its successor, the Ocean Drilling Program, researchers were able to compile increasingly complete and long records of the stable isotope history of the ocean basins.

Much of the long-term climate history that we know today is evident in the features of this early oxygen isotope compilation—the Paleocene and Eocene already were known to be vastly warmer than the more recent Plio-Pleistocene (and modern) times. This data laid the critical foundation for subsequent investigations, and empowered researchers to pin-point portions of the geologic record to the history of climate in much finer detail.

Ocean Drilling Program Leg 113 took the *JOIDES Resolution* drillship to the very southern part of the Atlantic near Antarctica (the Weddell Sea) at the beginning of 1987 to recover sediment cores that would enable researchers to reconstruct the Paleocene and Eocene climate history of the region at a level of detail unprecedented

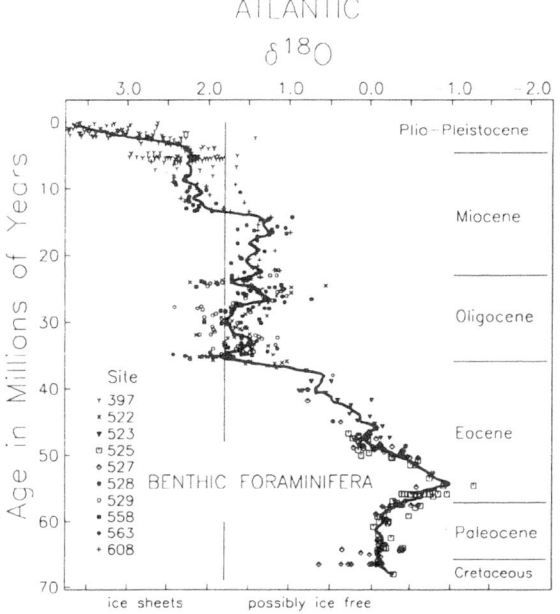

Fig. 6 Compilation of benthic foraminifera oxygen isotope data from ten drill cores spanning the same interval of geologic history as Fig. 1. From Miller et al. (1987).[13] Figure used with permission from Wiley. This figure is excluded from Creative Commons license

---

[13] Miller et al. (1987).

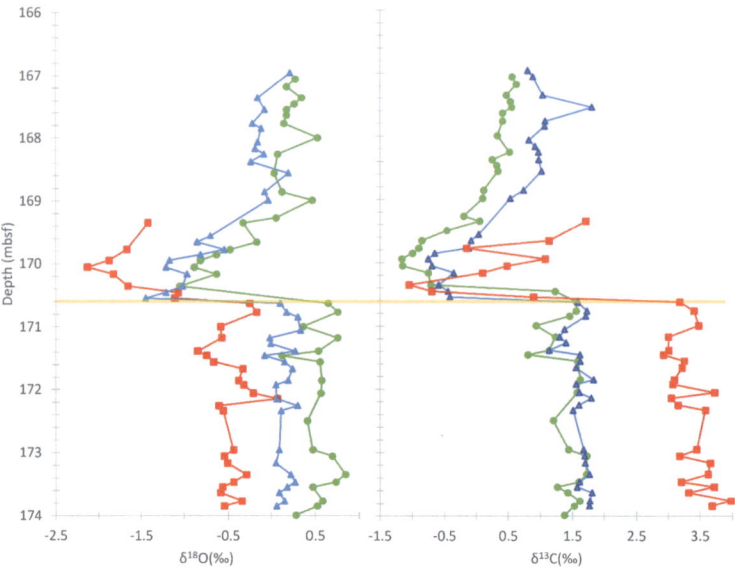

Fig. 7 Oxygen and carbon isotope data from Ocean Drilling Program Site 690 near Antarctica. These are the first data to indicate major changes in both isotope systems directly associated with the extinction of seafloor-dwelling foraminifera. Orange line marks the sediment depth at which the benthic foraminifera event was recorded. Replotted data from Kennett and Stott (1991)[14]

at the time. One of the most important studies to emerge from the expedition focused on investigating the benthic foraminiferal extinction event at the Paleocene-Eocene boundary by analyzing the stable isotopes at very high resolution to determine what environmental conditions may have contributed to the decline in species at that time. The results published in 1991 were astonishing.

Figure 7 is a graph of the very first high-resolution (closely spaced) stable isotope data generated from Ocean Drilling Program Site 690 recovered during Expedition 113 off of Antarctica (Maud Rise, Weddell Sea, 65°09′S, 01°12′W). This includes analyses from shells that grew in the surface waters (red), subsurface waters (blue), and at the seafloor (green). The most remarkable feature of these data is the large, abrupt, and simultaneous decrease in both carbon and oxygen isotopes in all three types of foraminifera. This means that the surface waters, subsurface waters approximately 100 m below the surface, and waters at the seafloor (~ 1900 m water depth) all responded in the same way and at the same time.

Even more remarkable was the fact that the rapid onset of the carbon and oxygen isotope decreases coincided with the largest extinction of benthic foraminifera of the past 90 million years (the orange line in both panels)! The abrupt onset of both the oxygen and carbon isotope decrease was followed by a more gradual increase, leveling off at a state slightly lower than the pre-event values.

A year after the publication of the rapid stable isotope changes recorded in the Antarctic drill core came the publication of another groundbreaking study, this time

---

[14] Kennett and Stott (1991).

from land-based data. At the same time that researchers sought to understand the environmental conditions surrounding the benthic foraminiferal extinction in the ocean, researchers also began applying the same stable isotope techniques toward exploring the environmental conditions that may have contributed to the mammal evolutionary events on land. But instead of analyzing the stable isotopes from fossil calcium carbonate shells, reconstruction of the terrestrial record required the analysis of fossil mammal tooth enamel and calcium carbonate nodules found in the layers of ancient soil (paleosol) in which the mammal fossils were preserved.

As shown in Fig. 8, there is a rapid and simultaneous decrease in the carbon isotope composition in both the tooth enamel and the carbonate nodule records, corresponding to the change in mammal evolution (the boundary between the Clarkforkian and Wasatchian fauna). It is important to note that the numerical age assigned to the Paleocene/Eocene boundary has been revised since the publication of this data in 1992.

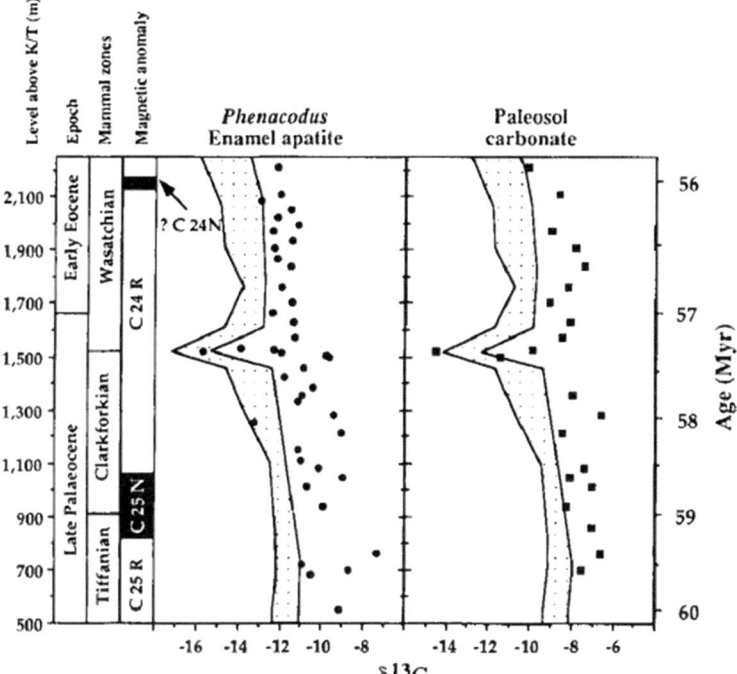

Fig. 8 Carbon isotope data from mammal tooth enamel and soil carbonate concretions indicating a prominent decrease associated with the major change in mammal evolution at the Paleocene-Eocene boundary. Level above K/T (m) refers to the height in the vertical section of exposed layers above the identification of the Cretaceous/Paleogene boundary. From Koch et al. (1992).[15] Used with permission from Springer Nature. This figure is excluded from Creative Commons license

---

[15] Koch et al. (1992).

4.3 Confirmation of the PETM as a Global Event ("Step" 4)

The identification of prominent changes in the carbon and oxygen isotope records associated with two major changes in the evolutionary record is remarkable in and of itself. But the fact that these events could be pinpointed to the same moment in geologic time on land and in the ocean generated tremendous excitement throughout the earth science community! Since the publication of these two groundbreaking data sets, these same isotope changes have been found in sediments from every ocean basin, at every range of water depth, and in plant and soil records worldwide (Fig. 9). Let's step back and reflect on what the abrupt changes in oxygen and carbon isotopes might reveal about the environment. Recall from Chap. 5, and from the discussion at the beginning of this chapter, that lower (lighter) oxygen isotope values recorded by foraminifera reflect warmer water temperatures . Thus, the spike in oxygen isotope

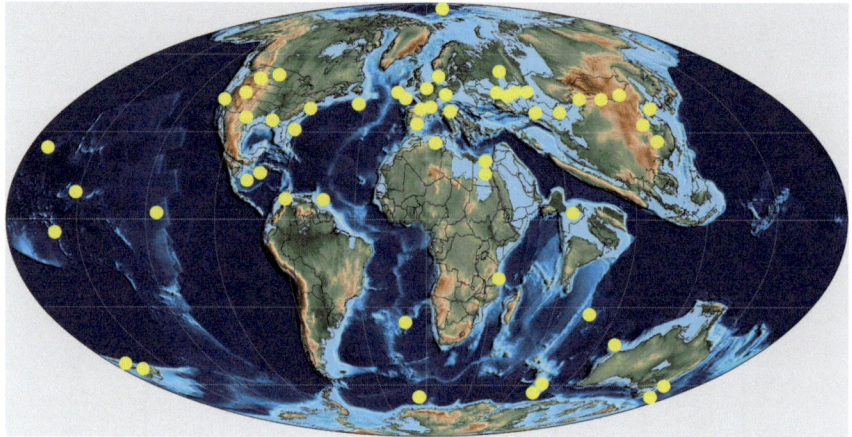

Fig. 9 Global distribution of localities, indicated on a map depicting the paleogeography of the planet 56 million years ago, known to record the paleontological and/or oxygen and carbon isotope changes during the PETM. Updated from McInerney and Wing (2011) and using Scotese (2021) map[16] with more recent localities drilled by IODP

---

[16] McInerney and Wing (2011) and Scotese (2021).

values indicates rapid warming followed by a return to "background" conditions (which during the Paleocene and Eocene already were significantly warmer than modern). Based on this evidence for warming, the event originally was termed the Late Paleocene Thermal Maximum, but a revision of the geologic time scale placed this event at the Paleocene/Eocene boundary. Hence, the event is now known as the Paleocene-Eocene Thermal Maximum (PETM). The coincident spike in carbon isotopes indicates that a substantial mass of isotopically light carbon was added to the ocean–atmosphere–biosphere system.

By now you likely have made the connection between the global warming and the change in the carbon cycle. Indeed, as suggested at the beginning of this chapter, the global warming during the PETM is related to the massive addition of an isotopically light source of carbon to the ocean–atmosphere–biosphere system. So far, the connection between a change in the carbon cycle and global warming during the PETM seems to resemble the current climate scenario. Such a release of carbon should have left its signature in the oceans in another manner: enhanced ocean acidification due to the mixing of CO_2 into the oceans. What is the geologic signature of ocean acidification? This question is addressed in the next section, and Box 2 explores acidification in the modern ocean.

Box 2 CO_2 and acidification in the modern oceans

CO_2 in the atmosphere mixes into the ocean waters where it combines with water to form the weak acid H_2CO_3 (carbonic acid). Carbonic acid readily dissociates (breaks) into its constituent bicarbonate (HCO_3^-) and hydrogen (H^+) ions (Fig. 10). You may be familiar with the concept of pH, which reflects the inverse concentration of hydrogen ions on a logarithmic scale. A pH of 7 is neutral, pH values lower than 7 are acidic (high concentration of H^+), and pH values higher than 7 are referred to as basic or alkaline (low concentration of H^+).

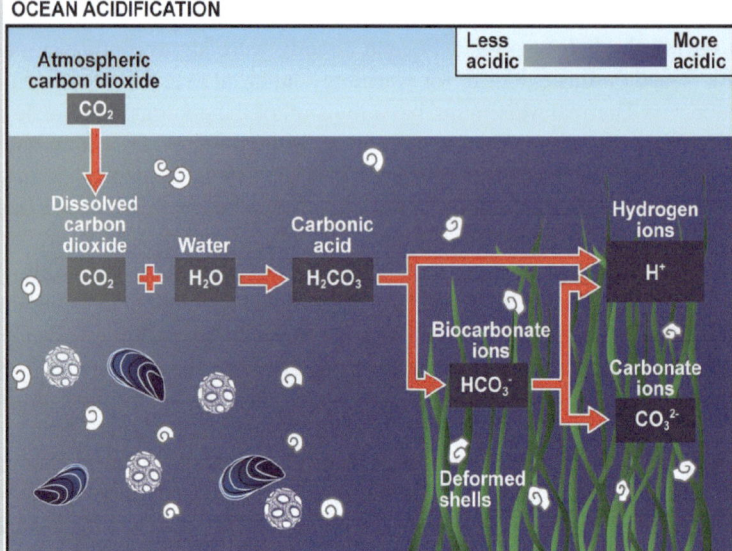

Fig. 10 Illustration of the reactions that occur when CO_2 dissolves in seawater. *Source* Birchenough et al. (2017).[17] U.K. Open Government License v3.0

What do you predict would happen to the pH of the oceans if CO_2 in the atmosphere suddenly increased? An increase in CO_2 in the atmosphere would result in an increase of CO_2 that mixes into the surface waters of the oceans. This increase, in turn, will yield an increase in H_2CO_3 that itself would cause increases in the concentration of HCO_3^- and H^+....connecting the dots would lead us to the prediction that an increase in atmospheric CO_2 would cause a decrease in ocean pH. And, in fact, this prediction already is bearing out—since the onset of the Industrial Revolution, the pH of surface ocean waters has decreased by 0.1 (Fig. 11). This may not seem significant, but it is important to recall that the pH scale is logarithmic—this 0.1 pH decrease actually reflects a 30% increase in acidity of the surface ocean.

---

[17] From Birchenough et al. (2017). Open Government License v3.0.

What are the consequences of ocean acidification? Many marine organisms are impacted by changes in the pH of their habitat. In particular, those organisms that rely on carbonate ion (CO_3^{2-}) to build calcium carbonate ($CaCO_3$) shells will suffer adverse effects from the decrease in CO_3^{2-} (which will combine with the extra H^+ due to increased CO_2 and hence not be available for $CaCO_3$ production). Currently, many species of coral and oysters are exhibiting stress related to decreased pH, as are plankton groups who use $CaCO_3$ for their shells. As pH levels continue to decrease, $CaCO_3$ shells will begin to dissolve.

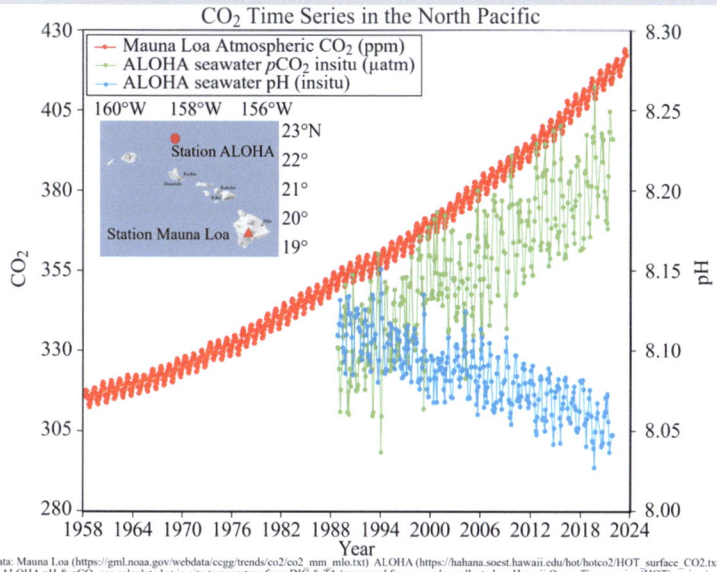

Fig. 11 Comparison of the trends in atmospheric CO_2 concentration measured at the Mauna Loa observatory (red line), the concentration of seawater CO_2 measured at the Aloha station offshore Hawaii (blue line), and the pH of the seawater at the Aloha station (green line). From: NOAA PMEL Carbon Program[18] used with NOAA PMEL permission. This figure is excluded from Creative Commons license

---

[18] From NOAA PMEL Carbon Program (accessed 2023, June 1). Hawaii Carbon Dioxide Time-Series. NOAA PMEL Carbon Program. https://www.pmel.noaa.gov/co2/files/co2_time_series_aloha_08-09-2023.jpg. Permission granted.

4.4 The Geologic Record of Ocean Acidification During the PETM ("Step" 5)

While investigating the nature of the benthic foraminiferal extinctions, researchers noted that some of the deep-sea drilling cores employed in the benthic foraminiferal investigations also showed evidence for a substantial decrease in the amount of calcium carbonate in the sediments around the time of the Paleocene-Eocene boundary. Typically, marine sediments rich in calcium carbonate are very light to white in color because the accumulation of the shells outpaces the accumulation of other sediment types. Color changes in the sediment are a useful indicator of changes in the accumulation of calcium carbonate shells. For sediments near-shore, darker colored sediment supplied by rivers may dominate the deposition resulting in a darker overall sediment color. But in the case of deep-sea sediments that accumulated relatively far from land in waters exceeding ~ 1000 m depth, terrigenous sediment is a minor component of overall deposition. Thus, changes in the sediment color generally correlate to *removal* of the calcium carbonate through dissolution.

In Fig. 12, this type of effect is immediately apparent. Intervals characterized by the dissolution of calcium carbonate tend to be beige to brown or darker shades of gray, reflecting the lack of shells leaving behind other types of sediment. In many cases carbonate shells disappeared completely and abruptly from the sediment layers, followed by a gradual increase back to previous levels. Carbonate dissolution of seafloor sediments was noted in nearly every deep-sea sediment record containing the PETM and now is one of the signature features of the PETM in marine records. The abrupt onset of dissolution and gradual recovery closely resembles the pattern of carbon isotope change in the same sediments (Fig. 12).

One of the most important characteristics of the geologic record of the PETM is that the geographic and bathymetric patterns of carbonate dissolution closely resemble the patterns that are predicted by theory. The intensity of dissolution is greatest in the deepest parts of the ocean, and less severe at shallower water depths. To appreciate this, let's compare two sediment records, classic DSDP Site 527 (one of the first to be studied for the benthic foraminiferal extinction, drilled very close to the same location as ODP Site 1262 in Fig. 12) and ODP Site 1209 (Fig. 13). The record at Site 527 was recovered from a modern water depth of ~ 4700 m and at the time of the PETM was situated at a water depth of ~ 3700 m.[19] The record at Site 1209 was recovered from a modern water depth of ~ 1200 m which also was its water depth at the time of the PETM. The intensity of the color change reflects the amount of calcium carbonate that dissolved from the sediment at the seafloor, and in

---

[19] After ocean crust is formed by volcanic activity, it typically contracts as it cools resulting in subsidence (sinking). This effectively means that the water depth at that location increases as the seafloor cools, so ocean crust that formed 60 million years ago originally occupied a shallower water depth and then subsided to its current water depth. In the example of Site 1262, the volcanic activity that formed the ocean crust upon which the sediments accumulated occurred about 75 million years ago, and over the course of time, cooling of that crust caused the water depth to increase as indicated in the text.

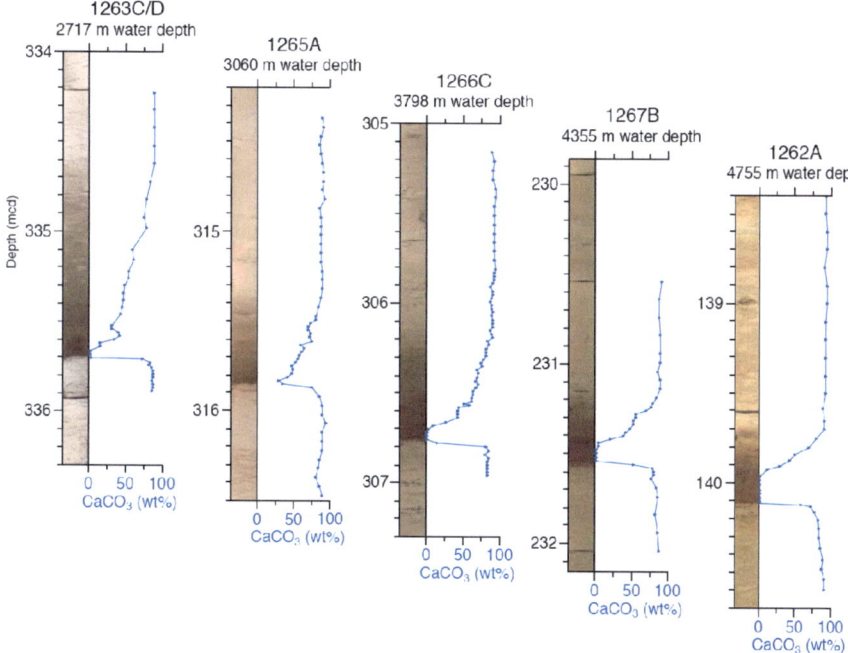

Fig. 12 Abundance of calcium carbonate in the sediments plotted alongside the photograph of the cores from Ocean Drilling Program Leg 208. From Zachos et al. (2005).[20] Used with permission from AAAS. This figure is excluded from Creative Commons license

turn reflects the greater susceptibility to dissolution of calcium carbonate at greater water depths (the degree of undersaturation of calcium carbonate increases in deeper water). While a thorough discussion of the chemistry of carbonate in seawater is far beyond the scope of this book, a very general analogy here is that the intensity of the pH decrease (increase in acidity) increases with water depth due to the pressure dependence of calcium carbonate saturation.

The prominence of calcium carbonate dissolution during the PETM should raise a question at this point: how is it possible to generate carbon isotope data from calcium carbonate when the calcium carbonate was dissolved so extensively? The answer is that the instruments that analyze the carbon isotope composition are highly sensitive and require very little mass of $CaCO_3$ to produce a very accurate and precise (and replicable) analysis. Fortunately, even when the majority of calcium carbonate has been removed from the sediment, a few shells or shell fragments remain that provide enough material for analysis.

---

[20] Zachos et al. (2005).

Fig. 13 Sediment cores from DSDP Site 527 (left panel) and ODP Site 1209 (right panel). Site 527 was located in a deeper part of the ocean than Site 1209 at the time of the PETM. In each photo, the core sections are laid out with the upper (younger) portion of the core at the left, progressing to deeper (older) portions to the right. Note prominent dissolution (i.e., dark color) in the second section of Site 527 and less intense dissolution in the first section of Site 1209 (indicated by arrows). Core images from IODP database[21]

4.5 Confirmation of Surface Ocean Acidification During the PETM ("Step" 6)

The realization that ocean acidification occurred in the deep ocean during the PETM was a very important step in testing the cause-and-effect relationship between the increase in carbon to the ocean–atmosphere–biosphere system and ocean pH. However, the ocean acidification data we have explored so far only provides us with the environmental situation that occurred in the deep ocean—the dissolution recorded in seafloor sediments *occurred at the seafloor*. However, the modern situation (Fig. 10) clearly indicates ocean acidification that is playing out in the surface waters. The geologic record of the PETM and other intervals of time contains clear indicators of ocean acidification at the sea floor, but we had no indication of how the

---

[21] Core photos for all DSDP-ODP-IODP cores can be accessed at: https://iodp.tamu.edu/database/coreimages.html.

ancient *surface* ocean may have responded to a release of carbon. In order to assess the pH response of ancient surface waters to an increase in CO_2, new proxies had to be developed.

Recall that the term "proxy" refers to a term or property (or person, in the example of voting representation) that can represent another. With few exceptions, our ability to reconstruct the different dimensions of ancient climate relies on proxies extracted from the geologic record (i.e., geologic archives). So far in this chapter, we have explored two main proxies: the oxygen isotope composition of foraminiferal calcium carbonate as a proxy for the temperatures in which the shells grew, and the carbon isotope composition of those same shells as a representation of the combination of carbon sources to the environment. How did we know that these isotope values could serve as proxies for such important components of the earth system? In the case of exploring and developing/testing proxies that employ the composition of foraminiferal shells, the strategy involves a combination of theory and experimentation. For example, the behavior of oxygen isotopes in response to changes in temperature can be predicted through physical chemistry. We then can test those predictions by growing live foraminifera in laboratory cultures and adjusting the water temperature and chemistry to determine how those changes translate to changes in the chemical composition of their shells. To complement the controlled laboratory culturing experiments, we also can collect modern foraminifera from the marine environment and compare the shell chemistry to direct measurements of the seawater from which they were collected.

One of the most important recent innovations in the field of paleoceanography was the introduction of a new way to reconstruct the pH of ancient seawater. Our existing proxies, such as isotopes of oxygen or carbon, carry no information about the pH of the environment. This new clever technique involves the chemistry of the trace element boron (B) in the calcium carbonate shells of foraminifera. Boron abundance and isotopic composition dissolved in seawater are controlled by the pH of the seawater. Although boron is not in the chemical formula of $CaCO_3$, the precipitation of any solid substance from a fluid in the natural environment (in this case $CaCO_3$ from seawater) incorporates minor to trace amounts of the other ions dissolved in that fluid. Careful testing indicates that the trace amounts of boron incorporated into foraminiferal shells records the same composition of the boron of the waters from which the shells grew. Figure 14 presents detailed boron records during the PETM. The top panel indicates the onset and progression of the PETM through the carbon isotope changes (with time on the x-axis proceeding from oldest on the left to youngest on the right). The middle panel is the boron isotope data, presented in the same "per mil" (‰) notation as the carbon and oxygen isotope data. Finally, the lower panel presents the boron abundance expressed as the ratio of boron to calcium. Both types of boron data recorded in planktonic foraminiferal shells sampled from several locations reveal substantial decreases in both the proportion of B to Ca and the boron isotopic composition, coincident with the change in carbon isotope composition. These decreases reflect a decrease in the pH of the surface waters in which those shells grew.

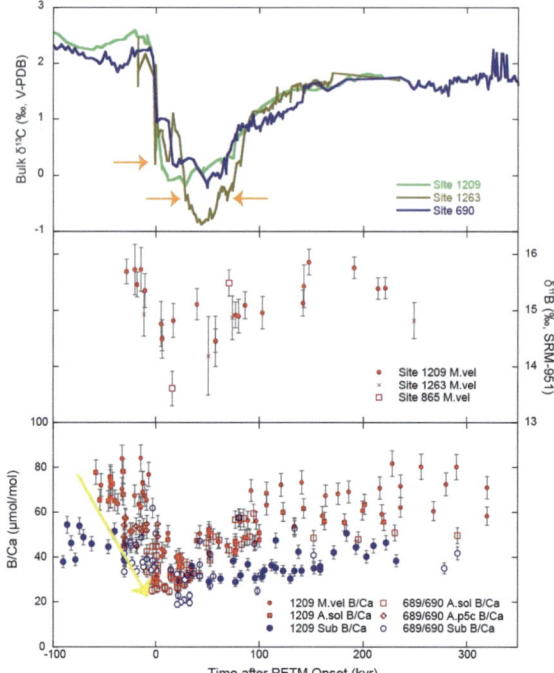

Fig. 14 Carbon isotope records acquired from bulk sediment analysis (top panel) compared to new boron isotope (middle panel) and B/Ca data from planktonic foraminifera (lower panel) from three drill sites indicating a decrease in sea surface pH coincident with the onset of the carbon isotope change during the PETM (yellow arrow in bottom panel). Orange arrows in top panel point to pulses of carbon isotope changes that followed the initial decrease in δ^{13}C. Modified from Penman and Zachos (2018)[22] and references therein

5 Rates of Onset and Total Duration of the PETM

Up to this point, we have discussed the progression of data collection and interpretation that has set the stage for understanding the geologic record of rapid global warming during the PETM, as well as using the PETM to test hypotheses regarding the effect of greenhouse gas increases on different portions on the atmosphere, ocean, and biosphere ("Steps" 1–6 in Table 1). Now we take on the challenge of determining the rate of onset and the total duration of the PETM. As a starting point, let's return to "Steps" 5 and 6. We found that dissolution of carbonate sediments at the seafloor and proxy-based data for surface water acidification confirm a critical prediction about the effects of a carbon release into the ocean–atmosphere–biosphere system. But, perhaps equally importantly, these two characteristics of the PETM also provide crucial constraints on the nature, timing, and duration of the carbon release.

---

[22] Penman and Zachos (2018).

5.1 How Abrupt Was the PETM? ("Step" 7)

To fully understand the causes, consequences, and utility of the PETM as an analog for current climate change, it is crucial that we identify the rate at which the carbon release happened. As you might imagine, decades of intense study have sought to definitively constrain the duration of the onset of the PETM. Yet there still remains considerable uncertainty for several reasons. For example, the original PETM records from terrestrial sections such as that presented in Fig. 8 indicate a more gradual change in the carbon isotope values than is apparent from the original marine sediment records (e.g., Figs. 7 and 9). Applying traditional geologic dating tools to the same event results in timing estimates that range from a few thousand years in the marine sections (geologically abrupt) to nearly 50,000 years in the terrestrial sections. The disparity in timing results from the challenge of using traditional geologic age indicators for this interval of time—the tie points pinned down with radiometrically dated numeric ages in this portion of the geologic time scale are spaced too far apart relative to the duration of the PETM to enable us to precisely establish the timing of the onset of the event.

Continuous accumulation of sediment in the deep-sea environment provides the opportunity to generate long-term, undisturbed records necessary for developing detailed chronologies. But sediments accumulate very slowly in this environment relative to more near-shore or terrestrial settings. For example, typical sediment accumulation rates for locations in the open ocean (with water depths of several thousand meters) characterized by high abundances of foraminiferal shells are 1–2 centimeters (cm) per thousand years!! This rate decreases substantially if the dominant sediment type is fine-grained clay (or if the shells dissolved)—in this case, a better approximation is on the order of millimeters (mm) per thousand years. In contrast, locations close to shore can experience sediment accumulation rates of cm per *year*. The onset of the PETM was characterized by a dramatic decrease in sediment accumulation rate due to the dissolution of seafloor carbonates, which is one of the reasons why it is so difficult to establish the rate of this event.

Let's return to Fig. 14 to examine more closely the nature of the carbon isotope records presented in the top panel. One of the major advances in our understanding of the PETM came through yet another refinement to the way in which we analyze the carbon isotope values. The top panel of Fig. 14 presents three carbon isotope records constructed using a different analytical strategy. The original data shown in Fig. 7 shows records produced by picking out 5–10 individual shells from distinct sediment layers spaced approximately 5–10 cm apart. The advance presented in Fig. 14 is that the samples were spaced much more closely (only 1 cm apart!) and instead of picking out individual shells, the analyses were made on the bulk sediment. So rather than extracting foraminiferal shells to separate out the surface and deep-water responses, this strategy gets a high-resolution snapshot of a mixture of all the different shell types in that sediment layer. The result is unprecedented detail into the shape of the carbon cycle changes, indicating that the initial and rapid decrease in carbon isotope values was followed by a few subsequent pulses (shown with orange arrows

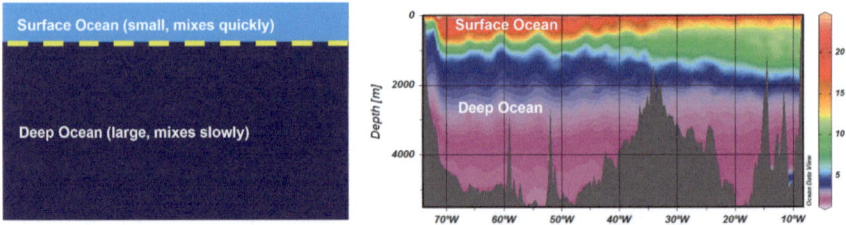

Fig. 15 Left: a simplified model of the ocean indicating the relative size and mixing rates of the surface and deep ocean, separated by the thermocline (yellow dotted line). Right: a cross section of the Atlantic Ocean near 35°N showing the distribution of water temperatures with depth. The yellow color provides an approximation of the depth of the thermocline. Right figure Atlantic Ocean profile courtesy Reiner Schlitzer, Alfred Wegener Institute; https://odv.awi.de/data/ocean/geosecs/ Ocean Data View figure used with permission from AWI. This figure is excluded from Creative Commons license

indicating the pulses in the brown curve in the top panel). This suggests that there was a rapid release of carbon that occurred in less than 1000 years, followed by a few subsequent releases prior to the recovery from the event.

The evidence for the release of carbon into the ocean–atmosphere–biosphere system occurring in several pulses, rather than one massive pulse, helps to explain the nature of the ocean acidification recorded during the PETM. Recall from Box 1 that if ocean acidification becomes sufficiently severe, it can impede the growth of $CaCO_3$ shells or even cause their dissolution in the surface ocean. What scenarios might produce such severe surface ocean acidification? The first thing we need to consider in order to answer this question is that the ocean itself is layered (as discussed in Chap. 2), with the relatively small surface layer separated from the immense deeper layer by a density interface most broadly represented as the "thermocline", in which the rapid change from warmer surface water to colder deeper waters creates a density barrier across which mixing is very slow (Fig. 15).

The rate at which atmospheric CO_2 can mix into the surface ocean is on the order of a decade while the rate at which the surface ocean mixes with the deep ocean is closer to 1000 years. Thus, you can imagine that a very large and rapid release of CO_2 into the atmosphere would have a relatively rapid impact on the pH of the surface ocean before that CO_2 could be mixed into the much larger deeper ocean (scenario A). In contrast, consider that same mass of CO_2 but released in several pulses spaced out in several thousand-year increments (scenario B). Each pulse could contribute to a decrease in surface ocean pH, detectable by proxy but not sufficiently severe to cause dissolution within the surface waters themselves. The temporal spacing of each pulse would allow time to mix a substantial fraction of that CO_2 into the deep ocean as well.

This latter scenario is reflected in a combination of the data we have examined so far. Glancing back at the right panel of Fig. 13, we have records of the PETM in some locations in which deep-sea dissolution only partially dissolved the shells that accumulated on the seafloor. This means that we have a record of the surface-dwelling shells preserved in the sediments (which is how we are able to apply the boron proxies!) that indicates some decrease in surface water pH, but not of the severity envisioned in scenario A described above. Surface water boron/calcium (B/Ca) data indicate that sea surface acidification conditions persisted for several tens of thousands of years. But, in contrast to the deep-water response, surface water acidification did not reach a level of severity that impeded shell formation or caused any extinctions. This implies that the combination of the masses of carbon released during any of the pulses, along with the rate at which each pulse occurred, was not sufficient to overwhelm the surface waters. Another way of expressing this is that the rate of carbon release was slow enough to allow oceanic mixing to transfer some of that additional carbon into deeper waters before surface waters could experience severe acidification (e.g., dissolution).

The combination of evidence that the PETM was caused by a pulsed release of carbon into the ocean–atmosphere–biosphere is an incredibly important finding and provides critical clues to identifying the source(s) of carbon involved, which in turn are critical for understanding what caused the initial release and subsequent pulses.

6 What Was the Source of Carbon Released During the PETM?

Now it is time to draw from the lines of evidence explored throughout this chapter, and to apply our understanding of Earth system processes explored throughout this book, to consider the potential sources of carbon released during the PETM. What observations are important to this discussion? For example, we already know that North Atlantic Igneous Province volcanic activity was contributing relatively large amounts of CO_2 into the ocean and atmosphere throughout the interval of time encompassing the PETM (yellow box in Fig. 3). We also have constrained the decrease in global carbon isotope values during the PETM to 3–4 per mil broken up into approximately 1–2 per mil pulses (Fig. 14).

To explore potential carbon sources, we can walk through the calculations similar to those set up in a study published in 1995.[20] It isn't important that we know the precise amount of carbon in each reservoir since we simply seek a general "proof of concept"—how much and what type of carbon might have produced the observed isotopic changes within a few thousand years. The study set up a simple equation to explore the parameters involved:

$$(M_{pre} + M_R) \cdot CI_{PETM} = M_{pre} \cdot CI_{pre} + M_R \cdot CI_R$$

where M_{pre} is the total mass of "exchangeable" carbon in the ocean–atmosphere–biosphere system pre-PETM, M_R is the mass of carbon released, CI_{PETM} is the carbon isotopic composition during the PETM, CI_{pre} is the carbon isotopic composition pre-PETM, and CI$_R$ is the carbon isotopic composition of the released carbon at the onset of the PETM. In very general terms, it is possible to achieve the PETM carbon isotope change with a range of M_R and CI$_R$ terms—a very negative CI$_R$ would require a smaller M_R term, and vice versa. This calculation assumes the transfer of released carbon occurred instantaneously and does not take into account the level of detail we now have regarding pulsed release or ocean acidification. For all practical (geological) purposes, the onset of the PETM of less than 1000 years fits this simplified set of assumptions.

To begin, we need an approximation for the M_{pre} term, which should include estimates of the masses of carbon in all of the exchangeable reservoirs of the atmosphere, ocean, and biosphere. While it is not possible to calculate these values for the planet 56 million years ago, we can still gain meaningful insight by applying estimates of the modern exchangeable reservoir. While estimates vary, the combined mass of carbon from inorganic sources in the atmosphere and oceans is ~ 30,500 × 10^{15} g of C, and the combined mass of organic carbon from land biota, soil matter, and within the oceans is ~ 3500 × 10^{15} g of C.[23] With a total estimate of 34,000 × 10^{15} g of C for the M_{pre} term, we can apply a value of 2 for the CI$_{pre}$ term, and a value of − 1 for the CI$_{PETM}$ term (from Fig. 14). Now we can approximate the mass of C released if we know some typical values of the $\delta^{13}C_R$ term! In the case of volcanic CO$_2$, we already discussed a value of -6 for the carbon isotope composition, so let's apply this to our calculation for an idea of how much mass of volcanically derived carbon would have been required to cause the entire PETM carbon isotope change.

This very quick and simplified calculation indicates that 20,400 × 10^{15} g of C would have been required to generate a ~ 3 per mil change in the total carbon reservoir at the PETM. Based on what we know of volcanic outgassing, an injection of 20,400 × 10^{15} g of C in less than 1000 years would be at least 20 times higher than known long-term rates of basaltic volcanism.[24] Not to mention that this would have nearly doubled the entire exchangeable carbon reservoir based on our estimated M_{pre} term. Both of these considerations render a simple, volcanic cause unrealistic.

As a foil to the incredibly large mass of relatively high carbon isotopic composition required for a volcanic mechanism, in 1995 a particularly intriguing hypothesis was proposed to explain the very rapid and dramatic carbon isotope decrease….methane[20]! It is important to remember that in 1995, we lacked the detail of the state-of-the-art carbon isotope curve, and the authors sought to explain what at the time was understood to be an instantaneous (~1000 years or less) and massive change in the global carbon cycle. They invoked the release of methane because of its very low carbon isotope signature (with a typical range of − 40 to − 65 per mil) and because it is stored at relatively shallow depths in the earth's crust in frozen soil (permafrost) and in the seafloor in a frozen form known as methane hydrates. (Methane gas associated with fossil fuel deposits is stored more deeply in the crust, but under certain circumstances also could be released suddenly). A sudden release of methane from any of these sources is consistent with the evidence for global

---

[23] For a discussion of these estimates see: Dickens et al. (1995).

[24] Dickens et al. (1995).

warming, given that methane itself is a potent greenhouse gas, and then rapidly converts to carbon dioxide in the presence of oxygen (in either the atmosphere or ocean).

Now solve the above mass balance equation to estimate the mass of C from methane required to produce a CI$_{PETM}$ of -1. Applying -40 to -65 for the M$_R$ term yields a range of ~ 1600 to 2600 × 10^{15} g of C, a mass consistent with an accessible proportion of methane that could be released from modern reservoirs of methane hydrate and soil permafrost methane. This hypothesis was so elegant and attractive because it could explain an unprecedented and simultaneous change in the global total carbon composition and global temperatures. For nearly a decade, it remained the default hypothesis to explain the PETM in spite of the fact that…it couldn't be confirmed with any geologic data!

We now know that the need to prove methane as the sole source of carbon released during the PETM was rendered unnecessary once it had been established that release of carbon had been pulsed. In fact, the carbon isotope data indicating a pulsed release opened up the idea that an initial release of C could have been followed by subsequent "leakage." This conceptual model is more consistent with the oxygen isotope records that indicate sustained warmth over nearly 100,000 years, as well as the sustained ocean acidification conditions. Furthermore, this model opened the door to a wider variety of carbon sources, and most likely, a combination of carbon sources involved that includes methane or buried organic carbon. While methane hydrate release presented a very appealing and elegant explanation, this source of carbon currently is not favored as one of the likely culprits for the initial, large pulse. The most plausible other potential sources of isotopically light carbon involve volcanic intrusion into subsurface deposits of organic-rich sediment (δ^{13}C of ~ -26) or reservoir of methane gas associated with fossil fuels released from a "pocket" in the earth's crust (~ -45), volcanic release of CO$_2$ (~ -6), and the oxidation of soil/sediment organic material (~ -26).[25] For example, although discounted initially, a growing body of data within the new constraints of the state-of-the-art isotope curves point to a strong role of North Atlantic Igneous Province volcanism in the PETM. An intense pulse of volcanism spanning 1 million years brackets the PETM,[26] and there is evidence of intense magmatic sill intrusion that may have caused the release of methane gas in the region now famous for the North Sea oil and gas deposits.[27] In other words, this potential trigger does not rely on the slow, long-term buildup of greenhouse gasses from the North Atlantic Igneous Province volcanism, but instead the chance, rapid release of trapped methane gas and/or buried organic carbon because of one or more of the volcanic episodes in this region.

Another intriguing aspect of the pulsed nature of the carbon isotope record is that the timing of the pulses may be related to orbital cycles—a topic you will learn more about in the next chapter.

---

[25] Panchuk et al. (2008).

[26] Storey et al. (2007).

[27] Svensen et al. (2010).

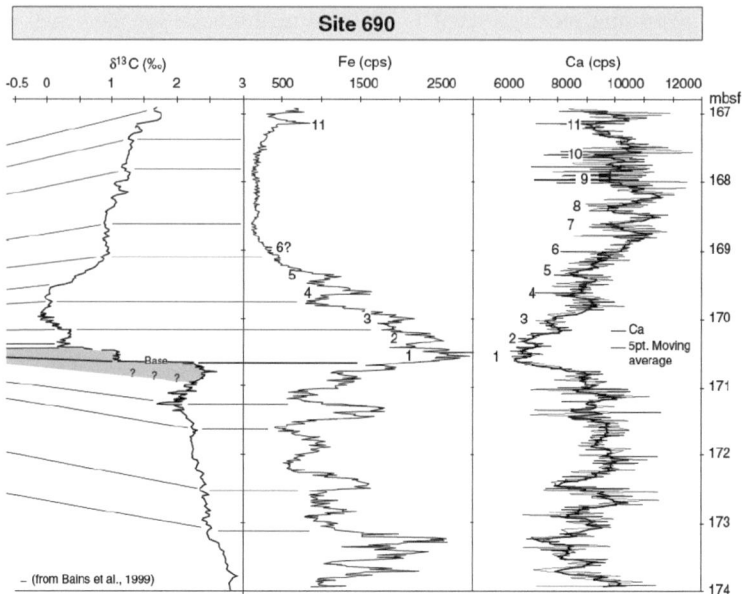

Fig. 16 Bulk carbon isotope data[28] shown with the elemental Fe and Ca data generated through scanned x-ray fluorescence. From Rohl et al. (2000).[29] Used with permission from the Geological Society of America. This figure is excluded from Creative Commons license

Figure 16 shows the high-resolution carbon isotope record introduced in Fig. 14 along with high-resolution Fe and Ca data. The Fe data in particular are very useful for identifying and correlating orbital cycles found within the sedimentary record. Intensive investigation of these and other PETM records suggests that the numbered cycles represent precessional (i.e., ~ 21,000 year) cycles. The first three of these cycles appear to correlate with pulses in the carbon isotope record, suggesting a connection between orbital precession and the release of light carbon. Identifying the causal connection is an area of ongoing research.

After the initial pulse of carbon release and associated warming, additional sources of carbon released in subsequent pulses may have been triggered by the PETM warming itself. It is important to consider that warming-related feedbacks may have resulted in pulses of carbon release from soil carbon and seafloor methane hydrates. For example, warming due to the initial pulse of carbon release could have promoted the dissociation of methane hydrates from sediment or soil reservoirs, with the methane contributing additional warming potential as a greenhouse gas. One of the big outstanding challenges in understanding the PETM is parsing the precise contribution of carbon from the various potential sources, including the carbon

---

[28] Bains et al. (1999).

[29] Rohl et al. (2000).

cycle feedbacks within the actively cycling ocean–atmosphere–biosphere system.[30] Resolving these uncertainties has substantial implications for our understanding of future warming and the associated warming-related feedbacks.

7 How Did the PETM End?

Nearly every type of record that spans the PETM demonstrates an exponential decay-shaped recovery curve indicating that once the pulsed injections of carbon from volcanism and other geologic sources ceased, the "excess" carbon was removed from the ocean–atmosphere–biosphere system over the course of the next 150,000–200,000 years, or potentially as long as ~ 300,000 years if estimated from counting precessional cycles (Fig. 16). This removal is corroborated by similar exponential decay recovery observed in the oxygen isotope records, indicating that peak PETM global warming was followed by cooling back to pre-PETM conditions (e.g., Figs. 3 and 7). Recall from the discussion in Chap. 7 the potential mechanisms for carbon removal from the atmosphere–ocean system and emplacement back into the lithosphere. The most plausible mechanisms involve enhanced weathering, or enhanced primary production that sequestered carbon either through burial of organic-rich sediments or within biomass. Limited evidence does exist for enhanced weathering of continental rocks coincident with the carbon isotope decrease.[31] However, model simulations of the carbon cycle response indicate that the recovery proceeded far more rapidly than possible if only weathering feedbacks had been operating.[32] The majority of marine sediment records do not indicate a substantial increase in the burial of organic matter, however this does not preclude such sequestration of organic carbon on land in soils or permafrost. The other possibility is that biological uptake and replenishment of carbon reservoirs such as methane that may have been released during the pulsed onset of the PETM contributed to the recovery.

8 Implications of the PETM for Current Climate Change: A Cautionary Tale

Decades of intensive investigation of the PETM are a classic example of the nature of scientific research, in which observational data lead to hypotheses, which can be tested as additional data is collected and new analytical methods are developed. For example, we thought the rapid release of methane hydrates was the definitive cause 25 years ago, but as data collection afforded increasingly detailed records (e.g., the pulsed record of carbon isotope change, and the measurable but not-too-severe

---

[30] Turner (2018).
[31] Ravizza et al. (2001).
[32] Bowen and Zachos (2010).

acidification of the surface ocean) it became clear that a single, massive release of methane was not the cause of the global carbon isotope change. A growing body of data instead favor pulsed volcanism as the likely trigger for carbon release,[33] however other critical details remain elusive as of the writing of this chapter: the source(s) of carbon released, why the pulsed releases ceased, and how the released carbon was removed from the ocean–atmosphere–biosphere system. We also have learned from the PETM that the biotic response to rapid warming can be complex, with some types of organisms suffering detrimental impacts while others experience rapid evolution.

Even with these remaining questions, several crucial aspects of the PETM are well-constrained and provide very robust tests of the dire predictions we make about current warming. It is eminently clear that the release of carbon recorded throughout the atmosphere, oceans, and biosphere caused an increase in atmospheric greenhouse gas inventories. The increase in greenhouse gas content, in turn, warmed the entire planet. The release of CO_2 also caused the predictable decrease in ocean pH as CO_2 was mixed from the atmosphere into the oceans. The rate at which the CO_2 was released into the ocean–atmosphere–biosphere systems, and subsequently mixed into the surface ocean, was sufficiently slow to prevent overwhelming acidification in the surface ocean (much of the CO_2 could be mixed into the deep ocean, where pronounced dissolution of calcium carbonate did occur at the seafloor).

The well-constrained details of the PETM provide alarming context for the current rate of global warming. Based on our ability to directly measure the rate of change from the geologic record, as well as clues extracted from the geologic record about the rate from the relative differences in surface and deep ocean acidification, the rate of carbon release at the onset of the PETM was substantially slower than the current annual average of carbon release. In fact, current estimates suggest that *modern rates of carbon injection into the atmosphere are nearly ten times higher than the annual rate during peak PETM carbon release.*[34] Thus, the consequences of carbon release during the PETM vastly underestimate the warming, acidification, and regional elements of environmental change that will result from the current crisis that is "unprecedented within the past 66 million years."[35]

Furthermore, some of the biggest predicted threats of climate change also are the most difficult to investigate from a geological perspective. Recall from Chap. 1 that regional changes in the hydrological cycle are one of the most immediate consequences of climate change, and with each new extreme weather event comes the associated question of whether such events are increasing in severity and frequency due to global warming. Some of these questions simply cannot be addressed from the geologic record either because the time scale over which such events occurs is too rapid to resolve with slowly accumulating sediments (think back to the typical rates at which open-ocean sediments accumulate), or because the geologic record may not afford the geographic coverage needed to assess regional-scale changes.

---

[33] Kender et al. (2021).

[34] Zeebe et al. (2016).

[35] Same as above.

There also is one incredibly important difference between the PETM and modern that requires elaboration: *ice*. Recall in our discussion of background climate during the Paleocene and Eocene that conditions already were vastly warmer than the modern, with little evidence of significant glaciers at the poles. Take a moment and consider the impacts of rapid warming on glaciers. One impact is rather intuitive—ice will melt. Glaciers form and build up when water, evaporated from the oceans, precipitates at high latitudes as ice and snow, and does not melt away during the warmer months. Over time, this process sequesters water in continental ice sheets (today on Greenland and Antarctica) resulting in a decrease in global sea level. When glaciers melt, the meltwater returns to the oceans consequently causing a rise in global sea level. As a quick geologic example, coastal flooding, and the enhanced threat of storm surges due to intensified coastal storms, is one of the most tangible threats of current climate change with the potential to impact billions of lives. According to the United Nations, approximately 10% of the world's human population lives in coastal areas that are less than 10 m above sea level, and 40% of the population resides within 100 km of a coastline.[36]

Glacial melting has the added challenge of the positive feedback involving changes in how much incoming solar radiation is reflected back out of the atmosphere. The term albedo refers to how "reflective" a portion of the earth's surface is to incoming solar radiation. The bright surfaces of ice and snow are highly reflective, hence portions of the Earth's surface that are covered in ice and snow reflect more (absorb less) incoming solar radiation than other types of groundcover (forest, soil, pavement). Melting high-latitude glacial ice contributes to a positive feedback in the sense that as the proportion of ice diminishes, the high-latitude groundcover increasingly absorbs more of the incoming radiation, serving to accelerate the warming even further. This feedback gives rise, in part, to the concept of a "runaway greenhouse." Of course, high-latitude warming also is predicted to thaw permafrost, allowing the methane stored in permafrost methane hydrate to escape to the atmosphere, contributing even more to the greenhouse gas inventory and fueling further warming.

The geologic record of the PETM provides a "best-case" example of the changes in the aftermath of an increase in the carbon content of the ocean–atmosphere–biosphere system. In the modern iteration of global change, human beings now are involved in both the cause and the consequences. Human beings aren't accustomed to thinking about processes that require more than a human lifetime to play out, but we must not only think about addressing the issues—we must act now. We don't have the luxury to sit back and allow the Earth system to recover over the next hundred thousand years.

---

[36] From: United Nations (2017). The Ocean Conference Fact Sheet. United Nations. https://www.un.org/sustainabledevelopment/wp-content/uploads/2017/05/Ocean-fact-sheet-package.pdf.

References

Anagnostou, E., John, E. H., Edgar, K. M., Foster, G. L., Ridgwell, A., Inglis, G. N., Pancost, R. D., Lunt, D. J., & Pearson, P. N. (2016). Changing atmospheric CO_2 concentration was the primary driver of early Cenozoic climate. *Nature, 533*, 380–384. https://doi.org/10.1038/nature17423

Bains, S., Corfield, R. M., & Norris, R. D. (1999). Mechanisms of climate warming at the end of the Paleocene. *Science, 285*, 724–727. https://doi.org/10.1126/science.285.5428.724

Birchenough, S., Williamson, P., & Turley, C. (2017). *Future of the sea: Ocean acidification* (19 p.). U.K. Government Office. https://assets.publishing.service.gov.uk/government/uploads/system/uploads/attachment_data/file/645500/Ocean_Acidification_final_v3.pdf

Bowen, G. J., & Zachos, J. C. (2010). Rapid carbon sequestration at the termination of the Palaeocene-Eocene Thermal Maximum. *Nature Geoscience, 3*, 866–869. https://doi.org/10.1038/ngeo1014

Burke, K. D., Williams, J. D., Chandler, M. A., Haywood, A. M., Lunt, D. J., & Otto-Bliesner, B. L. (2018). Pliocene and Eocene provide best analogs for near-future climates. *PNAS, 115*(52), 13288–13293. https://doi.org/10.1073/pnas.1809600115

Dickens, G. R., O'Neil, J. R., Rea, D. K., & Owen, R. M. (1995). Dissociation of oceanic methane hydrate as a cause of the carbon isotope excursion at the end of the Paleocene. *Paleoceanography, 10*, 965–971. https://doi.org/10.1029/95PA02087

Eldholm, O., & Thomas, E. (1993). Environmental impact of volcanic margin formation. *Earth and Planetary Science Letters, 117*, 319–329.

Friedlingstein, P., Jones, M. W., O'Sullivan, M., Andrew, R. M., Hauck, J., Peters, G. P., Peters, W., Pongratz, J., Sitch, S., Le Quéré, C., Bakker, D. C. E., Canadell, J. G., Ciais, P., Jackson, R. B., Anthoni, P., Barbero, L., Bastos, A., Bastrikov, V., Becker, M., ... Zaehle, S. (2019). Global Carbon Budget 2019. *Earth System Science Data, 11*, 1783–1838. https://doi.org/10.5194/essd-11-1783-2019

Gingerich, P. D. (2003). Mammalian responses to climate change at the Paleocene-Eocene boundary: Polecat Bench record in the northern Bighorn Basin, Wyoming. *Geological Society of America Special Publication, 369*, 463–478. https://doi.org/10.1130/0-8137-2369-8.463

Irving, E., & Wynne, P. J. (1991). The paleolatitude of the Eocene fossil forests of Arctic Canada in Tertiary Fossil Forests of the Geodetic Hills, Axel Heiberg Island, Arctic Archipelago. In R. L. Christie, & N. J. McMillan (Eds.), *Geological survey of Canada Bulletin* (Vol. 403, pp. 209–212).

Kender, S., Bogus, K., Pedersen, G. K., Dybkjaer, K., Mather, T. A., Mariani, E., Ridgwell, A., Riding, J. B., Wagner, T., Hesselbo, S. P., & Leng, M. J. (2021). Paleocene/Eocene carbon feedbacks triggered by volcanic activity. *Nature Communications, 12*, 5186. https://doi.org/10.1038/s41467-021-25536-0

Kennett, J. P., & Stott, L. D. (1991). Abrupt deep-sea warming, palaeoceanographic changes, and benthic extinctions at the end of the Palaeocene. *Nature, 353*, 225–229. https://doi.org/10.1038/353225a0

Koch, P. L., Zachos, J. C., & Gingerich, P. D. (1992). Correlation between isotopic records in marine and continental carbon reservoirs at the Paleocene-Eocene Boundary. *Nature, 358*, 319–322. https://doi.org/10.1038/358319a0

Kwon, E. Y., Timmermann, A., Tipple, B. J., & Schmittner, A. (2022). Projected reversal of oceanic stable carbon isotope ratio depth gradient with continued anthropogenic carbon emissions. *Communications Earth & Environment, 3*, 62. https://doi.org/10.1038/s43247-022-00388-8

McInerney, F. A., & Wing, S. L. (2011). The Paleocene-Eocene Thermal Maximum: A perturbation of carbon cycle, climate, and biosphere with implications for the future. *Annual Review of Earth and Planetary Sciences, 39*, 489–516. https://doi.org/10.1146/annurev-earth-040610-133431

McKenna, M. C. (1980). Eocene paleolatitude, climate and mammals of Ellesmere Island. *Palaeogeography, Palaeoclimatology, Palaeoecololgy, 30*, 349–362.

Miller, K. G., Fairbanks, R. G., & Mountain, G. S. (1987). Tertiary oxygen isotope synthesis, sea level history, and continental margin erosion. *Paleoceanography, 2*(1), 1–19. https://doi.org/10.1029/PA002i001p00001

Panchuk, K., Ridgwell, A., & Kump, L. R. (2008). Sedimentary response to Paleocene-Eocene thermal maximum carbon release: A model-data comparison. *Geology, 36*, 315–318. https://doi.org/10.1130/G24474A.1

Penman, D., & Zachos, J. C. (2018). New constraints on massive carbon release and recovery processes during the Paleocene-Eocene Thermal Maximum. *Environmental Research Letters, 13*, 105008. https://doi.org/10.1088/1748-9326/aae285

Ravizza, G., Norris, R. N., Blusztajn, J., & Aubry, M.-P. (2001). An osmium isotope excursion associated with the Late Paleocene thermal maximum: Evidence of intensified chemical weathering. *Paleoceanography and Paleoclimatology, 16*(2), 155–163. https://doi.org/10.1029/2000PA000541

Rohl, U., Bralower, T. J., Norris, R. D., & Wefer, G. (2000). New chronology for the late Paleocene thermal maximum and its environmental implications. *Geology, 28*, 927–930. https://doi.org/10.1130/0091-7613(2000)28%3c927:NCFTLP%3e2.0.CO;2

Saltzman, M. R., & Thomas, E. (2012). Chapter 11: Carbon isotope stratigraphy. In *The geologic time scale*. Elsevier. https://doi.org/10.1016/B978-0-444-59425-9.00011-1

Scotese, C. R. (2021). An Atlas of paleogeographic maps: The seas come in and the seas go out. *Annual Reviews of Earth and Planetary Sciences, 49*, 669–718. https://doi.org/10.1146/annurev-earth-081320-064052

Stidham, T. A., & Eberle, J. J. (2016). The palaeobiology of high-latitude birds from the early Eocene greenhouse of Ellesmere Island, Arctic Canada. *Scientific Reports, 6*, 20912.

Storey, M., Duncan, R. A., & Swisher, C. C. (2007). Paleocene-Eocene Thermal Maximum and the opening of the Northeast Atlantic. *Science, 316*, 587–589. https://doi.org/10.1126/science.1135274

Svensen, H., Planke, S., & Corfu, F. (2010). Zircon dating ties NE Atlantic sill emplacement to initial Eocene global warming. *Journal of the Geological Society of London, 167*, 433–436. https://doi.org/10.1144/0016-76492009-125

Thomas, E. (2007). Cenozoic mass extinctions in the deep sea: What perturbs the largest habitat on Earth? *Geological Society of America Special Paper, 424*. https://doi.org/10.1130/2007.2424(01)

Turner, S. K. (2018). Constraints on the onset duration of the Paleocene-Eocene Thermal Maximum. *Philosophical Transactions of the Royal Society A, 376*, 20170082. https://doi.org/10.1098/rsta.2017.0082

Westerhold, T., Marwan, N., Drury, A. J., Liebrand, D., Agnini, C., Anagnostou, E., Barnet, J. S. K., Bohaty, S. M., Vleeschouwer, D., Florindo, F., Frederichs, T., Hodell, D. A., Holbourn, A. E., Kroon, D., Lauretano, V., Littler, K., Lourens, L. J., Lyle, M., Pälike, H., ... Zachos, J. C. (2020). An astronomically dated record of Earth's climate and its predictability over the last 66 Million Years. *Science, 369*, 1383–1387. https://doi.org/10.1126/science.aba6853

Zachos, J. C., Röhl, U., Schellenberg, S. A., Sluijs, A., Hodell, D. A., Kelley, D. C., Thomas, E., Nicolo, M., Raffi, I., Lourens, L. J., McCarren, H., & Kroon, D. (2005). Rapid acidification of the ocean during the Paleocene-Eocene Thermal Maximum. *Science, 308*, 1611–1615. https://doi.org/10.1126/science.1109004

Zachos, J. C., Dickens, G. R., & Zeebe, R. E. (2008). An early Cenozoic perspective on greenhouse warming and carbon-cycle dynamics. *Nature, 451*. https://doi.org/10.1038/nature

Zeebe, R., Ridgwell, A., & Zachos, J. C. (2016). Anthropogenic carbon release rate unprecedented during past 66 million years. *Nature Geoscience, 9*. https://doi.org/10.1038/NGEO2681

Zhang, Y. G., Pagani, M., Liu, Z., Bohaty, S., & DeConto, R. (2013). A 40-million-year history of atmospheric CO_2. *Philosophical Transactions of the Royal Society A, 371*, 20130096. https://doi.org/10.1098/rsta.2013.0096

Debbie Thomas was born and raised in Cincinnati, Ohio but somehow decided at age 10 that she wanted to be an oceanographer when she grew up. After earning her bachelor's degree in Geological Sciences from Brown University, she earned her MS in Marine Sciences and her

PhD in Geological Sciences from the University of North Carolina—Chapel Hill. In 2004 she joined the faculty of the Department of Oceanography at Texas A&M University, fulfilling her childhood dream, and most recently is the founding dean of the College of Marine Sciences and Maritime Studies at Texas A&M University (located at the Galveston Campus). Debbie's disciplinary specialty of paleoceanography allowed her to apply her training in geological and marine science in a truly interdisciplinary manner. Debbie is the primary author of Chap. 8.

Kristen St. John Intent on being a political science major in college, Kristen's path changed after taking physical and historical geology as her science requirements. She's found that piecing together the stories that are preserved in climate archives of Earth's history is like working on a 10,000+ piece jigsaw puzzle. The challenges and rewards of figuring things out are much more fun—and solutions more attainable—when working with others. Synergistic research collaborations, and seeking ways to make geoscience more accessible and meaningful for students, have been guideposts for her career path. Her research focuses on reconstructing the history of land ice and sea ice by looking at clues in marine sediments. Kristen's degrees are from Furman University and The Ohio State University; she is a Professor at James Madison University. Kristen is the primary author of Chap. 1, a primary author of Chap. 2, and the author of Box 1 in Chap. 8.

Open Access This chapter is licensed under the terms of the Creative Commons Attribution 4.0 International License (http://creativecommons.org/licenses/by/4.0/), which permits use, sharing, adaptation, distribution and reproduction in any medium or format, as long as you give appropriate credit to the original author(s) and the source, provide a link to the Creative Commons license and indicate if changes were made.

The images or other third party material in this chapter are included in the chapter's Creative Commons license, unless indicated otherwise in a credit line to the material. If material is not included in the chapter's Creative Commons license and your intended use is not permitted by statutory regulation or exceeds the permitted use, you will need to obtain permission directly from the copyright holder.

Climate Cycles

Steven Clemens and Lawrence Krissek

Guiding Questions: What are climate cycles and what causes them? Where do we fit in a cycle?

1 Key Take-Away Points

- Science represents the accumulation of knowledge gained over many generations.
- Many aspects of the natural world are cyclical in nature, driven by the gravitational interaction among planetary bodies of our solar system. Cycles affect our lives as well as the climate system. Climate cycles are pervasive through geologic time, occurring in both Icehouse and Greenhouse climate states.
- Earth's orbit around the Sun has three primary characteristics, known as **eccentricity** (ellipticity of Earth's orbit about the Sun), **obliquity** (tilt angle of Earth's spin axis), and **precession of the equinox** (changes in the timing of the seasons associated with changes in the orientation of Earth's spin axis).
- Eccentricity, obliquity, and precession all change cyclically over long time-scales. Respectively, these cycles last 100,000, 41,000, and 23,000 years.
- The Sun is Earth's external energy source. The amount of solar radiation (watts per square meter) received at a given location on Earth changes significantly due to the 41,000-year obliquity and 23,000-year precession cycles. The 100,000-year-long eccentricity cycles have little direct impact on incident solar radiation.

S. Clemens (✉)
Department of Earth, Environmental and Planetary Sciences, Brown University, Providence, RI, USA
e-mail: steven_clemens@brown.edu

L. Krissek
School of Earth Sciences, The Ohio State University, Columbus, OH, USA
e-mail: krissek.1@osu.edu

- Energy from the Sun is absorbed and transferred among all components of Earth's climate system including the **atmosphere, hydrosphere, cryosphere, lithosphere**, and **biosphere** causing physical, chemical, biological, and isotopic changes that are preserved in many different geological settings, like sediments that accumulate beneath the ocean and in lake basins.
- Precession is the dominant control on summer insolation for all latitudes, and for winter insolation up to ~ 50° N latitude.
- Obliquity is an important component of changes in incoming solar insolation at high latitudes; times of lower insolation at high latitudes promote ice growth.
- The influence of eccentricity on insolation is indirect. For example, over the past 1.2 million years, energy input from the Sun (at 23,000 and 41,000 year cycles) has been modified by internal Earth-system feedbacks resulting in a dominant ~ 100,000 year-long cycle that characterizes the growth and decay of continental-scale (2000 m thick) **ice sheets** in the Northern Hemisphere.
- The 100,000 year-long cycles of growth and decay of the ice sheets are associated with the transfer of carbon (often in the form of **CO_2**) between the deep ocean and the atmosphere/biosphere. **Ice ages** (also called "**glacials**") were characterized by the storage of CO_2 in the deep oceans, causing atmospheric CO_2 to drop to ~ 180 parts per million by volume (ppmv) whereas interglacial intervals (times of low ice volume like now) were characterized by movement of CO_2 from the deep ocean into the atmosphere and biosphere, causing atmospheric CO_2 to rise to ~ 290 ppmv.
- Burning of **fossil fuels** (coal, oil, and natural gas) has caused CO_2 in our atmosphere to rise to > 420 ppmv (as of 2024), far higher than it ever was over the past one million years. Human-induced rise in CO_2 is altering the natural evolution of Earth's climate system; enough to prevent the next ice age from naturally occurring.
- In addition to the 100,000-year-long "short" eccentricity cycle, there is also a "long" 405,000-year-long cycle that is important at extremely long "deep time" time-scales (hundreds of millions of years). Whereas the timing of the other orbital cycles changes systematically over these extremely long time-scales, the 405,000-year cycle remains constant. This "long" eccentricity cycle is expressed in the sedimentary record through changes in sediment facies and sediment composition caused by changes in temperature, precipitation, sea level, and/or biological productivity. Perhaps most importantly, the constant duration of the "long" eccentricity cycles in these geological records provides time constraints on extremely ancient sediment sections.
- In addition to the orbital cycles, there is a much shorter Sun-induced cycle that also affects the insolation Earth receives. The solar power available to the Earth system at the 11-year sunspot cycle variation is a small fraction of the total solar output; very much smaller than orbital-scale changes responsible for the ice ages. Anthropogenic climate forcing has overwhelmed the influences of both orbital and solar cycles on Earth's heat budget.

2 Introduction: Clockwork-Like Cycles Are Everywhere in Our Lives

We may not think much about it but **cycles** impact all our lives to an extraordinary degree. Our sleep cycle is tied to Earth's 24-h (daily) rotation relative to the Sun; we generally sleep when the side of the Earth we are on faces away from the Sun (night) and are active when the side of the Earth we are on faces the Sun (day). We see the Moon's monthly cycle in the progression from the new Moon (when the side of the Moon facing Earth is shaded from the Sun) to the first quarter (half-Moon), full Moon (when the side of the Moon facing Earth is fully lit by the Sun), last quarter (another half-Moon), and back to new Moon. If you live on the coast, you know that local sea level rises to a maximum (high tide), falls to a minimum (low tide), and cycles again back to high tide every day, or sometimes even two times every day, like clockwork. These tidal cycles can range from insignificant to as much as 12 m (40 ft.). Earth's tidal cycles are driven by gravitational attraction of the Sun and Moon on ocean water; the highest tides occur when the Sun, Moon, and Earth align with one another, working together in terms of gravitational pull on ocean water. The very highest, so called "King Tides," occur when the Sun, Moon, and Earth align with one another and the Moon is closest to the Earth in its monthly orbit. Those of us who live in mid-latitudes experience seasonal cycles, spring into summer into fall into winter, and back into spring. The clockwork-like seasonal **cycle** happens because the Earth's rotational axis is tilted relative to the plane of Earth's orbit around the Sun, as will be discussed below. The timing of all these cycles is unique to Earth and is linked to the gravitational attractions that the Sun, the planets, and the moons of our solar system exert on one another (Fig. 1). Other planets in our solar system have their own unique cycles for similar reasons; for example, a day on Jupiter lasts only 10 h, unlike our 24 h-long day on Earth.

The Sun and the eight planets orbiting the Sun each exert a gravitational attraction on one another. The strength of these attractions and the orbital paths of the planets have been studied by astronomers, physicists, and mathematicians for hundreds of years and are so extremely well understood that the locations of each planet relative to all others can be accurately calculated millions of years into the future and millions of years back into the past. This chapter will focus on Earth's orbital cycles that take place over very long time-scales. Just like our daily, monthly, and seasonal cycles, Earth's orbital cycles are also driven by the relative orientations and gravitational attraction among the planetary bodies in our solar system, are also exceptionally well known, and can be calculated millions of years into the future and the past. The only difference is that Earth's orbital cycles are very long, with the shortest being 19,000 years long and the longest being 405,000 years long. These long orbital cycles are important to us because they regulate the amount of sunlight received by Earth and are responsible for very large, natural changes in global-scale climate. In particular, we will focus on the three orbital cycles that most impact Earth's climate, the 23,000-, 41,000-, and 100,000-year-long cycles. These orbital cycles are expressed in Earth's geological record as far back in time as scientists have been able to study. We will first

Fig. 1 Our solar system. The eight planets of our solar system orbit the Sun on slightly elliptical (almost circular) paths, with the asteroid belt occupying the orbital space between Mars and Jupiter. This figure is artistic in nature; in reality the planets rarely align as shown; the artist did so to show all of them in one close-up view. From NASA/JPL[1]

focus on the last million years and then on cycles found in records that are hundreds of millions of years old (so called "deep time"). From here on, large numbers like 100,000 will often be expressed in kiloyears (ky), so, for example, 100,000 years will be written as 100-ky.

Over the last million years, these orbital cycles were expressed dominantly in the form of 100-ky climate cycles known as "glacial-interglacial" or "ice age" cycles. In sections below, we will examine the ice age cycles in detail. Within each cycle, enough water was evaporated from the oceans to drop global sea level ~ 120 m (~390 ft.), moving all that water out of the ocean and storing it in the form of glaciers and ice sheets covering parts of Antarctica, North America, Europe, and Asia where it accumulated to thicknesses of up to 2000 m (Fig. 2). We will discuss and explore the ice ages from an historical timeline of their study, to impress upon you the idea that all scientific knowledge is gained over time, building on the work of previous generations of scientists over decades, centuries, and millennia.

It's important to know that these enormous ice age cycles are entirely natural, having nothing to do with humans. It's equally important to know that humans have disrupted these natural climate cycles. As we'll come to understand toward the end of the chapter, the most recent research indicates that even though Earth's orbital

---

[1] NASA/JPL (2018). Our Solar System Features Eight Planets, NASA Jet Propulsion Laboratory (accessed July 27, 2023), https://www.jpl.nasa.gov/images/pia11800-our-solar-system-features-eight-planets.

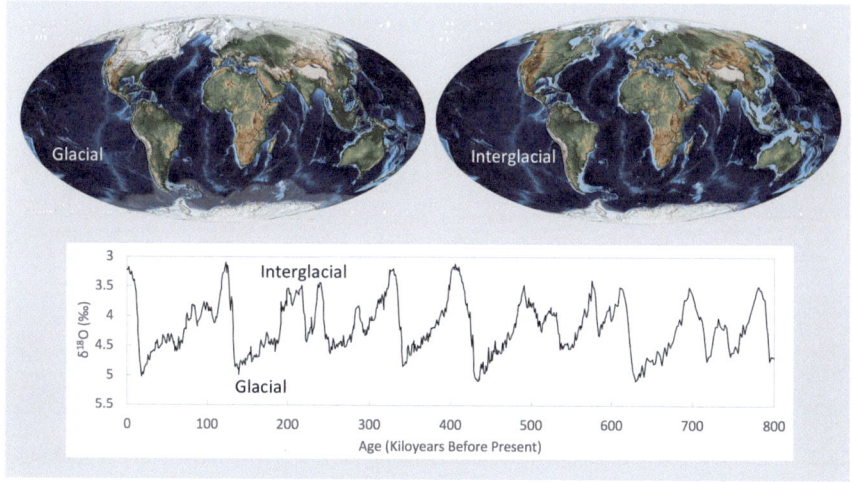

Fig. 2 Earth as it is in the modern interglacial (right) and during the last glacial ice age ~ 23,000 years ago (left).[2] Water evaporated from the oceans was stored as ice sheets on Antarctica, North America, Europe, and parts of Asia, causing sea level to drop ~ 120 m. Compare the shorelines of the two images to see now much more land was exposed. The graph at the bottom shows that glacial conditions (high ocean water $\delta^{18}O$ values) and interglacial conditions (low ocean water $\delta^{18}O$ values) have alternated with an approximately 100-ky cycle[3]

cycles will continue unabated into the future, it's extremely unlikely that another ice age will arise given the increasing amounts of CO_2 humans are injecting into the atmosphere.

3 The Great Ice Age Cycles

You learned in Chap. 2 that Earth's climate system is composed of five interacting components including the atmosphere, hydrosphere, cryosphere, biosphere, and geosphere. You learned that energy is exchanged among these five components, often through the transfer of heat, carbon, and water. You also learned that the Sun's radiative forcing is the ultimate source of the energy that flows through our climate system and that this energy can be stored and released over an enormous range of time-scales, from daily and weekly (we call this changes in our weather) to decadal and longer (we call this changes in climate). The glacial–interglacial cycles of the past one million years are an excellent example of how the Earth receives solar radiation (insolation) from the Sun and cycles it among the different components of Earth's climate system, sometimes storing it as ice sheets on the continents and

---

[2] Scotese and Wright (2018).
[3] Data from Lisiecki and Raymo (2005).

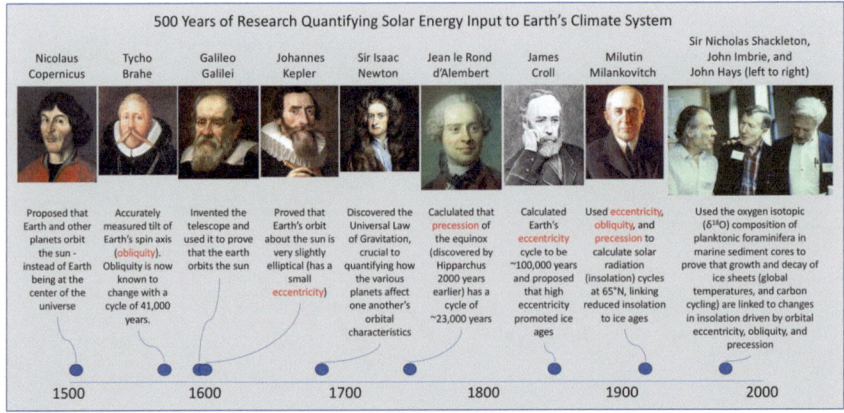

Fig. 3 Timeline of scientific milestones in understanding the role of solar forcing in climate change[4]

sometimes as water in the oceans, all the while using the carbon cycle to help make these transformations.

Our modern understanding of the links between the orbital cycles that drive changes in incoming solar radiation and the ice ages is only possible because of research that has taken place over the previous ~ 500 years. This previous research established how and why the amount and distribution of solar radiation received by Earth changes over time. As described in the next section we will see that it is due to changes in the shape of Earth's orbit around the Sun, the tilt of the Earth's spin axis, and the orientation of the spin axis. The observations and calculations that documented these orbital parameters represent the cumulative work of a great many scientists over hundreds of years, as far back as the works of Nicolaus Copernicus in the late 1400s.

3.1 Earth Orbital Cycles

All scientists stand on the shoulders of those who came before them; previous researchers who helped to establish the scientific basis of the fields in which they work. While the study of astronomy is thousands of years old, it was not until the past 500 or so years that technology advanced to the point that accurate measurements of orbital motions could be made. The timeline in Fig. 3 illustrates some of the important researchers and their contributions to understanding Earth's orbit about the Sun and its impact on climate.

These scientists came from all walks of life, from wealthy to poor, from classically educated at the world's leading universities, to the self-taught. One thing they all had

---

[4] First eight images are from https://commons.wikimedia.org. Shackleton, Imbrie, and Hayes photograph from Grove (2008).

in common was a deeply-held drive to understand the natural world. From our twenty-first century perspective, another aspect that all these researchers have in common is that they are all male, all Caucasian. The gender and ethnic homogeneity of this esteemed scientific group illustrates the historical opportunity gaps that have existed for many people, especially for women and people of color. An innate drive to better understand the natural world is, to a greater or lesser degree, common to us all, yet only recently are the sciences beginning to reflect and benefit from our diversity.

By the mid 1800s astronomers, physicists, and mathematicians (Fig. 3) had rigorously established the characteristics of Earth's orbital eccentricity, obliquity, and precession cycles. **Eccentricity** defines the ellipticity of Earth's orbit, ranging from 0.0002 (almost a perfect circle) to 0.06 (slightly elliptical) with cycles of 405-ky and 100-ky (Fig. 4a left). As well, the elliptical orbit itself changes orientation in space with a cycle of 112-ky; this is called **apsidal precession**. These cycles reflect the gravitational impact that various planetary bodies have on the ellipticity of Earth's orbit; in particular, the gravitational pulls of Saturn and Jupiter have the most impact on Earth's eccentricity cycles. Changes in eccentricity alone have a negligible impact on the amount of solar radiation reaching the Earth's surface but do play a large role in modulating (controlling) radiation cycles associated with precession of Earth's orbit.

Obliquity describes the tilt of Earth's spin axis relative to the plane of its orbit around the Sun (Fig. 4a middle). Tilt of the spin axis ranges from 22.1° to 24.5° and back to 22.1° every 41,000 years. The tilted nature of Earth's spin axis is thought to have been the result of a collision with another planet-size object over 4 billion years ago, knocking Earth's spin axis away from vertical. This obliquity (tilt) is the reason we have summers and winters here on Earth (Fig. 4b). The hemisphere tilted most directly toward the Sun experiences summer; at the same time, the opposite hemisphere is tilted away from the Sun and experiences winter. Over the course of the 41-ky obliquity cycle, the high latitudes receive increasingly more direct sunlight in summer as the obliquity increases, and then receive decreasing amounts of direct summer sunlight as the obliquity decreases. Thus, obliquity is an important component of changes in incoming solar insolation at high latitudes; times of lower insolation at high latitudes promote ice growth.

Axial precession reflects changes in the orientation of Earth's spin axis relative to the Sun and other stars (Fig. 4a right). For example, the spin axis now points at a star named Polaris, also known as the North Star. Because the direction of the spin axis changes very slowly in a circular manner (it precesses), it will point to other "North Stars" over time, eventually making its way back around to Polaris with a 26-ky cycle. The largest influences on precession are the gravitational attraction of the Sun and Moon. This axial precession, combined with eccentricity and apsidal precession of the elliptical orbit, result in a 23-ky precession cycle known as **climatic precession**; this determines the season during which Earth is closest to the Sun (perihelion) and furthest from the Sun (aphelion) over the course of its yearly orbit. The warmest

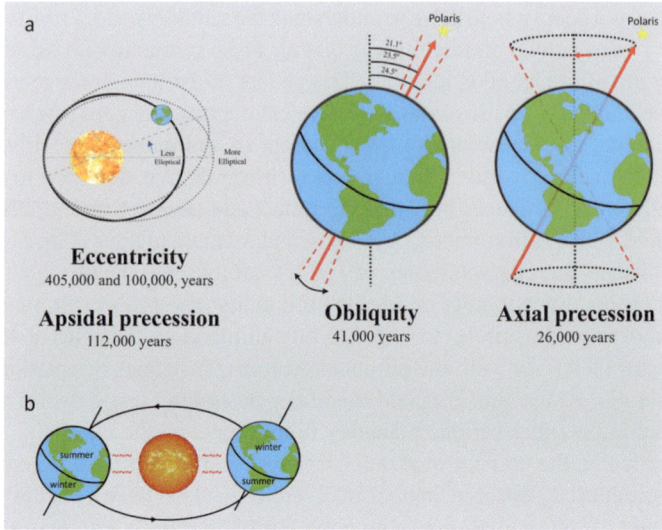

Fig. 4 Earth-orbital eccentricity, obliquity and precession. **a** (left) Earth's orbit about the Sun ranges from near circular to more elliptical (i.e., the eccentricity changes). These changes take place over 405,000- and 100,000- year cycles. At the same time, the eccentric orbit itself changes orientation in space with a cyclicity of 112,000 years (apsidal precession). Note that the eccentricity changes in this figure are greatly exaggerated for illustration purposes. If drawn to scale, the differences would be very hard to see. **a** (middle) Obliquity is the tilt of Earth's spin axis. It changes from 21.1° to 24.5° and back every 41,000 years. The obliquity is now such that the spin axis points at a star called Polaris, the current "north star." **a** (right) The spin axis also precesses over time, tracing out a cone shape such that it will point to different "north stars" over time, making its way back around to Polaris every 26,000 years; this is called axial precession. **b** Summers and winters. Northern Hemisphere summer occurs when Earth occupies the position on its yearly orbit around the Sun whereby the Northern Hemisphere is tilted toward the Sun and receives more direct sunlight. At the same time, the Southern Hemisphere is tilted away from the Sun, receiving less direct sunlight, causing winter. Six months later the Southern Hemisphere experiences summer, when the Southern Hemisphere is tilted toward the Sun. Although eccentricity does cause the Earth to be a little closer to and further away from the Sun during the course of the year (the figure is greatly exaggerated), this is not why Earth experiences summer and winter seasons. Figure drafted by Grace Berg

summers in a given hemisphere take place when Earth is nearest to the Sun on its elliptical orbit and the spin axis is tilted toward the Sun at the same time. Likewise, the coldest winters in a given hemisphere take place when Earth is farthest from the Sun on its elliptical orbit and the spin axis is tilted away from the Sun at the same time; these are times when ice growth is promoted. Now, Earth is at perihelion in January and is closest to the Sun in Northern Hemisphere winter and Southern Hemisphere summer.

These three fundamental components of Earth's orbit (eccentricity, obliquity, and climatic precession) combined can change the amount of solar radiation (insolation) received at a given location on Earth by ± 12% at orbital time-scales (100-ky, 41-ky, and 23-ky cycles). These orbitally-driven changes in insolation are the external source of energy supplied to Earth's climate system and greatly exceed the changes induced by the 11-year sunspot cycle that also influences climate on Earth (Box 1). In the following sections, we will explore how Earth's climate has changed in response to this external (orbital-related) input of energy; it's a story of taking this external energy and passing it through the various components of the Earth system (atmosphere, hydrosphere, cryosphere, biosphere, and geosphere).

3.2 The Milankovitch Hypothesis of the Ice Ages

By the early 1900s advances in the fields of physics and mathematics allowed Serbian scientist Milutin Milankovitch to combine accurate calculations of the eccentricity, obliquity, and precession parameters with Newton's discovery that the amount of solar radiation hitting the surface depends on the distance from the Sun and the angle at which the Sun's rays hit the surface. With these critical advances, Milankovitch was able to calculate and plot the amount of solar radiation reaching the Earth's surface over the past 600,000 years (Fig. 5). Milankovitch's painstaking calculations required years to complete. His work was interrupted by his participation in the first Balkan war and then again when he was taken prisoner during World War I. He nevertheless persisted, eventually putting forth the theory now known as the Milankovitch Theory of the Ice Ages; many scientists now refer to the cyclical (glacial-interglacial) nature of Earth's climate as "Milankovitch" cycles. Milankovitch's calculations suggested that changes in the amount of solar radiation received in the high latitudes (around 65°N) during summer was critical to the growth and decay of ice sheets. He postulated that larger amounts of solar radiation in high latitude summers would melt the winter snows entirely, preventing development of ice sheets. As well, he postulated that low amounts of radiation during high latitude summers would not be able to melt the winter snows, allowing it to accumulate over decades and centuries, growing into vast continental-scale ice sheets.

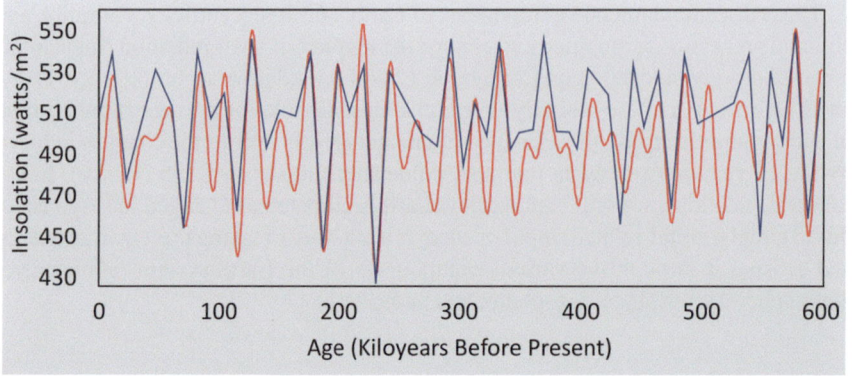

Fig. 5 Solar radiation over the past 600,000 years (watts/m^2 for summer at 65° north latitude). The blue curve was calculated by Milankovitch in 1938.[5] The red curve is the modern calculation by Laskar.[6] The match is quite good considering Milankovitch, doing all calculations by hand, could only produce one data point for every 12,000 years and the entire curve took years to complete. The Laskar curve has a data point every 1000 years; computers easily make these calculations in microseconds, for any latitude and season

Let's look into Milankovitch's hypothesis further, starting with an examination of incoming solar radiation. Examples of incoming solar radiation (insolation) are shown in Fig. 6. Summer insolation at all latitudes looks very similar to precession because precession is the dominant control on summer insolation. This is because precession determines where along the elliptical orbit Earth's spin axis is pointed directly at the Sun; the warmest summers happen when the spin axis is pointed most directly at the Sun and Earth is also closest to the Sun (perihelion) during its yearly orbit.

Precession is the dominant driver of winter insolation as well, up to about 50° north; as latitude increases obliquity contributes more and more as described earlier, because increased tilt yields more direct insolation at high latitudes. You can see (Fig. 6) that the December 45° N insolation curve has both precession and obliquity cycles embedded in it. At high latitudes during winter (e.g., December 65° N), obliquity is the most important driver. An important point here is that none of the insolation curves look like eccentricity. Note, however, that the precession curve fits like a glove into the eccentricity curve. This is because eccentricity determines the strength (amplitude) of precession. When eccentricity is high (a more elliptical orbit), the amplitude of precession is high and when eccentricity is low (more circular orbit) the amplitude of precession is low. Hence, the influence of eccentricity on insolation is said to be indirect. The relative influences of eccentricity, obliquity, and precession

---

[5] Redrawn from Milankovitch (1941).

[6] Laskar et al. (2004).

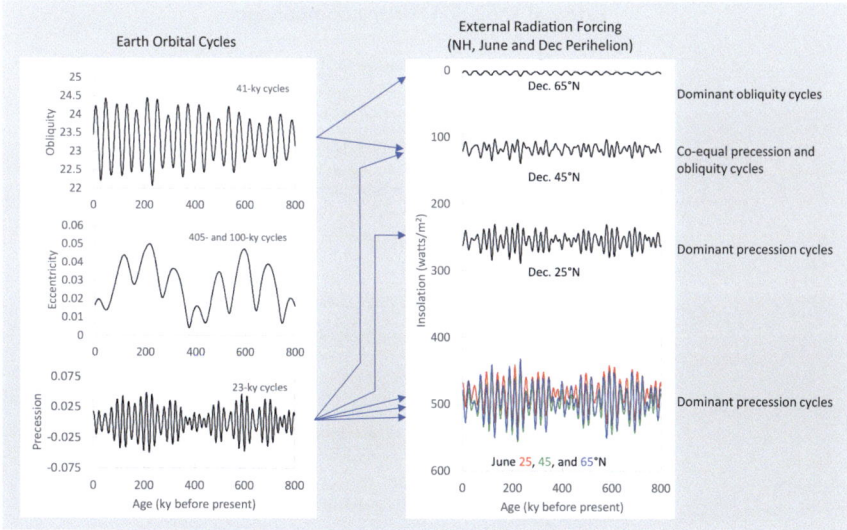

Fig. 6 Comparison among orbital cycles and solar radiation.[7] Left panel shows eccentricity, obliquity and precession components of Earth's orbit about the Sun over the last 800,000 years (800-ky). Right panel shows examples of incoming solar radiation (W/m^2) at various months and latitudes. Precession cycles are clearly evident in all but high latitude winter. Obliquity cycles are more evident at higher latitudes

are shown in Fig. 7 where summer insolation at 65° N is mathematically broken down (deconvolved) into its three main orbital components. Note the very small amount of insolation contributed directly by the 100-ky eccentricity cycle. It's important to note this because we'll shortly see, oddly enough, that the ice age cycles of the past million years are dominated by 100-ky cycles.

Milankovitch's remarkable hypothesis that summer insolation at 65° N is the main driver of the ice ages was published in 1941,[8] but was extremely difficult to prove for two simple reasons. The first is that each successive ice age destroys much of the land-based evidence of the previous one as ice sheets initiate in the polar latitudes and grind their way across the land surface, expanding toward low latitudes, and eroding the older surfaces. Second, although there was plenty of evidence left over from the last ice age (Fig. 8), which we now know started to melt back about 23,000 years ago, it was nearly impossible to date with the technology at the time. For these reasons, the Milankovitch hypothesis languished three decades until scientists started analyzing sediments from the deep sea.

---

[7] Laskar et al. (2004).
[8] Milankovitch (1941).

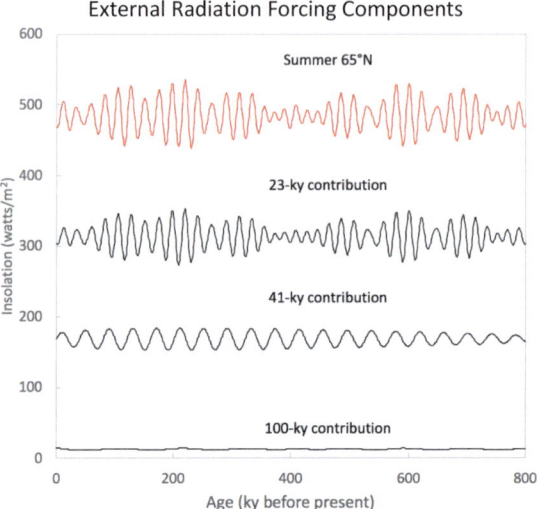

Fig. 7 Orbital components of 65° north summer insolation,[9] hypothesized to be the fundamental external forcing of the Ice Ages. Top curve shows the insolation coming into Earth at 65° north during June for the past 800,000 years. The bottom three curves show the contribution from the eccentricity (100-ky), obliquity (41-ky), and precession (23-ky) components. In other words, the sum of all three components yields the insolation curve above. It is clear from this analysis that most of the summer insolation at 65° north comes from precession, less from obliquity, and very little from eccentricity. We will see the opposite is true for global ice volume, CO_2, and sea surface temperature

3.3 Hays, Imbrie, and Shackleton Prove the Milankovitch Hypothesis

Unlike the land surface, where each new ice sheet scrapes clean much of the evidence of the previous ice sheet, sediments deposited on the sea floor typically continue to pile up over time, with the older sediment being buried beneath younger sediment. If these sediments pile up at a rate of 5 cm every 1000 years, then a 40 m deep pile of sediment will represent the past 800,000 years. A group of scientists in the mid 1970s analyzed the isotopic composition of fossils buried in well-dated marine sediments and were able to prove the Milankovitch hypothesis. Here's how that works.

As you learned in Chap. 5, the oxygen in water (H_2O) is composed dominantly of two isotopes (^{18}O and ^{16}O). When water is evaporated from the ocean, much of the heavy isotope (^{18}O) is left behind relative to the lighter isotope (^{16}O), which is transported in the water vapor and deposited on the continents as snow, eventually becoming ice sheets as the snow piles up and compacts. Because of this, the $^{18}O/^{16}O$ ratio (cast in delta notation as $\delta^{18}O$) of ocean water becomes larger as ice sheets grow and then smaller as that ice sheets melt and the light isotope (^{16}O) flows back into the

---

[9] Laskar et al. (2004).

Fig. 8 Evidence of the presence of past ice sheets. Top photo shows "glacial grooves," carved into limestone bedrock by the movement of the ice sheet (with other stones embedded in its base) over bedrock, acting like sandpaper. Ohio glacial groove photo used with permission from the Ohio Department of Natural Resources, Division of Geologic Survey. This image is excluded from Creative Commons license. Bottom photo shows a "glacial erratic" moved several 100 km from where it was picked up by the ice sheet. The erratic is located in Ohio and is composed of granite; the nearest surface exposures of granite, where this material could have been eroded, are several 100 km north in Ontario. The pole propped on the erratic is 2 m tall. Erratics can now be "**exposure dated**" to determine how long ago they melted out of the ice sheet, becoming exposed to cosmic rays. Glacial erratic photo by Larry Krissek

ocean. Amazingly, microscopic organisms known as "benthic foraminifera" live on the sea floor and make their skeletons out of calcium carbonate ($CaCO_3$), using the oxygen from ocean water (Fig. 9). Since these organisms only live for a very short time (a few weeks to a few years), many individuals are produced and their shells are preserved in seafloor sediments as the sediments pile up over geologic time. Analysis of the $\delta^{18}O$ in these fossil benthic foraminifera allows scientists to reconstruct the $\delta^{18}O$ of ocean water (and ice volume) through time. Similar organisms that live in the surface ocean (planktonic foraminifera) substitute magnesium (Mg) for Ca in

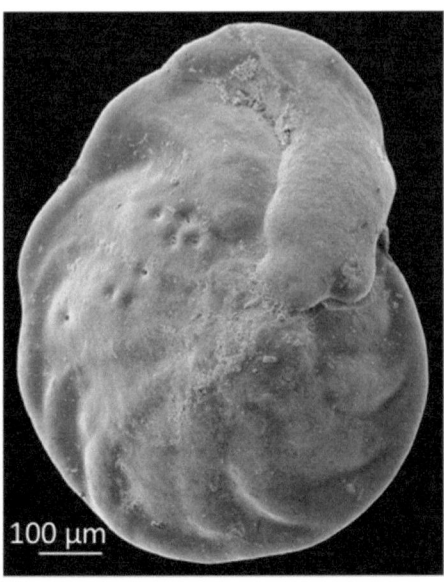

Fig. 9 Benthic foraminifer fossil *Planulina wuellerstorfi*. Made out of calcium carbonate ($CaCO_3$). It is the size of a sand grain. Photo by R. Mark Leckie

their $CaCO_3$ skeletal material when ocean temperatures warm, such that the Mg/Ca ratio can be used to reconstruct past sea surface temperatures (see Chap. 5).

Seeking to test the Milankovitch Theory of the Ice Ages, three marine geologists, John Hays, John Imbrie, and Nicholas Shackleton (Fig. 3) performed a mathematical analysis of the $\delta^{18}O$ in foraminifera isolated from a sufficiently long, well-dated marine sediment core. They verified that all three orbital cycles (100-ky eccentricity, 41-ky obliquity, and 23-ky precession) are embedded in the marine $\delta^{18}O$ record.[10] The modern version of the $\delta^{18}O$ (ice volume) record, compiled by marine scientists Lorraine Lisiecki and Maureen Raymo,[11] is shown in Fig. 10. The three smooth curves at the top of Fig. 10 show the $\delta^{18}O$ record mathematically separated (deconvolved) into its three main components. Divide 800-ky by the number of peaks (or valleys) in these curves and you will see that they represent the 100-ky cycles associated with eccentricity, the 41-ky cycles associated with obliquity, and the 23-ky cycles associated with precession. Combined, they will replicate the global ice volume curve (blue). Here's an important point; compare the 100-ky eccentricity, 41-ky obliquity, and 23-ky precession contributions to the external solar insolation forcing record in Fig. 7 to those of the ice volume record in Fig. 10. The 23-ky precession cycles dominate the energy input from the Sun whereas 100-ky eccentricity cycles dominate the ice volume record (and many other climate records as well). How can this be? To find out, let's examine other records, including carbon dioxide (CO_2) and Earth's surface temperature.

---

[10] Hays et al. (1976).

[11] Lisiecki and Raymo (2005).

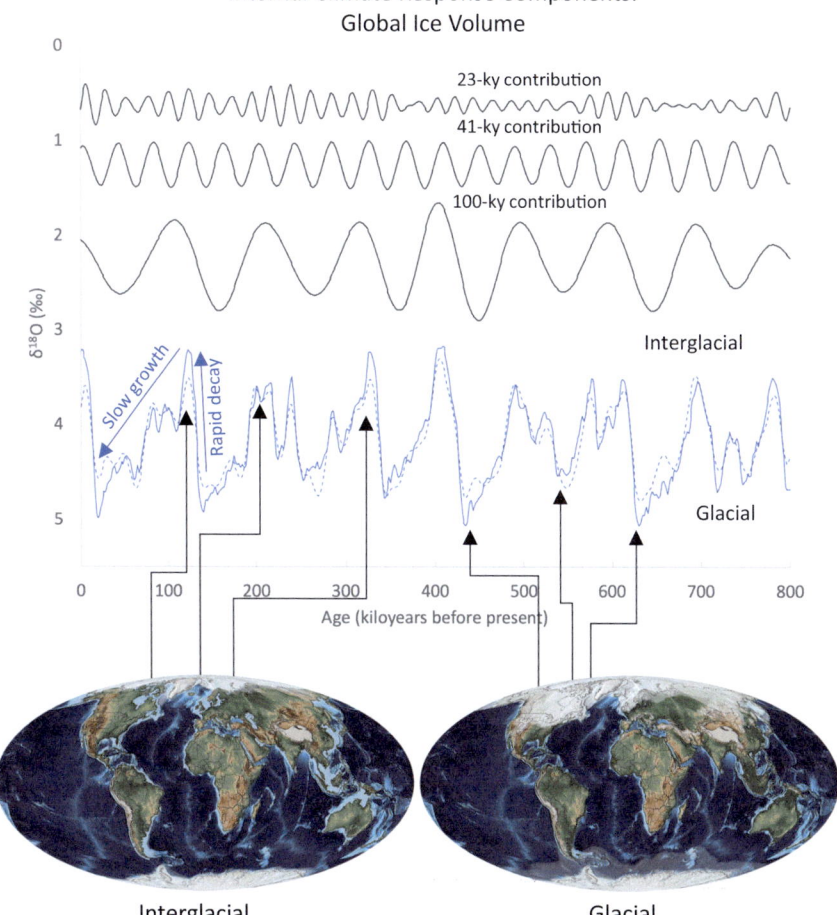

Fig. 10 Modern version of the marine $\delta^{18}O$ (‰ VPDB) curve is shown in blue.[12] Large values (4.5–5) indicate ice ages (glacials) and small values (3–3.5) indicate times similar to today (interglacials),[13] where Northern Hemisphere ice sheets are only found on Greenland. The orbital eccentricity (100-ky), obliquity (41-ky), and precession (23-ky) contributions are also shown. All three added together will yield a smoothed version of the marine $\delta^{18}O$ curve (dashed). Unlike the insolation records (dominated by precession), the marine $\delta^{18}O$ record is dominated by eccentricity. The "sawtooth" form of the ice age cycles is characterized by rapid decay and slow growth of ice sheets

---

[12] Lisiecki and Raymo (2005).

[13] Scotese and Wright (2018).

3.4 Ice Ages, Carbon Dioxide, and Surface Temperature

We just saw evidence that the 100-ky (eccentricity) cycle is the strongest cycle in the marine $\delta^{18}O$ record (our indicator of ice volume). That is odd, because we also just saw that the hypothesized solar radiation forcing (summer 65° N insolation) is dominated by 23-ky (precession) cycles. This means that the energy received from the Sun (at 23-ky cycles) is modified/stored within Earth's climate system, and ends up expressed mostly as 100-ky cycles where ice builds up slowly over time and then melts away quickly. As an analogy, think about a flashing light. Electricity is continually fed into a device called a capacitor which stores it until a sufficient amount of energy is accumulated (ice growth is at a maximum) and is then let go to create a flash (ice rapidly melts away) and the process repeats.

The climate system's internal response to external insolation forcing is strongly non-linear, like a capacitor; external energy from the Sun is stored, modified, and passed around within the hydrosphere, atmosphere, biosphere, cryosphere, and lithosphere. For example, consider the following sequence of events and follow along on Fig. 11. The sequence starts 128,000 years ago (on the right side of the figure) when Northern Hemisphere summer insolation at 65° N was strong, the only ice sheet in the Northern Hemisphere was on Greenland (like now), and the climate was warm. The Sun was continuously putting energy into the ocean surface (hydrosphere), heating it up (see Fig. 11; the blue insolation curve). Ocean water evaporated into the atmosphere (some of the Sun's energy was stored in the atmosphere in the form of water vapor). Shortly after 128,000 years ago insolation began to drop rapidly due to changes in obliquity and precession, reaching a threshold below which snowfall in winter did not melt during the following summer and ice sheets began to grow (blue dots on the ice volume record). Because it takes time to form an ice sheet, there was a lag between when the insolation starts to drop and the ice sheet begins to grow. From 128,000 years ago to 20,000 ago, the 23-ky precession insolation cycles became smaller (because eccentricity was becoming smaller) so there were longer intervals of time when ice growth was greater than ice melt (blue dots; ice volume growing) compared to times when ice melt was greater than ice growth (red dots; ice volume declining). As the ice sheet spread southward, the ice acted like a mirror, reflecting some of the Sun's energy back out to space, making it even easier for ice to grow, and it did so, reaching a maximum 20,000 years ago. In some places, the surface of the ice sheet was more than 2000 m above sea level. As a result, some of the Sun's energy that was transferred to the surface ocean, and then to water vapor in the atmosphere, was further transferred and stored in the mass of the ice; essentially storing potential energy in the mass of the (frozen) water now at elevations above sea level. Scaling up, it's difficult to conceive of the amount of energy stored in a continental-scale ice sheet 2000 m thick, such as the North American (Laurentide) ice sheet. In addition, the ice sheet weighed so much that it actually deformed Earth's crust (lithosphere) beneath it, creating a bowl-shaped depression in the Northern Hemisphere (where Hudson Bay is now located); this stored some of the Sun's energy in the lithosphere, not unlike storing energy in a compressed spring.

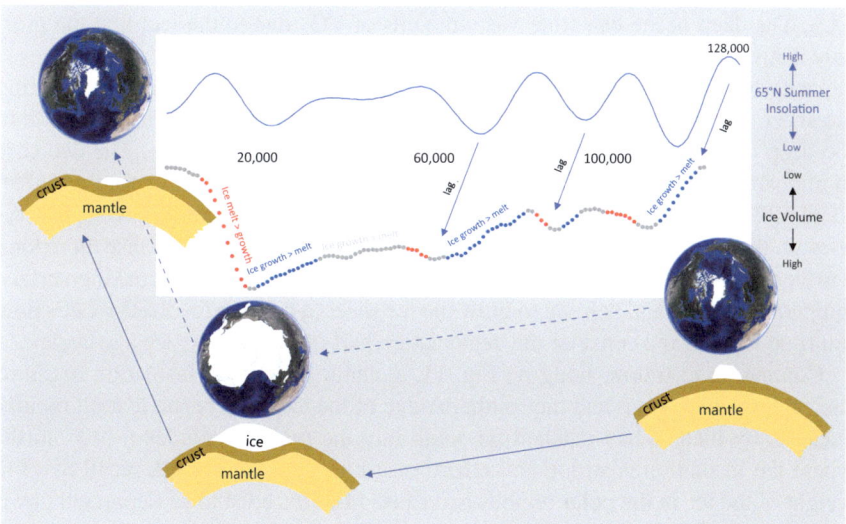

Fig. 11 100-ky cycle of the ice ages. Use this figure as you read the text to see how Northern Hemisphere summer insolation (solid blue line),[14] ice volume (dotted line),[15] and Earth's lithosphere (crust and mantle) interact to create the 100-ky ice age cycle.[16] Ice volume images are from Blue Marble[17]

Now let's take a short break from Fig. 11 to consider what was happening with CO_2 during this time from 128,000 to 20,000 years ago. As you learned in Chap. 4, past changes in atmospheric CO_2 can be reconstructed from measuring CO_2 content of air bubbles trapped in ice cores from Antarctica, where the ice accumulated over time has remained stable (it has not melted and reformed as in the Northern Hemisphere). Growth of the 2000 m high (Northern Hemisphere) ice sheet led to large changes in wind patterns which caused changes in global ocean and atmospheric circulation and rainfall patterns. These changes included, for example, erosion of the land surface and runoff of nutrients into the ocean, which caused an increase in surface ocean biological productivity stimulating **photosynthesis**. Increased surface ocean photosynthesis (productivity) removed CO_2 from the atmosphere and stored it as carbon in protist (e.g., phytoplankton such as coccolithophores and zooplankton such as foraminifera), plant, and animal tissue (the biosphere), exporting it to the deep ocean as the organisms died, sank, and decayed. This process increased the CO_2 content of the deep ocean, representing movement of the Sun's energy from the atmosphere into the biosphere, then to the hydrosphere in the form of deep ocean

---

[14] Laskar et al. (2004).

[15] Lisiecki and Raymo (2005).

[16] Abe-Ouchi et al. (2013).

[17] Google Earth Blue Marble: https://sos.noaa.gov/catalog/datasets/blue-marble-sea-level-ice-and-vegetation-changes-19000bc-10000ad/.

CO_2. The deep ocean can store vast amounts of CO_2 due to the fact that the pressure is high and the water is cold (only a few °C); both conditions lead to increased capacity to hold CO_2 dissolved in water. The important point here is that during ice sheet growth atmospheric CO_2 dropped because carbon was being moved into the deep ocean through this "biological pump" of the carbon cycle. The decrease in atmospheric CO_2 (a strong greenhouse gas) reduced its radiative forcing effect (remember that radiative forcing by greenhouse gases was discussed in Chap. 2). As a result, the heat energy retained in the climate system decreased (because more long wavelength radiation from the Earth's surface escaped into space), making surface temperatures drop and helping to grow the ice sheet (a positive feedback). Let's now return to Fig. 11 and consider the deglaciation part of the 100-kyr ice age cycle.

Continuing to follow along on Fig. 11, at about 23-ky ago, insolation began to rise (due to precession) and the southern edge of the ice sheet began to melt rapidly because, by then, it had reached far south into the middle-latitudes (~ 40° north) where the insolation is strong and effective at melting ice. As well, recall that the weight of the ice in the polar regions had caused Earth's crust to be depressed down into the mantle; in fact, parts of the bowl-shaped depression created by the weight of the last ice sheet are still springing back into shape today. Crustal depression is important because it lowered the top of the ice sheet down into the warm part of the atmosphere, helping the increased insolation to more effectively melt the ice. At the same time, decay of the 2000 m thick ice sheet caused changes in global-scale winds and ocean circulation patterns allowing the CO_2 stored in the high-pressure, cold, deep ocean to move upward into warmer, lower pressure shallower depths where water can no longer store that much dissolved CO_2. Hence CO_2 moved from the ocean back into the atmosphere, warming the atmosphere and helping to further melt the ice sheet (another positive feedback). Combined, these factors led to very rapid melting of the ice sheet; the last great Northern Hemisphere ice sheet melted away between 20,000 and 11,000 years ago, leaving only a small ice sheet on Greenland, the one that is there today. The entire sequence of events took about 100,000 years and illustrates how the external insolation forcing received by Earth, largely as 23-ky precession cycles, is transformed into dominant 100-ky ice age cycles by internal interactions of Earth's climate system. The global ocean $\delta^{18}O$ record (reconstructed from benthic foraminifera in marine sediments) and atmospheric CO_2 record (reconstructed from air bubbles trapped in ice cores) indicate that this cycling of ice volume and atmospheric CO_2 happened eight times in the past 800,000 years.

The slow growth and rapid decay of the Northern Hemisphere ice sheets is commonly referred to as "sawtooth" in form (see bottom panel of Fig. 10). According to the sequence of events described above, global records of past CO_2 and surface temperature should have characteristics similar to that of marine $\delta^{18}O$ (the ice volume record). Indeed, the 100-ky cycle and its "sawtooth" form are clearly evident in both types of paleoclimate records (Fig. 12). Like the global ice volume record, the 100-ky eccentricity component of globally averaged sea surface temperature and atmospheric CO_2 is dominant, with lesser contributions from 41-ky obliquity and 23-ky precession variability.

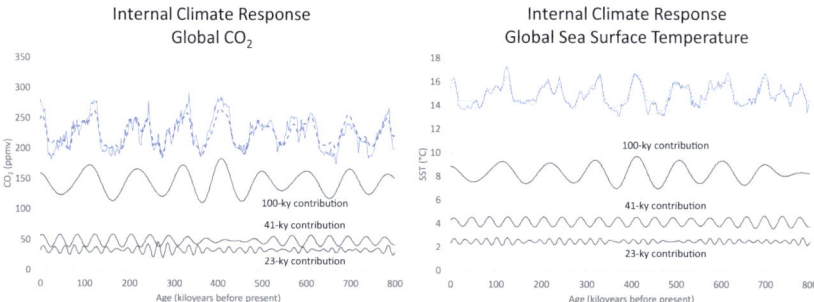

Fig. 12 Global records of greenhouse gas abundance[18] and sea surface temperature (SST)[19] over the past 800-ky. Top curves (solid blue) are global atmospheric carbon dioxide content (CO_2 in parts per million by volume, ppmv) and sea surface temperature (SST, °C). The orbital eccentricity (100-ky), obliquity (41-ky), and precession (23-ky) contributions are also shown. All three added together will yield smoothed versions of the CO_2 and SST curves (dashed blue). Unlike the insolation record, but similar to the ice volume record, these internal climate responses are dominated by 100-ky (eccentricity), with lesser amounts of 41-ky (obliquity), and 23-ky (precession)

3.5 Did 100-ky Variability Always Dominate Earth's Climate?

We just examined the last 800,000 years of Earth's climate history and found that although insolation received at Earth's surface has strong 23-ky cyclicity, its impact has been expressed in the climate system as strong 100-ky cycles in ice volume, atmospheric CO_2, and sea surface temperature. Was the system always this complicated, where energy comes in mostly at 23-kyr cycles and is transformed into 100-kyr cycles as it is passed around and modified by Earth's atmosphere, hydrosphere, cryosphere, lithosphere, and biosphere? The answer is no; there are many intervals in the past when ice sheets were less common (even absent; recall the discussion of Greenhouse times in Chaps. 6 and 7) and climate cycles looked much more like 23-ky insolation forcing. As an example, let's look at a 2-million-year-long record of changes in the Maldives island region (Indian Ocean) including the interval from 2 million to 1.2 million years ago, just prior to the development of the strong 100-ky ice age cycles we have been discussing. Figure 13 shows the elemental ratio of iron divided by potassium (Fe/K) measured on sediments from a drill site near the Maldives islands. This Fe/K ratio reflects changing input of fine-grained sediments eroded from the continents and delivered to the Indian Ocean by winds and river runoff. In this setting, river runoff is predominantly controlled by the amount of seasonal monsoon precipitation (recall that the Indian Ocean monsoon was discussed in Box 1 in Chap. 2). Large Fe/K values is a proxy for more humid conditions (increased rainfall) while low values indicate drier conditions.

---

[18] Luthi et al. (2008).
[19] Shakun et al. (2015).

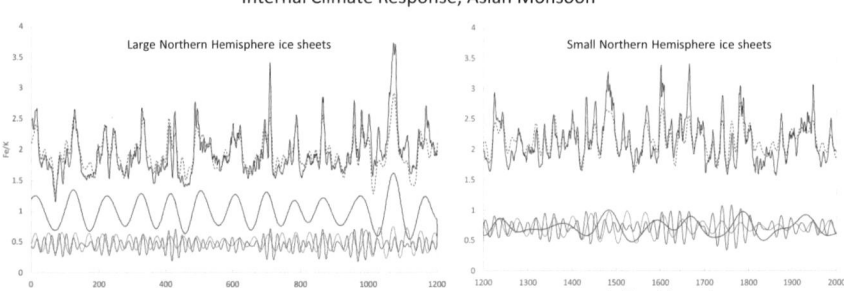

Fig. 13 Fe/K reconstruction[20] of Asian monsoon variability over the last 2 million years (2000-ky) separated into two intervals; 0- to 1200-ky (left panel) and 1200- to 2000-ky (right panel). The cyclical 100-ky (black), 41-ky (blue), and 23-ky (red) contributions are shown beneath the Fe/K record. All three combined yield a smoothed version (dashed) of the Fe/K record. When ice sheets were large, 100-ky variability dominated. When ice sheets were small, 41-ky and 23-ky ice variability dominated, similar to the obliquity and precession insolation forcing

Just like the ice volume, CO_2, and SST records, the interval from 1200-ky (1.2 million year ago) to modern is characterized by strong 100-ky cycles, with lesser contributions from the 41-ky obliquity and 23-ky precession cycles. This demonstrates that even tropical regions are impacted by high latitude ice sheets, which are linked to global-scale aridity (drying of the atmosphere). However this 2 million year record shows that before 1200-ka (1.2 Ma), the influence of 23-ky and 41-ky cycles was greater and the 100-kyr cycles were muted and less regular because Northern Hemisphere ice sheets were much smaller and less variable. Thus the influence of insolation was more direct and the energy received and the climate response were more similar. In this case, increased insolation caused by precession and obliquity are more directly expressed as 23-ky and 41-ky cycles in rainfall.

4 Orbital Cycles in "Deep Time" Sedimentary Records

Section 3 of this chapter emphasized the causes of orbital-scale variations during the past two million years, the effects of those variations on different parts of the Earth's climate system during that time, and the records of those effects as preserved in sedimentary proxies. It was noted, however, that the effects of orbital cycles have been recognized in older parts of Earth's geologic record. In this section, we will briefly examine a few of those older records in order to illustrate the influence of orbital variations on several components of the Earth's climate system during prolonged warm (greenhouse) and prolonged cold (icehouse) times in Earth's history that you were introduced to in Chaps. 6 and 7.

---

[20] Kunkelova et al. (2018).

For this discussion, we will transition from considering the last several million years of Earth history to considering the Phanerozoic (the last ~ 560 million years, composed of the Paleozoic, Mesozoic, and Cenozoic Eras). When this much longer time interval is considered, additional cyclic patterns and complexities are also observable as a result of several aspects of planetary interactions that have very little effect over the shorter span of a few million years or less. Two aspects of planetary interactions are particularly important for influencing orbital cycles in "deep time"[21] including:

- *The influence of additional gravitational interactions between Earth and other planets.* Gravitational interactions are responsible for the orbital cycles that have been the focus of earlier parts of this chapter. Infrequently, some of these gravitational interactions produce "chaotic behavior," which is when the normal patterns of certain orbital cycles are disrupted. Long-term interactions between the Earth and Mars have been identified as especially important for causing some episodes of "chaotic behavior." Chaotic behavior makes it difficult, for example, to trace the effects of orbital obliquity and precession continuously throughout Earth history. It is fortunate, however, that some planetary interactions have been demonstrated to be remarkably stable throughout Earth history, producing cycles often termed "orbital metronomes." One of the most commonly observed "orbital metronomes" arises because of interactions of the Earth with Jupiter and Venus, producing "long eccentricity" cycles with periods of 405,000- and 413,000-years[22] that have been observed in paleoclimate records throughout the Phanerozoic. In Paleozoic and Mesozoic paleoclimate records, the well-established duration of the long eccentricity "metronome" is very valuable because it provides a time-bound framework within which the shorter-duration orbital cycles can be defined more accurately than would be possible using only radiometrically-based dating methods.

- *The influences of "tidal dissipation" (sometimes called "tidal friction") and small changes in the distribution of Earth's mass through time.* Over very long time periods these factors have acted to slow the rate of Earth's rotation and lengthen the periods of the short-eccentricity, obliquity, and precession cycles through geologic time. For example, the principal period of obliquity cycles for the past 5 million years has been 41-kyr, whereas the principal period for obliquity was 34.5-kyr during the early Triassic (~ 245 Ma).[23]

---

[21] Hinnov (2018) and Huang (2018).
[22] Berger (1978) and Berger and Loutre (1991).
[23] Huang (2018).

4.1 Examples of Orbital Cyclicity in Older Paleoclimate Records

The body of literature produced by studies of paleoclimatic cyclicity is large[24] so we cannot synthesize all that information here. Instead, we will consider "deep time" examples from the greenhouse conditions of the Triassic and from the icehouse conditions of the Late Paleozoic, to illustrate the durable influence of orbital cycles on Earth's climate system.

4.1.1 Orbital Cyclicity in Greenhouse Time Paleoclimate Records from Triassic Rift Lakes of Eastern North America

One widely cited example of a "deep time" paleoclimatic record that clearly exhibits orbital cyclicity comes from Triassic-age sediments found in the Newark and Hartford Basins of eastern North America (present-day New Jersey, New York, and Connecticut).[25] It is an example you were first introduced to in Chap. 5; here we will explore it further. Equivalent sediments are found in similar Triassic-age basins in North Carolina, Virginia, Maryland, Pennsylvania, and the Canadian Maritime Provinces.

The Triassic was a global greenhouse time; atmospheric CO_2 concentration is estimated to have been at least several thousand ppm, with no evidence of large ice sheets at high latitudes on Pangea, the supercontinent that existed at that time.[26] Monsoonal atmospheric circulation, characterized by seasonal changes in wind directions and strong seasonal differences in precipitation, were an important component of the Triassic climate of Pangea, as interpreted from sedimentary proxy records[27] and as indicated by climate modeling.[28] The supercontinent of Pangea began to break up via rifting in the late Triassic, and the Triassic rift basins of interest to us were produced as North America separated from Europe and Africa during the early stages of opening what would become the present-day Atlantic Ocean. Those rift basins formed depressions that were relatively separate and completely—or almost completely—surrounded by higher land. This configuration produced a situation known as **internal drainage**, where water from the surrounding highlands drains into the basin but there is no major river flowing out of the basin, into the ocean. A similar situation of distinct basins with internal drainage produces the lakes in today's East African Rift zone, such as Lake Malawi and Lake Tanganyika. The size and water chemistry of a lake in an internally drained basin is very sensitive to the

---

[24] For example, see the references in the "Cyclostratigraphy" section of each chapter in Geologic Time Scale 2020 (2020), edited by Felix M. Gradstein, James G. Ogg, Mark D. Schmitz, and Gabi M. Ogg. Elsevier. https://doi.org/10.1016/B978-0-12-824360-2.00001-2.

[25] Olsen et al. (1996).

[26] Royer et al. (2004).

[27] Sellwood and Valdes (2006).

[28] Kutzback and Gallimore (1989).

Fig. 14 Repeating cycles of sedimentary rocks in the Newark Basin in Connecticut reflecting the influence of orbital cycles on the hydrological cycle and regional lake levels. From the bottom of the photo to the top, alluvial stream deposits grade into shallow lake lacustrine muds followed by deep lake shales and back again to lacustrine muds and alluvial deposits as the lake level rose and fell. Same as Fig. 5.3. Photos provided by Mark Leckie, figure by Adriane Lam

relative amounts of precipitation and evaporation in that basin, so past changes in lake size and water chemistry, which can be interpreted from the record of that lake's sediments, can serve as proxies for changes in the basin's water budget (Fig. 14). Since evaporation is driven by solar energy (through both air temperature and winds) and precipitation is determined by the local-to-regional water cycle, determining the water budget for a rift basin in the past provides valuable insights into the climatically important flows of energy and water in the atmosphere and hydrosphere at that time.

The Triassic sediments of the Newark and Hartford Basins (NHB) are a thick (5 km or more) sequence composed primarily of sandstones and mudstones, with lesser amounts of gravel-bearing sediments (i.e., conglomerates) and a few interbedded basalts.[29] The mudstones are the most abundant lithology in this sequence, and range in color from black and gray to purple-gray and red; some contain freshwater fossils and trace fossils of plant roots and reptile footprints, whereas others exhibit desiccation features (e.g., mudcracks and casts of evaporite minerals). The coarser-grained NHB sediments are interpreted to have been deposited on alluvial fans near the edges of the rift basins and by stream systems within the basins, whereas the finer-grained sediments (mudstones to fine-grained sandstones) primarily are interpreted to have been deposited in lakes. Characteristics of the lake sediments (e.g., color, grain size, bedding thickness, fossils, and sedimentary structures) have been interpreted as a proxy for water depth at the time of deposition, allowing the history of lake size and depth to be reconstructed through an estimated 20 My of the Late Triassic and earliest Jurassic. Depending on the water budget at that time, the lakes ranged from annual (i.e., the lake was present during the wet season but disappeared during the

---

[29] Olsen et al. (1996).

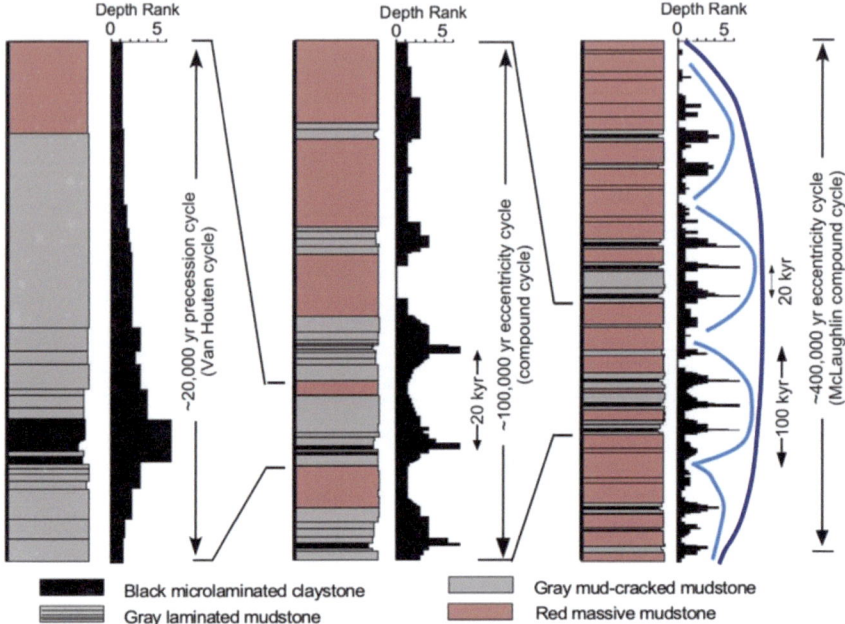

Fig. 15 Lithologic section reflecting one long eccentricity (405,000-year) cycle containing four short-eccentricity (~ 100,000-year) cycles each of which contains, five precession (~ 20,000-year) cycles. From Schlische (2022) modified from Olsen et al. (1989).[30] This figure is excluded from Creative Commons license

dry season) to perennial (i.e., persisting through multiple years), although even the perennial lakes varied seasonally in size and water depth.

Cyclicity in the NHB sediments has been recognized at several different distinctive thicknesses; cycles 3–6 m thick are interpreted to record variations between poorly oxygenated perennial lakes during wetter times and annual lakes under more arid conditions (also known as **playas**); these m-thick packages are themselves bundled into cycles ~ 12–18 m thick and ~ 60 m thick. Quantitative analysis of these cycles indicates that the shortest cycles had a period of ~ 20-ky, indicating a precessional control on the amount of monsoonal precipitation. The periods of the two longer cycles were ~ 100-ky (short-eccentricity) and 405-ky (long eccentricity), indicating an additional orbital influence on the magnitude of each precessional cycle (Fig. 15). Over the entire 20 My record, all of the orbital influences were modified by the gradual shift of the NHB through climatic belts from ~ 5° N to ~ 20° N by plate tectonics. Similar precession and long eccentricity periods have been identified in other Late Triassic records, including another monsoonal lake record from Germany[31] and a

---

[30] From Schlische (2022), https://eps.rutgers.edu/research/virtual-geo/projects/geology-of-the-new ark-rift-basin/1377-compound-cycles; Modified from Olsen et al. (1989).
[31] Bahr et al. (2020).

record of precession-driven cyclicity in marine productivity interpreted from chert/shale alternations in a deep-marine record in Japan.[32]

4.1.2 Orbital Cyclicity in Paleoclimate Records of the Late Paleozoic Ice Age (LPIA)

Global conditions during the Late Paleozoic icehouse (the Pennsylvanian [or Upper Carboniferous] and Permian) were quite different from those of the subsequent Triassic greenhouse world; the difference in overall climatic state has been explained by lower atmospheric CO_2 concentrations during the Late Paleozoic (due to carbon extraction and storage in widespread coal swamps) and differences in the geographic distribution of oceans and continents, which affected the poleward transport of heat by ocean currents. Despite this difference in overall climate state, however, orbital influences are still evident in records of different components of the Late Paleozoic climate system.

The Late Paleozoic is considered an icehouse time because it was marked by a long interval with widespread but locally/regionally sourced and temporally variable ice sheets. In this way, the distribution and behavior of ice sheets during the LPIA was different from the Cenozoic icehouse, which has been dominated by the large and persistent Antarctic and Greenland/Northern Hemisphere ice sheets. Evidence for these dynamic Late Paleozoic ice sheets is preserved across large portions of the landmass termed "Gondwana" (i.e., today's Antarctica, Australia, India, South America, and southern Africa). Although originally a separate landmass, Gondwana was incorporated into Pangea during the Late Paleozoic. In fact, the Late Paleozoic Ice Age (LPIA; ~335 to 260 Ma) was the longest-lived icehouse of the Phanerozoic. The LPIA's effects on cryosphere volume and global sea level change are estimated to have been somewhere between 30 and 100% those of the Pleistocene, although research on this topic continues.[33] Our examples of orbital cyclicity during the LPIA will consider two different types of sediments, deposited in different settings during the LPIA.

The first sediment type that can exhibit a record of orbital cyclicity was deposited in the tropical latitudes of Pangea during the LPIA, rather than in environments directly affected by glacial ice. These low-latitude deposits occur as distinctive sequences with repeated alternations of marine and nonmarine sediments, and are known by the general name **cyclothems**.[34] In some locations, the cyclothems alternate between coarse-grained and fine-grained detrital (i.e., clastic) sediments, whereas other cyclothems alternate between detrital mudrocks and limestones. Beginning with early studies of cyclothems almost a century ago, repeated fluctuations in sea level have been cited as a major control on the deposition of these alternating rock types. As a result, any cyclicity recorded by low-latitude cyclothems deposited during the LPIA can be used to evaluate the influence of orbital cycles on glacial fluctuations and

---

[32] Huang (2018).
[33] Montanez and Poulsen (2013).
[34] Fielding (2021).

Fig. 16 Coastal cliff exposure of part of a cyclothem at Tullig Point, western Ireland. The lower part of the cliff is composed of a thickening- and coarsening-upward succession of prodelta sediments, and is sharply overlain by a sandstone body representing river mouth bar deposits. Total thickness visible is 50 m. The sharp-based sandstone is 11 m thick. Figure used with permission from Elsevier. *Source* Fielding (2021).[36] This figure is excluded from Creative Commons license

associated sea level change (i.e., the influence of orbital cycles on the cryosphere and the hydrosphere).

LPIA cyclothems are widespread across the present midcontinent of North America, Europe, Russia, central Asia and China, and portions of northern Africa (Fig. 16). All of these occurrences exhibit the sedimentary alternations characteristic of cyclothems; many studies have hypothesized that those alternations record orbital cyclicity, based solely on qualitative analyses of the sedimentary patterns. In contrast, only a few cyclothem occurrences provide enough detailed age information for their potential orbital cyclicity to have been analyzed quantitatively; these include cyclothems from the Donets Basin in Ukraine and some of the cyclothems from the midcontinent of North America. Both of these exhibit the 405-ky long eccentricity cycle, with nested alternations that may record the ~ 100-ky short-eccentricity cycle.[35] Other cyclothem sequences with good age control have been analyzed quantitatively and do not exhibit orbital cyclicities, perhaps because geologic controls not affected by orbital cyclicities (such as variations in sediment supply driven by the effects of regional tectonics) were more important in those geographic regions during deposition.

The second sediment type that can exhibit a record of LPIA orbital cyclicity is marine limestones, which also were deposited in low-latitude settings of Pangea. These limestones exhibit repeated alternations in physical characteristics and fossil content that record repeated shallowing and deepening of the ocean, similar to the record of sea level change carried by the LPIA cyclothems. These limestone

---

[35] Aretz et al. (2020) and Fielding (2021).

[36] Figure 6E in: Fielding (2021).

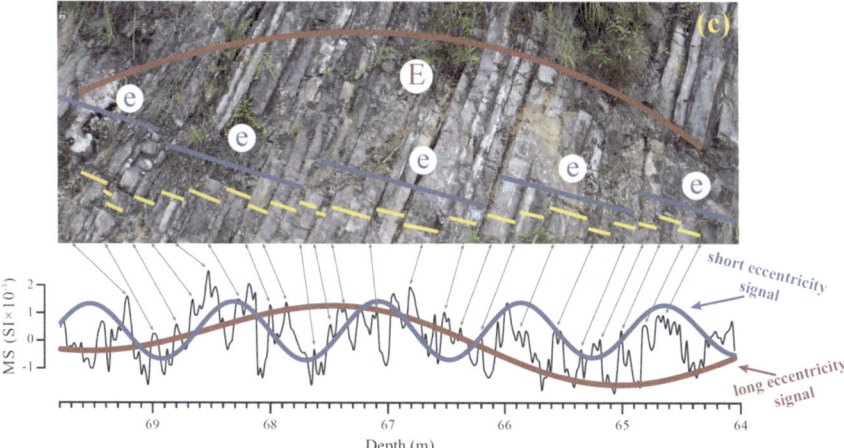

Fig. 17 Limestone cycles from the Luokun section, China. The magnetic susceptibility (MS) record reveals long- and short-eccentricity as well as precessional cycles. Figure used with permission from Elsevier. *Source* Fang et al. (2018).[37] This figure is excluded from Creative Commons license

sequences cannot be called cyclothems, however, because the entire limestone sequence was deposited in the ocean, and therefore lacks the alternation between marine and continental sediments that is characteristic of cyclothems.

Strong evidence for the influence of orbital cycles during the LPIA comes from a study of one of these marine limestone sequences, found in southern China (Fig. 17). In this study, the paleoclimatic proxy was the magnetic susceptibility of the sediment, which is a very sensitive and non-destructive measure of the abundance of land-derived material within the limestones. The supply of land-derived material to such a site of limestone deposition varies in response to changes in the amount of weathering on adjacent land, which varies with temperature, rainfall, and atmospheric CO_2 levels, as well as variations in the rate of sediment supply from that land, which varies with rainfall and distance to shore. Quantitative analysis of this record, which does have good age control, identified variations at the full range of orbital cyclicities, from long eccentricity (405-ky) through precession (19- and 15.9-ky). The researchers interpreted these results to indicate the importance of orbital forcing on LPIA cryosphere extent and associated sea level changes at their area of study.

---

[37] Fang et al. (2018).

5 Climate Cycles in Our Future

In Sect. 4 of this chapter, you learned how the observed cyclic patterns of terrestrial and ocean sedimentary records of the Mesozoic and Paleozoic Era exhibit orbital influences on Earth's climate during both greenhouse and icehouse times. In Sect. 3, you learned how scientists reconstructed the past 800,000 years of ice age cycles with great accuracy and now have a reasonable understanding of the interplay between external insolation and the cycling of carbon (in the form of CO_2) between the atmosphere and deep ocean. You learned that changes in insolation initiate the growth and decay of the ice sheets and that cycling of CO_2 between the deep ocean and atmosphere helps these processes along. The CO_2 feedback works in both directions; the movement of CO_2 out of the atmosphere and into the deep ocean during glaciation helps to promote ice sheet growth while the movement of CO_2 out of the deep ocean back into the atmosphere during deglaciations helps to rapidly melt the ice sheets.

So, what's next in terms of orbitally-driven climate cycles? Will there be another ice age? This is an interesting question given that Northern Hemisphere summer insolation has fallen significantly (naturally) since its recent peak 11,000 years ago. Obviously, the future of the ice ages cannot be found in existing terrestrial or ocean sediments. Instead, we turn to mathematical models of how the climate works in order to determine if another ice age is possible (you will learn more about these models in Chap. 10). Under natural conditions, the concentration of CO_2 in the atmosphere is around 180 ppmv during peak glacial times when much of it is moved into the ocean, rising to around 290 ppmv during peak interglacial times, when more of it resides in the atmosphere. Because humans continue burning fossil fuels (coal, oil, and gas), atmospheric CO_2 was at ~420 ppmv (as of 2024). Climate models indicate that this level of atmospheric CO_2 will delay the next phase of ice sheet growth by about 60,000 years.[38] In other words, with this extra CO_2 in the atmosphere, the next low insolation cycle will no longer be sufficient to initiate ice sheet growth.

> **Box 1 Sunspot cycles and Earth's climate**
>
> Much of this chapter examined orbital-scale changes in solar radiation received at Earth's surface (100,000-, 41,000-, and 23,000-year cycles) and the large climate changes in response to this energy input. In addition to these orbital-scale cycles, there is also a much shorter Sun-induced cycle (sunspot cycles) that affects the solar energy input that Earth receives. As described in Chap. 3, solar activity is extremely variable over interannual (i.e., two or more years) to centennial time-scales, and embedded within this variability is an 11-year cycle of increasing and decreasing number of sunspots and solar output.

---

[38] Ganopolski et al. (2016); Maslin (2020); and Milankovitch (1941).

Solar activity is measured by Total Solar Irradiance (TSI) and sunspot number. Sunspots are easily observed from Earth. Sunspots[39] are small regions (although they can be many times the side of the Earth) on the Sun's surface where the magnetic field is strongest. This inhibits the flow of hot gas from the Sun's interior to the surface, causing the surface to cool and darken at the location of the sunspots. At the same time, however, the area surrounding the sunspots (called faculae) become even hotter, so more sunspots actually indicate stronger TSI[40] (watts/m^2, integrated over all wavelengths). The number of sunspots[41] and TSI increase and decrease with a cycle of 11 years such that the Sun emits ~ 0.1% more radiation during sunspot maxima (when it is most covered with dark sunspots and faculae). The solar power available to the Earth system at the 11-year sunspot cycle variation is only $\sim$ 0.07% of the total solar output; very much smaller than orbital-scale changes responsible for the ice ages. Nevertheless, some recent climate change observations have been interpreted as being driven by these small changes in solar activity; the Little Ice Age (LIA), explored in Chaps. 3 and 4 (and described further below), is a good example.

The LIA was a period of cold conditions from 1550 to 1880 C.E. It was the final and most intense of a series of millennial-scale cooling events and glacial advances that took place within the otherwise generally warm Holocene epoch (11.7 ka-present). Glaciers advanced in Europe, North America, and South America. The coldest phase of the LIA was during a time when the amplitude of the sunspot cycle was very low called the Maunder Minimum (1645–1715 C.E.; Fig. 18); at this time average winter temperatures in Europe and North America were as much as 2 °C lower than at present. The Baltic Sea froze over, as did many of the rivers and lakes in Europe. The tree line and snowline dropped and glaciers advanced, overrunning towns and farms. The lack of sunspots meant that solar radiation was lower at this time suggesting that changes in solar (sunspot) activity contributed to these climate changes. Climate models suggest that changes in sunspot activity would have reduced average global temperatures by 0.4 °C at most, which cannot explain the ~ 2 °C regional cooling of the climate in Europe and North America. Other global to regional climate drivers must also have influenced the LIA, such as increased volcanic forcings (e.g., see Chap. 3, Fig. 1) and changes in internal climate oscillations (e.g., the North Atlantic Oscillation and the El Niño-Southern Oscillation) affecting regional patterns of heat and moisture.

---

[39] https://scied.ucar.edu/learning-zone/sun-space-weather/sunspots.
[40] Dewitte and Nevens (2016).
[41] Hoyt and Schatten (1998).

More recently, between 1900 and 1990 C.E. the amplitude of the 11-year sunspot cycles increased, as did the global average temperature (Fig. 18), leading some to conclude that solar activity was the primary cause of modern global warming, instead of the rapid increase in CO_2 caused by burning of fossil fuels. Evidence against this conclusion emerged between 1990 and 2020 C.E., when the amplitude of the solar cycle decreased dramatically whereas global surface temperatures continued to rise (along with CO_2; Fig. 18).

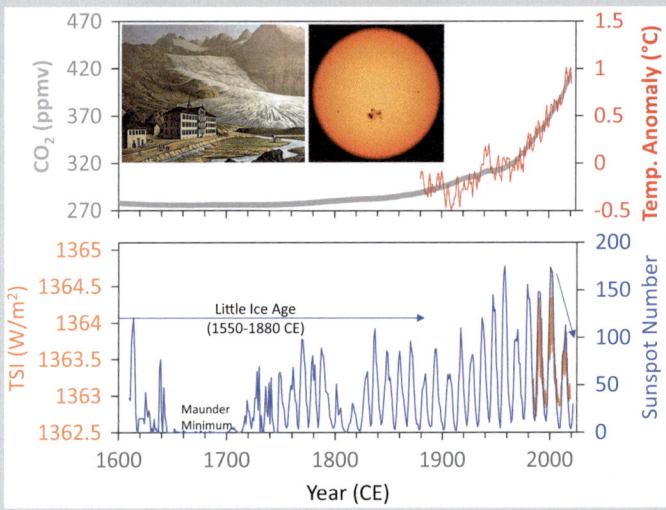

Fig. 18 Changes in solar activity, carbon dioxide and average temperature, 1600–2021 C.E. Lower graph: Total Solar Irradiance (TSI, in watts per square meter) and sunspot number, which are proxies of solar activity. The Little Ice Age and Maunder Minimum are labeled. Upper graph: atmospheric carbon dioxide concentration (in parts per million by volume)[42] and global mean temperature (degrees Celsius) relative to the 1951–1980 average.[43] Inset images: painting of Rhone Glacier titled "Gletsch mit Rhonegletscher 1870"[44] (left) and a large sunspot (right) that is 80,000 miles diameter (> 10× the diameter of Earth). Right image taken on October 23, 2014 by NASA Solar Dynamic Observatory[45]

---

[42] IAC Swiss data downloaded from https://www.co2.earth/historical-co2-datasets.

[43] https://data.giss.nasa.gov/gistemp/graphs/graph_data/Global_Mean_Estimates_based_on_Land_and_Ocean_Data/graph.txt.

[44] https://commons.wikimedia.org/wiki/File:Rhonegletscher1870.jpg; public domain work of art.

[45] NASA, https://www.nasa.gov/content/goddard/sdo-observes-largest-sunspot-of-the-solar-cycle/ Public domain.

Modern model simulations of twentieth century climate[46] include all of the major forcings (e.g., solar activity, volcanic activity, changes in greenhouse gasses and aerosols). These models indicate that anthropogenic forcing outgrew the solar forcing near the start of the twentieth century, and the divergence greatly accelerated during the second half of the twentieth century. Thus, modern global warming, largely driven by anthropogenic greenhouse gas emissions has overwhelmed any natural changes due to solar forcing. You will learn about how climate models are tools for understanding the Earth system in the next chapter. Chapter 10 will focus on how climate models work and what our existing climate models tell us about future climate change.

References

Abe-Ouchi, A., Saito, F., Kawamura, K., Raymo, M. E., Okuno, J., Takahashi, K., & Blatter, H. (2013). Insolation-driven 100,000-year glacial cycles and hysteresis of ice-sheet volume. *Nature, 500*(7461), 190–193. https://doi.org/10.1038/nature12374

Aretz, M., Herbig, H. G., Wang, X. D., Gradstein, F. M., Agterberg, F. P., & Ogg, J. G. (2020). Chapter 23—The Carboniferous period. In F. M. Gradstein, J. G. Ogg, M. D. Schmitz, & G. M. Ogg (Eds.), *Geologic Time Scale 2020* (pp. 811–874). Elsevier.

Bahr, A., Kolber, G., Kaboth-Bahr, S., Reinhardt, L., Friedrich, O., & Pross, J. (2020). Megamonsoon variability during the late Triassic: Re-assessing the role of orbital forcing in the deposition of playa sediments in the Germanic Basin. *Sedimentology, 67*, 951–970. https://doi.org/10.1111/sed.12668

Berger, A. L. (1978). Long-term variations in daily insolation and Quaternary climate changes. *Journal of the Atmospheric Sciences, 35*, 2362–2367.

Berger, A., & Loutre, M. F. (1991). Insolation values for the climate of the last 10 million years. *Quaternary Science Reviews, 10*, 297–317.

Dewitte, S., & Nevens, S. (2016). The total solar irradiance climate data record. *The Astrophysical Journal, 830*(1), 25. https://doi.org/10.3847/0004-637X/830/1/25

Fang, Q., Wu, H., Wang, X., Yang, T., Li, H., & Zhang, S. (2018). Astronomical cycles in the Serpukhovian-Moscovian (Carboniferous) marine sequence, South China and their implications for geochronology and icehouse dynamics. *Journal of Asian Earth Sciences, 156*, 302–315. https://doi.org/10.1016/j.jseaes.2018.02.001

Fielding, C. R. (2021). Late Palaeozoic cyclothems—A review of their stratigraphy and sedimentology. *Earth-Science Reviews, 217*, 103612. https://doi.org/10.1016/j.earscirev.2021.103612

Ganopolski, A., Winkelmann, R., & Schellnhuber, H. J. (2016). Critical insolation–CO_2 relation for diagnosing past and future glacial inception. *Nature, 529*, 200. https://doi.org/10.1038/nature16494

Gray, L. J., Beer, J., Geller, M., Haigh, J. D., Lockwood, M., Matthes, K., et al. (2010). Solar Influences on Climate. *Reviews of Geophysics, 48*(4).

Grove, A. T. (2008). The revolution in palaeoclimatology around 1970. In T. P. Burt, R. J. Chorley, D. Brunsden, N. J. Cox, & A. S. Goudie (Eds.), *The history of the study of landforms. Volume 4: Quaternary and recent processes and forms (1890–1965) and the mid-century revolutions* (pp. 961–1004). Geological Society.

---

[46] For example: Gray et al. (2010) and IPCC (2013, 2021).

Hays, J. D., Imbrie, J., & Shackleton, N. J. (1976). Variations in the Earth's orbit: Pacemakers of the ice ages. *Science, 194*, 1121–1132.

Hinnov, L. (2018). Chapter One—Cyclostratigraphy and astrochronology in 2018. *Stratigraphy & Timescales, 3*, 1–80. https://doi.org/10.1016/bs.sats.2018.08.004

Hoyt, D. V., & Schatten, K. H. (1998). Group sunspot numbers: A new solar activity reconstruction. *Solar Physics, 179*(1), 189–219. https://doi.org/10.1023/A:1005007527816

Huang, C. (2018). Chapter Two—Astronomical time scale for the Mesozoic. *Stratigraphy & Timescales, 3*, 81–150. https://doi.org/10.1016/bs.sats.2018.08.005

IPCC. (2013). In T. F. Stocker, D. Qin, G.-K. Plattner, M. Tignor, S. K. Allen, J. Boschung, A. Nauels, Y. Xia, V. Bex, & P. M. Midgley (Eds.), *Climate change 2013: The physical science basis. Contribution of Working Group I to the Fifth Assessment Report of the Intergovernmental Panel on Climate Change* (1535 pp.). Cambridge University Press.

IPCC. (2021). In V. Masson-Delmotte, P. Zhai, A. Pirani, S. L. Connors, C. Péan, S. Berger, N. Caud, Y. Chen, L. Goldfarb, M. I. Gomis, M. Huang, K. Leitzell, E. Lonnoy, J. B. R. Matthews, T. K. Maycock, T. Waterfield, O. Yelekçi, R. Yu, & B. Zhou (Eds.), *Climate change 2021: The physical science basis. Contribution of Working Group I to the Sixth Assessment Report of the Intergovernmental Panel on Climate Change* (pp. 3–32). Cambridge University Press. https://doi.org/10.1017/9781009157896.001

Kunkelova, T., et al. (2018). A two-million-year record of low-latitude aridity linked to continental weathering from the Maldives. *Progress in Earth and Planetary Science, 5*(1), 86. https://doi.org/10.1186/s40645-018-0238-x

Kutzback, J. E., & Gallimore, R. G. (1989). Pangaean climates: Megamonsoons of the Megacontinent. *Journal of Geophysical Research, 94*(D3), 3341–3357. https://doi.org/10.1029/jd094id03p03341

Laskar, J., Robutel, P., Joutel, F., Gastineau, M., Correia, A. C. M., & Levrard, B. (2004). A long-term numerical solution for the insolation quantities of the Earth. *Astronomy & Astrophysics, 428*. https://doi.org/10.1051/0004-6361:20041335

Lisiecki, L. E., & Raymo, M. E. (2005). A Pliocene-Pleistocene stack of 57 globally distributed benthic $\delta^{18}O$ records. *Paleoceanography, 20*(1). https://doi.org/10.1029/2004PA001071

Luthi, D., Le Floch, M., Bereiter, B., Blunier, T., Barnola, J. M., Siegenthaler, U., Raynaud, D., Jouzel, J., Fischer, H., Kawamura, K., & Stocker, T. F. (2008). High-resolution carbon dioxide concentration record 650,000–800,000 years before present. *Nature, 453*(7193), 379–382. https://doi.org/10.1038/nature06949

Maslin, M. (2020). Tying celestial mechanics to Earth's ice ages. *Physics Today, 73*, 48–53.

Milankovitch, M. (1941). *Kanon der Erdbestrahlung und seine Andwendungouf das Eiszeitenproblem* (I. P. f. S. Translations, Trans.). Koninglich Serbische Academie.

Montanez, I., & Poulsen, C. J. (2013). The Late Paleozoic Ice Age: An evolving paradigm. *Annual Review of Earth and Planetary Sciences, 41*, 629–656. https://doi.org/10.1146/annurev.earth.031208.100118

Olsen, P. E., Kent, D. V., Cornet, B., Witte, W. K., & Schlische, R. W. (1996). High-resolution stratigraphy of the Newark rift basin (early Mesozoic, eastern North America). *GSA Bulletin, 108*(1), 40–77. https://doi.org/10.1130/0016-7606(1996)108%3c0040:HRSOTN%3e2.3.CO;2

Olsen, P. E., Schlische, R. W., & Gore, P. J. W. (Eds.). (1989) Tectonic, depositional, and paleoecological history of the Early Mesozoic Newark Supergroup, Eastern North America. In *International Geological Congress, Guidebooks for Field Trips* T351 (174 p.).

Royer, D. L., Berner, R. A., Montanez, I., Tabor, N. J., & Beerling, D. J. (2004). CO_2 as a primary driver of Phanerozoic climate. *GSA Today, 14*(3), 4–10. https://doi.org/10.1130/1052-5173(2004)014%3c4:CAAPDO%3e2.0.CO;2

Scotese, C. R., & Wright, N. M. (2018). PALEOMAP Paleodigital Elevation Models (PaleoDEMS) for the Phanerozoic. *Zenodo*. https://doi.org/10.5281/zenodo.5460860

Sellwood, B. W., & Valdes, P. J. (2006). Mesozoic climates: General circulation models and the rock record. *Sedimentary Geology, 190*(1–4), 269–287. https://doi.org/10.1016/j.sedgeo.2006.05.013

Shakun, J. D., Lea, D. W., Lisiecki, L. E., & Raymo, M. E. (2015). An 800-kyr record of global surface ocean and implications for ice volume-temperature coupling. *Earth and Planetary Science Letters*, 426, 58–68. http://www.sciencedirect.com/science/article/pii/S0012821X1500 3404

Steven Clemens grew up in Texas, Colorado, Florida, New Jersey, Alaska, and Washington State. Each place offered new and different opportunities for exploration of the natural environment; from urban-influenced to untouched and natural, from mountains to oceans. These led to an interest in the sciences and in particular, the science that integrates all others, Earth Science. Steve is a Professor of Research at Brown University, studying how and why monsoon rainfall in Asia changes through time. He has planned, participated in, and led ocean drilling expeditions to the Arabian Sea, the South China Sea, the East China Sea, the Sea of Japan, and the Bay of Bengal. Each expedition is followed by years of research in the lab, reconstructing monsoon wind and rainfall patterns by analyzing changes in the chemical, physical, and biological properties preserved in the sea floor sediments and publishing the results so that other scientists can evaluate and build upon them. Nearing retirement, Steve looks forward to spending even more time on and in the ocean. Steve is a primary author of Chap. 9.

Lawrence Krissek became intrigued with the oceans while completing the Boy Scouts' Oceanography merit badge and a middle school science fair project, even though he lived far from the coast. He has been fortunate to be able to pursue that interest throughout his career, combining his initial focus on scientific research with more-recent efforts to improve earth- and climate-science literacy across the K-16 spectrum. Larry's research interests focus on the information about past climates carried by land-derived sediments deposited in the ocean, especially the history of past ice sheets. Larry has spent his academic career in the School of Earth Sciences (and its predecessors) at Ohio State University. Larry is a primary author of Chaps. 2 and 9.

Open Access This chapter is licensed under the terms of the Creative Commons Attribution 4.0 International License (http://creativecommons.org/licenses/by/4.0/), which permits use, sharing, adaptation, distribution and reproduction in any medium or format, as long as you give appropriate credit to the original author(s) and the source, provide a link to the Creative Commons license and indicate if changes were made.

The images or other third party material in this chapter are included in the chapter's Creative Commons license, unless indicated otherwise in a credit line to the material. If material is not included in the chapter's Creative Commons license and your intended use is not permitted by statutory regulation or exceeds the permitted use, you will need to obtain permission directly from the copyright holder.

Climate Models as Tools for Understanding Earth's Climate System

Shuang-Ye Wu

Guiding Question: How does a climate model work?

1 Key Take-Away Points

- Climate models are a numerical representation of the Earth's climate system based on physical laws, most importantly the conservation of mass, energy and momentum, and the ideal gas law.
- Climate models have different levels of complexity, from simple zero-dimension energy balance models to three-dimensional general circulation models with fully coupled atmosphere, ocean, and other components of the climate system.
- Although only the most sophisticated models can simulate the complete state of the climate, simpler models are effective tools frequently used to investigate different aspects of the climate system.
- Climate models can be run with different input conditions to investigate a wide range of scenarios of past, present, and future climate.
- The accuracy of climate models is evaluated by comparing model output with observed climate variables.
- Based on comprehensive evaluation, climate models are shown to be able to simulate well the observed climate conditions. The average of multiple models usually has better overall performance than any individual model.
- Based on the current climate models, the global temperature is likely to increase by 2.2–4.7 °C by the end of this century if we do not dramatically reduce our fossil fuel consumption. In addition, the models project a wide range of changes

---

S.-Y. Wu (✉)
Department of Geology and Environmental Geosciences, University of Dayton, Dayton, OH, USA
e-mail: swu001@udayton.edu

in the climate system, such as increased storminess, decreased ice coverage, and rising sea levels.
- Climate models provide valuable information about likely future climate change, which people can use to make informed decisions on mitigation and adaptation.

2 Introduction

In science, a model is a simplified representation of an object, a process or a system. Scientists have used models to describe and explain natural phenomena, and to make predictions of their future behaviors. Models can be physical, such as a fluvial model built in a laboratory to examine processes of sediment transportation and deposition. They can also be mathematical, using formulas to describe the processes involved in a phenomenon, simulate hypothetical scenarios, and make future predictions. These models are usually written in programming languages and run on computers. Given the sheer scale and complexity of the climate system, **climate models** can only be the latter kind. They are mathematical representations of the Earth's climate system, using numeric equations to describe fundamental physical, chemical and biological processes in the atmosphere, ocean, land surface, and cryosphere that affect our climate.

A frequently asked question about climate models is this: If we cannot reliably predict the weather beyond a week, how can we predict the climate in a hundred years? The answer starts with the distinction between weather and climate. **Weather** is the instantaneous condition of the atmosphere: its temperature, moisture (rain), air pressure, wind, etc. Weather is a chaotic system with limited predictability, because it contains many non-linear relationships, which can lead to a large amount of change even with small modifications of initial conditions. This high sensitivity to initial conditions means that even the slightest deviation of our measurements (which can be affected by instrument precision, location, time coverage, etc.) from the true state of the atmosphere can quickly lead to great divergence between the model predictions and the evolution of the real weather conditions. You likely have experienced this divergence by observing the rapid development of unpredicted local storms, such as afternoon thunderstorms during the summer in the North American midcontinent. **Climate**, on the other hand, is the average weather conditions over an extended period of time. As a rule of thumb in modern climatology, we need at least 30 years of weather data to derive a proper description of climate. Unlike day-to-day weather, climate is governed by a set of large-scale physical laws, most importantly, the distribution and transport of energy we receive from the Sun over the surface of the Earth, as well as many other factors described in Chap. 2. This average condition is more predictable than day-to-day weather, and this is why people go to Florida for spring break, not Alaska, even though there is no guarantee of beautiful warm sunny days every time they go. So, what do climate models predict? The climate models cannot predict the exact weather conditions of a particular day in the future, but they can predict the average conditions of a future period, such as mean values, range of variabilities, and probability of extreme weather events.

Climate models are primary tools to study **climate change,** the change of the long-term mean (i.e., average climate state) on which the short-term internal variabilities of weather are superimposed. They are used to test hypotheses regarding the mechanisms of climate change, to explain past climate changes recorded in the paleo-climate proxy data, and to make short-term and long-term predictions for future climate change caused by both natural variabilities (e.g., El Niño-Southern Oscillation, ENSO) and anthropogenic causes. Climate models provide the scientific basis for making national and international policies regarding climate change, and play a prominent role in the assessment reports of the **Intergovernmental Panel on Climate Change (IPCC)**. IPCC is an intergovernmental body of the United Nations dedicated to providing the world with objective, scientific information on the risks and impacts of anthropogenic climate change. Since its inception in 1988, the IPCC has periodically produced comprehensive **assessment reports (ARs)** on the scientific progress toward our understanding of the climate system and climate change (working group I), its impacts (working group II) and adaptation options (working group III). The most recent is the 6th assessment report (AR6) released in August 2021, and previous reports include the First (FAR) published in 1990, Second (SAR) in 1995, Third (TAR) in 2001, Fourth (AR4) in 2007, and Fifth (AR5) in 2013. In this chapter, we will draw extensive information from these reports, particularly the most recent Fifth and Sixth reports (AR5 and AR6), and you will see these abbreviations used in the following text and some of the figures.

This chapter aims to provide you with a first guide to climate models, focusing on their basic principles and structures (Sect. 3), the major types of climate models (Sect. 4), experiments commonly run on climate models (Sect. 5), methods of model evaluation (Sect. 6), and finally what our existing climate models tell us about future climate change (Sect. 7). Hopefully, by the end of the chapter, you will have a fundamental understanding of how climate models work and why they are such an important tool in studying climate change. This understanding could also help you make an educated assessment of the large amount of information provided by climate models.

3 What Is Inside a Climate Model?

The Earth's climate is primarily determined by a few fundamental properties: (i) how much solar radiation reaches the Earth; (ii) how much of the radiation is reflected by the Earth's atmosphere and surface; (iii) how much radiation is retained by the Earth's atmosphere; and (iv) how much radiation is absorbed by the Earth's surface (land and ocean). (The present-day globally averaged values for these properties were presented in the discussion of Earth's energy budget in Chap. 2.) Many processes could potentially affect these fundamental properties, hence changing the climate. These processes operate on vastly different spatial and temporal scales (Fig. 1). For

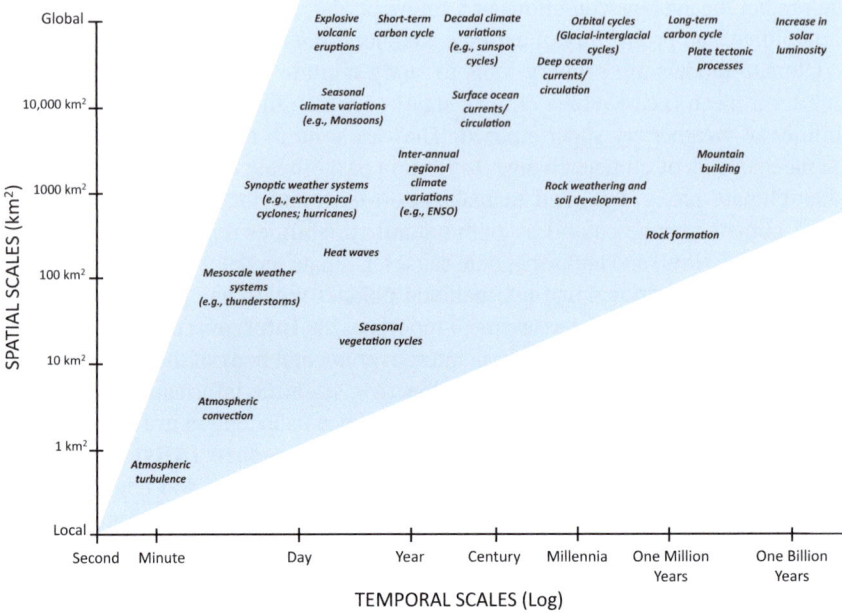

Fig. 1 Spatial and temporal scales of different climate processes. Figure by Kristen St John. Conceptual diagram based on similar representations in: Merritts et al. (1998) and in Commonwealth Science Industrial Research Organization–Commonwealth Science Industrial Research Organization[1]

example, the typical thunderstorm is ~ 20 km in diameter and lasts an average of 30 min. El Niño–Southern Oscillation (ENSO) is a recurring warming (El Niño) and cooling (La Niña) pattern in the eastern equatorial Pacific (see Chap. 3). These events occur on average every 3–7 years, and have global impact on the Earth's climate system. Plate tectonics and chemical weathering have global impacts on the climate, but operate on the time scale of millions to hundreds of million years (see Chaps. 6 and 7). Orbitally controlled changes in incoming solar radiation (insolation) also affect climate, playing a major role in glacial-interglacial cycles over tens to hundreds of thousands of years (see Chap. 9). In addition, these climate system processes interact with each other across different scales to enhance or weaken initial changes via positive or negative feedbacks, many of which have been discussed in previous chapters. A climate model cannot possibly capture all of the processes at work across all of these temporal and spatial scales, and is necessarily a simplified representation of the Earth's climate system with only a selected number of processes represented at different levels of sophistication. Therefore, a great number of climate models exist,

---

[1] Conceptual diagram based on similar representations in: Merritts et al. (1998); and Commonwealth Science Industrial Research Organization (accessed, 2023, May 26). *Climate Change Australia*. Commonwealth Science Industrial Research Organization. https://www.climatechangeinaustralia. gov.au/en/learning-support/climate-models/theory-and-physics/.

and they differ in terms of what processes are included and how these processes are represented in the model. The selection of processes and their representations is often guided by what scientific questions a specific model is designed to investigate. As a result, all models have their strengths and weaknesses, and there is not a single "right" or "wrong" model. With continued improvements in our understanding of the climate system and in computing power, climate models are constantly being improved by their ability to incorporate more processes with more details at higher spatial and temporal resolutions.

3.1 Input and Output

The purpose of climate models is to link climate forcings to climate states through basic laws of physics. **Climate forcings** are factors external to the climate system that affect the energy balance, hence the climate of the Earth, while a **climate state** is the characteristics of the climate at a particular time. Over the short time scales (years to millennia), the main climate forcings include solar radiation, atmospheric concentration of greenhouse gases (e.g., carbon dioxide (CO_2) and methane (CH_4)), atmospheric aerosols, volcanic eruptions, and land cover (e.g., vegetation and ice). Some of these forcings (e.g., solar radiation and volcanic eruptions) have their natural variabilities, and others are affected by human activities (e.g., CO_2 from burning of fossil fuels, aerosols from forest fires and land use change). The quantitative estimates of these forcings serve as the input to climate models. Data from direct observations of these forcings are often used as input for simulations of the recent past (such as the time periods described in Chap. 3). Over longer time periods (tens of thousands to hundreds of millions of years), other climate forcings also become important. These include rates of seafloor spreading and subduction (Chap. 6), rock formation and weathering (Chap. 7), and orbital cycles (Chap. 9). These additional relevant forcings are estimated from climate proxy data for simulations of past climate at longer time scales (such as the time periods described in Chaps. 4–9). For predictions of future climate, some of these forcings are estimated based on their past trend and variabilities. Forcings involving human activities are typically presented in a range of future scenarios based on different estimates of changes in population, technology, energy, and land use in the future. The input data could also be a set of hypothetical forcing conditions for models to simulate a wide range of "what-if" situations.

Different types of climate models produce different outputs. The most sophisticated climate models generate a comprehensive numeric description of the Earth's climate with an extensive list of characteristics, which can then be summarized at different time intervals, e.g., hourly, daily, and monthly, over the periods of simulation. The simulation period varies from decades to tens of thousands of years at various points in time (past, present, and future). The model outputs provide a detailed picture of the atmosphere (e.g., temperature, humidity, precipitation, atmospheric pressure, wind speed and direction, cloud cover), the ocean (e.g., temperature, salinity, acidity, and water velocity), the cryosphere (e.g., snow cover, sea ice, extent of glaciers and ice sheets), and the land surface (e.g., soil moisture and surface runoff). Based on these outputs, we can conduct further analyses on various aspects of the climate change and their potential impacts on natural environment and human society. For example, using the daily precipitation output for the last 30 years, we can establish the future changes in average precipitation amount, frequency, intensity, as well as changes in seasonal distribution, spatial variation, extreme storms, and storm tracks. Such information can be combined with other models (e.g., economic, ecological, and epidemiological) to further assess the diverse socio-economic and ecological impacts of precipitation change, such as the impact of extreme storms on infrastructure, changes of precipitation seasonality on agricultural production, and spatial changes of precipitation on vegetation, species range and the spread of vector-borne diseases.

One important analysis performed by most climate models is to estimate **climate sensitivity**, which is the amount of climate change (usually change in temperature) in response to changes in a particular forcing. The climate system usually takes time to respond to changes in climate forcings (see Chap. 2 for information about the response times of the "spheres" in Earth's climate system). Therefore, climate sensitivity refers to the change (e.g., in temperature) when the climate system has reached the new steady state (i.e., **equilibrium**). In the case of increasing greenhouse gas concentrations, in addition to the direct warming from the resulting change in energy balance, the climate model contains numeric representations that describe various physical/chemical/biological processes that may amplify or dampen the initial temperature change, such as water vapor feedback, ice-albedo feedback, and cloud feedback described in Chap. 2. Climate models take into account these feedbacks to calculate the final equilibrium temperature change due to a specific change in greenhouse gas concentrations. The equilibrium temperature will be different for different changes in the greenhouse gas concentrations.

3.2 *Basic Physical Laws*

Climate models simulate climate states by solving a set of fundamental equations derived through basic laws of physics. The most important laws are ones you may have learned in an introductory physics course, including the following:

a. **The conservation of mass** states that the mass of a closed system[2] is neither created nor destroyed, but remains constant over time.
b. **The conservation of energy** states that the energy of a closed system is neither created nor destroyed, but can change from one form to another.
c. **The conservation of momentum**: Momentum of an object is defined as the product of its mass and velocity. This law states that the total momentum of a closed system is neither created nor destroyed, but can change upon the action

---

[2] A closed system does not exchange matter with its external surroundings.

of forces as described by Newton's laws of motion. As momentum is a vector quantity with both magnitude and direction, it is conserved in all three directions (x, y, z).

d. **The ideal gas law** states the temperature (T), pressure (P), volume (V), and amount of gas (as a number of moles, n) are related in the following way (R is the ideal gas constant with the value of 8.314 J/K mol)[3]:

$$PV = nRT$$

Many aspects of the climate system can be interpreted from a small number of variables and their changes through time. The most important variables include temperature, wind speed and direction, atmospheric pressure, and atmosphere composition (such as greenhouse gas concentrations and atmospheric water content). Other variables can be derived from these basic variables. For example, precipitation can be derived from atmospheric water content and temperature. Indices for climate patterns, such as regional monsoons and El Niño, can be derived from wind and atmospheric pressure. In order to calculate these basic variables, equations are set up based on the basic laws of physics listed above. A common approach to modeling a complex system is to reduce it into boxes linked by flows between boxes. How we define our boxes depends on what we want to model at which level of detail. This gives rise to a hierarchy of climate models of various complexity so that we can explore different aspects of the climate system. At the simple end of the complexity spectrum, the entire Earth is treated as a single box in the case of the **zero-dimension energy balance models (EBMs)**, which is presented in Box 1. Toward the complex end of the complexity spectrum is the **three-dimensional coupled atmosphere–ocean circulation models (GCM)**, where the Earth is divided into a three-dimensional grid covering atmosphere, ocean, and land (Fig. 2). The set of important equations then has to be developed and solved for each box within that 3D grid. The climate models used in IPCC assessment reports are three-dimensional GCMs.

In any case, each box in the grid is assumed to be internally homogeneous with a uniform value for each property (e.g., temperature, water vapor content, atmospheric CO_2 concentration), but all boxes together represent the spatially varying state of the global climate. The conservation of mass, energy, and momentum (the basic physical laws mentioned above) dictate that nothing comes out of the "thin air." This means any change in a property of a box over time is the net change in the quantity of that specific property caused by either internal processes (those that occur within the box) or external processes (those that occur outside the box). This concept is called **continuity**. In this discussion of box models, "**production**" and "**destruction**" refer to internal processes that increase or decrease the quantity of a property; "**input**" and "**output**" refer to external processes that move a certain quantity of a property into and out of the box (Fig. 3). When this concept is expressed mathematically, it is called a **continuity equation**, which can be formulated for each of the basic properties or components (variables) of the climate system, including mass (mass

---

[3] Joules per kelvin per mole is a unit of energy per temperature increment per amount of substance.

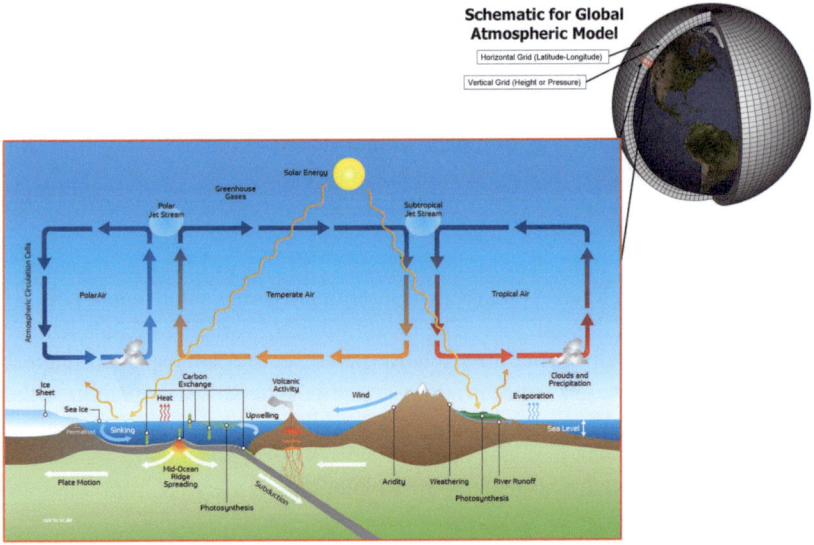

Fig. 2 Grid system for global climate models. *Source* Modified from Wikimedia Commons and inset figure from Koppers and Coggon (2020)[4]

continuity equation), momentum (equations of motion for each of the three dimensions), energy (the thermodynamic equation), and chemical components (e.g., water vapor and greenhouse gases). These equations are then solved for each grid box at each time step. Solving these equations often requires many calculations because highly complex equations such as those used in climate models usually do not have exact solutions, and the solutions can only be estimated through iterative numeric approximations. As a result, climate modeling often requires massive computational power and time.

For each climate variable, specific processes are involved in causing internal and external changes. Table 1 summarizes some examples of these processes. Figure 3b shows an example for water vapor content. The amount of water vapor in a grid box of atmosphere can be changed through several processes. Internally, if the grid box contains liquid water such as clouds (made of droplets of liquid water) or is located at the boundary with surface water (rivers, lakes, ocean, etc.), evaporation of such liquid water can add water vapor within the box. In this way, water vapor is "produced" through evaporation. On the other hand, when water vapor condenses to form liquid water (as clouds or precipitation), the amount of water vapor is decreased,

---

[4] Modified from Wikimedia Commons. 2012. https://commons.wikimedia.org/wiki/File:Global_Climate_Model.png. Public Domain, which was based on a figure from NOAA, . Inset diagram of climate system from Koppers, A., and Coggon (Editors; 2020). Exploring Earth by Scientific Ocean Drilling: 2050 Science Framework. Integrated Ocean Discovery Program, 132 pages, .

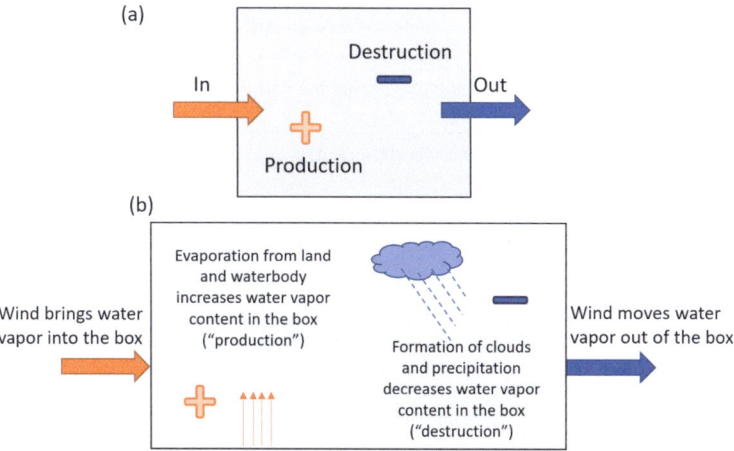

Fig. 3 Box model and the conceptual continuity equation (**a**). An example of water vapor content in a box model (**b**)

or "destroyed",[5] in the box. The water vapor content in the box can also be changed through external processes. Wind (the horizontal movement of the air) and convection (the vertical movement of the air) can bring extra water vapor from neighboring boxes into the target box (input), or move water vapor out of the target box (output). Consequently, the change of water vapor content in the box is the net effect of changes in the quantity of internal production and destruction and external input and output of water vapor. Another example is temperature, which is an indicator of total thermal energy (heat) in a grid box. Temperature within a grid box can change because of internal processes such as the absorption or release of latent heat during evaporation or condensation of water, decay of radioactive elements, chemical reactions, and radiative cooling and absorption within the box. Temperature change also occurs as a result of thermal energy exchange with neighboring grid boxes through various processes. For example, winds, convection, and turbulence can move warmer or cooler air into the box, hence changing the heat content of the box. Thermal energy can also be transferred directly through conduction and radiation. The movement of water (both vapor and liquid) also brings in or carries out additional thermal energy as latent heat, which is then released or absorbed through phase changes of water. The continuity equation for a specific variable (e.g., water vapor or temperature) thus incorporates the mathematical equations describing these processes. What processes are included and at what level of detail in these continuity equations thus determine

---

[5] The "destroyed" water vapor becomes liquid water, and adds to the total amount of liquid water, so that the overall mass of water in all forms is still conserved. Mass is moved from one form to another, but no water is really destroyed (or totally disappears).

Table 1 Examples of processes that change variable properties through production/destruction within a grid box and through input to/output from a grid box

Variable	Production/destruction within a box	Input to/output from an adjacent box
Mass	0 (mass is conserved)	Wind, currents, convection, turbulence
Water vapor	Condensation, evaporation, deposition to surface	Wind, convection, turbulence
Clouds (liquid water)	Condensation, evaporation, precipitation to surface	Wind, convection, turbulence
Atmospheric gas (CO_2, ozone, other pollutants, etc.)	Natural and anthropogenic emissions, chemical reactions, absorption by clouds and precipitation	Wind, convection, turbulence
Atmospheric aerosols	Volcanism, desert dust, anthropogenic emissions, deposition to surface	Wind, convection, turbulence
Temperature (thermal energy)	Latent heat, chemical reactions, radioactive decay, radiative cooling, radiative absorption	Wind, convection, turbulence, conduction, radiation, precipitation, etc.

the level of complexity and hence performance of climate models and the appropriate use of their output.

3.3 Climate Model Resolution

In a climate model where a three-dimensional grid is set up to represent the Earth's climate system, each grid box is often referred to as a cell. Although generally they mean the same thing, the term "box" is often used to describe conceptual models or models where the climate system is divided into very few boxes (e.g., four latitudinal boxes), whereas "cell" is used more often to describe the technical aspects of climate models, particularly those with three-dimensional grids. **Spatial resolution** refers to the horizontal and/or vertical size of a cell, and **temporal resolution** is the size of the time interval (i.e., time step) at which calculations are repeated. Spatial and temporal resolution are closely related. Higher spatial resolution also requires higher temporal resolution. Typical horizontal spatial resolution of present global climate models ranges from 1.5° to 3° of latitude and longitude, or around 150–300 km in the mid-latitudes. The spherical shape of the Earth means that grid cells based on longitude and latitude are not equally sized. They are the largest at the equator and get progressively smaller toward the poles. Alternative grid systems are sometimes used in recent climate models to avoid this problem. The grid systems also extend upwards in the atmosphere and downwards in the ocean (and soil in some models) with multiple vertical levels. Since the atmosphere and the ocean are so much thinner compared with the horizontal size of the Earth, the vertical spacing of these layers is much smaller than the horizontal spacing of the grid cells. Moreover, the vertical subdivisions are usually not at a uniform interval. The layers tend to be thinner near the Earth's surface, where vertical profiles of the atmosphere and the ocean show more rapid changes. For most recent models, the typical number of vertical layers is about 30–40 for the atmosphere and 30–60 for the ocean.

High-resolution models have more and smaller grid cells, and they can produce more detailed climate information, but they require higher computational power and longer computation time. Figure 4 shows the gradual increase of spatial resolution of climate models between the first and fourth IPCC assessment reports. The improvement resolves more details of the climate system, such as the topography of the land surface. Even though we want to resolve as much spatial detail as we can with the models, such improvement comes at high computational costs. For a relatively simple atmospheric model of 2.5° latitude–longitude resolution and 18 vertical levels, the atmosphere is divided into 128 (longitudinal) × 64 (latitudinal) × 18 (vertical) = ~ 150,000 grid cells. With a time step of 1 h, there are 24 × 365 = ~ 10,000 time steps per year. At each grid cell and each time step, there may be 10 variables, each with about 100 calculations. So, the simulation of climate for one year involves about 1.5 trillion calculations, which takes about 3 h on a 1 GHz computer. This is only an extremely *simple* atmosphere model, which just calculates changes in the atmosphere with a set of prescribed (thus fixed, non-changing) land and ocean conditions. The climate models typically used for climate research, e.g., those used for IPCC reports, are much more complex and run for much longer times. Typically, increasing the spatial resolution of a model by a factor of two requires around 10 times the increase in either the computing power or computing time. The computational intensity of sophisticated climate models such as GCMs means that they are usually run on very powerful supercomputers (Fig. 5).

Fig. 4 Increasing spatial resolutions of climate models used in the first through fourth IPCC assessment reports. Spatial resolution has continued to increase in subsequent assessment reports as well. Figure used with permission from the IPCC. Full IPCC caption for this figure can be found in the footnote *Source* IPCC.[6] This figure is excluded from Creative Commons license

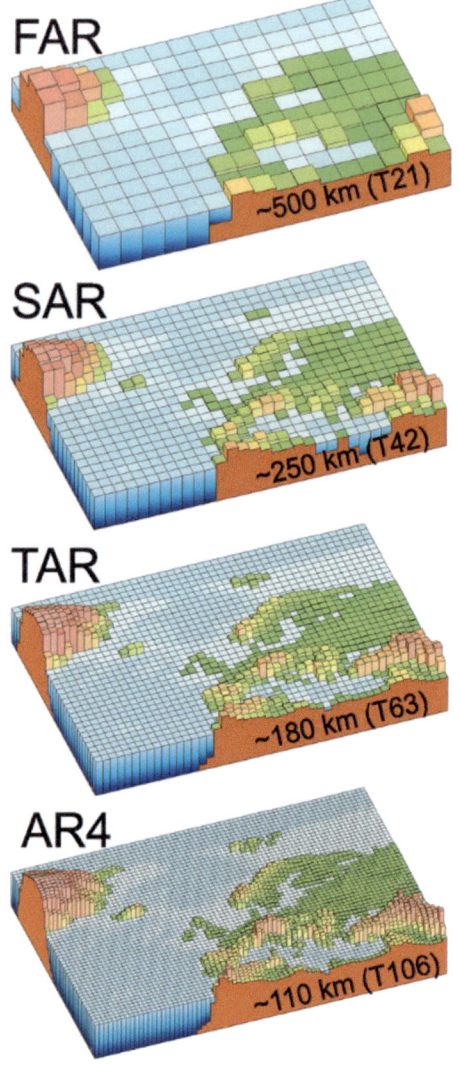

---

[6] From IPCC (2007). Figures 1–4. https://archive.ipcc.ch/publications_and_data/ar4/wg1/en/figure-1-4.html. Full caption from IPCC figure: Geographic resolution characteristic of the generations of climate models used in the IPCC Assessment Reports: FAR (IPCC, 1990), SAR (IPCC, 1996), TAR (IPCC, 2001a), and AR4 (2007). The figures above show how successive generations of these global models increasingly resolved northern Europe. These illustrations are representative of the most detailed horizontal resolution used for short-term climate simulations. The century-long simulations cited in IPCC Assessment Reports after the FAR were typically run with the previous generation's resolution. Vertical resolution in both atmosphere and ocean models is not shown, but it has increased comparably with the horizontal resolution, beginning typically with a single-layer

Fig. 5 National Oceanic and Atmospheric Administration's GAEA climate research supercomputer housed at Oak Ridge National Laboratory. *Source* NOAA[7]

3.4 Sub-grid Processes and Parameterization

Dividing the climate system into discrete boxes means that we can only obtain an averaged description of the real atmospheric conditions within a box (grid cell). The cell size is largely determined by feasible computational cost. However, many processes happen at scales smaller than the size of the cell. These processes cannot be directly described by equations based on laws of physics, but they can have a profound impact on the calculated climate variables. For example, the convective activities that form storm clouds often occur at the scale of 10 km or smaller, a process that cannot be resolved by most climate models. However, the formation of clouds profoundly changes the energy budget of the grid cell by releasing latent heat, reflecting incoming solar radiation and blocking the outgoing infrared radiation. In other situations, some processes in the climate system are not fully understood, making full mathematical representation impossible. Therefore, a reliable climate model has to account for these processes through approximation, a process called **"parameterization."** Such parameterizations aim to estimate the average or expected effect of unresolved variables on the resolved variables. One way to parameterize a process is to use the statistical relationship derived from empirical observations in the past. For example, we can first establish a statistical relationship between cloud

---

slab ocean and ten atmospheric layers in the FAR and progressing to about thirty levels in both atmosphere and ocean.

[7] National Oceanic and Atmospheric Administration (NOAA) (accessed 2023, May 25). GAEA. NOAA. https://www.noaa.gov/organization/information-technology/gaea.

cover and average temperature and humidity. Since temperature and humidity can be resolved (i.e., calculated directly) by the climate model, we then can use this established statistical model and the model outputs of temperature and humidity in the grid cells to calculate average cloud cover in the grid, even if we cannot simulate each cloud. Figure 6 provides a sample of important processes that are typically parameterized in global climate models. Parameterization is the greatest source of uncertainties in present climate models.

Fig. 6 Climate processes and properties that are typically parameterized within global climate models. Figure used with permission of The UCAR Comet Program. *Source* UCAR.[8] This figure is excluded from Creative Commons license

---

[8] The Comet Program. (n.d., accessed, 2023, May 25). UCAR. https://www.meted.ucar.edu/nwp/climate_models/media_gallery.php.

Box 1 Zero-dimension energy balance model

To help us begin to understand what happens inside a climate model, we are now going to examine some details of the simplest climate model, which is referred to as the "zero-dimension energy balance model." It is called "zero-dimension" because the Earth is treated essentially as a single point or box in space—it has no neighbors to interact with in any direction. As a result, any processes that occur on the Earth (i.e., within the box) are not considered. Fundamentally, this model lets us consider a simple question: given the amount of solar radiation received by the Earth, what should the Earth's average temperature be? Even though this model does not reproduce climate realistically, considering it still illustrates for us some basic principles of climate modeling.

The principle of energy balance indicates that, for the Earth's surface temperature to stay constant, the energy input to the Earth over a period of time has to be equal to the energy output from the Earth over that same time. (This is analogous to the situation needed to keep a constant balance in your bank account—the rate at which money comes in has to be the same as the rate at which money is spent). Among climate scientists, a common way to write this energy balance mathematically is

$$Q_a = Q_e \tag{1}$$

where Q_a is the total amount of solar energy absorbed by the Earth and Q_e is the total amount of energy emitted from the Earth into space (see Table 2 for variable definitions). Now let us use some principles of physics in order to calculate Q_a and Q_e, and then to calculate the average temperature of the Earth based on the amount of energy Earth emits to space. The calculation process is illustrated in Fig. 7, and explained in detail as follows.

Table 2 Summary of variables used in this box

Variable	Description	Unit
Q	Energy flow per unit time	W (Watts)
Q_a	Total energy absorbed by the Earth per unit time	W
Q_e	Total energy emitted by the Earth per unit time	W
F	Energy flux (energy flow per unit time per unit area)	W m^{-2} (Watts per square meter)
S_0	Solar energy flux that reaches the Earth, also referred to as the solar constant with the value of 1366 J m^{-2} s^{-1}, which is the same as 1366 W m^{-2}	W m^{-2}
F_e	Energy flux emitted by the Earth	W m^{-2}
A	Earth's planetary albedo (0.3)	Dimensionless
R	Earth's radius	m (meter)
T	Temperature	K (Kelvin)
T_e	Earth's effective radiating temperature	K
σ	Stefan–Boltzmann constant with a numerical value of 5.67×10^{-8} W m^{-2} K^{-4}	W m^{-2} K^{-4}

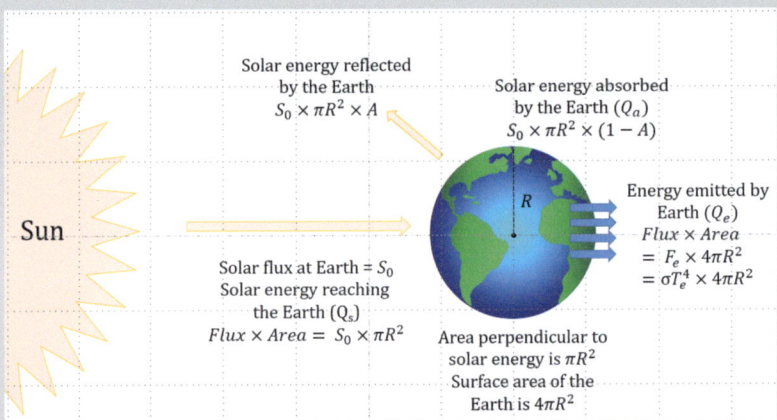

Fig. 7 Schematic diagram for the principle of energy balance

Total amount of energy emitted/received (Q) is calculated as the energy intensity multiplied by the effective area emitting/receiving such energy. Energy intensity is measured by energy flux (F), i.e., the amount of energy that passes through each unit of Earth's surface area oriented perpendicular to flow direction per unit time. We can express this mathematically as follows:

$$Q = \text{Flux} \times \text{Area} \tag{2}$$

According to the Stefan–Boltzmann law, the energy flux emitted by a black body is related to the fourth power of the body's absolute temperature as follows:

$$F = \sigma T^4 \tag{3}$$

where T is the temperature in kelvins and σ is the Stefan–Boltzmann constant with a numerical value of 5.67×10^{-8} W m^{-2} K^{-4}. A black body is something that emits and absorbs electromagnetic radiation with 100% efficiency at all wavelengths. Although a blackbody does not really exist, the celestial bodies such as the Earth and the Sun are often modeled as blackbodies. Therefore, the total energy emitted by the Earth can be calculated as the product of the energy flux based on the Earth's temperature and the Earth's total surface area. From geometry class, you may recall that the surface area of a sphere can be calculated using the formula $4\pi R^2$, where R is the sphere's radius. Therefore, the total energy emitted by Earth can be written mathematically as follows:

$$Q_e = \sigma T_e^4 \times (4\pi R^2) \tag{4}$$

where T_e is the Earth's temperature in kelvins,[9] and R is the Earth's radius in meters.

The energy flux from the Sun can be calculated from its surface temperature based on Eq. (3). The solar flux decreases with distance following an inverse-square relationship and is about 1366 W m^{-2} when it reaches the Earth's orbit. This number is often referred to as the solar constant and denoted as S_0. Determining the surface area affected at any one time by that flux is a bit more complicated than just using the surface area of the Earth. The surface area receiving that solar energy flux at any one time is limited to one half of the Earth, and the surface area effectively receiving solar radiation is limited even more because very little of Earth's surface is oriented completely perpendicular to that incoming solar radiation. (If you want to prove this to yourself, think about how often the Sun is positioned directly overhead where you live. All of the solar radiation is arriving perpendicular to the Earth's surface at that spot only when the Sun is directly overhead.). Therefore, the effective area receiving solar energy perpendicular to its flow direction is only the north–south cross-section of the Earth (i.e., the circle when the Earth is sliced with

---

[9] Zero (0) Kelvin is -273 °C and is -460 °F. The following websites are useful for temperature unit conversion (https://www.convert-me.com/en/convert/temperature/) and temperature increment unit conversion (https://www.convert-me.com/en/convert/temperature-inc/).

the plane of the cross-section passing through the center of the Earth), and the area of this circle is calculated as πR^2 (where R is the Earth's radius). Based on this, total incoming solar radiation can be written as:

$$Q_s = S_0 \times \pi R^2 \tag{5}$$

Not all incoming solar radiation is absorbed by the Earth, however. Some is reflected by clouds, ice sheets, snow cover, and other surfaces. The reflectivity of a surface is called its albedo, defined as the proportion of the incoming radiation that is reflected by the surface. A dark surface such as asphalt (0.05–0.2) has a low albedo, whereas a light surface like fresh snow (0.8–0.9) has a high albedo. Without describing all the details about the Earth's surfaces, we estimate an overall albedo for the Earth (A) at about 30%, and use it as a parameter of our model. Therefore, the total energy absorbed (i.e., not reflected) by the Earth is calculated as:

$$Q_a = Q_s(1 - A) = S_0 \times \pi R^2 (1 - A) \tag{6}$$

Equating the incoming energy (Q_a) with outgoing energy (Q_e), we get

$$S_0 \times \pi R^2 (1 - A) = \sigma T_e^4 \times (4\pi R^2) \tag{7}$$

Rearranging the equation, we get

$$T_e = \sqrt[4]{\frac{S_0(1-A)}{4\sigma}} \tag{8}$$

From the known values of S_0 (1366 W m^{-2}), A (0.3), and σ (5.67 × 10^{-8} W m^{-2} K^{-4}), we get

$$T_e \approx 255\,\text{K}$$

The result of this calculation represents the effective radiating temperature of the Earth, i.e., the temperature of a "black body" that would emit the same total amount of electromagnetic radiation as the Earth. The temperature of 255 K (− 18 °C, or ∼ 0 °F) is much colder than the Earth's actual average surface temperature of 288 K (or 15 °C, or ∼ 59 °F); the result of the model calculation is, however, very close to the temperature observed at the *top* of the atmosphere. This difference between the calculated temperature and the actual surface temperature (a difference of ∼ 33 K, or ∼ 33°C or ∼ 59 °F) occurs because we did not include the Earth's atmosphere, and particularly its greenhouse effect, in our model.

3.5 An Example of Climate Models—Energy Balance Model from Simple to Complex

The simplest type of climate model is the zero-dimension energy balance model. This model treats the entire Earth as a single homogenous box and determines its temperature based on the principle of energy balance: the amount of energy absorbed by the Earth is equal to the amount of energy the Earth radiates into space. Given the amount of solar energy that reaches the Earth and the Earth's average albedo, the model calculates the Earth's temperature to be 255 K ($-$ 18 °C, or ~ 0 °F). The details of this calculation are given in Box 1. This result, referred to as the Earth's **effective radiating temperature**, is much colder than the Earth's actual average surface temperature of 288 K (or 15 °C, or ~ 59 °F). The difference between this calculated temperature and the actual surface temperature (a difference of ~ 33 K, or ~ 33 °C or ~ 59 °F) occurs because the Earth's atmosphere, and particularly its greenhouse effect, is not included in the model.

The energy balance model can be extended to account for the greenhouse effect, however. This modification involves assuming that the Earth has a simple, one-layer homogenous atmosphere, which allows all incoming solar radiation to pass through it, but absorbs all infrared radiation from the Earth and reradiates that infrared radiation both into space and back to the Earth's surface. This is the "greenhouse effect." The model also assumes that the atmosphere behaves like a black body, has the Earth's albedo value, and has a different temperature than Earth's surface. This model leads to a mathematical situation with two energy balance equations (one for the Earth and one for the Earth's atmosphere) and two unknown variables (the temperatures of the atmosphere and the Earth's surface). Solving this modified model predicts that the Earth's atmosphere causes 47 K (or 47 °C or ~ 85 °F) of a natural greenhouse effect, which is close to—although somewhat larger than—the Earth's actual greenhouse effect.

While useful for demonstrating basic Earth system energy input and output relationships, this one-layer atmosphere energy balance model is not entirely realistic. In reality, the atmosphere does not absorb all of the Earth's outgoing infrared radiation. Radiation from the Earth's surface at certain wavelengths may pass through the atmosphere and escape to space. This will reduce the amount of the greenhouse effect due to the atmosphere. In addition, the atmosphere is not homogeneous; a more accurate model would further divide the atmosphere into many different layers and consider the energy balance for each layer. Despite the simplifications, however, this model does illustrate the basic nature of the greenhouse effect: when the atmosphere absorbs infrared radiation from the Earth's surface and reemits it that radiation warms Earth's surface to a temperature higher than the planet's effective radiating temperature, i.e., what the surface temperature would be without an atmosphere. This example shows that simple energy balance models are useful for investigating a wide range of questions regarding the planetary climate system as a whole, even though they are not meant to reproduce realistic climate states. The questions that can be examined with zero-dimension energy balance models include the magnitude of planetary climate sensitivity to feedbacks due to changes in ice and albedo, atmospheric water vapor, and the abundance of clouds.

4 What Kind of Climate Models Are There?

A wide array of climate models of varying complexity exists for different purposes. Complexity of climate models may vary based on their dimensionality, resolution, levels of parameterization and integration. The simplest form of climate model is the **zero-dimension energy balance model (EBM)** that we discussed above and in Box 1. This model treats the Earth as a single point and determines its surface temperature based on the energy balance of the Earth. This can be expanded horizontally to form a **one-dimensional EBM**, which uses simplified relationships to calculate the average temperature at each latitude band based on energy balance at each of these bands because of the latitudinal (also called zonal) distribution of energy. EBMs are useful for determining the effect of changes in solar radiation or the Earth's surface albedo on average Earth temperature.

Alternatively, the energy balance model can be refined vertically to form **one-dimensional radiative–convective models (RCMs)**. These models calculate the vertical profile of the global average temperature by incorporating the upward and downward energy transfer through layers of the atmosphere by radiation and convection. Radiation is the process of energy transfer through electromagnetic waves or particles. Convective energy transfer occurs when warm (hence less dense) fluids (e.g., air) spontaneously rise and cold fluids sink under the influence of gravity. For example, if you boil a pot of water on a stove burner, heat energy transfers from burner to the pot through radiative transfer, and from the bottom to the top of the water through convective transfer. The one-dimensional RCMs can determine the effects of varying greenhouse gas concentrations on the effectiveness of the Earth's ability to emit infrared radiation and hence the effect on surface temperature. They are also used to examine, for example, the effect of the ice-albedo feedback on global climate sensitivity.

The **two-dimensional statistical dynamical models (SDMs)** combine the zonal energy balance models with the vertical radiative-convective models. They incorporate zonally averaged energy and dynamics with a vertically resolved atmosphere. In addition, the SDMs simulate equator-to-pole energy transfer based on both theoretical and empirical relationships. Large-scale dynamic processes are parameterized in the model based on statistical relations, while energy diffusion is calculated based on the laws of motion as in an EBM. This combination gives this group of models the name "**statistical–dynamical.**" They are particularly useful for examining horizontal energy transfer.

The **general circulation models (GCMs)** fully represent physical processes in the atmosphere and ocean in three dimensions. They can simulate the atmosphere and the ocean separately in atmosphere general circulation models (AGCMs) or ocean

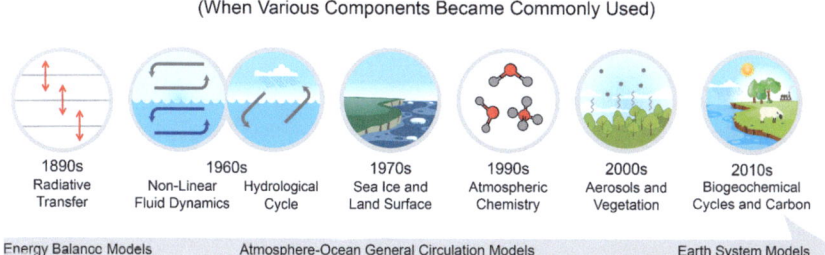

Fig. 8 Evolution of climate modeling. From USGCRP[10]

general circulation models (OGCMs), which can be combined into **coupled atmosphere–ocean general circulation models (AOGCMs)**. AOGCMs are the most sophisticated climate models available, and aim to incorporate as many processes as possible in order to produce a realistic picture of the evolution of the climate system in three dimensions. In addition to oceanic and atmospheric dynamics, newer generations of GCMs also include additional physical processes that affect the climate system such as sea ice and ice sheet dynamics, dust/sea spray, and sulfate aerosols. In particular, one group of models incorporates biogeochemical cycling in terrestrial and marine ecosystems and their feedbacks on the circulation. These models are often called the **Earth system models (ESMs)** (Fig. 8). As a special branch of GCMs, ESMs are becoming increasingly popular for examining ecological questions, such as how climate change will affect the seasonal timing of phytoplankton blooms, the geographic ranges of species, and the boundaries of terrestrial and marine biomes.

Even though the comprehensive GCMs (including ESMs) represent the most sophisticated climate modeling, that does not mean simpler models such as EBMs and RCMs are of no value. In fact, there is a large class of simpler, but valuable, climate models referred to as **Earth models of intermediate complexity (EMICs)**. These models are primarily used to investigate specific aspects of the Earth's climate systems at reduced computational cost. In these models, one or more aspects of the full climate system are either held constant or highly simplified so that we can examine specific processes at higher temporal and/or spatial resolutions or over long timescales that are not otherwise possible with the available computing resources. For example, some models (e.g., the CLIMBER models) have a highly simplified (e.g., parameterized) atmosphere in order to incorporate comprehensive ocean, sea ice and/or biogeochemistry models at reasonable computation time. Such models do not resolve any atmospheric features, such as daily weather patterns and regional atmospheric circulations, but offer performance in other components, such as temperature, humidity, and pressure, that is comparable to the outputs of more complex models.

---

[10] USGCRP (2017). Figure 4.3. https://science2017.globalchange.gov/. Public domain.

5 What Kind of Simulations Are Run on Climate Models?

Once a comprehensive climate model is built, it becomes a laboratory for scientists to run a great variety of experiments. Climate models are used to simulate past, present, and future climate changes, and to explore different "what-if" scenarios to determine the responses of climate systems to anthropogenic changes in climate forcings such as greenhouse gas emission, land surface changes, and geoengineering. Here we give a brief introduction to the most common experiments run by all major climate modeling groups.

5.1 Common Experiments of Climate Models

5.1.1 Pre-industrial Control Runs

For the pre-industrial control run, the climate model is run to simulate climate evolution for hundreds or thousands of years with fixed (non-evolving) pre-industrial conditions, including prescribed solar radiation, atmospheric concentrations or emissions of gases and aerosols, and unperturbed land surface. Control runs are often used to examine how well the model simulates natural variability of the climate system. The control runs are also used to identify "**model drift**," an artificial long-term trend sometimes generated by models even without changes in climate forcings. Model drift is problematic because it makes predictions less accurate as time progresses, and needs to be corrected.

5.1.2 Historical Runs

Historical runs simulate the climate of the recent past (since 1850) with observed climate forcings during this period, including solar forcing, atmospheric composition (including greenhouse gases) due to both anthropogenic and natural (e.g., volcanic) influences, emissions and concentrations of short-lived gases, natural and anthropogenic aerosols, and evolving land use. Historical runs are essential for model evaluation (see Sect. 6 for more details). A comparison between the model's "prediction" of the past climate (hindcast) and the observed climate record is the most direct way to evaluate the performance of a climate model.

Historical runs can also help us to determine the relative contributions of different climate forcings, in particular the natural versus human contributions. This is often referred to as climate change **attribution**, a process to scientifically establish the causal links between observed climate change and its many underlying mechanisms. Figure 9 shows the comparison of observed global mean temperature change from observed data (black line) to climate model historical simulations forced by natural forcings only (green) and those forced by both human and natural forcings (orange).

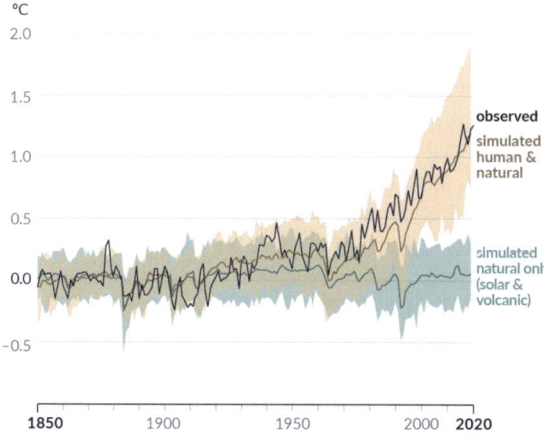

Fig. 9 Comparison of observed global mean temperature change (black line) to CMIP6 climate model historical experiments using anthropogenic and natural forcings combined (orange line), or natural forcings only (green line). The shadings indicate model spread (5%-95%). Figure used with permission from the IPCC. Full IPCC caption for this figure can be found in the footnote. *Source*: IPCC (2021).[11] This figure is excluded from Creative Commons license

The shadings (orange and green) represent the range of global temperature simulated by different models, whereas the solid lines (orange and green) represent the average of all models. In the "natural-forcing only" scenario (green line and shading), model simulations are forced by naturally evolving climate forcing conditions such as solar radiation and volcanic activity, while assuming constant pre-industrial levels for anthropogenic forcings such as greenhouse gas emissions and land use patterns. In the "human and natural forcings" scenario (orange line and shading), models are run with both natural and anthropogenic climate forcings, such as human emissions of greenhouse gases, land use change, anthropogenic aerosols, and their combined effects. The results show that natural forcing alone cannot account for the changes in global average temperature in the last several decades. Only when human contributions are also incorporated as climate forcings can we simulate the observed temperature change, confirming the causal links between human activities and the observed global warming.

---

[11] From: Figure SPM.1b in IPCC (2021a). Full caption from original IPCC figure: Changes in global surface temperature over the past 170 years (black line) relative to 1850–1900 and annually averaged, compared to Coupled Model Intercomparison Project Phase 6 (CMIP6) climate model simulations (see Box SPM.1) of the temperature response to both human and natural drivers (brown) and to only natural drivers (solar and volcanic activity, green). Solid colored lines show the multi-model average, and colored shades show the *very likely* range of simulations.

5.1.3 Future Projections

Future projections are model simulations of future climate based on estimated future conditions of climate forcings. Natural forcings are usually estimated based on their past behavior such as variability, trend, and cyclicity. However, human activities are much harder to predict, and the human contribution to greenhouse gas emissions is the greatest source of uncertainty for future predictions. Therefore, instead of prescribing one set of conditions, the common strategy is to develop a number of probable cases for the future, called **emission scenarios**, ranging from low to high, based on probable conditions of socio-economic development in the future, such as population, trade, total wealth, wealth distribution, energy use intensity, and energy sources. Each scenario provides a set of future conditions based on spatially explicit greenhouse gas emissions and land use changes. These conditions are then used as input to run climate models to establish future climate under each of the future emission scenarios. Some of the most important results from these future climate simulations are presented later in Sect. 7.

Each IPCC assessment report has specified its own set of future emission scenarios as the forcing input of all climate models, based on the most up-to-date research on possible future socio-economic changes. The most recent sixth report (AR6) focused on a set of future scenarios developed from a combination of two approaches. The first approach, known as the **Representative Concentration Pathways (RCPs)**, prescribes different levels of total increase of radiative forcing from greenhouse gasses and other sources for the year 2100; those increases are 1.9, 2.6, 6.0, 7.0, and 8.5 W/m^2. The other approach, termed the **Shared Socioeconomic Pathways (SSPs)**, is scenarios of future socioeconomic change, from which greenhouse gas emissions can be derived. There are five SSPs, and they are sustainability (SSP1), middle of the road (SSP2), regional rivalry (SSP3), inequality (SSP4), and fossil-fueled development (SSP5). A specific RCP can be achieved through different SSPs and vice versa, although some SSP-RCP combinations will be more plausible than others. AR6 provided assessment of the state of the climate for five future scenarios: SSP1-1.9, SSP1-2.6, SSP2-4.5, SSP3-7.0, and SSP5-8.5. These scenarios present a range of possible futures, from high emission scenarios in line with a business-as-usual world (SSP3-7.0, SSP5-8.5) to low emission scenarios based on aggressive mitigation measures (SSP1-1.9 and SSP1-2.6), with one middle scenario (SSP2-4.5). These scenarios describe different socio-economic pathways and expected radiative forcing increase due to human activities by 2100. Climate models are then run with those forcings as input so that we can examine the likely responses of the Earth's climate system under each of these scenarios.

5.1.4 Abrupt 4×CO$_2$ Runs and 1% CO$_2$ Runs

CO$_2$ runs examine how climate will respond to increasing CO$_2$ concentrations alone whereas all other forcings are fixed at pre-industrial levels. Abrupt 4×CO$_2$ runs impose an instantaneous quadrupling of atmospheric CO$_2$ relative to pre-industrial conditions, from 280 ppm to 1129 ppm, and then hold the CO_2 concentration fixed. 1% CO_2 runs increase CO_2 concentration by 1% per year until it reaches the same quadrupled CO_2 level, and then hold the CO_2 concentration fixed. These model runs allow us to isolate the effects of CO_2 from all other factors, and to compare the temperature change due to CO_2 with changes forced by other factors. In earlier IPCC reports, doubling CO_2 ($2\times CO_2$) experiments were more routinely run in climate models, as the future projections focused on the twenty-first century. However, later research into future emission scenarios by the IPCC and others suggests that CO_2 concentrations are likely to increase well beyond the doubling of pre-industrial level in the next couple of centuries, if no substantial reduction of fossil fuel consumption occurs. Therefore, quadrupling CO_2 experiments have been adopted to examine longer-term climate change.

5.1.5 Paleoclimate Runs

Climate models are also run to simulate climates of the past. Many climate models included in IPCC AR6 run paleoclimate simulations for five periods: 1000 years before the Industrial Revolution, the mid-Holocene warm period (6000 years ago), the last glacial maximum cold period (21,000 years ago), the last interglacial warm period (127,000 years ago), and the mid-Pliocene warm period (3.2 million years ago). These periods were chosen because they have relatively abundant proxy data to reconstruct past climate and environmental conditions. In addition, many of them were warm periods, which provide important analogues for our understanding of future global warming. There are also paleoclimate runs for other deeper time periods to examine particular trends (such as the long-term evolution of the carbon cycle and climate during the last 600 million years using the GEOCARB model[12]) and events (such as the Paleocene–Eocene Thermal Maximum (PETM) around 55 million years ago). Among the IPCC AR6 standard simulations, the runs for 1000 years before the Industrial Revolution impose evolving climate forcing conditions estimated from both theoretical relationships and climate proxy data, including solar radiation, volcanic eruptions, and atmospheric concentrations of greenhouse gases. The other simulations are forced with fixed solar radiations, greenhouse gas concentrations, and ice sheet extent, derived from geological and proxy data.

Special model runs are also conducted for other important periods, particularly times of rapid climate change. These runs help us understand possible mechanisms for large climatic changes that occurred in the past, as well as the responses of major components of the climate system such as sea level, ice sheet extent, and shifts in major circulation systems. One such period that draws much attention is the Younger Dryas (~12,000 years ago), a short period of rapid cooling that interrupted a warming trend as the Earth was transitioning out of the last Ice Age. Climate models are used to examine the possible causes for such a rapid return to short-lived glacial conditions (Box 2).

---

[12] More information about this model can be found at GEOCARB Geologic Carbon Cycle, University of Chicago (accessed 2023, May 25), http://climatemodels.uchicago.edu/geocarb/geocarb.doc.html.

Box 2 Multiple causes of the Younger Dryas cold period—a modeling perspective

The Younger Dryas (YD) was a cold period between 12,900 and 11,500 years before present (BP), characterized by an abrupt return to glacial conditions that interrupted the warming that started around 15,000 BP as the world was transitioning from the cold glacial to warm interglacial climate. It is one of the most well-established examples of abrupt and short-lived climate change, which has been detected in climate proxies across the Northern Hemisphere. Based on ice cores from Greenland, where the signal of the Younger Dryas is especially well-developed, the onset of this period was marked by a drop of 8 °C in less than 100 years. Toward the end of the period, temperature increased by 10 °C in several decades[13] (Fig. 10).

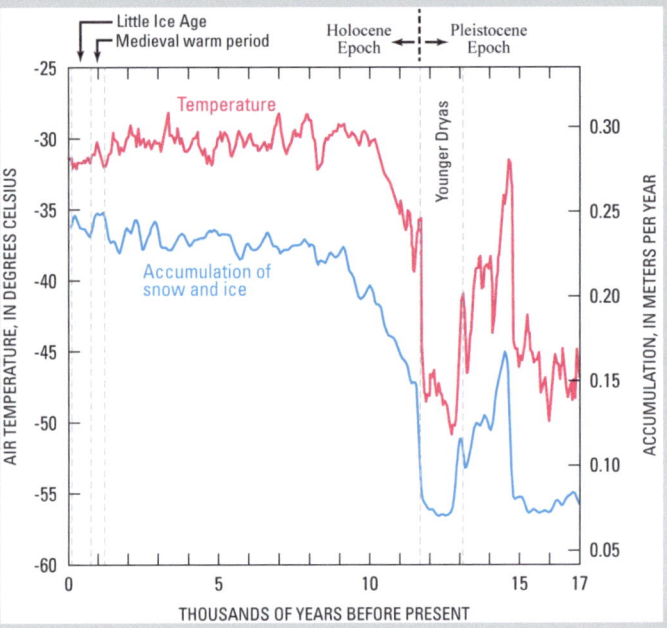

Fig. 10 Greenland surface temperature and snowfall over the last 17,000 years reconstructed from Greenland ice cores. From USGS[14]

---

[13] National Academies (2002).

[14] Figure from Williams, R.S., & Ferrigno, J.G., (2016). State of the Earth's Cryosphere at the Beginning of the 21st Century: Glaciers, Global Snow Cover, Floating Ice, and Permafrost and Periglacial Environments, USGS Professional Paper 1386–A, Chapter A-2, Fig. 35. https://pubs.usgs.gov/pp/p1386a/gallery2-fig35.html. Public domain.

Several causes have been suggested for the YD event. The most common hypothesis[15] involves the disruption of the thermohaline circulation of the ocean, also called the Atlantic Meridional Overturning Circulation (AMOC). During the transition from glacial to interglacial climates, warming caused massive melting of the land-based glaciers and ice sheets globally. The melting of the North America (Laurentide) ice sheet drained a large amount of fresh water into the North Atlantic. This may have weakened (or even stopped) the formation and sinking of highly saline and dense seawater, causing interruptions of the AMOC. As the AMOC is responsible for most northward energy (i.e., heat) transport in the Atlantic, its disruption could have led to a rapid return to glacial conditions. Climate model simulations suggest that the initial burst of fresh water input could have caused the sudden cooling, but was not sufficient to maintain the cooling conditions for over 1000 years. Instead, a constant input of fresh water was needed to sustain a weakened state of the AMOC, which could not have been supplied by meltwater under a cooling condition. Therefore, other causes are needed. An alternative hypothesis suggests that changes in topography and land configuration caused by ice sheet melting may have led to a northward shift of the atmospheric jet stream, bringing more precipitation to the North Atlantic.[16] As a result, the increased input of fresh water was enough to slow down the AMOC, leading to the onset of the YD. Another well-known hypothesis stipulates that the YD was initiated by an impact event, which greatly increased the dust level in the atmosphere and reduced the amount of solar radiation that reached the Earth's surface.[17]

Climate models have been widely used to investigate various hypotheses on the cause of YD cooling. For example, in a study by Renssen and others (2015),[18] a group of scientists used a climate model to investigate the importance of different forcings for the YD, such as a shutdown of the AMOC, a shift of the jet stream, and reduced solar radiation. They found that any single cause by itself was insufficient to create the YD conditions, and only a combination of different forcings could reproduce the conditions established by climate proxy data for the YD period. This study suggests greater complexity in the climate system that triggers abrupt climate changes such as the Younger Dryas.

The study by Renssen and others (2015) used the LOVECLIM model, a three-dimensional Earth system model of intermediate complexity with representations of atmosphere, ocean, sea ice, land surface (with vegetation), ice sheets, and carbon cycle. The model has coarser spatial resolution and simpler representations of physical processes (particularly in the atmosphere) than the

---

[15] Stocker and Wright (1991).
[16] Wunsch (2007).
[17] Firestone et al. (2007).
[18] Renssen et al. (2015).

comprehensive GCMs, but these simplifications allow faster computations to enable long-term simulations. The simplifications also facilitate multiple runs of the model (called ensemble) in order to assess the climate system's natural variability and/or its sensitivity to perturbations in certain parameters. Because of these advantages, LOVECLIM is often used in modeling past climate and has been successfully applied to simulate various past climates in many studies, including the last millennium, the last glacial maximum (LGM), the Holocene, the 8.2 ka cooling event, and the last interglacial. In this study, the model was first run with modern forcings and was able to simulate a present-day climate similar to observations. Through various diagnostic tests, LOVECLIM was found to perform similarly to other comprehensive GCMs in terms of important climate characteristics such as climate sensitivity, polar amplification, and the model's AMOC response to freshwater perturbations.

In order to establish the cause of the YD climate change, it was first necessary to establish a baseline climate condition. This was achieved by running the model with fixed pre-YD forcings (13,000 BP) for 5000 years to obtain an equilibrium climate state. These forcings included ice sheet extent, elevation, orbital parameters, and atmospheric concentrations of greenhouse gases (CO_2, CH_4, and N_2O) at 13,000 BP. These forcing conditions were constructed from various geological evidence and proxy data.

After the baseline was established, a series of model experiments were implemented by changing (perturbing) the reference climate state. These experiments were developed to test the existing hypotheses, including freshwater injections into the North Atlantic of various magnitudes and durations, a complete shutdown of the AMOC, and reduced radiative forcing (e.g., from increased dust levels or from reduced greenhouse gases). In each experiment, the model was run multiple times to generate an ensemble result over periods from centuries to millennia so that natural variability of the climate system could be examined. The resulting temperature changes were compared to proxy reconstructions, focusing on two key regions: Europe and the North Atlantic.

Based on the aggregated proxy data, the summer temperature of Europe was interpreted to have decreased by 1.7 °C and the annual sea surface temperature (SST) of the North Atlantic to have decreased by 2.4 °C during the YD period. The model simulations show that a sudden pulse of freshwater, 0.5–5 Sv (one million cubic meters per second) over a year, would produce a cooling of 0.6–0.9 °C over Europe and North Atlantic, weaker than the proxy estimates. A complete shutdown of the AMOC would reduce the summer temperature of Europe by 3.5 °C, and the annual SST of the North Atlantic by 4.3 °C, a much stronger cooling than the proxy estimates. A reduction of the radiative forcing by 10 W m^{-2} would reduce temperatures by about the same amount as the freshwater pulses, but over much wider areas, which is inconsistent with proxy data that indicate relatively mild conditions in the Southern Hemisphere during the YD. Therefore, none of these mechanisms by themselves could reproduce the YD conditions reconstructed from proxy data. Finally, a simulation was run with combined forcings of freshwater pulses (5 Sv over three years) plus a small reduction in radiative forcing (2 W m^{-2}), and produced temperature reductions and spatial patterns consistent with the YD changes interpreted from proxies.

5.2 Coupled Model Intercomparison Project (CMIP)

Previous sections in this chapter have shown that a wide variety of climate models exist, with different characteristics and different applications. Even a single climate model can have several versions with different features and parameterizations. The number of research centers around the world that are developing and running a wide variety of climate models has increased dramatically in the past several decades, so it has become necessary to establish a standard framework for all models so that their results can be compared. Model comparison is important for improving climate models and for quantifying model uncertainties, both of which ultimately will improve our understanding of past, present, and future climate change. For this purpose, the **Coupled Model Intercomparison Project (CMIP)** was first organized in 1995 by the Working Group on Coupled Modelling (WGCM) of the World Climate Research Program (WCRP). CMIP aims to provide a standardized framework for all coupled atmosphere–ocean circulation models, including model setup, input data, emission scenarios, experimental designs, output data, and format. Since its beginning, CMIP has gone through several phases, with each phase involving an increasing number of models of greater complexity and higher resolution running more sophisticated experiments. CMIP plays an important role in coordinating the modeling efforts that support the IPCC Assessment Reports. For example, the most recently completed phase, CMIP6, involved approximately 100 distinct climate models developed by 49 different modeling groups, and running a wide range of near-term and long-term experiments. These formed the foundation for climate projections presented in IPCC AR6. Figure 11 shows a comparison of observed (thick black line) and simulated (thick colored lines) global temperature changes from models in the third, fifth, and sixth CMIP phrases. A close examination of Fig. 11 and other results shows that the CMIP 6 models more accurately predict the global distributions of temperature and precipitation than the CMIP5 and CMIP3 models did. More significant improvements are found in the predictions of other variables, such as the vertical temperature, water vapor, and zonal wind speed distributions (see Sect. 6.3 for more details).

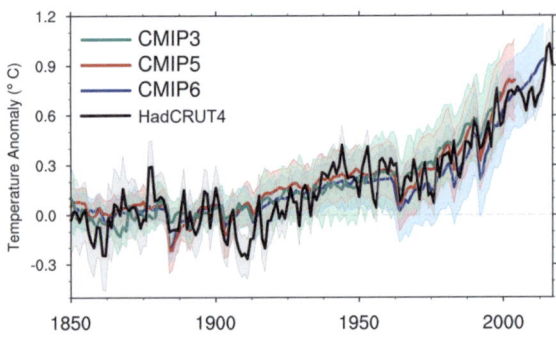

Fig. 11 Comparison between observed (thick black line) and simulated (thick colored lines) global temperature change from models in the third (CMIP3, green), fifth (CMIP5, red), and sixth (CMIP6, blue) phrases of the Coupled Model Intercomparison Project. Shadings indicate the range of model variability. *Source* Bock et al. (2020)[19]

6 How Do We Evaluate Climate Models?

Model evaluation is an effort to quantitatively establish how accurately climate models produce observed variables that describe the climate state, such as temperature, precipitation, atmospheric pressure, and wind. The most direct way to evaluate models is to compare modeled variables (i.e., the model output) with observed measurements. Climate models routinely simulate the climate of the recent past based on the observed forcing factors (e.g., solar radiation, greenhouse gas concentrations, volcanic eruptions, and land cover), and produce modeled output for climate variables for the past (such as temperature, precipitation, pressure, and wind), which are then compared with observational data of these climate variables. However, observational data come with their own errors and uncertainties. A good understanding of these limits on observational data is fundamental for effective model evaluation.

6.1 Observational Data

As described in Chap. 3, observational data come from many different sources. The most direct observational data on land come from tens of thousands of weather stations at the Earth's surface, which measure essential climate variables such as temperature, humidity, pressure, and wind speed and direction. These station data provide the longest observational time series available, with some stations going back as early as 1850, although the majority of stations started in the 1950s. The disadvantage of station data is their uneven spatial distribution, with major gaps in regions that are sparsely populated. In addition, station time series often contain inhomogeneities (i.e., changes that are not related to climate) with causes such as

---

[19] From Fig. 2 in Bock et al. (2020). Creative Common License.

instrument upgrades, station relocations, or changes in the surrounding environment (e.g., urbanization and major land use change). Data on vertical profiles through the atmosphere generally are limited, and primarily provided by radiosondes launched on weather balloons. Climate observation over the ocean is carried out by ships, buoys, and stationary platforms. The bulk of the data come from ships and drifting buoys, which do not have fixed locations, hence do not provide long time series at specific locations. Satellites provide the first global coverage of the Earth's climate conditions. They are equipped with sensors that measure the intensity of radiation in a wide range of wavelength bands, from which many climate parameters can be derived. Satellites provide data over large areas at high spatial and temporal resolutions. However, most satellite data began in the 1990s, hence, they do not provide long time series. In addition, the raw satellite measurements must be converted into the desired climate variables, a process that introduces additional uncertainties.

In order to overcome the deficiencies of individual data sources and to fill data gaps, various data assimilation schemes have been used to combine datasets for better spatial and temporal coverage. Some data assimilation approaches are based on sophisticated statistical methods. An alternative approach is to combine observational data with climate models to provide a complete numerical description of the recent climate. This approach is called **climate reanalysis**. Global reanalysis datasets cover all locations on Earth at high spatial and temporal resolutions, and they can extend back decades or more.

Both climate models and climate reanalysis use observation data, but in very different ways. Climate models use observations only as the starting conditions for model runs. Once the model starts, the simulated climate is allowed to evolve by itself. On the other hand, climate reanalyses assimilate observations during the whole period simulated. At each time step, observational data are used to constrain and adjust model calculations. Such assimilated data products, based on either reanalysis models or statistical methods, have several distinct advantages for evaluating climate models. They provide complete coverage for all climate variables, even in places and times that have no observations. In addition, unlike station-based data, which only measures the conditions of a small area adjacent to the station, the assimilated datasets provide area-average values for each grid cell. This is more appropriate for comparison with climate model output in order determine the skill of climate model simulations.

6.2 Model Evaluation Approaches

To evaluate climate models, we compare simulated values of climate variables (i.e., the output of the climate model) with estimates based on observational data. Comparison can be made in many different ways, depending on the climate variables, temporal and spatial scales, and the statistical properties of the variables. Since its third report, the IPCC assessment reports have increasingly relied on a set of standard statistical measurements for model comparison, called **performance metrics**,

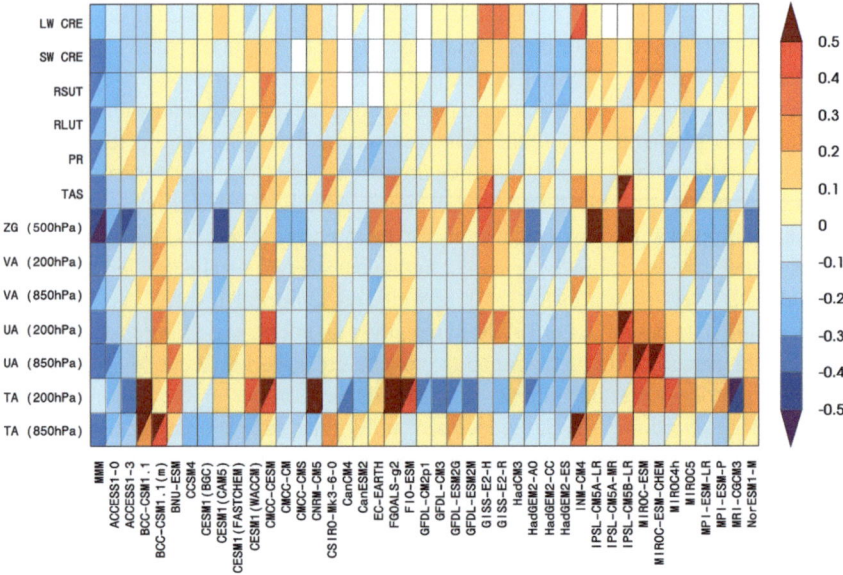

Fig. 12 Climate model performance matrix: relative error measurements for 40 climate models and the multi-model mean (MMM) (columns) in simulating 13 climate variables (rows). The number (and color) indicates the percentage greater (positive) or smaller (negative) than the median error of all model results. Figure used with permission from the IPCC. Full IPCC caption for this figure can be found in the footnote. From IPCC (2013).[20] This figure is excluded from Creative Commons license

to provide an objective and quantitative assessment of each model's ability to simulate the real climate. Performance metrics are developed to evaluate accuracy for the average climate state and seasonality, as well as short- and long-term climate variabilities. The use of standard metrics simplifies synthesis and visualization of model performance, and enables the quantitative assessment of improvements in climate models over time. For example, Fig. 12 shows the relative errors for a set of common climate variables simulated by different climate models (IPCC AR5). The number shown for each climate model indicates the percentage (0.5 means 50%) more (positive) or less (negative) of its output for that climate variable compared to the median value of all model errors. Therefore, the blue shadings indicate better performance than average, whereas the yellow–red shadings indicate worse performance. This figure provides a useful overall portrait of the relative performance of all models, showing that some models perform consistently better than others (i.e., a column containing only blues), and that simulations are more consistent for some climate variables than others (i.e., a row containing only pale yellows and blues). Moreover, the multi-model mean (MMM) performs better than any individual model.

---

[20] Figure 9.7 in IPCC (2013). Full caption from original figure: Relative error measures of CMIP5 model performance, based on the global seasonal-cycle climatology (1980–2005) computed from

Over the years, the ensemble methods have been used more and more frequently to explore the uncertainty in climate model simulations. Instead of evaluating individual models, the ensemble approach evaluates a group of models at the same time so that we can examine the average and range of the simulated climate variables in order to determine model uncertainties. Many factors contribute to model uncertainties, such as internal variability in the climate system, uncertainties in boundary conditions, differences in parametrization schemes and parameter values, and differences in model formulations and structures. The ensemble approach can help quantify model uncertainties and compare the relative contribution from each major source of that uncertainty.[21]

One of the most important outcomes of the systematic evaluation of climate models is the estimation of the range of uncertainty inherent in climate model simulations and predictions. These efforts have generally found that (1) no one single model performs well in all climate variables; (2) no one test evaluates all aspects of the models; and (3) the model group mean (after excluding unreasonable outliers) outperforms any one model. As a result, it is common practice for people to use the mean of multiple models to project future climate change.

6.3 How Well Do Climate Models Simulate the Observed Climate of the Recent Past?

The most sophisticated climate models simulate the complete state of the atmosphere and ocean. Many of the simulated variables can be evaluated against the increasing amount of observational data, including temperature, humidity, wind, precipitation,

---

the historical experiments. Rows and columns represent individual variables and models, respectively. The error measure is a space–time root-mean-square error (RMSE), which, treating each variable separately, is portrayed as a relative error by normalizing the result by the median error of all model results (Gleckler et al., 2008). For example, a value of 0.20 indicates that a model's RMSE is 20% larger than the median CMIP5 error for that variable, whereas a value of –0.20 means the error is 20% smaller than the median error. No colour (white) indicates that model results are currently unavailable. A diagonal split of a grid square shows the relative error with respect to both the default reference data set (upper left triangle) and the alternate (lower right triangle). The relative errors are calculated independently for the default and alternate data sets. All reference data used in the diagram are summarized in Table 9.3 [of IPCC report].

[21] There are generally two types of ensembles: Multi-model Ensembles (MMEs) and Perturbed Parameter (or Physics) Ensembles (PPEs). The MMEs are generated from simulations of many different models. Sometimes all members of a particular MME are treated equally. This could be problematic because some climate models share model components and hence are more similar to each other than others. More often statistical techniques are used to assign different weights to different climate models to create a weighted MME in order to reduce redundancy in shared biases. MMEs allow us to examine the uncertainties related to different formations and structures among climate models. Unlike MMEs, PPEs are generated by multiple runs of a single climate model. At each run (i.e., an ensemble member), parameters in the model's parametrization schemes are slightly changed (i.e., perturbed). This allows us to compare uncertainties associated with different parameters and parameterization schemes, and determine the main drivers of parametric uncertainty across the ensemble.

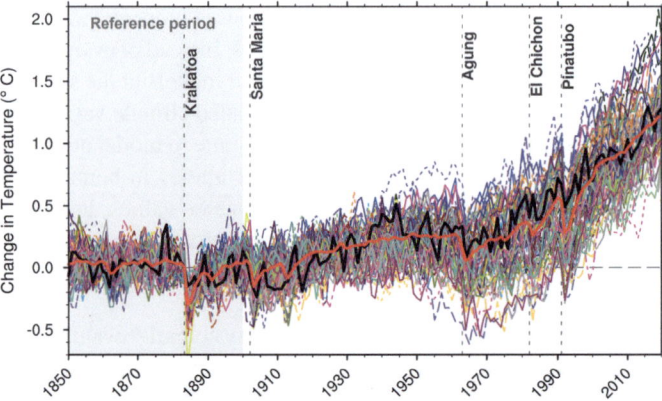

Fig. 13 Observed and simulated time series of the global surface air temperature (GSAT) change with respect to the average of the reference period of 1850–1900. *Source* Bock et al. (2020)[22]

cloud cover, energy fluxes, etc. In Sect. 5.1, we showed that by incorporating both natural and anthropogenic forcings climate models can reconstruct global mean temperature changes accurately for the past 150 years, although more uncertainties exist for reconstructions of local temperature changes. In this section, we will explore more details about the ability of climate models to simulate the climate of the recent past. This discussion will focus on surface temperature and precipitation, the most routinely evaluated variables. Figure 13 presents a direct comparison of the global surface temperature variations for 1850–2010 from observed data (HadCRUT5, thick black line) and the individual CMIP6 models (thin colored lines) as well as the multi-model mean (thick red line). Results from the individual models are not important, except for showing the range of results from the different models. Despite moderate model spread (within 0.5 °C), the CMIP6 mean of global surface temperature (thick red line) follows the observational record (thick black line) fairly closely (mostly within 0.2 °C), showing the gradual warming trend during the twentieth century that becomes steeper during the recent decades. Even with the strong match between the mean simulated and observed temperature trends, it is important to note that simulated temperatures may not always track the observations because the climate system contains random internal variabilities. The model mean has noticeably smaller interannual variations than those in the observations, and this is largely because the simulated variability of individual models is smoothed by averaging. Notice too that the climate model simulations are in good agreement with observations during short-term cooling forced by episodic volcanic eruptions (indicated by vertical dashed lines, e.g., the Pinatubo eruption in 1991), although the models tend to overestimate the cooling response to such events. This strong aerosol forcing in several models also results in lower simulated temperatures during the mid-to-late twentieth century, when air pollution was more severe globally.

---

[22] From Fig. 1c in Bock et al. (2020). Creative Commons license.

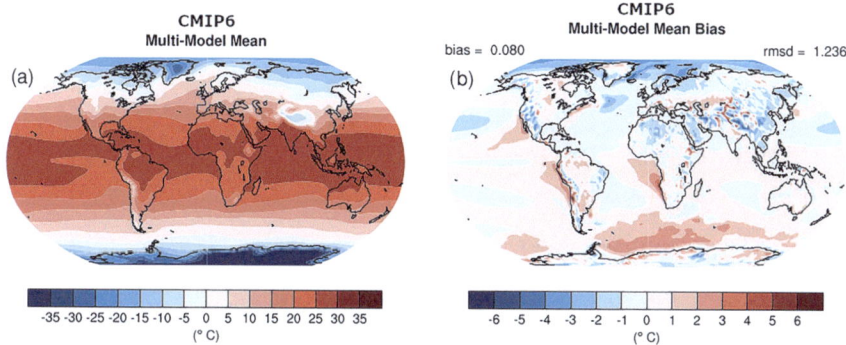

Fig. 14 **a** Multi-model mean surface temperature; **b** the difference between multi-model mean and ERA5 reanalysis data. Values close to 0 °C (pale colors) indicate regions with no significant biases. *Source* Bock et al. (2020)[23]

In addition to the global average surface temperature, the models are also evaluated for their ability to simulate the spatial (or geographic) distribution of temperature. Figure 14a presents the multi-model mean of the annual mean surface air temperature (at 2 m above the ground surface) simulated by the CMIP6 models, and Fig. 14b shows the difference between the multi-model means and the ERA5 reanalysis, which is one of the best estimates of local temperatures based on observational data. The deviation of the model output from the observed value is referred to as model **bias**, and it can be positive (i.e., overestimation) or negative (underestimation). The multi-model mean simulates well the equator-to-pole temperature gradient, and model bias is generally less than 2 °C in most places. Largest biases are most common in polar regions, high elevation regions (e.g., the Tibetan Plateau and the American Cordillera), and ocean upwelling regions (e.g., off the west coasts of Africa and South America). Such biases can arise from the difficulties that climate models have in resolving complex processes, as well as uncertainties in the observational data, particularly in regions where observational data are scarce.

Let us look at a different example of spatial evaluation results, this time of modeled precipitation. Figure 15a presents the multiple model mean of annual precipitation rate (mm day^{-1}) for the period 1980–2005, and Fig. 15b shows the difference between model simulations and the precipitation reanalysis from the Global Precipitation Climatology Project (GPCP), the best estimate of past observed precipitation. The model simulates well the large-scale spatial distribution patterns of global precipitation, and the biases are largely less than 10%, within the uncertainties of observations. Climate models tend to overestimate precipitation in the tropical region over the ocean, but underestimate inland precipitation across a range of latitudes.

---

[23] From Fig. 3a and b in Bock et al. (2020). Creative Commons license.

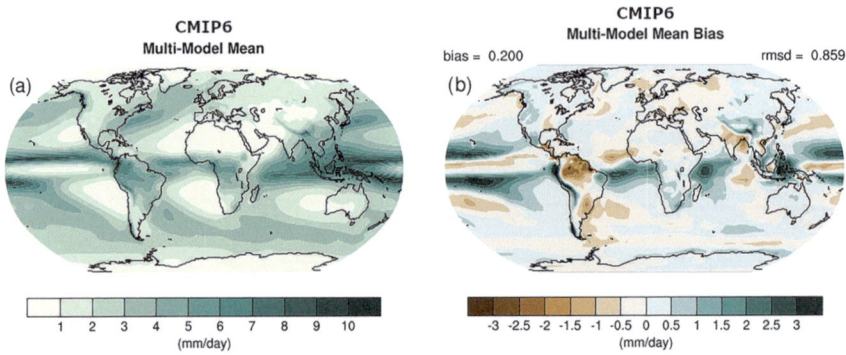

Fig. 15 a Multi-model mean surface precipitation rate; **b** multi-model mean bias when compared to the Global Precipitation Climatology Project (GPCP) observation data. Values close to 0 mm/day (pale colors) indicate regions with no significant biases. *Source* Bock et al. (2020)[24]

7 What Do Climate Models Tell Us About Future Climate Change?

Once the climate models are tested and shown to be reliable in simulating the observed past climate, we can use them to project possible future climate changes based on different socio-economic scenarios. This information is essential for developing effective mitigation and adaptation strategies to deal with future climate change. This section provides an overview of climate models' future projections.

Since human processes are the most difficult to project, the most recent IPCC assessment report (AR6) includes five possible greenhouse gas emission scenarios: SSP1-1.9, SSP1-2.6, SSP2-4.5, SSP3-7.0 and SSP5-8.5. As mentioned in Sect. 5.1, these scenarios were developed by integrating different future socioeconomic projections (SSPs) with a target increase of radiative forcing (RCPs) at the end of the twenty-first century. In doing so, these emission scenarios also represent a wide range of mitigation scenarios and policy actions. SSP1-1.9 and SSP1-2.6 are "very low" and "low" emission scenarios. Both require that carbon dioxide emissions start declining by 2020 and reach zero before 2100 (by 2050 for SSP1-1.9 and by 2075 for SSP1-2.6). These two emission scenarios also require significant reductions in other greenhouse gases, such as methane and nitrous oxide. If actual conditions were to follow these two SSP1 scenarios, global temperature rise would likely be kept below 2 °C by 2100, the upper limit of the Paris Climate Agreement goal (see Chap. 1). SSP2-4.5 is an "intermediate" scenario, which requires that carbon dioxide emissions start to decline by 2050 without reaching net zero by 2100. SSP3-7.0 and SSP5-8.5 are "high" and "very high" emission scenarios, with carbon dioxide emissions rising throughout the twenty-first century. The total greenhouse gas emissions would double

---

[24] From Fig. 4a and b in Bock et al. (2020). Creative Commons license.

by 2100 under SSP3-7.0, and triple by 2075 under SSP5-8.5. Future CO_2 concentrations can be established for these scenarios using a carbon cycle model. They range from 393 (SSP1-1.9) to 1135 ppm (SSP5-8.5) by 2100 (Fig. 16).[25] These scenario and concentration data are then used as input for global climate models to produce future climate conditions.

7.1 Future Changes in Surface Temperature

Based on the CMIP6 ensemble of climate models, global mean temperatures are expected to continue to rise over the twenty-first century under all future scenarios. An increase of 2.2–4.7 °C (relative to 1995–2014) is projected by the end of the century if we do not dramatically reduce our fossil fuel consumption (SSP3-7.0). Figure 17 shows the spatial distribution of annual surface temperature change under two different scenarios for mid-century (2041–2060) and late-century (2081–2100) relative to the recent past (1995–2014). In general, warming is greater on land than over the ocean, and it is greater in polar regions than in low latitudes. Assuming no

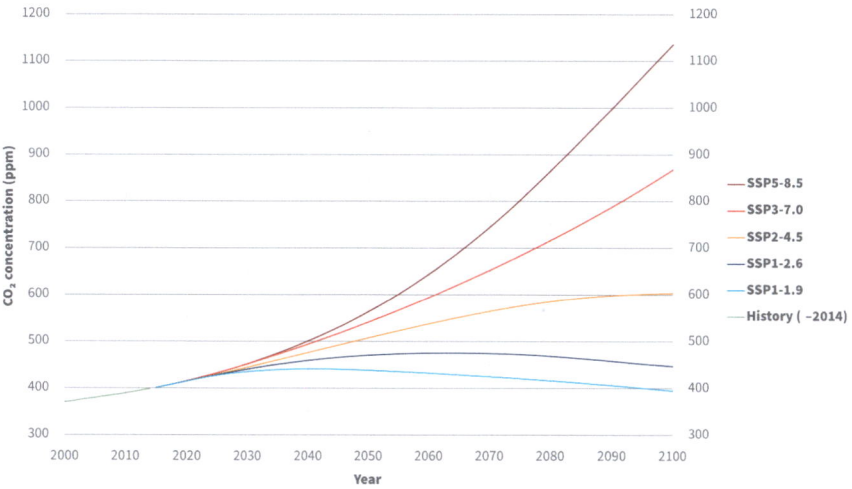

Fig. 16 Atmospheric CO_2 concentrations by SSP scenarios across the twenty-first century, projected by MAGICC7, a reduced complexity climate-carbon cycle model. From Wikimedia Commons, modified from Meinshausen et al. (2020)[26]

---

[25] Meinshausen et al. (2020).

[26] From Wikimedia Commons (accessed August 1 2023). Atmospheric CO_2 concentration by SSP across the twenty-first century. https://en.wikipedia.org/wiki/Shared_Socioeconomic_Path ways#/media/File:Atmospheric_CO%E2%82%82_concentrations_by_SSP_across_the_21st_cent ury.svg. Creative Commons Attributions-Share Alike 4.0 International license. It is modified from Meinshausen et al. (2020).

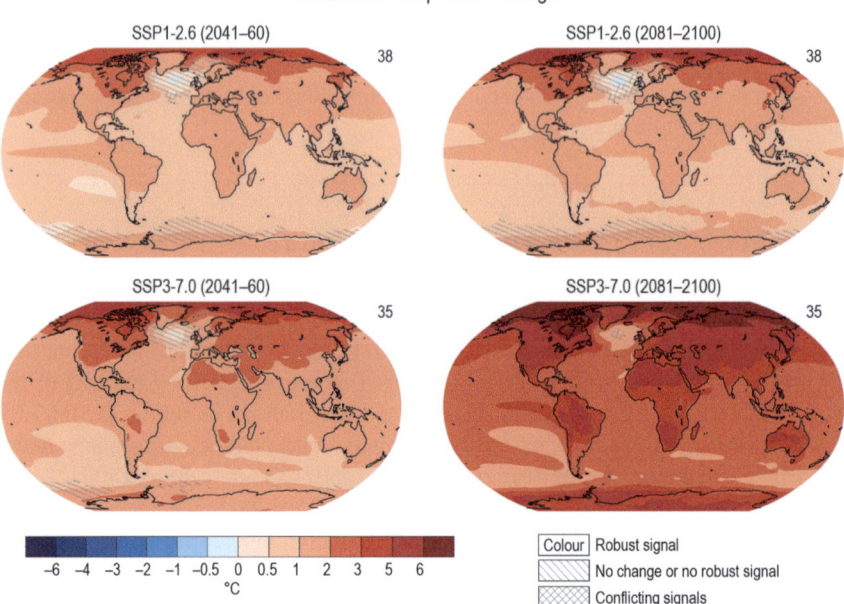

Fig. 17 Spatial distribution of projected changes in annual mean temperature for the mid and late twenty-first century relative to the reference period of 1995–2014 for the low (SSP1-2.6) and high (SSP3-7.0) emission scenarios. Diagonal lines indicate regions with no significant change. Figure used with permission of the IPCC. Full IPCC caption for this figure can be found in the footnote. *Source* IPCC (2021).[27] This figure is excluded from Creative Commons license

significant reduction in the Atlantic Meridional Overturning Circulation (see Box 2), the Arctic region is likely to experience the largest amount of warming in the future (as it has in the recent past), largely due to the positive feedback from reduced ice and snow cover.

---

[27] From Fig. 4.19 in IPCC (2021b). Original IPCC figure caption: Mid- and long-term change of annual mean surface temperature. Displayed are projected spatial patterns of multi-model mean change in annual mean near-surface air temperature (°C) in 2041–2060 and 2081–2100 relative to 1995–2014 for (top) SSP1-2.6 and (bottom) SSP3-7.0. The number of models used is indicated in the top right of the maps. No overlay indicates regions where the change is robust and likely emerges from internal variability, that is, where at least 66% of the models show a change greater than the internal-variability threshold (see Sect. 4.2.6) and at least 80% of the models agree on the sign of change. Diagonal lines indicate regions with no change or no robust significant change, where fewer than 66% of the models show change greater than the internal-variability threshold. Crossed lines indicate areas of conflicting signals where at least 66% of the models show change greater than the internal-variability threshold but fewer than 80% of all models agree on the sign of change. Further details on data sources and processing are available in the IPCC chapter data table (Table 4.SM.1).

7.2 Future Changes in Precipitation

Global warming has the potential to enhance the hydrological cycle and alter many key characteristics of precipitation, such as its quantity, frequency, intensity, and duration. Several important changes are expected. First, higher temperature enhances evaporation, as well as the moisture-holding capacity of the atmosphere at a rate of about 7% per °C based on physical laws. Therefore, global precipitation is expected to increase to balance the enhanced evaporation, but great uncertainty still exists in how precipitation will change locally. Second, with more moisture in the atmosphere, when storm systems develop, the increased humidity is expected to lead to heavier rainfall events. In addition, increased moisture content in the atmosphere also enhances latent heat in storm systems and thereby is expected to increase their intensity. As a result, global warming could lead to an increase in heavy precipitation events at the expense of light and moderate precipitation. Third, assuming no significant changes in the global atmospheric circulation pattern, increases in evaporation and humidity mean more moisture will be transported from the arid subtropics to the humid tropics and into the storm tracks at higher latitudes. This could cause wet areas to become wetter and dry areas to become drier. However, global warming could also potentially alter large-scale circulation patterns, leading to shifts of the world's dry and wet regions. The onset of many of these changes already has been confirmed by both observations and climate models.

Future changes in precipitation simulated by CMIP6 models indicate that overall, precipitation is expected to increase more over land than over the ocean. The increase is more significant for higher emission scenarios and further into the future. Figure 18 shows the global distribution of precipitation change simulated by CMIP6 models for the mid- and late century under the low- and high-emission scenarios. These models project an increase of global precipitation at 1–3% for every degree (°C) increase in surface temperature for most scenarios at the end of the twenty-first century, but precipitation change is likely to exhibit substantial spatial variation. High latitude regions are very likely to get wetter due to both the increased humidity of the warmer atmosphere and increased transport of water vapor from lower latitudes by the end of this century. In the mid-latitudes, many arid and semi-arid regions are likely to experience a decrease in precipitation, whereas many humid regions are likely to see an increase in precipitation by the end of this century. Based on the model simulations, warming is also likely to increase the contrast in annual mean precipitation between dry and wet regions and between wet and dry seasons in most places.

7.3 Future Changes in the Cryosphere

Climate models project significant future changes in many aspects of the cryosphere. The current Arctic sea ice cover is at its lowest since at least 1850, and probably for the past 1000 years. CMIP6 simulations indicate that the Arctic sea ice cover will continue to shrink and thin in the twenty-first century as global temperature increases

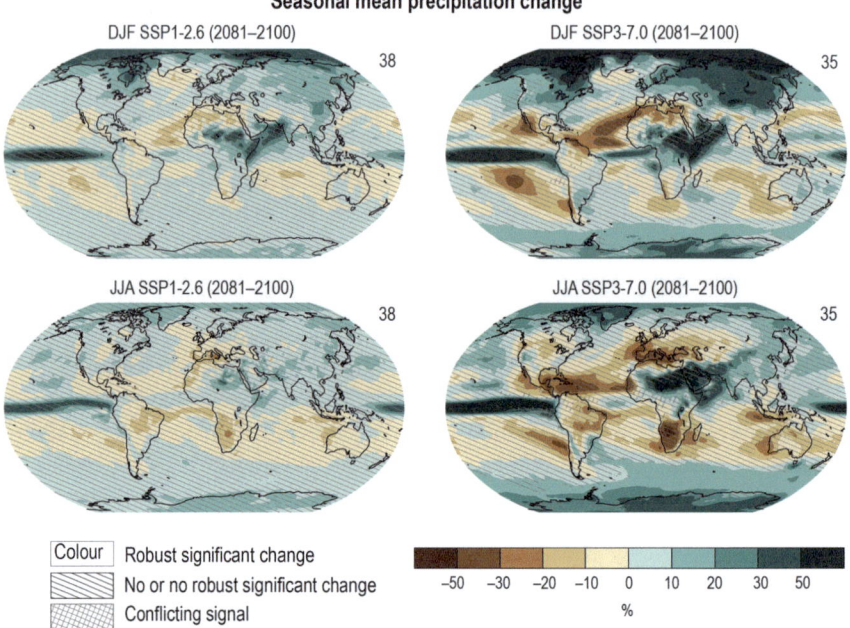

Fig. 18 Spatial distribution of projected changes in seasonal precipitation (%) for the late twenty-first century relative to the reference period of 1995–2014 under the low and high emission scenarios. "DJF" refers to December, January, and February; "JJA" refers to June, July, and August. Figure used with permission of the IPCC. Full IPCC caption for this figure can be found in the footnote. *Source* IPCC (2021).[28] This figure is excluded from Creative Commons license

(Fig. 19), and on average the Arctic will be practically ice free in September by the end of the twenty-first century under the medium (SSP2-4.5), high (SSP3-7.0) and very high (SSP5-8.5) emission scenarios. Sea ice extent and volume in the Antarctic is also expected to decrease, but with greater uncertainties. Although reduction in sea ice does not contribute to sea level rise, it does play an extremely important role in the global climate system, because reduced sea ice provides a strong positive

---

[28] From Fig. 4.24 in IPCC (2021b). Original IPCC caption: Long-term change of seasonal mean precipitation. Displayed are projected spatial patterns of multi-model mean change (%) in (top) December–January–February (DJF) and (bottom) June–July–August (JJA) mean precipitation in 2081–2100 relative to 1995–2014, for (left) SSP1-2.6 and (right) SSP3-7.0. The number of models used is indicated in the top right of the maps. No map overlay indicates regions where the change is robust and likely emerges from internal variability, that is, where at least 66% of the models show a change greater than the internal-variability threshold (Sect. 4.2.6) and at least 80% of the models agree on the sign of change. Diagonal lines indicate regions with no change or no robust significant change, where fewer than 66% of the models show change greater than the internal-variability threshold. Crossed lines indicate areas of conflicting signals where at least 66% of the models show change greater than the internal-variability threshold but fewer than 80% of all models agree on the sign of change. Further details on data sources and processing are available in the IPCC chapter data table (Table 4.SM.1).

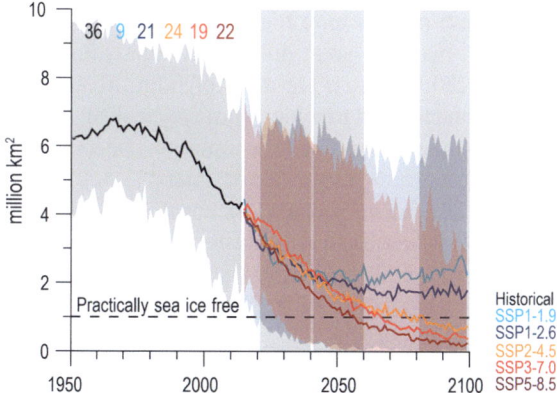

Fig. 19 Projected summer Arctic sea ice extent (September) by the end of the twenty-first century under five emission scenarios. The numbers near the top show the number of model simulations used for each scenario. Figure used with permission of the IPCC. Full IPCC caption for this figure can be found in the footnote. *Source* IPCC (2021).[29] This figure is excluded from Creative Commons license

feedback to global warming through albedo changes. A significant reduction in sea ice extent, particularly in summer, causes less energy to be reflected back to space, and hence more energy being absorbed by the Earth's surface, leading to even more warming. This feedback has already led to significantly more warming in the Arctic region than the global average.

A global increase in temperature will also lead to continued reduction in the mass of ice sheets and glaciers (Fig. 20), with significant consequences for sea level. Globally, mountain glaciers have been retreating since the start of the Industrial Revolution, and at the fastest rate in at least the last 2000 years. The 2021 IPCC report summarizes historical changes in the cryosphere and future predictions.[30] Between 1901 and 2018, glacial retreat contributed 67 (42–93) mm to the observed sea level rise. This is going to continue over the twenty-first century. CMIP6 models project a continental ice-volume reduction of 18% under the low (SSP1-2.6) and 36% for high (SSP5-8.5) emission scenarios. Over the period 1992–2020, when satellites have made continuous and frequent observations, Greenland likely lost 4890 Gigatons (Gt) of ice, contributing 13.5 mm to global sea level rise. At the same time, the Antarctic Ice Sheet has lost 2670 Gt, contributing 7.4 mm to sea level rise. Using a group of climate models that specializes in simulating the dynamic responses of ice sheets to climate change, the Ice Sheet Model Intercomparison Project (ISMIP6), IPCC AR6 projects a continued decrease in the Greenland and Antarctic ice sheets under all

---

[29] From Fig. 4.2c in IPCC (2021b). Original IPCC figure caption specific to 4.2c: September Arctic sea ice area, September averages. The curves show averages over the CMIP6 simulations, the shadings around the SSP1-2.6 and SSP3-7.0 curves show 5–95% ranges, and the numbers near the top show the number of model simulations used. Results are derived from concentration-driven simulations.

[30] IPCC (2021c). See also the full AR6 report: IPCC (2023).

emission scenarios. These models have higher confidence in simulating mass loss caused by surface melt, but have great difficulties simulating abrupt changes due to ice sheet instability resulted from warming of both the atmosphere and ocean water. As a result, the models tend to underestimate ice sheet mass loss in response to climate warming. Disintegration of ice sheet margins becomes more likely under high emission scenarios, causing greater uncertainties.

Global warming will also impact global snow cover and cause a dramatic decrease in permafrost extent. Snow cover is affected by both precipitation and temperature. Sometimes the changes in temperature and precipitation can have opposite

Fig. 20 Photos of the Helheim Glacier, in SE Greenland. Helheim is one of Greenland's largest glaciers and one that has retreated and accelerated considerably since 2000. As a result, the terminus of the glacier (*upper photo*) retreated and upstream ice (which feeds the glacier) thinned. The glacial terminus is sticking up 80 m above the surface and extends 600 m below the surface. The gray water in front of the terminus is a turbulent, muddy meltwater plume at the glacier's edge. In the foreground, there is a mix of iceberg pieces, known as ice mélange. The blue puddle is surface meltwater on top of an iceberg sitting in the mélange. A larger view of the Helheim Glacier is shown in the *lower photo*. It is about 5 km between the mountains in the foreground and in the distance. The horizontal "bathtub ring" shown by a color change on the distant mountains is a mark left by the glacier when it was larger in the recent past and during the Little Ice Age. Photos courtesy of James Holte and Fiamma Straneo, University of California, San Diego, Scripps Institution of Oceanography. This figure is excluded from Creative Commons license

effects. Although higher temperatures can reduce snow cover, increased precipitation resulting from warming can lead to an increase in snow cover. CMIP6 models generally project a decrease in the snow-covered area of the Northern Hemisphere at a rate of approximately − 8% per 1 °C increase of global temperature until the snow cover disappears. The disappearance of Northern Hemisphere snow cover is projected to occur at about + 2 °C increase from the recent past (1995–2014) global mean temperature for July and August. However, the models do not exhibit high confidence in these projections, because snow processes are highly simplified in global climate models. Permafrost responds to both temperature increases and changes in snow cover, because both affect the underlying soil conditions. Permafrost regions have experienced widespread increases in ground temperatures over the past decades. For each additional 1 °C of warming (up to + 3 °C change from the 1995-2014 baseline), CMIP6 models project a 25% decrease in volume of permafrost. However, these decreases may be underestimated due to an incomplete representation of relevant physical processes in the climate models. Similar to the effect of reduced sea ice cover, changes in snow cover and permafrost would have significant feedback impacts on the climate. Reduced snow cover decreases the planet's reflectance and causes more solar radiation to be absorbed, reinforcing the initial warming. In addition, permafrost contains carbon stored in frozen plant and animal remains. As temperatures warm, these remains decompose, releasing greenhouse gases such as methane and carbon dioxide. This positive feedback loop will again serve to amplify the initial warming.

7.4 Future Changes in Global Sea Level

Global warming causes sea level to rise for two main reasons: thermal expansion of a warming ocean (water expands as it warms) and increased melting from land-based ice such as glaciers and ice sheets. Historical and future prediction of global sea level change are summarized in the IPCC AR6 report.[31] Over the period 1901–2018, global sea level rose by 0.2 (0.15–0.25) m, with an average rate of 1.7 mm per year. The rate of increase accelerated to 3.7 mm per year from 2006 to 2018. Between 1993 and 2018, thermal expansion and increased melting of mountain glaciers, the Greenland ice sheet and the Antarctic ice sheets contributed 42%, 21%, 15%, and 8%, respectively, to global sea level rise. Sea level will continue to rise in response to a warming climate, but the response is projected to be slower than for many other components of the climate system. This means sea level change could be small over the twenty-first century, but sea level will continue to rise for hundreds and thousands of years after all greenhouse gas emissions are stopped before reaching a new equilibrium state. Based on CMIP6 models, global mean sea level is projected to rise by 0.38 m under low (SSP1-1.9) and 0.77 m under high emission (SSP5-8.5) scenarios by 2100 (Fig. 21). However, these projections are dominated by relatively

---

[31] IPCC (2021c). See also the full AR6 report: IPCC (2023).

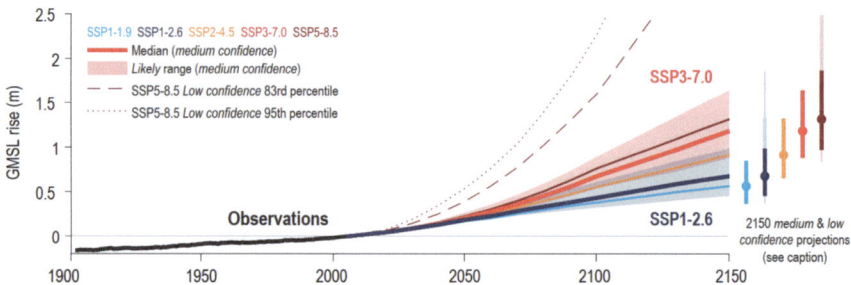

Fig. 21 Observed and projected future global mean sea level (GMSL) rise under various emission scenarios, relative to a 1995–2014 baseline. Solid lines show median projections. Shaded regions show likely ranges for different emission scenarios. Dotted and dashed lines show low confidence projections. Bars at right show likely ranges in 2150. Used with permission of the IPCC. Full IPCC caption for this figure can be found in the footnote. *Source* IPCC (2021).[32] This figure is excluded from Creative Commons license

well-understood processes such as surface melting of land-based ice. Under higher scenarios, much greater uncertainties exist due to the increasing likelihood of more abrupt changes caused by processes that are less understood and/or less quantified, such as increasing instability of ice sheets. In a low-likelihood, high-impact storyline and a high emissions scenario, such ice-sheet processes characterized by great uncertainty could drive sea level rise of as much as 5 m by 2150. Over a much longer period, the eventual sea level rise is projected to be 2–3 m if global warming is limited to 1.5 °C (SSP1-1.9), up to 6 m if warming peaks at 2 °C (SSP1-2.6) and 19–22 m if warming peaks at 5 °C (SSP3-7.0). Given that more than one billion people currently live on land below 10 m above sea level, future sea level rise is going to have significant socio-economic impacts.

Global warming will cause profound changes in many aspects of the Earth, including the physical and chemical properties of the atmosphere and the ocean, the carbon and other biogeochemical cycles, and consequent ecosystem changes. Warming also will affect different regions in different ways. Climate models are essential tools to investigate these present and future changes, and to help people make informed decisions about mitigation and adaptation strategies under a rapidly warming climate.

---

[32] From Box TS.4a, Fig. 1 in IPCC (2021d). Original caption of IPCC Box TS.4a Fig. 1 [https://www.ipcc.ch/report/ar6/wg1/figures/technical-summary/box-ts-4-figure-1/]: GMSL change from 1900 to 2150, observed (1900–2018) and projected under the SSP scenarios (2000–2150), relative to a 1995–2014 baseline. Solid lines show median projections. Shaded regions show *likely* ranges for SSP1-2.6 and SSP3-7.0. Dotted and dashed lines show respectively the 83rd and 95th percentile low confidence projections for SSP5-8.5. Bars at right show *likely* ranges for SSP1-1.9, SSP1-2.6, SSP2-4.5, SSP3-7.0 and SSP5-8.5 in 2150. Lightly shaded thick/thin bars show 17th–83rd/5th–95th percentile *low-confidence* ranges in 2150 for SSP1-2.6 and SSP5-8.5, based upon projection methods incorporating structured expert judgement and marine ice cliff instability. *Low confidence* range for SSP5-8.5 in 2150 extends to 4.8/5.4m at the 83rd/95th percentile.

Box 3 Changes of extreme precipitation in the midwestern United States

Climate model outputs consistently indicate that global warming is likely to increase overall precipitation. In particular, heavy precipitation could increase preferentially, so that more intense and frequent extreme rainfall events become more common. Such a shift toward heavy rainfall events also is a common conclusion of climate models, as well as a result of analyses of observed rainfall data at the global scale. However, there is great spatial variation within this average pattern. Therefore, detailed regional studies are often necessary to provide useful information for local decision-making (e.g., stormwater management, construction of flood control structures).

Present global climate models typically produce results with a spatial resolution of 150–300 km. This level of spatial resolution of GCMs is often insufficient for establishing detailed local future climate projections, particularly for a climate variable with great spatial heterogeneity such as precipitation. The process of deriving information on local or regional scales from the global climate scenarios generated by GCMs is called **downscaling**. There are two main approaches to downscaling. Dynamic downscaling involves a nested regional climate model (RCM) driven by boundary conditions generated by GCM output. Statistical downscaling first derives statistical relationships between the large-scale climatic state and local variations from historical data, and then employs such relationships to develop detailed future projections from the coarse GCM output.

Over the U.S., precipitation change has not been uniform, and is not predicted to be uniform in the future. For example, the relatively dry western U.S. is getting even drier, whereas the relatively wet eastern U.S. is getting wetter. With increasing precipitation, the change is also different for rainfall events of different intensities. Here we give an example of using regional climate models to project average and extreme precipitation in the Midwest United States, an agriculturally important region where climate change could have a profound impact. This study uses an ensemble of six RCMs driven by four GCMs over the domain of North America under the North American Regional Climate Change Assessment Program (NARCCAP). These regional models typically have a spatial resolution of 50 km. To evaluate these models, the RCMs were run retrospectively; the model predictions of daily precipitation during 1971–2000 were compared to the daily precipitation data of 204 climate stations in the region. The results of this comparison were also used to correct the biases in the climate model simulations. Future daily precipitation then was simulated for the period 2040–2070 (mid-century) under a medium high CO_2 emission scenario similar to SSP3-7.0. The model results were then subjected to statistical analysis in order to determine the present and future magnitudes of storms of various **return intervals**. Return interval is a relative measure of event intensity. For example, a storm with a return interval of

20 years (called a 20-year storm) typically occurs only once every 20 years. Longer return intervals mean more extreme (higher intensity) storms.

Figure 22 shows the percentage change of mean precipitation over the region, and Fig. 23 shows the percentage change for extreme storms with various return intervals. These results indicate that the Midwest U.S. region is likely to see a precipitation increase in the range of 3–15% by the mid-twenty-first century, with an average precipitation increase of ~ 7%. The rates of increase are likely to be higher for extreme precipitation events than for average precipitation; for example, the increase is predicted to range from 12% for the 1-year storms to 20% for the 20-year storms. Spatially, the northern part of the study area could see a larger increase in both mean precipitation and extreme precipitation events, likely due to increased evaporation of the Great Lakes from a combination of higher temperature and less ice cover under a warmer climate.

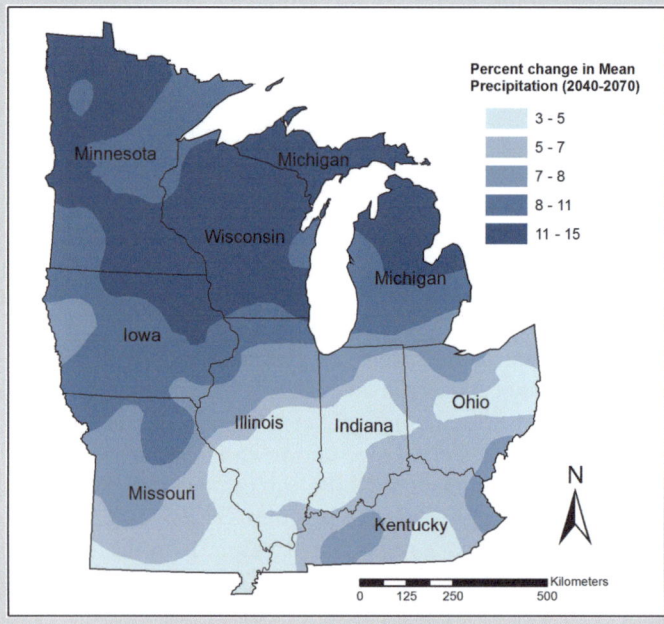

Fig. 22 Projected percent change in mean precipitation in the Midwest U.S. From Wu (2012)[33]

---

[33] Wu (2012).

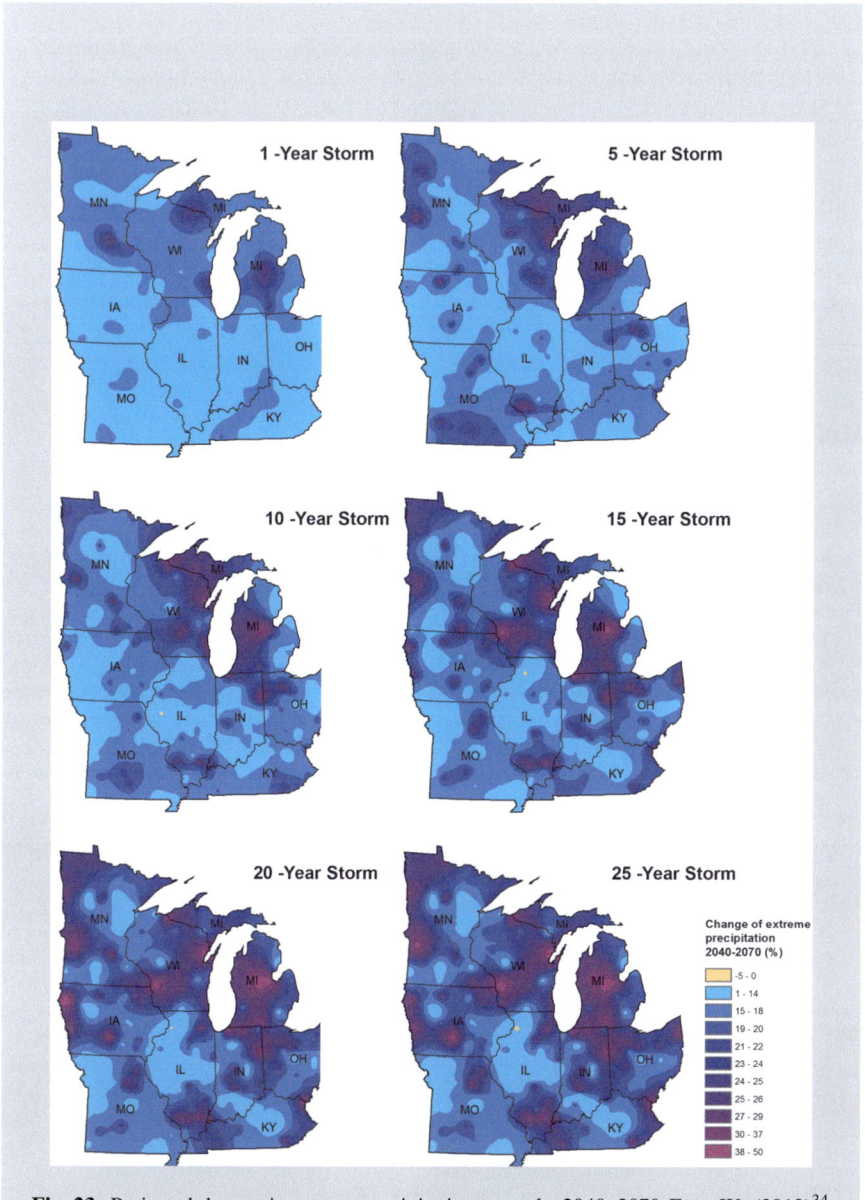

Fig. 23 Projected changes in extreme precipitation events by 2040–2070. From Wu (2012)[34]

---

[34] Wu (2012).

> At the time this chapter was written, record-breaking rainfall brought devastating flash floods and landslides to Missouri, Kentucky, and other parts of the Midwest United States (late July 2022). The flash floods in Kentucky killed at least 37 people. Flooding is the most expensive and one of the most fatal natural disasters in the U.S. The increase in precipitation in general, and the disproportionate increase in extreme rainfall in particular, has caused and will continue to cause devastating socioeconomic consequences under a warming climate.

References

Bock, L., Lauer, A., Schlund, M., Barreiro, M., Bellouin, N., Jones, C., Meehl, G. A., Predoi, V., Roberts, M. J., & Eyring, V., (2020). Quantifying progress across different CMIP phases with the ESMValTool. *Journal of Geophysical Research: Atmospheres, 125*(21), e2019JD032321. https://doi.org/10.1029/2019JD032321

Firestone, R. B., West, A., Kennett, J. P., Becker, L., Bunch, T. E., Revay, Z. S., Schultz, P. H., Belgya, T., Kennett, D. J., Erlandson, J. M., Dickenson, O. J., Goodyear, A. C., Harris, R. S., Howard, G. A., Kloosterman, J. B., Lechler, P., Mayewski, P. A., Montgomery, J., Poreda, R., ... Wolbach, W. S. (2007). Evidence for an extraterrestrial impact 12,900 years ago that contributed to the megafaunal extinctions and the Younger Dryas cooling. *Proceeding of the National Academy of Science (PNAS), 104*, 16016–16021. https://doi.org/10.1073/pnas.070697710

Gleckler, P., Taylor, K., & Doutriaux, C. (2008). Performance metrics for climate models. *Journal of Geophysical Research: Atmospheres, 113*, D06104.

IPCC. (2007). In Core Writing Team, R. K. Pachauri, & A. Reisinger (Eds.), *Climate change 2007: Synthesis report. Contribution of Working Groups I, II and III to the Fourth Assessment Report of the Intergovernmental Panel on Climate Change* (104 pp.). IPCC.

IPCC. (2013). In T. F. Stocker, D. Qin, G.-K. Plattner, M. Tignor, S. K. Allen, J. Boschung, A. Nauels, Y. Xia, V. Bex, & P. M. Midgley (Eds.), *Climate change 2013: The physical science basis. Contribution of Working Group I to the Fifth Assessment Report of the Intergovernmental Panel on Climate Change* (1535 pp.). Cambridge University Press. https://www.ipcc.ch/report/ar5/wg1/

IPCC. (2021a). Summary for policymakers. In V. Masson-Delmotte, P. Zhai, A. Pirani, S. L. Connors, C. Péan, S. Berger, N. Caud, Y. Chen, L. Goldfarb, M. I. Gomis, M. Huang, K. Leitzell, E. Lonnoy, J. B. R. Matthews, T. K. Maycock, T. Waterfield, O. Yelekçi, R. Yu, & B. Zhou (Eds.),*Climate change 2021: The physical science basis. Contribution of Working Group I to the Sixth Assessment Report of the Intergovernmental Panel on Climate Change* (pp. 3–32). Cambridge University Press. https://doi.org/10.1017/9781009157896.001

IPCC. (2021b). Chapter 4. In V. Masson-Delmotte, P. Zhai, A. Pirani, S. L. Connors, C. Péan, S. Berger, N. Caud, Y. Chen, L. Goldfarb, M. I. Gomis, M. Huang, K. Leitzell, E. Lonnoy, J. B. R. Matthews, T. K. Maycock, T. Waterfield, O. Yelekçi, R. Yu, & B. Zhou (Eds.), *Climate change 2021: The physical science basis. Contribution of Working Group I to the Sixth Assessment Report of the Intergovernmental Panel on Climate Change* [J.-Y. Lee, J. Marotzke, G. Bala, L. Cao, S. Corti, J. P. Dunne, F. Engelbrecht, E. Fischer, J. C. Fyfe, C. Jones, A. Maycock, J. Mutemi, O. Ndiaye, S. Panickal, & T. Zhou, 2021: Future Global Climate: Scenario-Based Projections and Near-Term Information. In Climate Change 2021: The Physical Science Basis. Contribution of Working Group I to the Sixth Assessment Report of the Intergovernmental Panel

on Climate Change] (pp. 553–672). Cambridge University Press. https://doi.org/10.1017/9781009157896.006
IPCC. (2021c). In V. Masson-Delmotte, P. Zhai, A. Pirani, S. L. Connors, C. Péan, S. Berger, N. Caud, Y. Chen, L. Goldfarb, M. I. Gomis, M. Huang, K. Leitzell, E. Lonnoy, J. B. R. Matthews, T. K. Maycock, T. Waterfield, O. Yelekçi, R. Yu, &and B. Zhou (Eds.), *Climate change 2021: The physical science basis. Contribution of Working Group I to the Sixth Assessment Report of the Intergovernmental Panel on Climate Change.* Cambridge University Press, Cambridge (in press). https://doi.org/10.1017/9781009157896
IPCC. (2021d). Technical summary. In V. Masson-Delmotte, P. Zhai, A. Pirani, S. L. Connors, C. Péan, S. Berger, N. Caud, Y. Chen, L. Goldfarb, M. I. Gomis, M. Huang, K. Leitzell, E. Lonnoy, J. B. R. Matthews, T. K. Maycock, T. Waterfield, O. Yelekçi, R. Yu, & B. Zhou (Eds.), *Climate change 2021: The physical science basis. Contribution of Working Group I to the Sixth Assessment Report of the Intergovernmental Panel on Climate Change* [D. Chen, M. Rojas, B.H. Samset, K. Cobb, A. Diongue Niang, P. Edwards, S. Emori, S. H. Faria, E. Hawkins, P. Hope, P. Huybrechts, M. Meinshausen, S. K. Mustafa, G.-K. Plattner, & A.-M. Tréguier, 2021: Framing, Context, and Methods. In Climate Change 2021: The Physical Science Basis. Contribution of Working Group I to the Sixth Assessment Report of the Intergovernmental Panel on Climate Change] (pp. 147–286). Cambridge University Press. https://doi.org/10.1017/9781009157896.003
IPCC. (2023). In Core Writing Team, H. Lee, & J. Romero (Eds.), *Climate change 2023: Synthesis report. A report of the Intergovernmental Panel on Climate Change. Contribution of Working Groups I, II and III to the Sixth Assessment Report of the Intergovernmental Panel on Climate Change.* IPCC (in press).
Meinshausen, M., Nicholls, Z. R., Lewis, J., Gidden, M. J., Vogel, E., Freund, M., Beyerle, U., Gessner, C., Nauels, A., Bauer, N., Canadell, J. G., Daniel, J. S., John, A., Krummel, P. B., Luderer, G., Meinshausen, N., Montzka, S. A., Rayner, P. J., Reimann, S., … Wang, R. H. (2020). The shared socio-economic pathway (SSP) greenhouse gas concentrations and their extensions to 2500. *Geoscientific Model Development, 13*(8), 3571–3605.
Merritts, D. J., De Wet, A., & Menking, K. (1998). *Environmental geology: An Earth system science approach*, Figure 3.2 (550 p.). Freeman. ISBN-10: 0716728346.
National Academies. (2002). *Abrupt climate change: Inevitable surprises*. Available at National Research Council. 2002. *Abrupt climate change: Inevitable surprises*. The National Academies Press. https://doi.org/10.17226/10136
Renssen, H., Mairesse, A., Goosse, H., Mathiot, P., Heiri, O., Roche, D. M., Nisancioglu, K. H., & Valdes, P. J. (2015). Multiple causes of the Younger Dryas cold period. *Nature Geoscience, 8*(12), 946–949.
Stocker, T. F., & Wright, D. G. (1991). Rapid transitions of the ocean's deep circulation induced by changes in the surface water fluxes. *Nature, 351*, 729–732.
USGCRP. (2017). In D. J. Wuebbles, D. W. Fahey, K. A. Hibbard, D. J. Dokken, B. C. Stewart, & T. K. Maycock (Eds.), *Climate science special report: Fourth national climate assessment* (Vol. I, 470 pp.). U.S. Global Change Research Program.
Wu, S. Y. (2012). Projecting changes in extreme precipitation in the midwestern United States using North American Regional Climate Change Assessment Program (NARCCAP) regional climate models. *Greenhouse Gases: Emission, Measurement and Management, 337*, 354.
Wunsch, C. (2007). Abrupt climate change: An alternative view. *Quaternary Research, 65*, 191–203.

Shuang-Ye Wu often considers herself an accidental scientist. Working for the National Environmental Protection Agency in China after her Master's degree in linguistics, she became fascinated by the multi-faceted and interconnected nature of climate science. She went on to obtain her Master's and Doctoral degrees in geosciences from the University of Cambridge, focusing on the impact of climate change on the hydrological cycle both in the past and future. She has worked in the Department of Geology and Environmental Geosciences at the University of Dayton since

2004, and is currently a professor and the Chair of the Department. Her research uses both historical proxies and observations and climate models to examine the changing patterns of precipitation, floods, droughts under different climate drivers, and how such changes affect vegetation and land cover. Shuang-Ye is the primary author of Chap. 10.

Open Access This chapter is licensed under the terms of the Creative Commons Attribution 4.0 International License (http://creativecommons.org/licenses/by/4.0/), which permits use, sharing, adaptation, distribution and reproduction in any medium or format, as long as you give appropriate credit to the original author(s) and the source, provide a link to the Creative Commons license and indicate if changes were made.

The images or other third party material in this chapter are included in the chapter's Creative Commons license, unless indicated otherwise in a credit line to the material. If material is not included in the chapter's Creative Commons license and your intended use is not permitted by statutory regulation or exceeds the permitted use, you will need to obtain permission directly from the copyright holder.

Large-Scale Climate Interventions: Carbon Dioxide Removal and Solar Radiation Management

Walker Raymond Lee, Douglas MacMartin, and Amanda Borth

Guiding Question: Is anthropogenic climate change fixable?

1 Key Take-Away Points

- It is very likely that carbon emissions will cause global warming to exceed the 1.5 °C threshold prescribed by the 2015 Paris Climate Agreement. Based on recent emissions trends, the planet is on track for 2–4 °C of warming by the year 2100.
- The surface temperature of the Earth is determined largely by the planet's energy balance: the fluxes of radiation into and out of the Earth system. The two main components of the energy balance are the shortwave energy from the Sun and the longwave energy emitted by the Earth.
- Carbon emissions warm the planet by reducing the amount of longwave radiation that can leave the planet, and this is causing surface temperatures to rise.
- Any proposed attempt by humans to modify the Earth's energy balance in order to cool the planet would be a climate intervention or geoengineering. Climate intervention proposals fall into two broad categories: carbon dioxide removal (CDR) and solar radiation management (SRM).

W. R. Lee (✉)
National Center for Atmospheric Research, Boulder, CO, USA
e-mail: walkerl@ucar.edu

D. MacMartin
Sibley School of Mechanical and Aerospace Engineering, Cornell University, Ithaca, NY, USA
e-mail: dgm224@cornell.edu

A. Borth
Consortium for Science, Policy & Outcomes at Arizona State University, Washington D.C., USA
e-mail: amanda.borth@asu.edu

© The Author(s) 2025
K. St. John and L. Krissek (eds.), *Climate Change*,
https://doi.org/10.1007/978-3-031-82869-0_11

- CDR proposals aim to remove carbon dioxide from the atmosphere through photosynthesis or through chemical reactions that produce carbonate compounds.
- CDR directly offsets carbon emissions, and at a large enough scale, CDR could reduce or eliminate global warming. However, the technologies are immature and often compete with each other (e.g., for land), and there is currently no technology that can be implemented economically at the necessary scale.
- SRM aims to alter the Earth's energy balance by scattering particles into the atmosphere. Stratospheric aerosol injection (SAI) and marine cloud brightening (MCB) would reduce the amount of sunlight reaching the surface, while cirrus cloud thinning (CCT) would increase the amount of longwave energy leaving the atmosphere.
- SRM would cool the planet, but it would not directly counteract the mechanism by which carbon emissions warm the planet, and therefore "global warming + SRM" would not produce the same climate state as "reduced greenhouse gas emissions". SRM would have unintended and possibly unwanted consequences, which could include changes to circulation and precipitation patterns and impacts to ecosystems.
- The three proposed methods of SRM are understood to varying degrees, but any of them would introduce a host of physical, political, and ethical complications.
- Substantially more research is required to determine the roles that CDR or SRM could play in responding to climate change.
- To make informed decisions, it is essential that entities with decision-making power, such as governments and corporations, engage with publics, stakeholders, and communities directly impacted by climate intervention decisions.

2 Introduction: Earth's Energy Balance

The 2015 Paris Climate Accords established the goal of limiting global warming above pre-industrial temperatures to well below 2 °C, and ideally to below 1.5 °C. However, it is a near certainty that this goal will not be achieved. The international community's pledged emission cuts are well below what is needed to meet these targets, and current trajectories place us on track for 2–4 °C of warming by the end of the century. Even if emissions are reduced, warming is proportional to greenhouse gas (GHG) concentrations in the atmosphere, not to the rate at which we produce them. Therefore, even if we were to completely cut GHGs out of society by mid-century, it wouldn't erase global warming—it would only stop the problem from getting worse. Even if emissions were zeroed tomorrow (an obviously impossible feat!), the current (at the writing of this chapter) concentration of atmospheric CO_2 is 421 ppm, and the current NOAA Annual Greenhouse Gas Index, which factors in other greenhouse gasses, measures over 500 ppm of CO_2 equivalent. The pre-industrial concentration was approximately 280 ppm, and each part per million represents more than 7 billion tons of CO_2; therefore, we are more than 1000 gigatons of CO_2 above the pre-industrial equilibrium, and these emissions could still very well cause temperatures to rise above the 1.5 °C target. In simpler terms, no matter what we do in the future, our past emissions may have already committed us to at least 1.5 °C of warming. In fact, in 2024 the planet's average temperature already breached the 1.5 °C threshold.

So, where does that leave us? If there is a possibility—even a small one—that emission cuts will not be sufficient to prevent unacceptable consequences of climate change, it may be worth consideration to look into **climate intervention**, also called *climate engineering* or geoengineering: deliberately modifying the Earth's climate system on a planetary-scale through artificial means. Figure 1 shows a simple radiative flux diagram for the Earth: the Earth receives energy, largely in the form of *shortwave* visible radiation, from the Sun, and radiates that energy back into space in the form of *longwave* infrared radiation. According to the First Law of Thermodynamics, if the energy fluxes received and emitted by the planet are approximately equal, then the planet will remain in *radiative equilibrium,* and Earth's average temperature will remain constant. However, GHGs trap infrared energy inside the planetary system; since GHG concentrations are increasing, incoming energy now exceeds outgoing energy, and Earth's average temperature has begun to rise. The idea behind climate intervention is to artificially correct this imbalance, through methods that generally fall into one of two categories: **carbon dioxide removal (CDR)** and **solar radiation management (SRM)**. As shown in Fig. 1, CDR aims to directly offset global warming by removing GHGs from the atmosphere, resulting in less longwave radiation trapped and more emitted back to space. Conversely, SRM aims to decrease the amount of shortwave energy reaching the surface to balance the decrease in outgoing longwave radiation under global warming. In this chapter, we discuss both ideas, including feasibility, cost, logistics, and ethics (Table 1).

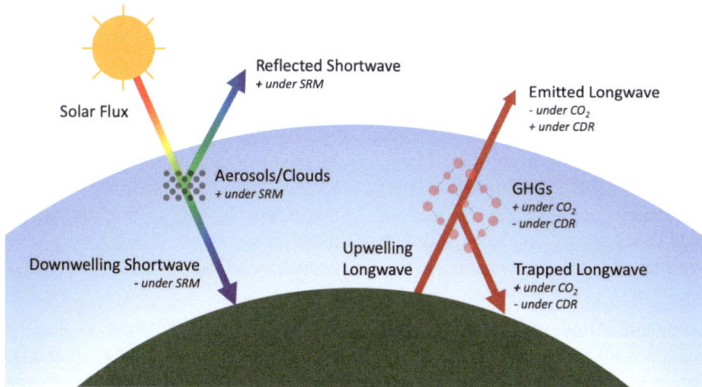

Fig. 1 Energy balance of the Earth system. "+" and "−" indicate how energy fluxes would change under warming (denoted "CO_2"), carbon dioxide removal (denoted "CDR"), and solar radiation management (denoted "SRM"); "+" denotes that the flux will increase under the denoted circumstances, whereas "−" denotes that the flux will shrink

Table 1 Overview of CDR and SRM methods, with brief, qualitative comments for each on the potential to impact the climate, primary limitations, cost relative to other methods of climate intervention, and current level of scientific understanding

Method	Potential impact and primary limitations	Relative cost at present or near future	Scientific understanding
Carbon dioxide removal (CDR)			
Forest management: expansion and management of forests to increase CO_2 uptake	Low—limited by forest area, demand for wood and land, and slow rate of carbon uptake	Very low	Very high—managing forests is not fundamentally challenging
Coastal blue carbon: management of coastal ecosystems to increase CO_2 uptake	Very low—limited by available coastal area, possible future sea level rise, and rate of carbon uptake	Very low	Very high—managing wetlands is not fundamentally challenging
Bioenergy with carbon capture and storage (BECCS): energy production through plant biomass while capturing any carbon produced, resulting in a net-negative emissions process	Moderate—limited by available land area and demand for plant use for food and fiber production	Moderate to very high; likely increases with scaling	Moderate—limited by understanding of the chemical reactions that govern the biomass burning and carbon capture process
Direct air capture (DAC): chemical reactions that directly remove carbon from the ambient air	High—competes for land area and power draw, but could theoretically be implemented anywhere; captured carbon also must be stored	Extremely expensive	Moderate—advancements are needed to reduce cost, but the technology does exist at present
Geological carbon sequestration: storage of captured carbon deep underground in stable rock formations	High (effectively unlimited storage capacity), but needs to be coupled with a mode of carbon capture	Modest—actual sequestration is not expensive, but mode of carbon capture could be	Moderate to good—other waste has been stored in a similar manner for decades
Carbon mineralization: induced reactions to induce "weathering," in which CO_2 bonds with a reactive mineral	Unknown—low scientific understanding limits ability to estimate potential for captured CO_2; enormous ranges for different minerals and reactions	Unknown—cost could be low or high depending on rate of scientific progress; large ranges for different minerals and reactions	Very low—large pool of possible minerals, each with many relevant chemical reactions

(continued)

Table 1 (continued)

Method	Potential impact and primary limitations	Relative cost at present or near future	Scientific understanding
Solar radiation management (SRM)			
Stratospheric aerosol injection (SAI): scattering of reflective particles into the middle atmosphere to reflect sunlight	Very high—injection rates can be tailored to achieve desired global cooling, but political and ethical concerns are enormous; will affect everyone on the planet in some way	Likely low relative to other methods of climate intervention	Low to moderate—fundamental mechanism is clear, but uncertainties present in climate response
Marine cloud brightening (MCB): seeding of low ocean clouds to brighten them and reflect sunlight	Likely moderate—can be scaled, but bounded by available ocean area and susceptibility of clouds to artificial brightening. More localized, but political and ethical concerns still present	Likely low	Low—aerosol-cloud interactions are extremely challenging to represent in models
Cirrus Cloud Thinning (CCT): seeding of high clouds that trap radiation to thin them	Uncertain, but likely very low—bounded by availability and susceptibility of cirrus clouds	Likely low	Very low—unclear exactly which clouds can be targeted, and how to actually produce a cooling effect instead of a warming effect

Note that this table is intended to provide a snapshot of the current state of the science; impact, cost, feasibility, and limitations for some of these methods could change, possibly very rapidly, depending on investments, technological advancements, and the political landscape. Adapted from NASEM 2019, Table S1[1]

3 Carbon Dioxide Removal (CDR)

3.1 Introduction

Since the root cause of modern global warming is the increased concentrations of GHGs in the atmosphere, the most straightforward response would be to reduce those concentrations. While this can primarily be achieved by reducing emissions (i.e., mitigation), it could also be achieved by artificially increasing the rate at which they are removed. Figure 2 shows a simplified version of the carbon cycle: the

---

[1] National Academies of Sciences, Engineering, and Medicine (2019).

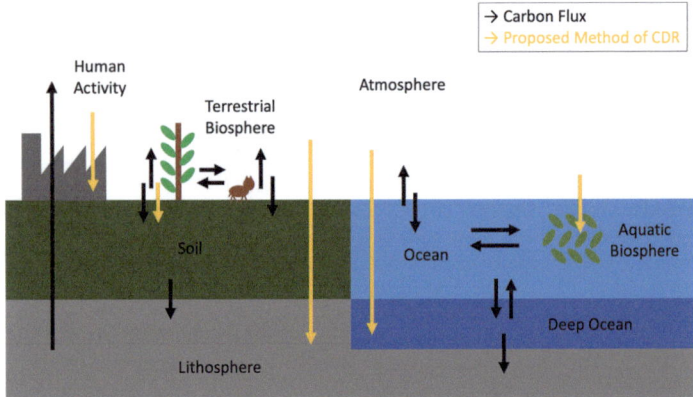

Fig. 2 A simplified version of the carbon cycle. Black lines represent fluxes in the carbon cycle, and gold lines represent potential paths for increased carbon uptake or storage through CDR

pathways through which CO_2 and other carbon molecules are naturally exchanged between the atmosphere, living matter, the soil, the ocean, and the lithosphere. Carbon dioxide in the atmosphere is naturally removed by plants through photosynthesis and is also stored in the soil. Buried organic matter can be incorporated into carbon-containing sedimentary rocks over thousands of years. Carbon dioxide in the atmosphere diffuses into the surface ocean (recall the discussion of *ocean acidification* in Chap. 8), and CO_2 reaches the deep ocean via advection from the surface ocean and the decomposition of marine organic matter. However, many of these processes are very slow compared to the rate at which humans emit carbon dioxide (as of the early 2020s, humans emit more than 50 Gt of CO_2 each year). Gold arrows in Fig. 2 denote pathways for carbon uptake or storage which could be artificially enhanced, as well as additional artificially constructed methods for long-term carbon removal. Such processes are commonly called *carbon dioxide removal* (CDR), or *carbon capture and storage* (CCS). In this section, we provide brief overviews of some of the proposed methods of CDR, as well as an overall assessment of prospects for the future of CDR technology.

3.2 Methods of CDR

Broadly speaking, there are two main ways to remove carbon dioxide from the atmosphere: through photosynthesis, and through chemical reactions with alkaline materials (a group of chemical compounds that react naturally with CO_2) to form carbonate compounds. Carbon captured through photosynthesis is generally stored in soils and plant matter, while carbon captured as carbonate compounds can either be left in mineral form or extracted and stored elsewhere, such as deep underground. These possibilities have given rise to several methods of CDR; here, we give a brief introduction to each. Some categories described below combine both a mode to capture carbon from the atmosphere and a mode to store it in a permanent or semi-permanent state, while others only capture carbon and must be combined with another mechanism to sequester it, or vice versa. In evaluating individual methods of CDR, as well as prospects for CDR as a whole, important aspects to consider include cost (usually expressed in dollars per ton of CO_2), capacity to capture and/or sequester CO_2 (commonly expressed in gigatons per year), expected lifetime of sequestered CO_2 (if applicable), practical barriers to implementation, and public perception.

3.2.1 Forest Management

Perhaps the most well-known natural pathway for the removal of carbon dioxide from the atmosphere is uptake by plant life through photosynthesis and storage in plants and soils. This category of methods aims to increase uptake through this pathway through the planting of new forests (*afforestation*), the replanting of previously deforested areas and the regeneration of damaged forests (*reforestation*), and other forest management practices designed specifically to enhance the ability of woodland areas to take in and store carbon (*improved forest management,* or IFM). Forest management (Fig. 3) is generally perceived by the public as a "natural" mode of CDR, and as such it tends to have widespread approval among both laypersons and experts. Compared to other modes of CDR, forest management also has the potential for extremely low cost, with estimates generally in the range of $18–20/ton, and one estimate even as low as $1/ton. Additionally, there are enormous benefits to forest management practices, both for biodiversity and conservation efforts and directly for humans, such as improved water quality. However, forest management as a method of CDR faces severe scaling limitations; both existing forests and potential new forest areas face competition for land use, and care needs to be taken to ensure that new forest land is developed ethically. The potential capacity for increased carbon storage is small, likely on the order of 10 Gt/yr, and a total storage of more than 100 Gt total by the end of the century could require unfeasibly large areas of land. Additionally, the growth of a forest is a slow process, and it would likely take as long as a hundred years for any new forest to sequester appreciable amounts of carbon. Finally, because carbon sequestered by plants is part of the natural carbon cycle, additional carbon storage through forest management is not as permanent as through other methods of CDR.

3.2.2 Coastal Blue Carbon

The world's mangroves, marshes, and other wetland regions hold huge quantities of carbon; the maintenance, upkeep, and management of these regions for the purpose of increasing their carbon uptake and storage is referred to as **coastal blue carbon** (Fig. 4). As a mode of CDR, coastal blue carbon has many similarities to forest management, but the prospective advantages and disadvantages of coastal blue

Fig. 3 Reforestation—the replanting of depleted forests—has many ecological benefits, including increased capture and storage of carbon dioxide. Left: a United States Forest Service worker plants seeds in a burned area.[2] Center: a 21-year-old plantation of red pine trees in Ontario, Canada.[3] Right: a tropical tree nursery in Vichada Department, Colombia.[4] *Sources* Wikimedia Commons

carbon are respectively more extreme. Coastal blue carbon is likely one of the least expensive mechanisms to enhance carbon capture and storage; efforts in the United States would likely cost between $0 and $20 per ton, and worldwide efforts would likely cost no more than $100/ton. Just as with forest management, the preservation of wetland regions also has hugely beneficial side effects toward biodiversity and sustainability. However, efforts in this area are limited by total wetland area; coastal blue carbon will likely only be able to sequester a very small amount of additional carbon, with projections for global efforts ranging from 5 Gt/yr to as little as 0.1 Gt/yr. Additionally, global-warming-driven changes such as increasing temperatures and sea level rise are eroding wetland ecosystems. Therefore, coastal blue carbon efforts could be undone by future climate change; as sea levels rise, the coastal wetlands will migrate inland, but in any regions where sea level rise is faster than coastal wetland migration, the wetlands will be lost, and any efforts to manage those wetlands would be lost with them.

3.2.3 Biomass Energy with Carbon Capture and Storage (BECCS)

The concept of a carbon-negative power plant, in which plant matter is converted into biofuel and the resultant carbon waste is captured and sequestered, is referred to as **biomass energy with carbon capture and storage**, or BECCS (Fig. 5). There are several possible modes for the conversion of plant matter into energy, including

---

[2] Wikimedia Commons, https://commons.wikimedia.org/wiki/File:Replanting_a_burned_area_on_the_Idaho_Panhandle_NF_(39730798234).jpg. Public domain.

[3] Wikimedia Commons, https://commons.wikimedia.org/wiki/File:RedPinePlantation.JPG. Photo by Padraic Ryan, Licensed under the Creative Commons Attribution-Share Alike 3.0 Unported, 2.5 Generic, 2.0 Generic, and 1.0 Generic license.

[4] Wikimedia Commons, https://commons.wikimedia.org/wiki/File:Hasta_luego_001.JPG. Licensed under the Creative Commons Attribution 3.0 Unported license.

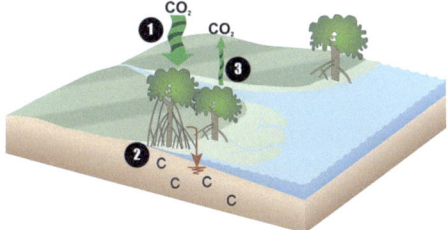

Fig. 4 Carbon moves in and out of coastal wetlands. Plants take up carbon through photosynthesis (1), and as plants break down in tidal-water-covered soil, carbon is stored beneath the surface (2). Other carbon is stored in roots, leaves, and branches, while some escapes back into the atmosphere through respiration (3) *Source* NOAA[5]

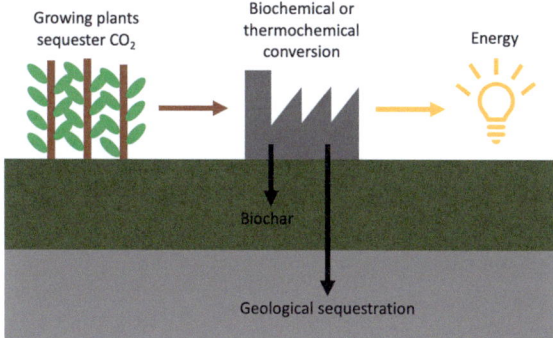

Fig. 5 Process diagram for Biomass Energy with Carbon Capture and Storage (BECCS), in which biomatter is converted into energy through biochemical or thermochemical reactions. Waste CO_2 is captured and sequestered, either as compressed CO_2 gas or as carbon-rich biochar, resulting in a process with net-negative emissions

biochemical pathways (such as conversion using yeast or bacteria), and thermochemical pathways (controlled heat-based conversions such as *gasification*, the conversion of carbon-rich materials into combustible gasses, and *pyrolysis*, the decomposition of organic substances into volatile compounds and high-carbon residue called *biochar*). Each of these processes produces energy alongside carbon waste in the form of either a waste stream of pure or nearly-pure CO_2 or a carbon-rich solid product. If the waste stream is captured and sequestered (see, for example, Sect. 3.2.5, "Geological Carbon Sequestration"), or the solid carbon-rich waste is returned to the soil, the amount of carbon captured by growing plants through photosynthesis can exceed the amount of carbon emitted through agricultural, production, and transportation processes, resulting in a net removal of carbon from the atmosphere.

---

[5] NOAA, "Coastal Blue Carbon," https://oceanservice.noaa.gov/ecosystems/coastal-blue-carbon/. Public domain.

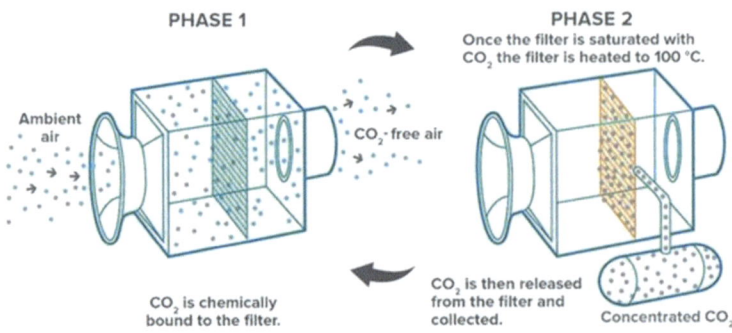

Fig. 6 Overview of a chemical process through which CO_2 is "scrubbed" from the ambient air and isolated for storage. *Source* ACS Publications[6]

Currently, the major factor limiting the implementation and scaling-up of BECCS is cost; depending on the source of fuel, method of conversion, method of storage, and other factors, cost estimates range from $15 to $400/ton. If the technology evolves to the point where BECCS can be scaled up profitably, the limiting quantities for capacity will be the future availability of land and biomass; because these quantities are uncertain, estimates for global capacity vary substantially, ranging from little over 1 Gt/yr of CO_2 to over 70 Gt/yr of CO_2. Realistically, competition for land use will likely limit global BECCS capacity to under 5 Gt/yr.

3.2.4 Direct Air Capture (DAC)

The direct "scrubbing" of CO_2 from the open atmosphere via chemical reaction is called **direct air capture**, or **DAC** (also called *direct air carbon capture and storage,* or *DACCS*). Current DAC technology uses large fans to draw air across chemicals that bond with CO_2; the CO_2 is subsequently recovered by breaking the bonds with high temperatures or other chemical reactions, after which it can be sequestered (Fig. 6). The most significant present-day barrier to the implementation of DAC is cost; currently, DAC costs about $600/ton, and while it is likely that operating costs can be reduced to $200–300/ton in the next few years, significant global investment and larger scales are likely necessary to bring prices down to $100/ton. If DAC becomes economically feasible, scaling up will be limited by infrastructure, energy draw, and land area, as the process requires both electrical energy to power the fans and thermal energy to recover captured carbon. This necessitates either land area to provide carbon-free energy (i.e., wind or solar), or increased DAC output to account for the additional carbon released by other sources of energy. There are currently few estimates for a theoretical upper bound on the capacity for DAC to capture and sequester carbon, but achieving 5 Gt/yr on a global-scale would require a significant global undertaking.

---

[6] Beuttler et al. (2019). Used under the CC-BY 4.0 Creative Commons license.

3.2.5 Geological Carbon Sequestration

Among the most promising prospective locations for the storage of carbon captured through BECCS or DAC is deep underground. Such geologic carbon storage is shown in Fig. 7. Different layers of sedimentary rock in the Earth's crust have different properties; some layers are impermeable, and others have interconnected pores that can hold fluids such as brine, oil, and natural gas. A naturally occurring sequence of layers such that a permeable porous layer is located directly beneath an impermeable layer is conducive to the storage of CO_2. At such a location, a well can be drilled such that supercritical CO_2 (i.e., CO_2 in a fluid state because it is held at high temperatures and pressures) is injected into the permeable porous layer, and the impermeable layer above serves as a "cap" that prevents the injected carbon from escaping. This method for disposing of unwanted waste fluids is not unique to carbon dioxide and has been used for over a hundred years for disposal of other waste fluids. Statistically, geological sequestration is stable, with leakages occurring in approximately 1 in 1000 cases. The injection of supercritical CO_2 has also been used to enhance the extraction of oil or gas from their subsurface reservoirs. Based on the physics of the sequestration, decades of research experience and data collection from waste disposal and CO_2-enhanced oil and gas recovery, and numerical modeling, it is estimated with high levels of confidence that fluids stored more than a kilometer beneath the surface in this manner can remain trapped for tens of thousands of years or longer. Additionally, over long periods, the high pressure of supercritical CO_2 makes it more chemically prone to react with underground rocks to form new carbon-bearing minerals (e.g., calcite, $CaCO_3$), enhancing the expected storage lifetime. Estimated storage costs are on the order of $10/ton of sequestered CO_2, but this does not include the cost of the actual carbon capture, or of transportation to the sequestration site. Estimates for global storage capacity are on the order of thousands to tens of thousands of gigatons of CO_2.

3.2.6 Carbon Mineralization

As discussed in Chap. 7, the process of weathering silicate minerals uses carbon dioxide. When calcium or magnesium is also available (in silicate minerals or dissolved in water), the products will include carbonate compounds (see Chap. 7, Eqs. 7.1–7.4). **Carbon mineralization** refers to an enhanced or engineered means of inducing such chemical reactions, which can be used as a mode of carbon capture, a mode of carbon storage, or both. This mineralization can be initiated "naturally" by introducing high concentrations of CO_2 to reactive minerals (as in direct air capture), or it can be artificially accelerated in a high-pressure or high-temperature reactor ("*ex-situ*"; Fig. 8) or by injecting CO_2-rich fluids deep underground into the presence of such minerals ("*in-situ*," as with geological carbon sequestration). Igneous or metamorphic silicate rocks of the appropriate composition can be used for carbon mineralization. These include basalts, the most abundant volcanic rock

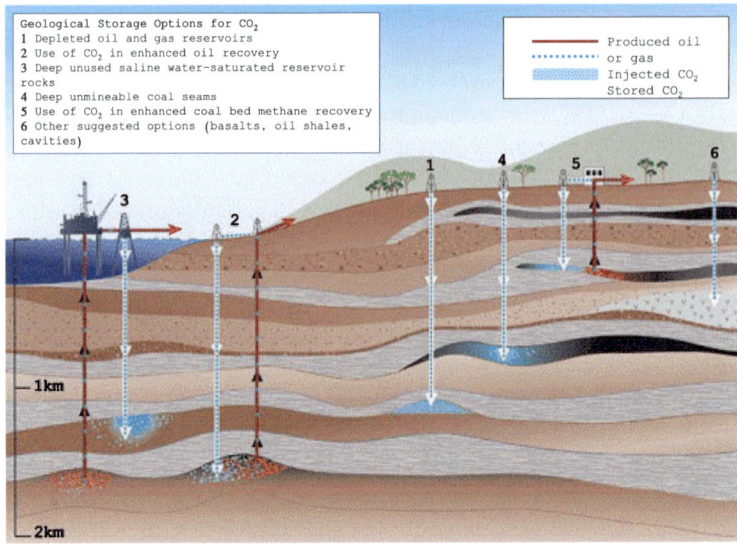

Fig. 7 CO_2 can be injected into wells drilled underground that tap into porous, permeable rock layers, where it can remain for thousands of years or longer when trapped by overlying layers of impermeable rock. *Source* Modified from Virginia Department of Energy[7]

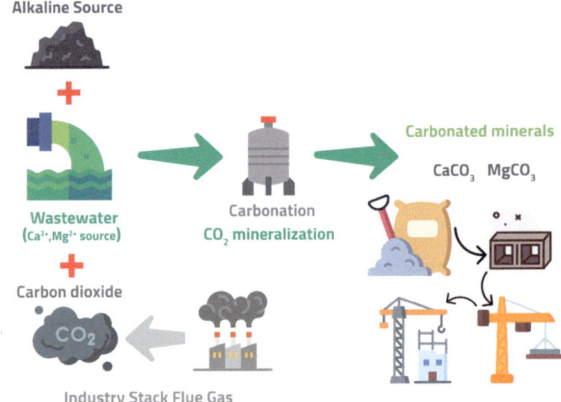

Fig. 8 Combining CO_2 with reactive minerals, especially those containing calcium or magnesium, can be used to store CO_2 in the form of carbonate compounds. These compounds could then be used, for example, in the construction of buildings. *Source* MDPI[8]

in the Earth's crust. As a result, there exists a huge number of possible pathways through which mineralization can be accomplished.

Because of the sheer variety of mineralization processes, it is difficult to give a single estimate for cost or capacity that encompasses the entire category. Across a

---

[7] Virginia Energy, Carbon Sequestration. https://energy.virginia.gov/geology/CarbonStorage.shtml. Public domain.

[8] Alturki (2022). Used under the open-access CC BY Creative Commons license.

dozen different in-situ, ex-situ, and ambient methods, actual or expected costs range from $10 to $500/ton, and estimates for storage capacities range from less than 1 Mt/yr to 10 Gt/yr or more. Ambient weathering using industrial or mining waste as reactants is relatively inexpensive but slow, and while the quantity of available reactants might be enough for storage of up to 3 Gt/yr, the slow reaction speed may limit actual capacity to less than this. Storage via ex-situ mineralization provides more opportunities for direct observation and control of the mineralization process than in-situ geological sequestration of supercritical CO_2 but must compete based on price. Ex-situ storage is currently much more expensive, and while in-situ mineralization has huge potential capacities on the order of tens of thousands of gigatons, this category faces challenges in the understanding of reaction kinetics and identification of reaction sites.

3.3 Practical, Logistical, and Societal Considerations of CDR

The amount of global warming caused by anthropogenic CO_2 emissions is roughly proportional to the concentration of CO_2 that humans have added to the atmosphere. Therefore, as long as humans continue to add more CO_2 to the atmosphere than natural pathways can remove, the Earth will continue to warm. In order to achieve a world without warming, humanity will need to transition to a society with *net-zero emissions*, in which either no GHGs are produced or any GHG emission is balanced by removal such as CDR. A society with no GHG emissions is near-impossible because it will be very difficult to remove the dependence of certain industries, such as air travel, on fossil fuels. As such, some amount of CDR is necessary for a net-zero future, and therefore necessary for a world without global warming.

The question remains: how much CDR is possible, and will it be enough? This topic is one of the fundamental debates surrounding responses to climate change. In its 2019 report,[9] the National Academies of Science, Engineering, and Medicine discuss the emission cuts required to limit global warming to less than 2 °C by 2100 (Fig. 9). The figure shows two pathways; the upper "business as usual" pathway represents a future with little mitigation. Emissions peak around 2080 and decline thereafter, but the world still emits over 60 Gt/yr by 2100. In contrast, the lower (blue line) pathway, necessary to limit global warming to under 2 °C, has sharp emissions cuts beginning as soon as possible, and the implementation and ramping up of CDR throughout the century: about 5 Gt/yr by 2050, about 10 Gt/yr by 2075, and about 15 Gt/yr by 2100. As discussed above, some sources of GHG emission cannot be eliminated, but these are balanced by CDR, reaching a net-zero society by around 2075 with net-negative emissions thereafter. Based on the current state of each of the broad CDR categories discussed above, the report estimates a global capacity for about 10 Gt/yr of CO_2 removal. Therefore, on the surface, there is hope that CDR might just be humanity's ticket out of the global warming mess: implement

---

[9] National Academies of Sciences, Engineering, and Medicine (2019).

CDR at full global capacity by mid-century, and technology will mature to the point where 15 Gt/yr is doable by the end of the century. Unfortunately, the reality is significantly more complicated. Firstly, there is a large gap between calculating a potential of 10 Gt/yr and actually implementing the technologies to achieve this rate of CDR. Consider the Paris Climate Agreement: the nations of the world pledged to reduce their emissions by an amount sufficient to limit global warming to less than 1.5 °C, but current assessments indicate that wealthy nations are far from on track to see those promises through. Even the Inflation Reduction Act of 2022, the largest climate change legislation ever passed by the U.S. Congress, will likely not be sufficient to cut U.S. greenhouse gas emissions in half by 2030.[10] At $100/ton, 10 Gt/yr of CDR amounts to $1 trillion a year, and there is no evidence that the international community will prioritize those costs any more than they have the costs of mitigation and sustainability. Secondly, the true global capacity for CDR is not as simple as summing up the potential contributions from each individual method. Many of the modes of CDR compete with each other for resources, especially land area: afforestation and reforestation require dedicated land to grow forests, BECCS requires land to grow biomass, and DAC requires land for infrastructure and power supply. Additionally, there will always be uncertainty in estimating how CDR will compete for resources with other needs as global population and subsequent demand for food and energy continue to increase. Care needs to be taken to ensure that CDR is implemented ethically and sustainably (see Chap. 12 for a discussion on ethical considerations). Finally, the "true" amount of CDR needed to prevent unacceptable consequences of global warming may well be more than 5–10 Gt/yr by mid-century, as given past trends, a sharp decrease in emissions in the next few years is far from guaranteed, and limiting warming to under 2 °C may not be enough to prevent the worst impacts.

In summary, the research and implementation of CDR are absolutely necessary if the worst consequences of global warming are to be abated. However, the technology is far from the state of readiness necessary for this to happen, and even if the technology does mature quickly enough, there is no guarantee that it will be implemented. The current state of CDR is perhaps best summed up by this September 2021 headline from Business Insider[11]: *"The world's biggest carbon-removal plant just opened. In a year, it'll negate just 3 seconds' worth of global emissions"*.

---

[10] A report by the Rhodium Group estimates that the Inflation Reduction Act will cut U.S. GHG emissions by 32–42% by 2030 (https://rhg.com/research/climate-clean-energy-inflation-reduction-act/); Princeton University's REPEAT Project estimates cuts of approximately 42% (https://repeatproject.org/docs/REPEAT_IRA_Prelminary_Report_2022-08-12.pdf); and the think tank Energy Innovation Policy & Technology estimates cuts of 37–41% (https://energyinnovation.org/publication/modeling-the-inflation-reduction-act-using-the-energy-policy-simulator/). However, changing priorities in the U.S. federal government may result in less funding and support for climate change mitigation efforts tied to the Inflation Reduction Act.

[11] Woodward (2021).

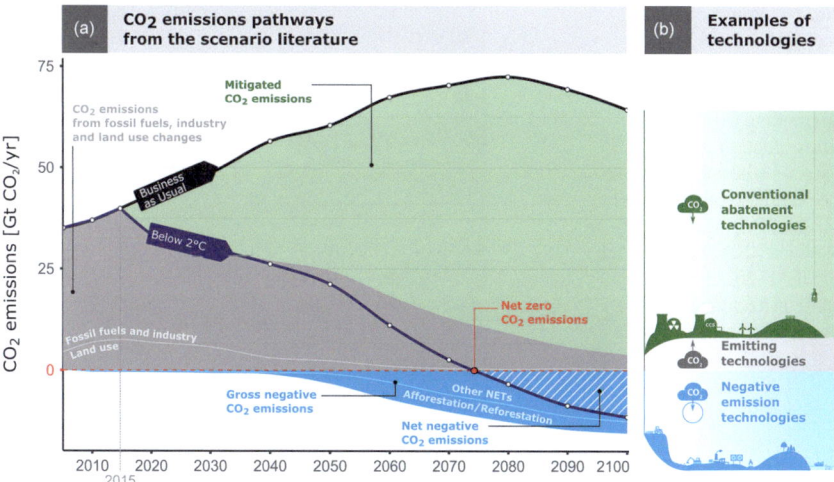

Fig. 9 CDR (blue), by itself, would not be enough to limit global warming to below 2 °C. However, with enough emissions cuts (green), CDR could balance those emissions which could not be eliminated (gray), achieving a "net-zero" society. *Source* Fuss et al. (2018), Environmental Research Letters, Creative Commons License[12]

4 Solar Radiation Management (SRM)

4.1 Introduction

The only permanent, sustainable solution to global warming is to tackle the root of the problem by reducing the amount of CO_2 in the atmosphere. Unfortunately, this is easier said than done. While no analogy is perfect, cutting emissions is like taking your foot off the gas when the car in front of you suddenly screeches to a halt: it would be silly not to, but it likely won't solve the problem by itself. In the case of global warming, the car is already going fast enough that taking one's foot off the accelerator won't prevent a crash; CO_2 levels are high enough that our emissions may well have already committed us to warming above 1.5 °C and possibly above 2 °C. The next step would be to hit the brakes, analogous to artificially removing carbon dioxide from the atmosphere. However, CDR needs to be scalable to gigatons per year, not have severe local impacts (like competition for food and clean water), and be cheap enough to be viable; as discussed in the previous section, right now there is no single approach that meets all of these requirements. The challenge is that, while there is hope, we cannot know today how rapidly these ideas will be developed, and whether they will be developed in time to prevent the worst impacts of climate change. In other words, using the car analogy, it is uncertain whether taking our

---

[12] Fuss et al. (2018). Creative Commons license.

Fig. 10 Proposed SRM methods include stratospheric aerosol injection (SAI), marine cloud brightening (MCB), and cirrus cloud thinning (CCT). Used with permission from the National Academies of Science, Engineering, and Medicine (NASEM). *Source* NASEM (2021).[13] This figure is excluded from Creative Commons license

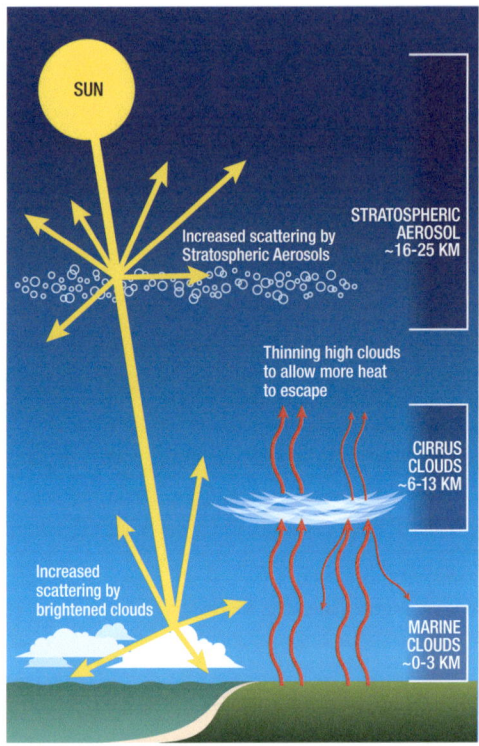

foot off the gas (i.e., cutting emissions) and hitting the brakes (i.e., employing CDR strategies) will be sufficient to avoid an accident.

One possible supplement to emission cuts and carbon dioxide removal is to reflect a small amount of incoming sunlight back to space in order to deliberately cool the planet. This concept is commonly called **solar radiation management** (**SRM**; Fig. 10), but has also been referred to as *solar radiation modification, sunlight reflection methods, climate engineering, solar geoengineering,* and *solar climate intervention*. At the time of writing this chapter, global average temperatures are approximately 1.2 °C warmer than pre-industrial levels; using SRM alone, offsetting all of this warming would require reflecting a little less than 1% of incoming sunlight back into space, cooling the climate for as long as the SRM effort is sustained. This could be done in several ways, including *stratospheric aerosol injection* (SAI), *marine cloud brightening* (MCB), and *cirrus cloud thinning* (CCT).

SRM would reduce global warming, but it does not deal with the underlying problem of increased CO_2 levels. As such, following the car accident analogy, SRM is analogous to seatbelts and airbags: it doesn't slow the car down, but it may reduce

---

[13] National Academies of Sciences, Engineering, and Medicine (2021). Reproduced with permission from the National Academy of Sciences, Courtesy of the National Academies Press, Washington, D.C.

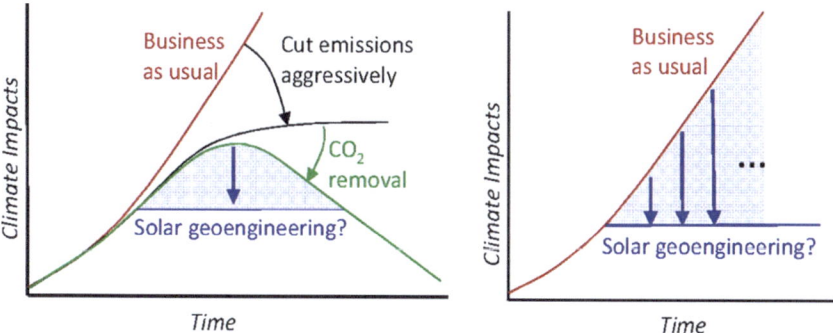

Fig. 11 A "contextual framing" for the use of SRM. The left panel demonstrates the role of SRM as part of a multi-faceted response to climate change, in which solar geoengineering "flattens the curve" to offset climate impacts until mitigation (represented by the "cut emissions aggressively" arrow) and CDR (represented by the "CO_2 removal" arrow) can be scaled up to the necessary levels. The right panel demonstrates why SRM on its own does not "fix" global warming, as without addressing the underlying problem of increased CO_2 levels, global warming—and the amount of solar geoengineering needed to offset that warming—would increase indefinitely. Used with permission of PNAS under author re-use policy. *Source* PNAS.[14] This figure is excluded from Creative Commons license

the loss of lives and livelihoods in the event of a crash. As such, SRM should be considered only as a possible supplement to aggressive mitigation, and not as a substitute or excuse to delay it. For the same reason, SRM could only ever be a temporary endeavor. As demonstrated by the "napkin diagram" (Fig. 11), the amount of SRM required to reduce impacts scales with the amount of global warming we wish to offset, and as shown in the panel on the right, without cutting emissions and/or CO_2 removal, the amount of SRM needed to stave off climate impacts would keep increasing. However, as shown in the left panel, when used as a supplement to mitigation and CDR, solar geoengineering could act as a temporary salve to lessen the worst impacts of climate change. In this section of the chapter, we provide an overview of what is known about these approaches and the context for considering them.

4.2 Methods of SRM

There are three primary methods of large-scale climate intervention commonly classified as SRM (Fig. 10): stratospheric aerosol injection (SAI), marine cloud brightening (MCB), and cirrus cloud thinning (CCT). In this section, we provide a broad introduction to each of them, and in the next section, we discuss similarities and differences among the three.

---

[14] MacMartin and Kravitz (2019). Author permission to re-use according to https://www.pnas.org/about/rights-permissions.

Other planetary-scale methods of reflecting sunlight have been proposed, such as changing the reflectivity of land surfaces, but these methods are unlikely to be capable of a reduction of the required magnitude; for example, achieving a 1% reduction in sunlight would require a reflective area of about 8 million km^2, roughly equal to the area of the continental USA. Similarly, a space-based mirror between the Earth and Sun would need an area of approximately 2 million km^2; for example, this would require constructing an average of 100 km^2 a day for the next 50 years. This chapter does not further discuss either of these ideas. Additionally, a number of small-scale methods, such as the scattering of tiny glass beads on the surface of Arctic sea ice to increase reflectivity, have been proposed, but these would have regional, smaller-scale impacts, and this chapter is focused on the planetary-scale methods of SRM.

4.2.1 Stratospheric Aerosol Injection (SAI)

Some volcanic eruptions are powerful enough that their gasses and ash are ejected into the stratosphere, a layer of the atmosphere stable enough that particles can remain aloft for a year or longer. Sulfur dioxide (SO_2), one of the products of volcanic eruptions, reacts with OH and combines with water vapor to form tiny droplets, called *sulfate aerosols*. Sulfate aerosols scatter sunlight, reflecting some of it back to space, producing a net cooling effect on the planet (Fig. 12); the 1991 eruption of Mt. Pinatubo in the Philippines was sufficient to cool the entire planet by about 0.4 °C for ~ 15 months (until the sulfate aerosols settled out of the atmosphere). The effect of aerosols on Earth's energy budget was discussed in Chap. 2, and the record of variations in forcing due to volcanic aerosols over the past 2000 years was discussed in Chap. 3. The premise of **stratospheric aerosol injection (SAI)** is to replicate this cooling effect by artificially introducing sulfur dioxide into the stratosphere, and maintaining it by replenishing the aerosol layer at regular intervals. The amount of sulfur would depend on the goal; roughly speaking, injections of 10 Tg of SO_2 per year would produce about 1 °C of cooling, and the amount can be scaled based on the desired goal (although larger injections are somewhat less efficient because the aerosols are more likely to clump together). Hypothetically, aerosols other than sulfur could be used, but the majority of SAI research to date has focused on sulfate aerosols because they are found naturally in the stratosphere and elsewhere due to volcanic eruptions, meaning that they are relatively well understood from a climate modeling perspective, and the behavior of sulfate aerosols in climate models can be validated against observations of volcanic activity.

SAI would not be an exact replica of a volcanic eruption; rather than deploying in a single burst, SAI would be continuous, likely deployed by aircraft specially designed to carry heavy payloads from the surface into the stratosphere and requiring repeated flights to replenish the aerosol layer. Furthermore, the location(s) of injections are important. In the stratosphere, the air is rapidly mixed across longitudes (i.e., in the east-to-west direction), and the *Brewer-Dobson circulation* (pictured in Fig. 13) carries air slowly toward the poles. Therefore, the choice of injection longitude will

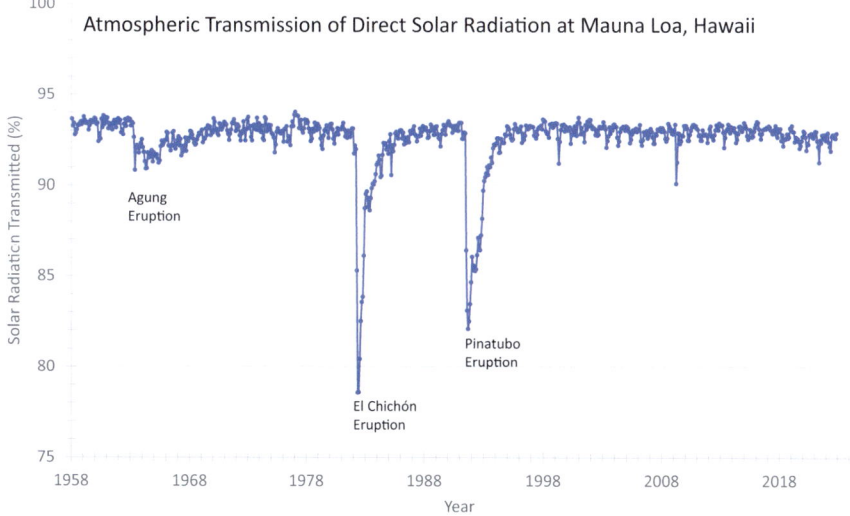

Fig. 12 Percent of solar radiation at the top of the atmosphere that reaches the surface in Mauna Loa, Hawaii. Sufficiently powerful volcanic eruptions eject material into the stratosphere, which can remain aloft for years and reflect enough sunlight to cool the surface. Data from NOAA[15]

have little influence, but injecting at different latitudes will have very different effects on the surface climate. Injecting near the equator means the aerosols would spread over the globe and provide relatively uniform cooling (a "global" strategy), whereas injecting at higher latitudes, such as 60° N (an "Arctic" or "high-latitude" strategy) would preferentially cool the poles. As such, an Arctic strategy could be used to preserve sea ice, ice sheet mass, and permafrost with relatively less effect on lower latitudes. However, it is not possible to cool only the Arctic (or any single region of the planet) without affecting the rest of the world; since cooling high latitudes would draw heat from the tropics toward the preferentially cooled poles, there would still be some effect on the tropics.

It can be said with certainty that SAI would cool the planet, and it would very likely reverse many of the other consequences of global warming, including the melting of snow and ice, increased extreme heat, and precipitation extremes stemming from the ability of warmer air to carry more moisture or increased evaporative demand. However, the magnitude and distribution of cooling—and of these other effects—are less certain. Adding more sulfate aerosols to the stratosphere would also introduce other complications; for example, SAI would increase sulfur pollution with proportional adverse effects on human health, and sulfate aerosols in the polar stratosphere also contribute to the destruction of stratospheric ozone, which shields the surface from harmful UV sunlight. However, the only way to predict the effects of SAI

---

[15] Data from National Oceanic and Atmospheric Administration (NOAA), https://gml.noaa.gov/grad/mloapt.html. Public Domain.

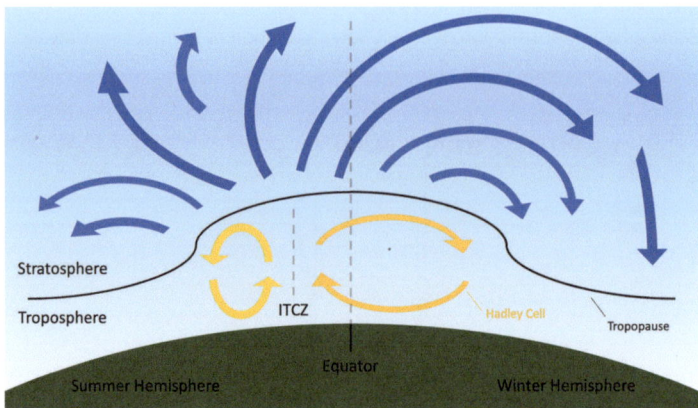

Fig. 13 Simplified diagram of the Earth's circulation in the troposphere (the layer of the atmosphere closest to the surface) and the stratosphere (the next-closest layer); the boundary between them is called the tropopause. Particles in the troposphere are deposited out quickly, but the stratosphere is more stable, and particles can remain aloft in the stratosphere for much longer; this makes SAI feasible

(without actually doing it) is to use a climate model. Full-scale, global field experiments (i.e., actually injecting SO_2 into the stratosphere and observing the results) are impossible because, at the point, where a field experiment is large enough to produce a detectable change to the climate, it is no longer just an experiment, but an actual deployment. However, models are, of course, not perfect, and they use many approximations and simplifications. Furthermore, volcanic eruptions, which are commonly used to validate aerosol behavior in models, are not perfect analogs for SAI, and the data we do have from large volcanic eruptions are imperfect. For example, the total amount of SO_2 injected into the stratosphere by the eruption of Mt. Pinatubo in 1991 (the most recent eruption that ejected sufficient sulfur into the stratosphere to cool the planet by a detectable amount) is uncertain because the satellite instruments were saturated (i.e., there was too much sulfur for the instruments to measure). As such, there will always be uncertainties inherent to SAI that will never be resolved prior to a decision to deploy. Some of these physical uncertainties—the exact amount and distribution of cooling, for instance—may be managed to some extent by adjusting the injection quantity and location over time to push toward desired outcomes; but there will always be "unknown unknowns" representing impacts or interactions that nobody has considered, and negative impacts on the environment, ecosystems, or communities may not be discovered until after deployment. However, there are also significant unknowns associated with *not* deploying SAI (and potentially allowing the climate to warm to unprecedented levels). While uncertainty cannot be eliminated, SAI research—if done ethically, with public, stakeholders, and community input (see Box 1)—might reduce uncertainty to the point where the risks and benefits of deploying SAI can be weighed against the risks and benefits of not deploying it (see Chap. 12 for a discussion on ethics).

The most immediate challenge to SAI implementation is lofting the material, almost certainly involving aircraft. At the equator, the tropopause is 17 km above the surface, and the desired injection altitude for low- and mid-latitude injections would probably be at least 20 km. The deployment vehicle would need to carry a payload to that altitude and maintain that altitude while that payload is dispersed; currently, there exists no aircraft capable of doing this. However, studies suggest that overcoming this obstacle would not be fundamentally difficult for a state or organization with sufficient resources. A strategy focused on the Arctic might be easier; the tropopause is lower at high latitudes, and an injection at 60°N (for example) could inject at 15 km. This could possibly be done with existing aircraft, but they would likely still require heavy modification, and development of new planes might still be desired depending on the scope of the operation. Estimates for the cost of an SAI deployment range from billions to tens of billions of dollars, depending on the latitude (and hence the altitude) of injection and on the amount of cooling provided.

4.2.2 Marine Cloud Brightening (MCB)

Under certain conditions, smokestack pollution from ships at sea causes vessels to leave behind bright "tracks" of clouds that can persist for days (Fig. 14).[16] This happens because aerosols in the ship's exhaust serve as cloud condensation nuclei: droplets of water coalesce around the aerosols, forming cloud particles. When aerosols contribute to cloud formation in this way, the clouds are "brighter" because the same amount of water vapor is distributed across more and smaller droplets, resulting in a higher cloud surface area and thus higher reflectivity. This mechanism, known as the *Twomey Effect*, is the principle behind **marine cloud brightening (MCB)**: by deliberately seeding marine skies with cloud condensation nuclei, which could be done by spraying salt water into the atmosphere, we could brighten existing clouds, reflecting more sunlight back into space to cool the planet. The clouds most susceptible to this kind of intervention are located over the ocean; additionally, the surface of the ocean is darker than the surface of the land, meaning that the contrast in albedo between brightened clouds and the surface will be greatest over the ocean. Finally, spraying salt water into clouds over land could be harmful to vegetation. As such, cloud brightening over land is probably not feasible.

MCB is less well understood than SAI. The primary reason for this is the low level of understanding of how clouds and water droplets interact with aerosols; laboratory

---

[16] Recently, an unintended test of solar radiation management occurred related to ship cloud tracks. Pollution regulations set by the United Nations' International Maritime Organization in 2020 resulted in an 80% decrease of ship-sourced sulfur pollution. Air quality improved worldwide. However, because the number of highly reflective ship track clouds (like those in Fig. 14) was reduced, less solar radiation was reflected back to space, which in turn meant more heat remained in the troposphere and surface ocean, contributing to global warming. This unintended 'experiment' is the opposite of marine cloud brightening (MCB), which would intentionally increase the number of highly reflective clouds to reduce solar radiation in order to cool the planet. See summary in Voosen (2023).

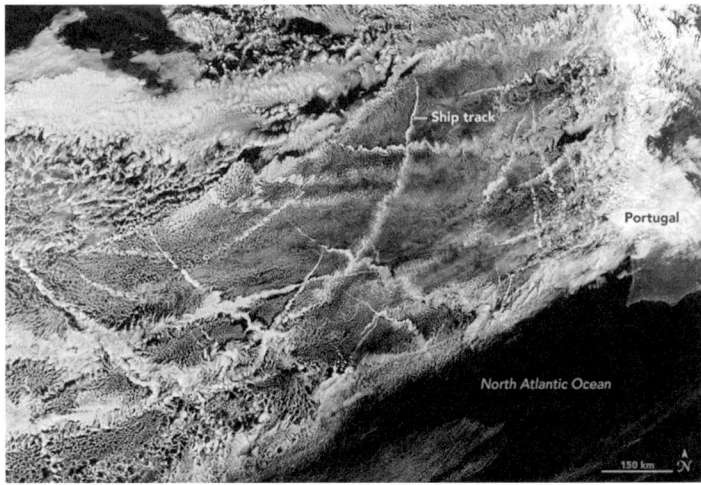

Fig. 14 Aerosols in ships' exhausts cause "tracks" to form, which are visible in this satellite image. Marine cloud brightening (MCB) would use the same principle to cool the planet. *Source* NASA[17]

observations of these interactions at micro-scale cannot easily be translated to the complexity of the real world at climate-wide scales, and cloud-aerosol interactions are difficult to model because of the complexity of the physics and chemistry as well as the range of scales involved. While there are areas and conditions where MCB will "work," in the sense of increasing the amount of reflection, there are also meteorological conditions where adding aerosols might have the opposite effect (by causing a cloud to rain out earlier, for example), and the exact conditions where adding aerosols increases reflectivity versus decreasing it are not well understood. Ultimately, it is unclear today where, when, and how much we can increase cloud albedo. As such, the maximum cooling effect achievable through MCB is not certain. Humans emit significant aerosol pollution; current estimates place existing forcing due to worldwide cloud-aerosol interactions at -0.9 W/m^2 (a net cooling effect; refer to Chap. 2 for more discussion on radiative forcing imbalances), but with considerable uncertainty.

MCB could be achieved using salt water, reducing concerns associated with the material itself (relative to sulfur, which is typically assumed for SAI). While MCB would be implemented only 1–3 km above the surface, which means that it could be deployed by ships, the technology needed to deploy would still involve challenges. Developing spray nozzles that generate the right size of salt aerosols is challenging, and the number of vessels needed and the extent to which their spray strategy might need to be finely tuned based on local conditions remains unclear. Unlike SAI, MCB could be applied locally; while the stratospheric aerosols of SAI would spread globally and can persist for a year, MCB could be applied over a smaller region, meaning

---

[17] NASA Earth Observatory, https://earthobservatory.nasa.gov/images/91608/signs-of-ships-in-the-clouds. Public domain.

MCB could be either a "regional" approach or a "global" approach. However, as heat is transported through the atmosphere, the resultant cooling would not be local even in the former case, and there would still be transboundary effects; this would be analogous to turning on the air conditioning in your house, but leaving the windows open—cool air would escape through the windows, and warmer air would come inside. Another concern is that while MCB is implemented over the ocean, a strong intervention close to shore could have transboundary effects large enough to affect the subsequent distribution of rain on land, with follow-on impacts on agriculture, or cause larger-scale circulation changes. Additionally, the right type of cloud only exists over perhaps 10% of the planet's surface, and so if the goal is global cooling, a stronger effect would be needed in those smaller areas. This could lead to more spatially heterogeneous outcomes, though the details are not yet clear.

Unlike SAI or CCT, MCB can be deployed by ships, and the technological barriers to create ships designed to spray seawater into clouds would be relatively low, even considering the need to develop spray nozzles that can produce the desired droplet size distribution. However, for a global-scale deployment, a large number of ships would be required, along with the facilities and people to maintain them. There is little to no data for estimates of the cost of an MCB deployment, but it would likely not be more expensive than an SAI deployment of a similar scope.

4.2.3 Cirrus Cloud Thinning (CCT)

Low, warm clouds—most clouds—have a net cooling effect on the atmosphere: they emit approximately the same amount of longwave radiation as they absorb, and therefore "trap" little upwelling longwave radiation, and they reflect some incoming sunlight. However, high, cold cirrus (ice) clouds that form in the upper troposphere (6–13 km), primarily in the mid-to-high latitudes, emit much less longwave radiation than they absorb and reflect little sunlight, and are therefore a net warming influence on the Earth (Fig. 15). The goal of **cirrus cloud thinning (CCT)** is to artificially thin or remove cirrus clouds, thus lessening their warming influence. Therefore, CCT is not technically a method of "solar" radiation management because it seeks to influence longwave radiation rather than shortwave. Nonetheless, it is commonly grouped with other methods of solar geoengineering because it is a proposed planetary-scale method of cooling the surface and because many of the practical, logistical, political, and ethical considerations are similar to those of MCB and SAI.

Cirrus clouds can form in two ways: homogeneous freezing and heterogeneous freezing (see Fig. 16). Heterogeneous freezing occurs where there are sufficient ice nuclei, typically dust. In the absence of sufficient ice nuclei, homogeneous freezing can occur; this requires colder temperatures and higher relative humidity and results in smaller ice crystals than heterogeneous freezing. The relative fraction of cirrus clouds that form via homogeneous freezing is not certain, but it could be the dominant mechanism in many places. By introducing *ice-nucleating particles* (INPs) in regions where homogeneous freezing occurs, we could cause a larger fraction of cirrus clouds to form through heterogeneous freezing. The ice particles could grow larger, and

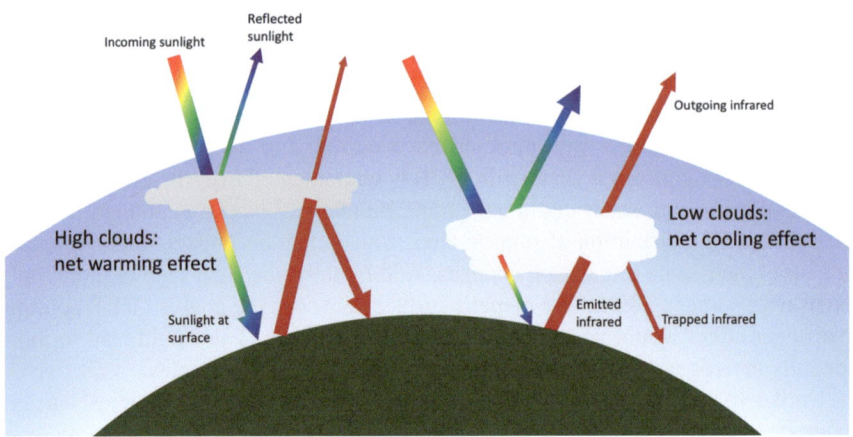

Fig. 15 Low clouds (right) have a net cooling effect on the atmosphere: they reflect sunlight and trap little outgoing infrared radiation. However, some high clouds (left) trap outgoing infrared radiation and reflect little sunlight, with a net warming effect. Thinning these high clouds could cool the planet

there would be fewer of them; the same amount of water spread over less surface area would trap radiation less effectively, and more infrared radiation could escape. Additionally, the heavier particles would fall out of the atmosphere more quickly, reducing cloud lifetime. The result would be fewer, thinner cirrus clouds, and a net cooling effect on the planet.

As discussed previously, CCT seeks to modify longwave radiation instead of shortwave (in contrast with SAI and MCB); since longwave radiation is the same component of the planetary energy budget that is affected by greenhouse gasses, this could be a potential advantage of CCT over the other methods of solar geoengineering. That said, CCT would still not have the same spatial or seasonal pattern of radiative forcing as well-mixed atmospheric greenhouse gasses, and would thus still lead to changes in regional temperature and precipitation patterns. More critically, CCT is the least well-understood of the three solar radiation management methods considered here. CCT is based on the assumption that cirrus clouds form primarily through homogeneous freezing, but the extent to which this is true is an open question in atmospheric science. Thus far, CCT has only been explored in climate models, as there is no natural analog for it; in climate models, the percent of natural cirrus that form homogeneously varies hugely depending on the physics, even in the same model. Due to these uncertainties, there is a possibility that an attempt at CCT would actually *warm* the planet; overseeding—supersaturating a region where no more cirrus would naturally form—would result in the formation of additional cirrus rather than thinning them. Additionally, the mechanism of CCT itself is not well understood; more information is needed regarding the concentration, temperature, and supersaturation dependence of cirrus ice-nucleating particles. Finally, the logistics of CCT deployment are also unclear, i.e. what material, particle size, and

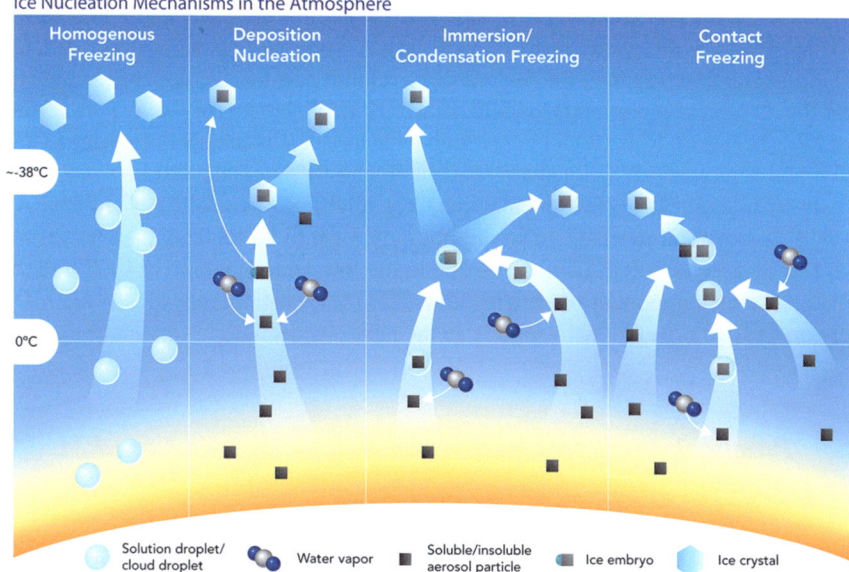

Fig. 16 Water–ice clouds are extremely complex. In areas where cirrus clouds form through homogeneous freezing (left column), introducing particles (gray squares in the diagram) could encourage clouds to form through heterogeneous freezing (right three columns) instead, causing larger and fewer cloud droplets (difference in size is not represented in the diagram) that would fall out of the atmosphere more quickly and trap less outgoing infrared radiation. *Source* Wikimedia Commons[18]

dispersion mechanism would be most effective, and how well a deployment could target regions in which the heterogeneously-formed cirrus clouds that were created replaced homogeneously-formed cirrus clouds that otherwise would have occurred, rather than forming new cirrus clouds where there otherwise would have been none. As such, one cannot say confidently how well CCT would work, or what the climate effects would be if it did.

The overall potential of CCT to reduce global warming is not large. It is estimated that cirrus clouds currently warm the planet by 5 W/m^2 (perhaps 2–3 °C; see Chap. 2), and this provides a hard upper bound on the change CCT could bring about. CCT would be most effective at high latitudes, but cirrus are most common at low latitudes, and thus the maximum achievable effect of CCT would be a reduction of about 2 W/m^2 (perhaps 1 °C). The effect would be spatially diverse, and little is known about how the climate or ecosystems would respond. Due to the uncertainties discussed above, it is unclear what the realistically achievable effect is, and the predicted effect differs between climate models.

The largest barrier to deployment is arguably the state of the research. Setting aside the relatively low upper bound of realistically achievable forcing, more needs to be

---

[18] Wikimedia Commons, https://commons.wikimedia.org/wiki/File:Ice_Nucleation_Mechanisms.png. Used under the Creative Commons Attribution-Share Alike 4.0 International License.

known about the formation of cirrus clouds to ensure that an attempt at CCT does not end up warming the planet instead of cooling it. The deployment altitude for CCT would be much lower than for SAI, as cirrus clouds form in the upper troposphere, and the amount of material that would need to be deployed is likely much lower than for SAI. However, little research has been done as to the mechanism by which CCT might be deployed (i.e., piloted aircraft, autonomous drones, etc.), or how much this might cost. As such, while implementing CCT might be practically easier than SAI, a CCT deployment mechanism would not only have to be implemented but would have to be designed first. Ultimately, due to the low understanding of CCT and the relatively small amount of forcing it could achieve, we cannot realistically expect to rely on CCT to offset global warming in the next few decades.

4.3 Practical, Logistical, and Societal Considerations of SRM

Besides the infrastructure and scientific barriers described in the previous section, other major considerations regarding deployment are political, cultural, and ethical. All forms of SRM are extremely contentious, and wildly varying opinions about SRM can be found across the general public, environmental groups, and climate scientists. Since any meaningful SRM deployment would affect everyone on the planet, ideally, everyone on the planet would get a say. However, given that we cannot even agree on whether SRM should be researched (or, in some cases, that global warming is a problem at all!), the odds of reaching a global consensus one way or the other are near zero, at least in the near term. As global warming continues to worsen, it is likely that SRM will become a greater topic of discussion on the international stage. Regardless, the only way a nation or group of nations could deploy is if they believe the benefits of SRM outweigh the physical, ecological, and political risks. On the other hand, given that SRM can be deployed from more or less anywhere in the world, once somebody decides to deploy SRM, it is very difficult to stop them barring direct military intervention. Political and economic sanctions could be implemented, but a backer may be big enough to weather the political storm, or vulnerable enough to climate change that they have nothing else to lose.

Cost is a lesser consideration. Most estimates for SAI (there is little cost estimate data for MCB and little to none for CCT) fall in the range of single billions to tens of billions of USD per year. This would likely be relatively easy for most developed nations and international unions to afford (there is debate over whether smaller countries or even wealthy individuals or corporations could independently conduct SRM—see below). Note that the costs of SRM would include not only the direct costs of manufacturing, materials, and dispersal but also budgets for monitoring and analysis. Additional costs could include compensation for nations or indigenous communities disproportionately impacted by SRM, extra money budgeted for healthcare due to impacts from SAI (for example, sulfate particle deposition or ozone depletion), or even legal costs should the effort to deploy be challenged in court.

The question of who could deploy SRM is largely answered by who could overcome the political, financial, and technological barriers discussed above. The low direct costs raise the question of whether even moderately small countries or wealthy individuals could deploy. Any deployment that actually affected climate (as opposed to a symbolic effort) would involve a sustained effort lasting for years. For SAI, for example, this would presumably involve many aircraft flights a day. As a last resort, this would be simple to shut down unless there was at least the implicit backing of a major power. Even a wealthy individual needs access to airbases and a supply chain, and maintaining these would also require implicit backing of the nation that operations were taking place in. For high-latitude strategies, aircraft exist today that could reach sufficient altitude; anyone could buy used business jets. For lower-latitude strategies, the tropopause is much higher, and as described above, no existing aircraft could be used and a new aircraft would be required. Designing and building the aircraft itself is not particularly difficult. However, there are relatively few global suppliers of aircraft engines that could operate at the required altitudes. Developing a new engine from scratch would take years and could be beyond the capabilities of all but a relatively small number of countries. The technology behind MCB or CCT would present few challenges and could be deployed at least at small scale by nearly any country. For MCB, any effort large enough to have a global influence would involve operations in international waters.

If someone chose to deploy SAI, it would certainly be detected. Long before the stratospheric aerosol levels were large enough to actually affect the climate, they would be large enough to be easily detected above the background variability of the stratosphere. Furthermore, the scale of the operation required may be small enough to be economically viable, but it would still be a large enough operation that the regular flights would be readily observed. A small-scale MCB deployment within territorial waters of an individual country might more easily be kept secret, but it would also not have global ramifications. Any MCB effort large enough to have global ramifications would likely also be detectable quickly due to the sheer number of vessels required.

One final consideration discussed here is referred to as *termination shock*. If a deployment of SRM was abruptly terminated, then the climate would rapidly warm back toward roughly the same state that it would have been in had SRM never been deployed, with most of the warming occurring over a few years. The consequences of that rapid warming are generally assumed to be worse for ecosystems than the gradual, long-term warming that would have occurred had SRM never been deployed. Termination is clearly a risk, but how much of a risk depends on how much cooling is being done; in scenarios that gradually ramp up cooling, it might be decades after deployment before it becomes too risky to suddenly stop. Even so, if after many decades of deployment, some physical climate impact became evident that justified stopping deployment, then it is of course possible to gradually ramp down the amount of cooling rather than suddenly terminating it. Regardless, the threat of termination shock is an additional risk that must be accounted for in a discussion of SRM. Additionally, if one country develops the technology for SRM, the prospect of termination shock provides a strong incentive for other countries to do so too.

5 Where Should We Go from Here? A Discussion on Climate Intervention Research

While we know that SRM would cool the climate, these methods would not perfectly cancel the effects of increased greenhouse gasses: some climate impacts would be reduced, others might not be, and there would be new risks introduced, including both purely physical effects as well as ecological, ethical, political, and governance concerns. Is it less risky to consider some limited amount of SRM, or to not consider it? This is not a straightforward question, as we need to consider not only physical impacts and uncertainties (e.g., the effects on the climate; ecological impacts; and impacts on human lives, health, and livelihood), but also the human dimension—how could humans manage, govern, and implement an SRM strategy in an ethical, just, and fair way? And, how will we react to it? The evidence (to date) from climate modeling suggests that if only the physical impacts were considered, it is *plausible* that SRM would reduce risk for most (and possibly all) people on the planet; however, "plausible" is an inadequate level of certainty for risks of this magnitude, and further research is needed to better understand the physical and ecological impacts of SRM. These uncertainties are compounded by complications from the human dimension, such as the potential for conflict over whether and how to deploy SRM or the potential for "moral hazards"[19] (e.g., will SRM lessen our resolve to cut emissions and invest in mitigation and CDR technology?). These complications increase the need for research, including research that prioritizes engagement and partnership with the stakeholders and communities that would be most affected by the potential outcomes.[20]

Just as one would never talk about chemotherapy without talking about cancer, or airbags without talking about car accidents, the risks of SRM must be considered not against the world as it is today, but against the risk of future climate change. The litany of concerns about climate change is well-known from observational (Chap. 3) and paleoclimate records (Chaps. 4–9), including droughts, forest fires, heat waves, stronger hurricanes, and increased sea level rise. As the Earth's climate is pushed away from the regime that we are familiar with, in addition to the impacts getting worse, the uncertainty also gets worse. For example, Arctic permafrost contains vast amounts of stored organic material; as this permafrost thaws, the organic material is released as carbon dioxide and methane, and we don't know how quickly this would accelerate climate change. We also don't know, for example, how much global warming might trigger the collapse of the ice shelves that hold back Antarctic ice and prevent the catastrophic sea level rise caused by melting that ice. In total, our ability as an international community to limit global warming and its consequences

---

[19] Tsipiras and Grant (2022).

[20] Prioritizing engagement and partnership with diverse stakeholders to benefit society and address Earth systems-related problems at community, state, national, and international scales is envisioned as a key characteristics of "next generation" Earth systems science research, as described in: National Academies of Sciences, Engineering, and Medicine (2022).

to "acceptable" levels through mitigation and CDR alone requires four things, none of which are certain right now:

- *First,* it requires sufficient political will in every major GHG-emitting country in the world, sustained for multiple decades. Popular discourse often assumes both that this is the only requirement, and that if we want this to be true hard enough, then it will happen (as an aside, the use of the word "we" here glosses over several significant assumptions. The question of who "we" represents, and who has agency, is a recurring theme here).
- *Second,* it requires technology that, while plausible, doesn't yet exist. A truly sustainable economy requires net-zero emission, and while we know how to cut our emissions dramatically with existing technology, we don't yet have the scalable energy storage options required for 100% of our electricity to come from renewables, and it is difficult to remove the dependency of some sectors (like aviation) on fossil fuels. As discussed above, we also have not yet demonstrated scalable CDR approaches required to reach net-zero emissions. Net-zero solutions also need to be cheap enough to enable deployment not just in the developed world, but to allow the rest of the world to develop economically.
- *Third,* we need reasonable luck on climate sensitivity: in other words, there is uncertainty in the remaining carbon budget—how much additional CO_2 can be emitted—in order to limit warming to 1.5 °C or 2 °C above pre-industrial conditions. According to the IPCC's 6th Climate Assessment Report,[21] typical emissions scenarios that are described as being sufficient to stabilize below the Paris Agreement upper limit of 2 °C are not guaranteed to meet that goal. For example, the odds that emissions scenarios expected to limit warming to 1.8 °C would still lead to warming above 2 °C may be as high as 1 in 3, and those odds with these consequences would be unacceptable in any other realm of life.
- *Fourth,* we need to gamble that the climate impacts from a given temperature are acceptable; there is no certainty that staying below 2 °C or even the 1.5 °C goal of the Paris Agreement will be sufficient to avoid significant risks. For example, there are already increases in extreme heat, flooding, and drought, and it is unclear how much warming will lead to even more severe cascading impacts (sometimes called "tipping points") such as the collapse of Antarctic ice shelves and subsequent catastrophic sea level rise, or when, how much, and how quickly carbon will be released from melting permafrost.

While we can certainly hold out hope that all four of these conditions will be met, they are clearly not guaranteed, especially given that two of these four aren't even under human agency. This leaves us facing the reality that there is a significant possibility of a substantial and sustained overshoot of global temperature targets; in layman's terms, there is a real chance that we will miss the goal of limiting global warming to 1.5 °C or even 2 °C, that we will miss the goal by a lot, and that temperatures will remain over the target for a long time before technology evolves to the point where we can reduce them. Furthermore, even if we make the 1.5 °C target,

---

[21] IPCC (2021).

it is not guaranteed that the resulting impacts will be "acceptable". Different people can reasonably disagree about the probability of avoiding significant damages, but given the uncertainties involved, it is impossible to pretend that the probability is small.

Currently, not enough is known about any of the SRM methods described above to make informed decisions about their use, and further research is needed. The questions on the table today are not the hard ones of whether (and how, and where, and when) to deploy, nor the equally challenging question of how to make that decision, or even who gets to decide. Rather, the current question is whether or not to conduct additional research into SRM, and how that research should be governed. The U.S. National Academies of Science, Engineering, and Medicine (NASEM) released a report in 2021 concluding that further research is warranted and making recommendations on research governance, and advising that the U.S. should create a research program. Note that as laid out in that report, while an international research program and international governance framework would be preferable, a U.S. body can only legitimately make recommendations for actions taken by the U.S. The recommended research agenda includes not only research into the physical impacts on climate, humanity, and ecology, but into the human dimensions of the challenge as well. If some form of SRM were ever implemented, the ideal scenario would be one in which every country agreed and cooperated. This aspiration seems unlikely today (though the 2015 Paris Agreement does provide some precedent). While the ideal world will likely continue to be just that—if we lived in an ideal world, we would have already addressed CO_2 emissions, and we wouldn't be faced with this conversation—it is appropriate that in parallel with the research, we must think about how decisions involving these ideas might get made. One aspect of this is simply capacity-building: making sure that people in every country understand the issues and how these decisions might affect them.

While the idea of SRM has been discussed at the international level, no nation or group of nations currently has an official stance on SRM or SRM research. Likewise, different groups, cultures, and communities have differing stances on SRM. While some work has been done to examine public perceptions of SRM, it is difficult to measure public attitudes accurately due to low levels of awareness; existing measures have attempted to capture very general perceptions of what the public thinks in order to offer some guidance on perspectives beyond those of researchers and decision-making elites. Perception of SRM (and SRM research) is largely influenced by perception of the risks posed by climate change compared to the risks posed by SRM. Opinions on SRM research are based on trust of the people involved (e.g., scientists and policymakers), the trade-offs involved (e.g., the seriousness of climate change and the difficulty of mitigation and CDR), and beliefs about humanity's role in the natural world.[22] Those who are especially vulnerable to climate change (such as those living in the Global South, or Indigenous peoples in the Arctic) may also

---

[22] Raimi (2021).

have different views on SRM research than those living in wealthy, developed countries.[23] The idea of manipulating nature is a common theme in international survey responses regarding SMR. Many people express discomfort at the idea of humanity deliberately modifying the world on a global scale in such a way, a perspective which could stem from either practical concerns or more spiritualistic reasons.[24,25] Broadly speaking, respondents tend to oppose imminent SRM deployment, tentatively oppose field research, and more broadly support laboratory and modeling research.[26] As the technology evolves, public discourse surrounding it will likewise evolve, and continuing to understand public perceptions of SRM will be critical in the years to come (see Box 1). These differences in beliefs and identities will likely present challenges to cooperation that will be difficult to overcome. As an example, a research group from Harvard University prepared for a small field experiment in northern Sweden in the spring and summer of 2021. The Stratospheric Controlled Perturbation Experiment, or SCoPEx, planned to release a small amount of sulfur from a weather balloon and fly instruments through it to better understand the properties of sulfate aerosols and improve representations of SAI in climate models.[27] However, the indigenous Sámi people were not consulted as the experiment was first being planned, and the Sámi objected to this experiment based on their deeply-held cultural beliefs that humans should learn to live with their environment rather than manipulate it. As a result of these differences in perspectives between the indigenous community and the researchers, the experiment ultimately was postponed. This series of events demonstrates that even at the local and regional levels, such differences in views can pose challenges to agreement on any action related to SRM, and national and international cooperation will likely be much harder. These events also demonstrate the need to prioritize engagement and partnerships with stakeholders throughout the research process and thus work to co-develop knowledge.[28]

There is clearly a risk that use of SRM (and even the awareness that SRM is possible) could lead to less emphasis on emission reduction and CDR—the "moral hazard" mentioned above. If there is mitigation deterrence, it could conceivably result in worse outcomes in the long term; this is perhaps the primary reason why some are opposed to SRM, or even research into SRM. The opposite attitude is also possible—that is, some people may react with, "Wow, if that's what we're going to be

---

[23] Sugiyama et al. (2020).

[24] Visschers et al. (2017).

[25] Raimi et al. (2019a).

[26] See [20].

[27] SCoPEx, Keutsch Group at Harvard, https://www.keutschgroup.com/scopex.

[28] The prioritization of engagement and partnerships with diverse stakeholders to benefit society and address Earth systems–related problems at community, state, national, and international scales is a key characteristic of "next generation" Earth system science. See: National Academies of Sciences, Engineering, and Medicine (2022).

faced with if we don't cut emissions, we better cut emissions faster".[29] Discussions of the moral hazard[30] of climate intervention are also fraught with ethical questions associated with the idea of "we", as if humanity were a single actor, which it is not. The people who suffer the most from climate change impacts (and thus might benefit from SRM), aren't the same people who might use SRM as an excuse to keep emitting. Additionally, there are ethical questions of withholding knowledge because we don't trust future people to use it. Ultimately, while there is the possibility that an SRM deployment would reduce short-term incentives and motivation to reach net-zero emissions, the choice of "emissions cuts, CDR, or SRM" is not a binary one, any more than one needs to choose between wearing seat belts and using the brakes. There is no evidence that an immediate worldwide halt to SRM research[31] would substantially accelerate progress in mitigation or CDR, or that such a halt would be the difference between a world with unacceptable climate consequences and one without.

Decisions about climate intervention, as well as climate mitigation and adaptation, are fraught with complexities. Framing our individual and community-based decisions on a foundation of science, technology, and ethical reasoning can help. Several geoscience groups have put forth position statements that rest on such a foundation.[32] Thus far, chapters of this book have provided you with key elements to understand the science and emerging technology. That last foundational piece—climate ethics—is the topic of the next (and final) chapter in this book.

> **Box 1 Including multiple perspectives on climate intervention strategies**
>
> By Amanda Borth, Consortium for Science, Policy & Outcomes at Arizona State University.
>
> **Thought Exercise and Primer**
>
> Please consider what you learned from this chapter alongside your own personal experiences and perspectives to analyze the following fictional scenario that is informed by real-world events.
>
> The country you live in has passed historic legislation to update the country's infrastructure. This includes significant funding for building carbon dioxide removal (CDR) facilities across your country. Government agencies have

---

[29] Raimi et al. (2019b).

[30] See Chap. 12 for a discussion of ethics and climate change decision making.

[31] A worldwide halt was in fact raised in a 2021 open letter signed by 400 scholars, https://www.solargeoeng.org/non-use-agreement/open-letter/.

[32] See the American Geophysical Union (AGU) Position Statement on Climate Intervention, which was adopted by AGU in 2018 and reaffirmed in 2023, https://www.agu.org/science-policy/position-statements/climate-intervention. See also the AGU Ethical Framework for Climate Intervention, a white paper (report) released in 2022 that can serve as a code of conduct to guide the ethics of climate intervention, https://www.agu.org/ethicalframeworkprinciples.

worked with universities and private companies to determine what types of CDR are ready to be implemented and what locations have the right geologic and environmental conditions for these CDR methods. Based on this, they are proposing to put a direct air capture with carbon capture and storage (DACCS) facility in your town. However, the national government wants to make sure that the DACCS facility serves your community's needs in addition to national-level goals. To do so, the government and project developers are holding a dialogue at your local science center, where you get to learn about the proposed DACCS project, share your views on it, and be a part of the decision-making process. During this, they ask you the following questions:

- What are your community's priorities?
- What are your hopes and concerns for the proposed DACCS project?
- What previously unheard perspectives in your community need to be reflected in decisions about the project?
- What role would you like the national government to play in developing this project? How about the role of other community members, large corporations, start-ups, community organizations, etc.?
- Who do you trust to oversee the project's decision-making process? This can include people from your own community or outside of it.
- How do you want your community involved at various stages of the project (conceptualization, construction, monitoring, etc.)?

This thought exercise displays the crux of engagement for climate intervention strategies (CIS) decision-making. The idea is to include voices like yours in the decision-making process around CIS, so your perspectives can guide responsible decision-making. While the concept of engagement and how to do it well continues to evolve over time, this textbox offers a definition and insights into this topic to help you think about the role engagement might play in current and future CIS decisions.

Engagement in CIS Decision-Making is any activity by which an entity with decision-making power—like companies or governments—justly fosters **publics'**, **stakeholders'**, and/or **communities'** (Fig. 17) attention and action to elicit and include their perspectives in decisions about the research, development, deployment, and/or monitoring of CIS.

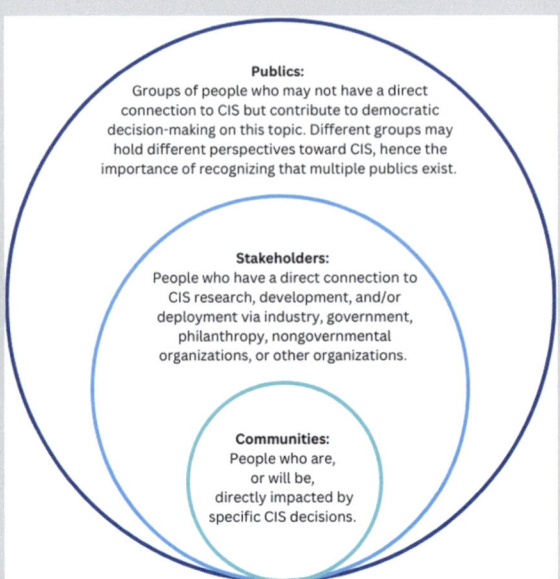

Fig. 17 Concept diagram of the nested relationships among public groups, stakeholders, and communities. Definitions and rationale for this diagram are adapted from NASEM[33]

Note that public groups, stakeholders, and communities are not mutually exclusive terms nor definitive, as indicated by the nested circles in Fig. 17. Rather, this is one framework to help determine how different people may relate to CIT in a given context and what their roles may be in engagement.

Why is decision-making on climate intervention strategies ripe for engagement?

As you've learned in this chapter, the research, development, deployment, and monitoring of CIS come with complexities, uncertainties, and risks that look very different across cultures, CIS methods, and levels of governance (i.e., local, state, regional, national, and international). The decisions made about CIS and how they are enacted have very real impacts on people's livelihoods, the environment, etc. Engagement is one of many tools in our toolbox to help guide decision-making on very complex topics like CIS.

---

[33] National Academies of Sciences, Engineering, and Medicine. (2016). Gene Drives on the Horizon: Advancing Science, Navigating Uncertainty, and Aligning Research with Public Values. The National Academies Press. https://doi.org/10.17226/23405

Over the past decade, there has been an emerging interest among philanthropies, private companies, governments, nongovernmental organizations, social science researchers, and physical science researchers in conducting engagement to support CIS decision-making. These interests stem from many intentions, including, but not limited to, making 'better' decisions, not repeating past harms, addressing climate change, enhancing community resilience, advancing democratic ideals, aligning hyper-global and hyper-local goals, and others. And these interests are rooted in many values such as equity, justice, transparency, responsibility, and accountability.

Engagement has already been used as a decision-making tool by governments, nongovernmental organizations, and researchers to gather insights from groups of people and to inform decisions about CIS. However, just like CIS is emerging, so are understandings of CIS engagement. As you finish reading, build on what you learned from this textbox. How would like to see CIS engagement happen, and when?

References

Alturki, A. (2022). The global carbon footprint and how new carbon mineralization technologies can be used to reduce CO_2 emissions. *Chemical Engineering.*

Beuttler C., Charles, L., & Wurzbacher, J. (2019). The role of direct air capture in mitigation of anthropogenic greenhouse gas emissions. *Frontiers in Climate, 1*(10). https://doi.org/10.3389/fclim.2019.00010

Fuss, S., Lamb, W. F., Callaghan, M. W., Hilaire, J., Creutzig, F., Amann, T., Beringer, T., de Oliveira Garcia, W., Hartmann, J., Khanna, T., Luderer, G., Nemet, G. F., Rogelj, J., Smith, P., Vicente Vicente, J. L., Wilcox, J., del Mar Zamora Dominguez, J., & Min, J. C. (2018). Negative emissions—Part 2: Costs, potentials and side effects. *Environmental Research Letters, 13*(6), 063002. https://doi.org/10.1088/1748-9326/aabf9f

IPCC. (2021). Summary for policymakers. In V. Masson-Delmotte, P. Zhai, A. Pirani, S. L. Connors, C. Péan, S. Berger, N. Caud, Y. Chen, L. Goldfarb, M. I. Gomis, M. Huang, K. Leitzell, E. Lonnoy, J. B. R. Matthews, T. K. Maycock, T. Waterfield, O. Yelekçi, R. Yu, & B. Zhou (Eds.), *Climate change 2021: The physical science basis. Contribution of working group I to the sixth assessment report of the intergovernmental panel on climate change* (pp. 3–32). Cambridge University Press. https://doi.org/10.1017/9781009157896.001. https://www.ipcc.ch/report/ar6/wg1/downloads/report/IPCC_AR6_WGI_SPM.pdf

MacMartin, D. G., & Kravitz, B. (2019). Mission-driven research for stratospheric aerosol geoengineering. *Proceedings of the National Academy of Sciences (PNAS), 116*(4), 1089–1094. https://doi.org/10.1073/pnas.1811022116

National Academies of Sciences, Engineering, and Medicine. (2016). *Gene drives on the horizon: Advancing science, navigating uncertainty, and aligning research with public values.* The National Academies Press. https://doi.org/10.17226/23405

National Academies of Sciences, Engineering, and Medicine. (2019). *Negative emissions technologies and reliable sequestration: A research agenda.* The National Academies Press. https://doi.org/10.17226/25259

National Academies of Sciences, Engineering, and Medicine. (2021). *Reflecting sunlight: Recommendations for solar geoengineering research and research governance*. The National Academies Press. https://doi.org/10.17226/25762

National Academies of Sciences, Engineering, and Medicine. (2022). *Next generation earth systems science at the national science foundation*. The National Academies Press. https://doi.org/10.17226/26042

Raimi, K. T. (2021). Public perceptions of geoengineering. *Current Opinion in Psychology, 42*, 66–70. https://doi.org/10.1016/j.copsyc.2021.03.012

Raimi, K. T., Wolske, K. S., Hart, P. S., & Campbell-Arvai, V. (2019a). The aversion to tampering with nature (ATN) scale: Individual differences in (dis)comfort with altering the natural world. *Risk Analysis, 40*(3), 638–656. https://doi.org/10.1111/risa.13414

Raimi, K. T., Maki, A., Dana, D., & Vandenbergh, M. P. (2019b). Framing of geoengineering affects support for climate change mitigation. *Environmental Communication, 13*(3), 300–319. https://doi.org/10.1080/17524032.2019.1575258

Sugiyama, M., Asayama, S., & Kosugi, T. (2020). The north-south divide on public perceptions of stratospheric aerosol geoengineering? A survey in six Asia-Pacific countries. *Environmental Communication, 14*(5), 641–656. https://doi.org/10.1080/17524032.2019.1699137

Tsipiras, K., & Grant, W. J. (2022). What do we mean when we talk about the moral hazard of geoengineering? *Environmental Law Review, 24*(1), 27–44. https://doi.org/10.1177/14614529211069839

Visschers, V. H. M., Shi, J., Siegrist, M., & Arvai, J. (2017). Beliefs and values explain international differences in perception of solar radiation management: Insights from a cross-country survey. *Climatic Change, 142*(3), 531–544. https://doi.org/10.1007/s10584-017-1970-8

Voosen, P. (August 2, 2023). We're changing the clouds. An unintended test of geoengineering is fueling record ocean warmth. *Science, 381*, 467–468. https://www.science.org/content/article/changing-clouds-unforeseen-test-geoengineering-fueling-record-ocean-warmth

Woodward, A. (2021, September 9). The world's biggest carbon-removal plant just opened. In a year, it'll negate just 3 seconds' worth of global emissions. *Business Insider*. https://www.businessinsider.com/carbon-capture-storage-expensive-climate-change-2021-9

Walker Raymond Lee grew up in Minnesota, where he spent lots of time outdoors with the Boy Scouts of America and became passionate about the environment. As an engineering student at university, he studied fluid mechanics and heat transfer and worked on renewable hydropower research. He expected to continue on a similar path in graduate school, but during his first week of orientation at Cornell, he was fascinated by a presentation on climate intervention research, and he immediately changed tracks to join the group. In 2023, he completed his Ph.D. on climate model simulations of solar geoengineering, in which aerosols (tiny droplets or solid particles) are scattered into the middle atmosphere to reflect sunlight and mitigate the impacts of global warming. Walker is now a postdoctoral fellow in the Climate & Global Dynamics research group at the National Center for Atmospheric Research, where he continues to study climate intervention. Walker is a primary author of Chap. 11.

Douglas MacMartin grew up in Ottawa, Canada, spending all of his summers outside, developing a love of nature. Not knowing what he wanted to do, he originally trained as an aerospace engineer, with a Bachelors in Engineering Science at the University of Toronto, a Ph.D. at MIT, followed by six years in industry at United Technologies. After a break that included hiking in New Zealand and Nepal, and diving in Australia and Thailand, he left industry for Caltech in 2000, determined to do something to help with climate change. A chance encounter in 2004 introduced him to solar geoengineering (or solar radiation modification). After initially thinking this was just a curious academic idea, he gradually realized both that it may actually be helpful to limit damages from climate change, and that his background as an engineer was essential for thinking about a problem that isn't just studying how the climate works, but rather how one might deliberately intervene in the climate system. He has been at Cornell University since 2015, focused on better understanding what the options are for solar geoengineering, what the effects would be if it were ever deployed, and the risks and uncertainties. Douglas is a primary author of Chap. 11.

Amanda Borth 16-year-old Amanda was a climate change denier. But in a determined attempt to avoid taking high school Physics, she enrolled in Advanced Placement Environmental Science: a choice that forever changed her career path. Since then, Amanda has dedicated her studies and work to climate change governance along with public participation and community engagement in science. While receiving her BA in International Studies, concentrating on Global Environmental Policy, at American University (AU), she worked as a Museum Educator at the Smithsonian National Air and Space Museum and held internships related to science policy. She then worked as a Project Coordinator for the Institute for Carbon Removal Law and Policy at AU before entering her PhD program in Communication at George Mason University (GMU). Specializing in science and climate change communication during her Ph.D., Amanda worked as a Graduate Research Assistant for GMU's Center for Climate Change Communication and focused her research on public engagement in carbon removal governance. Post-Ph.D., Amanda is now an Associate Research Professional at Arizona State University's Consortium for Science, Policy & Outcomes, where she focuses on data collection and analysis for Participatory Technology Assessments of climate intervention strategies. Amanda is the author of Box 1 in Chap. 11.

Open Access This chapter is licensed under the terms of the Creative Commons Attribution 4.0 International License (http://creativecommons.org/licenses/by/4.0/), which permits use, sharing, adaptation, distribution and reproduction in any medium or format, as long as you give appropriate credit to the original author(s) and the source, provide a link to the Creative Commons license and indicate if changes were made.

The images or other third party material in this chapter are included in the chapter's Creative Commons license, unless indicated otherwise in a credit line to the material. If material is not included in the chapter's Creative Commons license and your intended use is not permitted by statutory regulation or exceeds the permitted use, you will need to obtain permission directly from the copyright holder.

Going Beyond the Science: Climate Ethics

Greg Hitzhusen and Jill Schneiderman

Guiding Question: How do climate change and ethics intersect?

1 Key Take-Away Points

- In addition to the sciences, the humanities are essential for responding to complex, transdisciplinary global issues like climate change.
- Scientific information is not sufficient to lead to action; in fact, thinking that is a logical error. A values premise is needed to complement a scientific premise in order to create sufficient motivation for action.
- Combining scientific information and values can lead different people to opposite conclusions about climate change, depending on the values and related interpretations of history, so it is important to understand the range of values that exist to better understand how a culture might address climate change.
- There are six distinct views about climate change in America: alarmed, concerned, cautious, disengaged, doubtful, and dismissive.
- Well-funded and sophisticated public relations efforts have been successful at creating debate about climate science and generating denial of climate science and resistance to climate policy in the US. This influence has affected the landscape of views in America about climate change.

G. Hitzhusen (✉)
School of Environment and Natural Resources, The Ohio State University, Columbus, OH, USA
e-mail: hitzhusen.3@osu.edu

J. Schneiderman
Department of Earth Science, Vassar College, Poughkeepsie, NY, USA
e-mail: schneiderman@vassar.edu

© The Author(s) 2025
K. St. John and L. Krissek (eds.), *Climate Change*,
https://doi.org/10.1007/978-3-031-82869-0_12

- One illustration of how ethical ideas play into climate science has been the debate over the naming of a geological epoch the "Anthropocene". The science-based decision has been made to not define such a formal epoch on the geologic timescale. An accompanying broad public debate has had little to do with the scientific criteria for naming a geological epoch, but much to do with the debate about what is the moral/ethical point that such a name communicates.
- There are many ethical concerns that relate to climate change, but three primary ones are harm, justice, and concern for future generations.
- An important skill in ethical reasoning that helps people reach ethical and moral decisions in life is being able to deal with uncomfortable conversations. It is also helpful to consider underlying assumptions about "what counts as knowledge" when trying to understand the views of others.
- It is important to think about and develop our own values and ethical responses to climate change and its effects; it is also important to develop positive and constructive ways to dialogue with others who may—or may not—share those values.
- There are many points of consensus about climate ethics, and justice/fairness is one of the big ones that many communities are responding to; meanwhile, a wide array of value bases is serving to motivate action.
- While there is great urgency to generate more sufficient action, there are many encouraging examples of initiatives and actions being taken around the world to address climate change.
- Because it takes scientific information and values for any action on climate change, the challenge for you, as a user of this book, is to work on clarifying the values you have related to climate change and to connect those values to the scientific information you've learned. This connection can provide you with the foundation for developing *your* view of climate change and motivating *your* climate-related choices and actions.

2 Introduction

The preceding chapters have painted a compelling picture of many of the scientific and technical issues surrounding climate change and climate change interventions, and have provided a discussion of many of the factors that must be considered if we are to solve the climate challenges of our day. Indeed, many types of factors are important—science, technology, politics and policy, economic, information communication, cultural, and others. This chapter will focus on the complementary ethical factors that affect our individual and collective responses to climate change. Ethics and values underlie and influence all of these other factors, and many people feel that we won't solve environmental problems like climate change until we solve the human problems that influence our choices and cultures.[1] The role of **environmental humanities** is particularly important for highlighting these principles because the humanities is the realm in which ethical ideas are studied most closely. So in order to practically and effectively address the realities and risks revealed by climate science, we will need to go beyond science and consider climate ethics and the ways that different values intersect and affect responses to climate change.

2.1 Road Map for This Chapter

Because this chapter explores some of the perspectives that ethics can contribute to our understanding of and response to climate change, we need to start by thinking about what it means to think ethically and morally, as individuals and in groups, and to discuss values and how ethics and science relate to each other. We'll examine how polar opposite views on climate (skeptics v advocates) might be contextualized in the wider spectrum of views about climate change, and how this complex landscape of views might nuance our understanding and communication about climate change. We'll consider how the "Anthropocene" concept is also an example of how science and ethics intersect.

With these contextual and background points in place, we'll look at some of the main ethical concerns related to climate change (like harm, justice, and concern for future generations), and explore ways to engage in ethical reasoning to get beyond political and cultural clashes related to climate change. We'll consider eight key questions of ethical reasoning and apply them to climate change. We will also discuss different epistemological views (i.e., views about the nature of knowledge) that influence how people approach climate debates. This will help us address the phenomenon of climate denial, and better diagnose why some climate conversations fail; it will also help us account for the uncertainties that make climate change a challenging topic to address.

Finally, we'll explore some fruitful examples of climate dialogue and understanding—examples from social, governmental, religious, and cultural sources—that can help point a direction for ongoing responses to climate change. We end by inviting reflection on climate action and providing some sources of hope in addressing the climate challenges of our time.

---

[1] Richard Baer, the first American academic to teach a course focusing on environmental ethics (in 1966) argued for the role of the humanities in solving environmental problems in: Baer (1976).

3 Intersection of Ethics, Morals, and Values with Climate Change

3.1 How Do We Approach Questions of Should?

Climate change undoubtedly raises many ethical questions; after all, the primary question that we ask in ethics is "how *should* we live?"...and the risks related to climate change suggest that we should make significant changes to policies, energy systems, economies, and personal behaviors. One mistake that previous generations of scientists and environmental thinkers have made, however, is assuming that better science will drive better policies and this will suffice to solve societal problems like climate change. Gus Speth, founder of the Natural Resources Defense Council, former Dean of Yale's School of Forestry and Environmental Studies, and the first advisor to a U.S. president (Jimmy Carter) on climate and energy issues, once believed this, but more recently he has reflected on the complexity of issues like climate change, and the insufficiency of science and policy alone to deal with such challenges. Speth concludes that where he once thought that issues like pollution, biodiversity, and climate posed the biggest challenges to planetary thriving, he now thinks that pride, apathy, and greed are larger barriers to environmental solutions.[2] Speth notes that science tends not to deal with things like pride, apathy, and greed, which are fundamentally moral and ethical factors. If Speth and others are right, we will need to more carefully and intentionally address important moral and ethical dimensions of climate change if we want our scientific, legislative, and other social responses to be more effective in addressing climate change and its impacts on our world.[3]

3.2 An Ethics Primer

How do one's morals and ethics affect one's judgment of "should"? What do we mean by ethics? Or morals? What do we mean by values? These terms have a fair amount of overlap in meaning and usage, but they also each refer to particular aspects of the ethics landscape (Fig. 1; see also Table 1, which provides a glossary of ethics terms used in this chapter). **Values** generally refer to the social principles, goals, and standards held or accepted by an individual, class, group, or society; we often think about values as being personal to individuals or groups who hold them, and we think

---

[2] Speth made comments similar to these at various events, apparently first at Melhana Plantation in Georgia in late 2006 and at several other venues (heard in person by the author) in the 2010s. One citation from 2013 is noted in the following: NC Interfaith Power & Light (2017, October 25). *The environmental crisis is not environmental. It is spiritual.* NC Interfaith Power & Light. https://ncipl.org/environmental-crisis-not-environmental-spiritual/.

[3] More discussion here: Hitzhusen, G. (2019). Chapter 3: Defining our terms and direction. Religion and Environmental Values in America. https://ohiostate.pressbooks.pub/enr3470/chapter/defining-our-terms/.

of values as the precepts upon which we base our behavior. Values are learned, by experience and through culture. The values we hold may not align with the larger groups to which we belong and can change over time. Often that change comes from greater attention to ethics or ethical reasoning. **Ethics** relates to how people evaluate what is right and wrong; "ethics" is also the term we use for the study of moral and ethical thinking and judgment. The related term, "morals," is often used interchangeably with "ethics," but **morals** more specifically refers to views (sometimes religiously held) and principles we hold on different issues about what is right and wrong, while ethics tends to refer to our larger system of organizing our moral values to understand right and wrong across multiple issues. Charles Darwin in his *Origin of Species* described humans as "the moral species"—the species that wonders about, talks about and argues about what is right and wrong. It seems natural for humans to ponder what is right and wrong; indeed, any notion that humans should do anything—like pollute the environment or not—is a question of morals, ethics, and values. In relation to climate change, it has been noted that scientific information alone cannot lead to action about climate change, because humans act on the basis of values. Therefore, any information we gain from science about the state of the climate and the planet must be complemented by values that tell us why we should do anything about those facts of climate change. Fundamental ethical principles like harm, justice, and concern for future generations provide key examples. Humans tend to agree that it's wrong to cause unnecessary harm, that it's wrong to cause injustice, and that we should avoid disadvantaging future generations. These are values that prompt humans to take action about climate change, because climate change is causing harm, is a source of environmental injustice, and puts future (and present!) generations at risk. Values like these may seem so basic as to be taken for granted, but without such values in place to spur our actions, there is no logical reason to act on climate change.

3.3 Why Isn't Science Sufficient to Address Climate Change?

History has shown us that science alone has not been sufficient to guide individual, community, and political action to address climate change. Witness the public confusion and political and cultural contention about climate science in the United States, despite what some have called the most robust scientific consensus[4] ever reached on any issue. Communicating more science has not tended to reduce public confusion: even though 97 to > 99% of peer-reviewed science agrees with the consensus,[5] the

---

[4] Two sources that highlight this consensus include: NASA (accessed 2023, May 23). Scientific Consensus: Earth's Climate Is Warming. https://climate.nasa.gov/scientific-consensus/ and NASA (accessed 2023, May 23). How Do We Know Climate Change Is Real? https://climate.nasa.gov/evidence/.

[5] Note the following sources: Anderegg et al. (2010); Cook et al. (2016); Lynas et al. (2021); Powell, James Lawrence (15 November 2012); also note that Katharine Hayhoe and others conducted a study on the 3% of climate literature that apparently doesn't agree with the consensus, and found that all

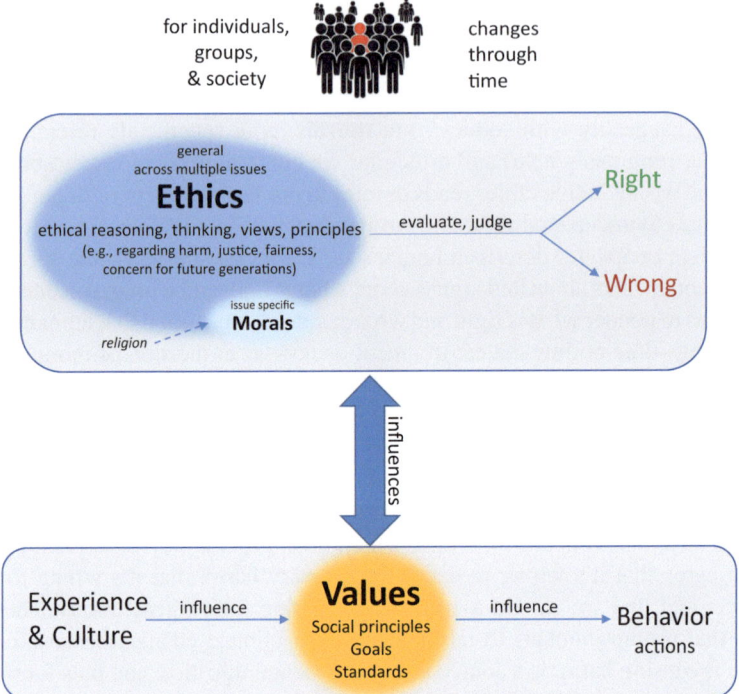

Fig. 1 The Ethics Landscape for individuals, groups, and society. The landscape changes through time. An individual's ethical views and/or values may not always align with those of their immediate larger group. This can create conflict and stall action. Strategies for dealing with this situation are provided later in the chapter. Diagram by Kristen St. John

American public long remained split on whether *they think* scientists agree about climate change.[6] The American public is also split on climate policy support, with the growing divide on whether climate change action should be a priority for the U.S. President and Congress gaping by 2020 (see Chap. 1, Fig. 11) and remaining wide at the writing of this book.[7] Americans have been slow to financially support much change—as of 2018, ~ 60% of Americans would pay a mere $1 a month to fight

---

of the 3% could either not be replicated, or contained theoretical or computational errors—thus, they conclude that it is closer to 100% of published climate science literature that supports the consensus (see: Foley, K.E. 2017, September 5). Those 3% of scientific papers that deny climate change? A review found them all flawed. *Quartz.* https://qz.com/1069298/the-3-of-scientific-papers-that-deny-climate-change-are-all-flawed/ and the study: Benestad et al. (2015).

[6] A 2021 Yale Climate study found that only 57% of respondents believed that "most scientists think global warming is happening": Marlon et al. (2022).

[7] See Tyson et al. (2023).

Table 1 Glossary of ethical terms discussed in this chapter

Ethics	Systems of thought for *evaluating* what is right and wrong; the *study of* moral and ethical thinking and judgment.
Morals	Views and principles people hold on different specific issues about what is right and wrong.
Values	The social principles, goals, and standards held or accepted by an individual, class, group, or society, upon which behavior is often based.
Differential vulnerability	That observation that some subjects (e.g., poor and vulnerable people and communities) *disproportionately* suffer the burdens of climate change.
"Anthropocene"	A term intended to signify the current interval of anthropogenic global environmental change, the precise significance of which is debated. Not officially defined on the geologic timescale.
Epistemology	A sub-field of ethics that focuses on questions of what it means to know something, and examines differing views of what counts as knowledge.
Absolutism	The belief in absolute principles that are understood to be true at all times and in all cultural situations.
Ethical skepticism	The belief that no ethical beliefs are certain, and thus all ethical beliefs are unjustified.
Fallibilism	The view that regardless of what we believe about whether things can be absolutely true or not, *we can be mistaken about what we think we know is true.*
Relativism	The belief that all truth is relative—that whatever is "true for you" is true for you, and whatever is "true for me" is true for me—making truth relative to each individual.
Scientism	The belief that science is sufficient to answer all meaningful questions about reality.
Precautionary approach	An ethical and political perspective that promotes enacting protective policies when there is a reasonable chance of risk or harm to the public.
Phronesis	Greek term for practical wisdom, or knowledge about *how* to live ethically.

climate change, but only 28% were willing to pay $10 per month (and 68% would be opposed to paying $10 per month).[8] That's a low commitment to actual change; on the other hand, perhaps because of the Biden administration's renewal of interest in addressing climate change, as of 2021, 31% of Americans were willing to pay $100 per month. As priorities change with new government administrations, this figure may be very different now and in the future. August 2022 legislation passed by the U.S. Congress (albeit with support from only one political party), which included $369 billion in incentives and measures to address climate change, may influence the public support moving forward should it be fully implemented and not overturned by subsequent policy makers. Notably, though, this legislation was not crafted to require the U.S. to reduce greenhouse gas emissions, but rather to provide credits and incentives for Americans to invest in energy-efficient or clean energy alternatives to carbon-intensive energy sources. Interestingly, after decades of inaction, the first

---

[8] The following article: Rainey (2019), highlights the findings of: Leiserowitz et al. (2019).

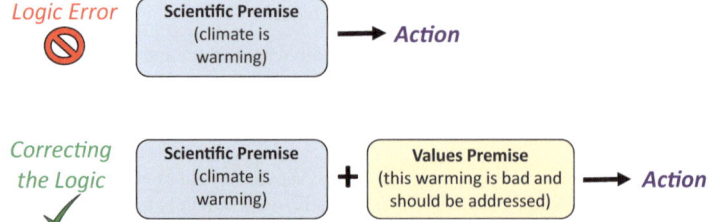

Fig. 2 Ethicist K. D. Moore explains that it is a logical error to think that scientific information alone will lead to action on climate change. A values premise must complement a scientific premise to lead to action. Note, however, that values premises are not always as easy to agree on, and a complex host of reasons can influence the expression of different values. Figure by Kristen St. John

major climate legislation to pass in the United States was noteworthy for its avoidance of creating any penalties for carbon emissions and opted instead to provide rewards for low-carbon alternatives. Such legislation seems to therefore capitalize on some underlying human motivators for change, in resonance with the concepts we will explore in this chapter.

A great explanation of why science alone is not sufficient to enact change comes from ethicist Kathleen Dean Moore, who explains[9] how scientific premises need values premises to avoid a logical error in the assumption that scientific information will lead to changes in policies and behaviors (Fig. 2). Moore's explanation highlights why we should not expect science alone to solve problems like climate change. Science addresses questions about what is, what was, and what is likely to be in the future, based on an understanding of physical, chemical, and biological principles; ethics answers questions about what *should be*. For example, science tells us that different people have different skin colors because of differing melanin levels in their skin. Ethics tells us that we should not discriminate against others on the basis of skin color and that if some people are systematically discriminated against because of their skin color, that is unjust and changes should be made to reduce such discrimination. History tells us that such changes will not come easily. Likewise, science tells us that modern climate is changing due to anthropogenic factors, predicts which actions can reduce greenhouse gas emissions, and identifies the actions that are likely to help mitigate the worst effects of climate change. Ethics explores whether and why we should take action to do anything about climate change. History has so far shown us that any changes in behavior to mitigate the effects of climate change will not come easily. Part of the challenge in reaching agreement about action is that people don't always agree on the values premise, even if they reach the point of agreeing on the scientific premise.

As obvious as it seems to anyone who places a high value on protecting the natural world that current facts about climate change demand action, often those values and

---

[9] This point is summarized here: Moore, KD, & Nelson, MP (2018). Ethics, Chapter 10 (https://open.oregonstate.education/climatechange/chapter/climate-ethics/) in Schmittner, A. *Introduction to Climate Science*. Oregon State University. https://open.oregonstate.education/climatechange/.

their related ethical tenets are not made explicit. And, often we fail to consider whether environmental values are as universal as the facts of climate science. It might be that many people do not hold a high enough value for solving environmental problems to decide to take action based on climate science. Many people (especially people of means who hold positions of influence in economically developed nations) value the security and comfort they have gained from the economic and environmental patterns of the status quo. And, the value they place on retaining that security may well override whatever value they place on the environment. And, if too often those values aren't made explicit or we are told not to discuss ethical issues because they are contentious or "unscientific," it seems unlikely that we will get better at addressing the moral dimensions of issues like climate change.

So what does it look like when we begin to be more explicit about the values involved in climate change? Let's start with a simple example. If we filled in the values premises of a typical stand-off about climate change, it might look something like this:

Climate Advocate: *climate science (scientific premise) shows us that we are changing the climate primarily because of land use practices and our burning of fossil fuels, and I value (values premise) keeping the climate within "safe" bounds for the sake of ecological integrity and avoiding the disturbances to human and non-human life that climate change brings. I think this goal of reducing the risks of climate change is more important than maintaining the status quo. We should therefore reduce GHGs!*

Climate Skeptic: *measurements may show us that climate is changing (scientific premise), but history also shows us that climate has changed a lot in the past, and I value (values premise) the comfort and security I have now due to the economy made possible by burning fossil fuels. The prospect of changing that economy scares me, and it might disrupt society enough to cause harm. Many environmental activists who get all excited about climate change seem like hippies and socialists who just want to disrupt society and destroy businesses. So I'd rather not change things if we don't have to.* (Fig. 3)

As the examples above illustrate, the same climate science matched with different values (and various possible interpretations of history) leads to very different calls for action. But, the situation is of course much more complex than this. There is a wide landscape of views about these questions.

The Yale Program on Climate Change Communication and the George Mason University Center for Climate Change Communication have described and continue to monitor six distinct and predominant views about climate change in America, what they call the "Six Americas." Their research has shown that American views on climate change can generally be divided into six main stances: alarmed, concerned, cautious, disengaged, doubtful, and dismissive. The predominance of each of these views from 2010 to 2021 is shown in Fig. 4.

These data show that there has been a major increase in the "alarmed" since 2015, and that group was the largest portion of the landscape in 2019 and 2021. This 12-year landscape of views suggests why the simple conflict between climate advocates and climate skeptics sketched above only touches on some of the intersections between

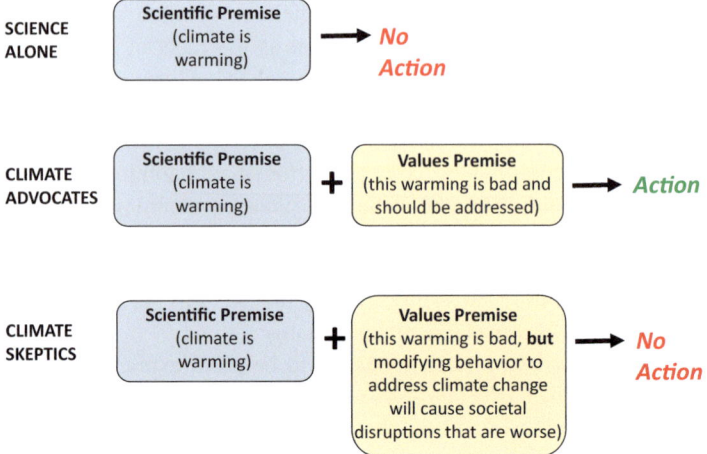

Fig. 3 Competing values can lead to different action outcomes. Figure by Kristen St. John

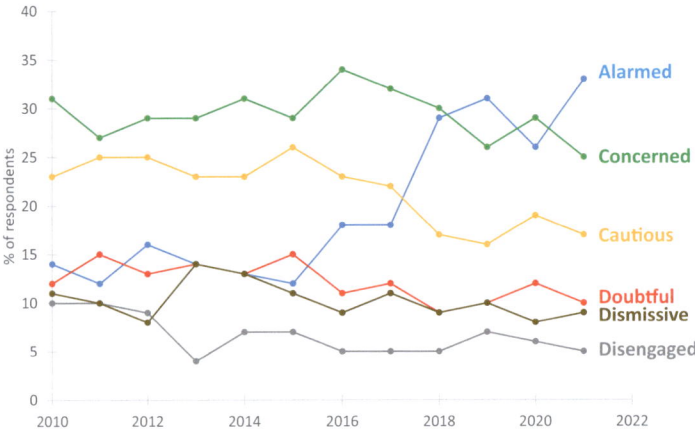

Fig. 4 The proportions of the American public professing different views of climate change from 2010 to 2021. Figure by Kristen St. John, from data in reports accessible from Yale Climate Communication[10]

climate science and ethics. If we were to imagine the science and values premises through time across this landscape, we would gain a more complete picture of the complex interplay between scientific information and ethical views related to climate change. The Yale study noted above briefly describes these six views across the landscape in this way:

---

[10] https://climatecommunication.yale.edu/about/projects/global-warmings-six-americas/.

- The **Climate Alarmed** agree with the scientific consensus and voice support for immediate action.
- The **Climate Concerned** agree with the scientific consensus and are generally supportive of plans to address climate change.
- The **Climate Cautious** agree with the scientific consensus and are cautious about climate change but not necessarily driven to act.
- The **Climate Disengaged** aren't sure one way or another about the climate consensus, and aren't concerned enough to engage one way or another.
- The **Climate Doubtful** have doubts about whether the scientific consensus is true, and are unlikely to take action.
- The **Climate Dismissive**, interestingly, have been shown to understand climate science[11] more accurately than the Climate Alarmed, but are skeptical and dismissive of solutions offered by climate advocates.

This range of views makes it easier to see why even if many people are aware of climate change and are concerned about it, their concern may not add up to much action across society. We might also consider adding three more views to the mix:

- The **Climate Denial Operative** understands that (and has been hired because) the facts of climate science suggest that the highly profitable fossil fuel industries of today may not be able to maintain their economic dominance, and thus works to generate confusion about climate science (just like nicotine addiction denial operatives did for the tobacco industry in the past) to delay public action.[12] They therefore enhance the role that doubt plays in diluting action by people across the spectrum described above. Or they may delay changes to energy systems by promoting the appearance of "carbon neutrality" that fails to reduce carbon emissions.[13]

---

[11] To some extent, those with higher science literacy are better at using scientific claims to support their existing values, so climate skeptics may have higher awareness of climate science than some alarmists, but use it to support their skepticism about taking action on climate change. See: Kahan et al. (2012).

[12] Commentary and research on climate denial can be found in Robert Brulle's study on "Institutionalizing Delay": Brulle (2013); additional commentary on the moral dimension of climate denial can be found here: Hitzhusen, G. (2019). Chapter 9.2 Faith and Skepticism: A Matter of Morality. Religion and Environmental Values in America. https://ohiostate.pressbooks.pub/enr3470/chapter/9-2/; a very detailed account of some examples of climate denial and its similarities to tobacco industry tactics is found in Oreskes and Conway (2011). Frontline's 2022 three-part video series titled "The Power of Big Oil" is another source of reporting about this issue: Frontline (2022). The Power of Big Oil. PBS.

[13] Commentaries like the following highlight that there are various types of misinformation, beyond just denying climate change, that delay action towards reducing climate change; promoting "carbon neutrality" that doesn't reduce carbon emissions or "clean coal" that isn't clean might be examples: Schendler (2021); Stapczynski et al. (2021); Wright, C., & Nyerg, D. (2021, October 13). Planetary Challenges: The Fossil-Fuel Industry, Climate Change, and the Disruption of the World. In Kipping et al. (2021).

- The **Climate Advocacy Lobbyist** agrees with the scientific consensus, and is aware of the massive funding behind the climate denial system,[14] so works to highlight the immoral aspects of these industries and to publish the truth about how the denial system is spreading misinformation to delay action.
- The **Climate Scientist** agrees with the scientific consensus, but as a scientist tends not to take a position about advocacy, and resists stretching data to say more than it should; however, as a person or citizen, many scientists find it hard not to support action on climate change because of the implications of the science.

The sum total of these various and intersecting values and scientific premises, amplified by how these issues are reported in the popular news and social media (as noted in Chap. 1), has been lots of confusion about what to do about climate change, and a weakened political will to make related changes. Attending to ethics can help bring previously ignored values premises into climate conversations, which can catalyze action, but values themselves are complex and contentious. And even when values are shared, a perennial human issue is that knowing what we "should" do doesn't mean we are good at actually doing it (many of us, for instance, instinctively resisted our parents when they told us to "eat our vegetables" or do other things that are "good for us," and recall from above how little money Americans who are concerned about climate change are actually willing to pay to solve it). Add to this the fact that values can also be manipulated—climate denial operatives (funded by fossil fuel industry supporters) have successfully generated doubt and opposition to climate science through disinformation campaigns, and thus their greed and pride have fueled campaigns to encourage apathy in the culture at large. Some climate ethicists, like Donald Brown, have wondered whether the climate denial actions of companies that profit from carbon pollution amount to a new kind of crime against humanity,[15] which seems an even stronger ethical judgment; others might note that the confusion of factors reflects the tragic patterns of human behavior that ethicists, prophets, and playwrights have highlighted for millennia. In addressing ethical aspects of climate change, we should not expect easy results, but we can work to better understand the nuances and complexities of how science and ethics intersect. If we want to clarify our own values related to climate change and enhance our ability to discuss these values with others (including those who might not share our same values) we can also consider additional aspects of ethical reasoning, and consider some of the approaches to discussing ethics highlighted below.

---

[14] See for example: Fischer (2013).

[15] See the following sources of this commentary: Brown (2010a, 2010b).

3.4 How Might the Concept of the "Anthropocene" Frame Our Ethical Concerns About Climate Change?

One way of thinking about climate change is through the concept of the "Anthropocene", which alludes to the pivotal impact that human beings have on the planet. In science, as in many issues of work and life, the devil is in the details. The essay in Box 1 is an example of that. It is a glimpse of the behind-the-scenes controversies and multiple perspectives related both to scientific data and to ethical concerns that exist in what at first look may seem so simple—giving something a name.

> **Box 1 The "Anthropocene": critiques from the environmental humanities**
>
> By Jill S. Schneiderman, Vassar College, Poughkeepsie, NY
>
> The decision whether to recognize and name the "Anthropocene" as a new epoch of geologic time is fundamentally a stratigraphic one; it is a decision about whether there is a distinct geologic signal of human impact on the planet, and a decision about when such impact first began.[16] Making these determinations is challenging and there are widely-varying perspectives. In July 2023, the Anthropocene Working Group of the International Commission on Stratigraphy (ICS) recommended that the start of an Anthropocene epoch be marked by the deposition of plutonium sometime between 1950 and 1954 in Crawford Lake, a deep waterbody near Toronto, Canada.[17] However, in March 2024 a panel of the ICS voted against this proposal.[18] Regardless of whether this (or a different) proposal is ultimately approved, the term "Anthropocene", intended to signify the current interval of anthropogenic global environmental change, has captivated not only the imaginations of geoscientists, but of scholars outside the geosciences, the media, policy makers, and the general public as well.[19] Some geoscientists have questioned the stratigraphic basis for such a division of time, and that view was reflected by the negative vote of the ICS panel, while humanities and social science scholars have heeded only rarely the stratigraphic debate and instead have focused their attentions on nomenclature.[20]

---

[16] See: Zalasiewicz et al. (2008); Autin and Holbrook (2012); Waters et al. (2014); Yang (2017); Zalasiewicz et al. (2015); 6; Finney and Edwards (2016); The Anthropocene Working Group (2019). Working group on the Anthropocene—Results of binding vote by WG. Subcommission on Quaternary Stratigraphy, http://quaternary.stratigraphy.org/working-groups/anthropocene/; Zalasiewicz et al. (2019) and Voosen (2022).

[17] Witze (2023).

[18] Voosen (2024).

[19] See: Zalasiewicz et al. (2008); Bonneuil and Fressoz (2015); The Economist (2016); Farrier, D. (2016). Farrier and Aeon (2016); Moore (2016); Yang (2017a, 2017b).

[20] Schneiderman (2015) and Crist (2016).

These scholars' primary critique has been that the appellation, "Anthropocene", "crystallizes human dominion, corralling the already-pliable-in-that-direction human mind into viewing our master identity as manifestly destined, quasi-natural, and sort of awesome".[21] In other words, according to such naysayers, the name feeds our human hubris. What's more, they claim that "Anthropocene," a term designed solely by geoscientists, implicates *all of humankind* in enacting environmental change on the planet rather than recognizing the disproportionate effects of industrialized nations on, for example, atmospheric change, sea level fluctuations, and soil erosion.[22] These scholars have proposed names such as Capitolocene—for the rise of capitalism, Plasticene—for the predominance of plastic on the planet, Petrolcene—for dependence on fossil fuel and Androcene—to reflect the relative power of men in decision-making processes, among others.[23] The coining of such neologisms has led geoscientists to question whether the "Anthropocene" has become a matter of either pop-culture or politics.[24] Nevertheless, despite the ICS panel's recent vote against establishing a formal Anthropocene epoch, it seems likely that discourse surrounding identification and characterization of the "Anthropocene" will endure within and outside of the geosciences.

Geoscientists should acknowledge that critiques from outside our subject area have some merit. As we hope your reading in this chapter on ethics and climate change has already demonstrated, all disciplines benefit from engaging with perspectives from outside our subject area. The term "Anthropocene" does indeed totalize all human actions into one 'human activity' generating a single 'human footprint' on Earth. Truly, populations of less industrialized countries have had little if any role in the Great Acceleration—changes in the Earth systems driven by the global economic system. We should take seriously this observant point. We should connect with the cultural criticisms by recognizing "Anthropocene" as a problematic name and wholeheartedly participate in a broad conversation whether, and how, to appropriately name this epoch—a conversation that incorporates both scientific and cultural perspectives.

Jill Schneiderman's overview of the various considerations and ethical overtones that attend the discussion of the "Anthropocene" highlights the complexity and nuances of how ethics intersects with science. Clearly, there are ethical stakes in this otherwise scientific naming debate. On the one hand, the "Anthropocene" has been evaluated via the basic criteria for designating stratigraphic intervals of geological time, and has been found wanting—at least for now. New junctures in the geological record may deserve to be named, and the geologic argument has revolved

---

[21] Crist (2016).

[22] Bonneuil and Fressoz (2015).

[23] Haraway (2016).

[24] Autin and Holbrook (2012) and Finney and Edwards (2016).

around whether the "Anthropocene" really qualifies as a unit of geologic time (the presence of man-made materials—atomic and other industrial fingerprints—in the geologic record provided the evidence supporting an argument for naming a new unit of geologic time). But, much of the force of the discussion of the "Anthropocene" seems to be that it raises up human impact on the planet as ethically relevant— significant, weighty—impactful enough to distinguish a geologic interval, which is perhaps more a moral point than a geological one.[25]

In terms of the ethical point of naming the current epoch as a recognition of human impacts on the planet, the interesting question is not *if* we should name a new epoch, but rather *what*, really, should we name it to get the moral commentary right? If human impacts deserve to be blamed, but the moral force of that blame directs us away from those who are most blameworthy and generalizes the blame to all humans, doesn't our moral imprecision then contribute to obscuring the causes and thus confusing our responses to climate change? The concept of the "Anthropocene", with its geologic/scientific debate and its ethical implications and connotations, exemplifies one intersection of science with climate ethics. Scientific matters clearly tread on moral ground, and the coherence and fitness of a scientific name here seem to hinge on the precision of the moral connotations of the name in question. This is an important cultural conversation. With a similar interest in moral precision, we can also consider a number of primary ethical concerns directly related to climate change.

3.5 What Are the Main Ethical Concerns Related to Climate Change?

Some fundamental ethical concerns about climate change have already been introduced, such as **differential vulnerability** as discussed in Chap. 1, but there are several primary ethical concerns that are highlighted by climate ethics.

One way to look at the main tenets of climate ethics is to consider some of the fundamental ethical principles that relate to climate issues: harm, justice, and concern for future generations are three key areas of moral/ethical concern that cut across many contemporary issues (Fig. 1). The simple facts that climate change causes or multiplies *harm*[26] (to humans and to other creatures), produces negative impacts that are *unjust or unfair* (differential vulnerability means poor and vulnerable people and communities disproportionately suffer the burdens of climate change),

---

[25] Some of this debate is addressed here: Zalasiewicz et al. (2011). It is also interesting to consider other science/ethics intersections and how they've been handled, such as the tendency toward renaming features and/or ocean currents that used to be named for male European explorers or scientists, but now are more commonly referenced using geographic names (e.g., the shift from "the Sea of Cortez" to "the Gulf of California", or "the Humboldt Current" to "the Peru-Chile Current").

[26] Chapter 1 refers to climate change as a "threat multiplier", meaning it makes existing threats worse. For example, extreme weather and water shortages increase the risk of civil unrest and political instability. So some of the harm isn't just because "climate change causes X"; rather it is because climate change has cascading consequences that can increase other existing risks.

Table 2 Moral reasons to care about climate change summarized from the insights of global moral leaders (Moore & Nelson, 2010)

Summary of moral reasons to care about climate change	
For our children's sake	For human survival
For the sake and love of Earth and all life	For moral integrity and virtue
Because life is beautiful	Because life is a gift, and therefore we have an obligation of gratitude
Because we have an obligation to be good stewards	Because honor is due to great, evolved life systems
Out of obligation to other creatures and diversity	Out of the duty of compassion to prevent suffering
Because people have a right to their cultures, which are being destroyed	For justice

and creates negative impacts on *future generations* (human actions today degrade the environment that future generations will inhabit) are some of the most important reasons why people should be (and are) morally concerned about climate change.

Recall K.D. Moore's point (Fig. 2) that in order for our scientific knowledge about climate change to result in any action, a values premise is also needed. For many, the fact that climate change causes harm and injustice, and impinges on future generations is enough ethical justification to take action in response to climate risks. K.D. Moore and Michael Nelson provide an expanded list of moral-ethical reasons why people care about climate change (Table 2). They invited more than 100 moral leaders from around the world to answer the question: "Is climate change a moral issue?" and the answer was overwhelmingly "yes." In their book *Moral Ground* (2010), Moore and Nelson highlight the main ethical reasons to care about climate change that were voiced by these moral leaders from around the world.[27]

These (Table 2) are weighty moral concerns, and some might speak more strongly to some people than others. Additionally, just because people acknowledge a moral concern doesn't mean it will be easily solved. Consider important concerns in the U.S. about racial justice that have been a focus since the abolition of slavery in the nineteenth century and the civil rights movement of the 1960s; only recently, as the Black Lives Matter movement (founded in 2013) has gained momentum, have many non-black people become aware of and begun to act to address these issues, and even so, efforts to address systemic racism and racial equity have a long path ahead. So too we expect that adequately addressing climate change will involve a long process of working out conflicting interests and changing behaviors.

---

[27] Moore and Nelson (2010). In a similar vein, UNESCO adopted a *Declaration of Ethical Principles in relation to Climate Change* in 2017. Another good commentary on addressing ethics to address climate change is found here: Somerville (2008). In addition to these fundamental moral concerns, there are a wide range of specific and even more technical ethical considerations that ethicists debate in relation to climate change. Idil Boran and Corey Katz's article on climate justice in the Routledge Encyclopedia of Philosophy (Boran & Katz, 2017) highlights some of these deeper discussions related to justice and governance; Simon Caney's entry on climate justice in the Stanford Encyclopedia of Philosophy (Caney, 2020) similarly highlights a wide range of ethical concepts, for readers interested in a deeper dive into scholarly discussions of climate ethics.

Ethical issues are complex by nature and tend to involve strongly held beliefs that can be tricky to navigate in polite conversation. In fact, one of the most important things that discussing the ethics of climate change can do is help us understand and disentangle the various (and sometimes conflicting) values and ethical claims that are relevant to climate change. Discussions of ethical issues require the ability to deal with discomfort and uncomfortable conversations; this is a life skill worth fostering. Dealing with questions of "how to have a reasonable and reliable and respectful discussion about moral/ethical issues" (after all, climate change dialogue has been contentious, and moral/ethical issues are notoriously debatable) is an important next step in considering the ethics of climate change, and may just be a prerequisite to having adequate discussions about climate change in the first place.

4 Integrating the Ethics Premise with the Scientific Premise to Motivate Climate Change Action

4.1 How Can We Engage in Ethical Reasoning to Wrestle with Complex Issues and Get Beyond Opinions, Political Agendas, Special Interests, and Social Media Bias?

Undoubtedly, good climate science will remain essential, and effective climate policy will require adroit policy-making, but the connection between these two will remain influenced by ethics and values, for good or ill. Recall how K.D. Moore introduced the idea that taking action (including legislation) on climate change requires both scientific and values premises. She also recommends that beyond just reading about climate ethics, *conversations* about climate change are the most important way to engage in thinking about climate ethics.[28] Moore suggests a discussion format where different values are drawn out, expressed, and put in dialogue in a respectful process where opposing views can be considered fairly, and conversation partners can come to better understand the perspective of those whose policy preferences differ.[29] One example of this approach is described in Box 2.

---

[28] Katharine Hayhoe's TED Talk makes a similar point: Hayhoe (2018). Note also: Goldberg et al. (2019).

[29] Moore describes a process of dialogue that centers on bringing people together to ask whether we have a moral obligation to leave a world as rich in possibilities as our own. She suggests discussing this in pairs and small groups and then with the whole group, not to rank ideas, but to share what we each deeply believe, as a basis for moving forward respecting what one another deeply believes in, which tends to provide common ground (e.g., people care about children, about our health, about security in an uncertain world, about relations with something bigger than ourselves, about obligations to God). While we may not all agree about these, we can honor them in each other because they are honestly held beliefs. And, we can then move forward in action to honor our obligations to the future based on these beliefs (see: More, 2010).

> **Box 2 Eight key questions of ethical reasoning**
>
> A helpful framework for discussing climate ethics is to consider the "Eight Key Questions" of ethical reasoning outlined by the James Madison University project on ethical reasoning.[30] The process starts by asking:
>
> What key questions should I (we) use to evaluate the ethical dimensions of a situation?
>
> - **Fairness**—How can I (we) act justly, equitably, and balance legitimate interests?
> - **Outcomes**—What possible actions achieve the best short- and long-term outcomes for me and all others?
> - **Responsibilities**—What duties and/or obligations apply?
> - **Character**—What actions help me (us) become my (our) ideal self (selves)?
> - **Liberty**—How do I (we) show respect for personal freedom, autonomy, and consent?
> - **Empathy**—How would I (we) act if I (we) cared about all involved?
> - **Authority**—What do legitimate authorities (e.g., experts, law, my religion/god) expect?
> - **Rights**—What rights, if any (e.g., innate, legal, social) apply?
>
> These eight questions are designed to open up the ethical concepts and values that may be important to a given issue. There are different ways to organize the eight questions—you might see the questions in two categories: those related to avoiding harm and those related to seeking to do good—or you might think about the questions in three categories: those that are outcomes-based, duty-based, or virtue-based.[31] Regardless of how you might conceptualize the questions, it is important to note that the questions are not designed to give us the "right" answers to all ethical questions, but they can help us to understand the choices that others make, and they can help us clarify our own ethical choices. It is also important to note that not all issues will necessarily strongly invoke all eight questions, but addressing any ethical issue by evaluating at least three of these questions will help uncover and clarify some of the complexity of moral issues, and the moral dimensions of contemporary concerns like climate change.[32]

---

[30] James Madison University, Ethical Reasoning In Action, The Eight Key Questions: https://www.jmu.edu/ethicalreasoning/8-key-questions.shtml

[31] These categories correspond to ethical theories of consequentialism, deontology, and virtue ethics.

[32] For more information on the 8 Key Questions program, visit: https://www.youtube.com/watch?v=e72iASGLVfA&t=2s; a similar ethics decision-making framework has recently been developed by a multi-stakeholder team and Cornell University for wildlife management decision-making, see: Decker et al. (2019).

A helpful exercise is to answer the eight key questions in reference to climate change. One set of responses (in the form of further questions) might look like this:

Fairness: How fair is it that Africans have contributed less than 2% to global greenhouse gas emissions, but they are bearing the brunt of climate drought in the form of lost agricultural livelihoods and refugee-ism?[33] How fair is it that one person's way of getting power/electricity creates pollution for another person? How fair is it that some climate change solutions privilege states like California that already has lots of solar power, and disadvantage states like Ohio, Pennsylvania, and West Virginia that have traditionally been fueled by coal?

Outcomes: What outcomes are we facing from sea level rise, wildfires, heat waves, and enhanced drought and storms because of climate change? What outcomes will we face if development and positive health outcomes provided by energy from fossil fuels are reduced to curb GHG emissions?

Responsibilities: What responsibilities do we have to future generations who may suffer greatly from a warmer planet? What responsibilities do industry leaders have to assure the public that they are not illegally polluting or cutting corners on emissions regulations just to make a profit? What responsibilities do government environmental regulating agencies have to decrease GHG emissions?

Character: What sort of person runs a business that pollutes the Earth and harms people and wildlife to make a profit? What kind of person demonizes an entire political party for supposedly caring less about climate change, and in the process contributes to further polarization about climate change due to identity politics?

Liberty: How is the freedom of the poor and of future generations affected by the damages being done by climate change? How is the freedom of drivers of inefficient vehicles impinged upon by regulations to reduce greenhouse gas emissions?

Empathy: Who is really being most harmed by climate change impacts? Do you know someone who is being harmed? Can you name them? What should we do to help climate refugees? Pacific Islanders leaving their homes because of salt water inundation? Poor farmers in Africa whose crops are failing due to

---

[33] See, for instance: Williams (2022).

drought? Farmers in the Southwestern and Western United States experiencing drought?

Authority: Does some higher spiritual authority want us to stand by while creation is destroyed? Would a loving and just God approve of current climate injustices? Why are climate scientists sounding such a strong alarm about climate change? What are the goals of international climate negotiations? Should climate morality be enforced by law so people are forced not to emit greenhouse gases, or should people be allowed to do the right thing on their own?

Rights: Do future generations have a right to clean air, clean water, and climate cycles that don't lead to ecological and economic devastation and destruction? Do poor urban residents have a right not to have their electricity bill go up because of a transition to clean energy? Do the rural poor have a right not to have their land and livelihood destroyed by drought?

The moral of the story here is that it is helpful to evaluate ethical questions from multiple perspectives, and in some cases it may only be by doing so that we can gain sufficient understanding to make decisions and change behaviors, together with others, in addressing tough issues.

There are additional perspectives on the philosophy of science that can help lay a clearer foundation for discussing ethics. The sub-field of ethics known as **epistemology** focuses on questions of what it means to know something and examines differing views of what counts as knowledge (see Table 1 for a glossary of ethical terms). When conflicts arise between moral viewpoints or in relation to complex issues (imagine a face-off between a pacifist and a gun rights advocate, or recall the clash between the climate advocate and the climate skeptic sketched above; Fig. 3), sometimes opponents default to different epistemological beliefs, but only rarely are these made explicit, with the result that opponents simply talk past each other, destined to remain opposed through a lack of understanding. For instance, someone who believes in **ethical skepticism** (that no ethical beliefs are certain, and thus all ethical beliefs are unjustified) would be unlikely to be motivated by someone else's moral beliefs about what is wrong with anthropogenic climate change—not because climate science isn't compelling, but because they are predisposed to doubt ethical claims.

Or sometimes in the face of conflicting ideas, opponents will invoke the epistemological perspective of "**relativism**," the belief that "whatever is true for you is true for you," while "whatever is true for me is true for me" (to each their own, eh?). For example, frustrated by the different views of a climate advocate, a climate skeptic might just say: "well, climate change is true for you, but it's not true for me"… and walk away from the conversation. As philosopher Allen Wood makes clear, such a response is unsatisfying for several reasons.[34] First is that anthropogenic climate change doesn't seem to be the sort of thing that could be true for some people and false for others: climate is either changing because of human actions or it is not. So this move seems to side-step the real issue. And indeed, philosophically, this move is disingenuous because relativism itself is self-refuting: in asserting the view of relativism (that all truth is relative to the believer), the relativist is asserting that relativism is true (for everyone), a view that is impossible for the relativist. Furthermore, what would be the point of debating different interpretations of truth (whether climate is changing, for instance) if truth is just a matter of what any given person prefers to believe? If opponents in an argument invoke relativism to deal with the conflict, or generally think that truth is relative, then it is no wonder that confusion results, and we reach no greater understanding of the issue at hand. Such confusion about truth may be why some people aren't even interested in examining what is true about climate change or not, or generally don't care to worry about supposed obligations they have to "the planet" or to the common good. If everyone is free to make their own truth, then we are free to ignore the truths of climate science and any obligation to climate action. Such views tend to weaken political will towards climate action (or strengthen political will against change!), so discussing the shortcomings of relativistic arguments can help us treat others' views more respectfully.

Epistemology helps us sort things out in another way, too. If relativism is a shaky basis for knowledge, many people opt instead for various forms of "**absolutism**" or the belief in absolute principles that are understood to be true at all times and in all cultural situations. This fares well for fairly settled moral intuitions, like the belief that murder is wrong, or that torturing children in front of their parents for no good reason is wrong, or that allowing willy nilly lying is probably a poor idea for any community.[35] But trouble, or at least confusion, arises amid the complexities of many moral issues—often there are multiple overlapping and possibly conflicting moral principles at play in a debate (the Eight Key Questions in Box 2 help ferret those out), and climate change is no exception.

This is why Kathleen Dean Moore's recommendation for climate ethics is to start with a discussion between people to identify our honestly held and deepest beliefs, which people tend to honor in each other, in part because such beliefs often reflect more common ground than typically seen with politically polarized issues. Our more deeply held values tend to be similar in contrast to our political agendas. In the U.S., for example, it seems correct to infer that Republicans and Democrats alike care for their children, and prefer good health, and desire security. In fact, one of the quintessential differences between conservatives and liberals may be based on a deeper shared value. Liberals tend to prefer *policies* to create positive social change, and conservatives tend to emphasize *individual integrity* as a

---

[34] Wood (2002). For further discussion, see a related chapter in: Hitzhusen (2019b).

[35] Dr. Richard Baer made these points and developed this set of epistemological frames during his career teaching environmental ethics at Cornell University. The "Religion, Ethics, and the Environment" course he began teaching in 1974 (derived from the environmental ethics course he taught at Earlham College starting in 1966) was the source of these ideas about epistemology.

basis for social well-being—but it seems that both want socially good outcomes that provide freedom, health, fairness, and justice for our communities. In an important study, Graham et al. (2009) examined liberal and conservative moral preferences and discovered that liberals tend to emphasize limiting harm and maximizing justice as the most important moral goals, while conservatives tend to focus on loyalty, respect, and purity/sanctity.[36] Subsequent research on environmental concerns showed that liberals respond to harm/care messages while conservatives resonate with purity/sanctity messages.[37] The tragedy is that these honorable goals get lost during climate change debates, where liberals claim that conservatives don't care about the environment or justice, and conservatives claim that liberals don't care about loyalty, respect, etc. Thus by failing to see and respect the deeper values we each hold, we create a clash of values and lose the common basis for moving forward that Moore promotes. This is a classic illustration, typical of our times, of how an inability to deal well with ethics derails an otherwise promising path forward and creates political and social division when there could be common ground.

Another classic example of conflict where there could be complementarity, especially in the United States, is the history of tension between religion and science. Typically, scientists may worry that some religious people resist science because of a preference for faith over science. But in these arguments about science, faith, and truth, it has been much less common to examine the view of *scientism*, which is *the belief that science is sufficient to answer all meaningful questions about reality*.[38] Modern science, which seems to be the most powerful and effective process humans have ever devised for gaining knowledge about the physical universe, is not the same thing as *scientism*, which is a faith-based view held by some scientists, though not held by others. Scientific methods do not have the capacity to discern what is true or false morally (Is racism wrong? Is equality a worthy principle? Science cannot measure or test these moral ideas). Ironically, those who claim that science can answer every meaningful question apparently can't scientifically answer the question of "whether science can answer every meaningful question," and thus can't verify their own most basic truth. Here as before, the point is not to pass judgment per se on those who believe in scientism (to each their own, eh?), but in cases where a scientist opposes a climate skeptic because they don't "believe in science,"[39] or criticizes an opponent for holding some religious view that is unscientific, those seem to be claims that issue from scientism and not science, and it should be no surprise that they give rise to conflict and resistance. If some scientists criticize people who don't share their belief in scientism, that's no different than a religious fundamentalist criticizing

---

[36] Graham et al. (2009). More recent research has linked this study with additional dimensions of morality and politics in the United States: Strupp-Levitsky et al. (2020). This study showed liberals responding to pleas for compassion and justice, and conservatives to values of loyalty, authority, purity, and patriotism: Wolsko et al. (2016).

[37] Feinberg and Willer (2013).

[38] Scientism is the belief that all valid knowledge is science, and thus anything real or worthwhile to know is known best by science; See Hutchinson (2011).

[39] The underlying premise is that saying "belief" implies science is more like faith…thus scientism.

someone for not believing the same thing they do; this clearly confuses the boundaries of science, and actually generates a reason for some religious people to oppose "science". What comes off as resistance to science is more likely just resistance to "scientism," and so long as some scientists insist on equating science with the beliefs of scientism that confusion will remain. It is also likely that some scientists could simply be more careful with their words. A frustrated reply about someone "not believing in science" might simply reflect exasperation about the low level of trust some people have for science; such a reply from a scientist might simply mean they think science deserves more credit given the robust methods that are designed to provide reliability in scientific results. So reducing the proselytizing of scientism and being more careful with our words can help eliminate some needless forms of opposition to science.

Another thing that causes division where there could be collaboration on addressing challenging issues, including climate change, is an aspect of various absolutisms. Modern science, in contrast to superstition and guesswork, is underlain by theoretical frameworks and robust methodologies that help us uncover how the world works, how atoms and matter are structured, how CO_2 traps heat in the atmosphere, how diseases are caused. This can lead to assumptions among the public-at-large that knowledge, to be reliable, true, or real, must be "certain", as in "scientifically proven" or "absolutely true." But most scientists, epistemologically, don't practice absolutism, but rather they recognize the uncertainty in science and aim to be transparent about what those uncertainties are, and their level of confidence in their findings (e.g., as is done in IPCC reports, including in their summaries for policy makers). This perspective is therefore something more like **fallibilism**, which is the notion that regardless of what we believe about whether things can be absolutely true or not, *we can be mistaken about what we think we know is true.* Indeed, most science proceeds with this assumption, since new evidence can always come along to overturn what we previously thought was true (Does the Sun revolve around the Earth, or the other way around? Does burning come from phlogiston?). And in fact, most of us likely make many daily decisions upon which our life depends with a fallibilist understanding of truth and certainty. We can never be 100% certain that our car, or an elevator ride, won't crash us to our death, but we trust our lives to something less than 100% certainty and proceed.[40]

If we think about it, religion and faith also rely on fallibilist notions. What does it mean to have "faith" in something if you know it with 100% certainty? That's not really faith. God is not provable by science, but that doesn't stop a majority of the world's people from identifying with a faith tradition, and many people from living their lives on the basis of their faith. Similarly, we can't technically "prove" that our

---

[40] The abrupt change in the Centers for Disease Control (CDC) guidance in May of 2021 about wearing masks to protect against COVID-19 spread is a good example of fallible knowledge. It also highlights that once the threat of COVID-19 was shown to be less dangerous than driving a car, society moved to recognize non-mask-wearing as an acceptable danger for those vaccinated and not at high risk. COVID-19 restrictions that would be 100% certain not to spread COVID would likely leave everyone paralyzed (unable to do anything for fear of possible infection) for the rest of our lives.

parents love us (they could just be lying when they say they do), but many people know well that they are loved despite lacking absolute certainty. Gravity sure seems real, even if it is an invisible force that is not well understood, and we cannot know for *certain* that it won't fail tomorrow[41]; nonetheless, engineering, aviation, and pretty much all physical reality proceed with reasonable faith (or expectation) that gravity will remain, regardless of what we think about it. Why all this practical, scientific, and religious fallibilism seems to get forgotten when we argue about politics is perhaps a mystery (absolutism seems to reign in political rhetoric more than in other realms!), but the point is this: when we demand absolute certainty (say, in climate science before climate policy is made) before we can proceed, we are pretending to be devoted to "certainty" more than any of us really are, and that also tends to obstruct taking action. We usually don't act only when we have certainty, so why stop ourselves from implementing policy and taking action because of the rhetoric of uncertainty that's generated around climate change? Overcoming this obstacle is sometimes referred to as taking a **precautionary approach**, and seeing these often hidden barriers (like the appeal of "certainty") can sometimes help remove them. Some of these considerations help explain why the IPCC creates a *range* of future scenarios (as discussed in Chap. 10)—scenarios like these are prudent because of this type of uncertainty.

A great example that combines many of the above insights is the climate communication work of Katharine Hayhoe, a climate scientist who has become known as a great communicator of the moral values involved in climate change (Hayhoe has been named one of TIME's 100 Most Influential People and FORTUNE's 50 world's greatest leaders). Hayhoe has particularly keen insight into some of the complex ethical questions related to climate change: her husband is an evangelical pastor, and they have co-published a book[42] about the challenges of understanding and responding to climate change in their congregation and community. She also has been involved in cutting-edge social science and environmental psychology research to understand the barriers to climate communication and action, and why climate science communication has sometimes increased, not decreased, climate skepticism. In a 2020 presentation[43] about communicating climate science, Hayhoe notes that for most people, it's not the head but the heart that determines how they will act in response to issues like climate change. This means that values and beliefs will play a key role, so when climate science is assumed to be the only message necessary, climate change action has stalled. She counsels focusing on climate impacts

---

[41] Perhaps the laws of physics will wobble this afternoon, perhaps a new big bang type event emerging next door will suddenly insert highly unusual phenomena into our physical environment tomorrow…

[42] Hayhoe and Farley (2011).

[43] Conversations on the Politics and Science of Climate Change in the Buckeye State 10.13.20 EPN_part1, https://www.youtube.com/watch?v=PotWFl9hYYA&t=3877s.

here and now and on *why* we care, rather than highlighting more and more global climate statistics. Hayhoe notes that for most Americans and Canadians and indeed for most people in the world—about 80% of the world's people self-identify with a religion—faith-based values and beliefs are a primary lens informing their reasons for the heart[44] (their why). For herself as a Christian, she has talked about *why* we should care about climate change in many forums[45]—including that climate change is creating many negative impacts, and that we are called to love others and be good caretakers of the planet, and to serve others with compassion. She notes that climate change is exacerbating and multiplying risks, which are disproportionately affecting the most poor and vulnerable of the Earth. But she has also recently highlighted that a key missing piece in climate action is that people don't *talk* about climate change very much.[46] And if we don't talk about it, we don't care about it, and if we don't care about it, we won't likely take action. And too often, climate discussions are more like debates or diatribes. Much like Kathy Dean Moore's advice about productive discussions, Hayhoe recommends not starting with differences (bold climate science facts often are divisive and contentious between political ideologies), but rather finding common ground, bonding with conversation partners, and then exploring common actions you can take.

4.2 What Are Some Fruitful Lines of Dialogue About the Key Points in Climate Ethics?

With the above background on thoughtful ethical reasoning and discussion, let's take a look at some of the diverse ethical perspectives related to climate change that are being expressed around the world. These perspectives can be the foundation for fruitful lines of dialogue about climate ethics and ultimately climate action. The perspectives include consensus statements on climate change from world leaders, businesses, and religions that complement consensus statements by scientists (e.g., IPCC Reports; US National Climate Assessments), and arguments that climate justice and concern for future generations are ethical imperatives.

---

[44] See Hayhoe's comments here: Hayhoe, K. (2015). Climate Change: Faith and Science - Katharine Hayhoe. YouTube, https://www.youtube.com/watch?v=td_OB9BKgIs.

[45] Including this sermon: Hayhoe, K. (2019). Christians and Climate Change with Dr. Katharine Hayhoe. The Meeting House. YouTube. https://www.youtube.com/watch?v=UOJuHpeWoPE&t=1034s.

[46] Hayhoe (2021).

4.2.1 Climate Change Statements Across Cultures

So much attention has been given to climate change that there are official statements published by most nations, industries, religions, organizations, and local communities. One global statement of interest is the United Nations Educational, Scientific and Cultural Organization's (UNESCO) *Declaration of Ethical Principles in relation to Global Climate Change*.[47] Drafted in 2017, this statement brings together principles from decades-worth of previous statements and negotiations and highlights the following ethical principles: prevention of harm, a precautionary approach, equity and justice, sustainable development, solidarity, and scientific knowledge and integrity in decision-making. An interesting example of how businesses are championing professional standards that are responsive to climate change is the Science-Based Targets Initiative (SBTi), which involves large companies committing to industrial-scale emissions reductions in response to the Paris Climate Agreement.[48] Another example of a globally impactful statement comes from the religious sector: Pope Francis's environmental encyclical letter, *Laudato Si'* (2015), addressed all people on Earth and implored everyone to work together in defense of our common home.[49] Francis noted climate change as one of the most critical environmental issues of our time and stated that it will require radical change (an "ecological conversion") and a dialogue of everyone to adequately respond. Francis noted that in order to address climate change, we will also need to move towards phasing out fossil fuels, and he highlighted issues of justice and solidarity.

Laudato Si' highlights a larger trend: there are many climate change statements from religious groups around the world. These faith-based statements (Table 3) typically refer to the scientific consensus on climate change, and then focus on various moral dimensions of the climate crisis, be they harm, justice, future generations, power and greed, solidarity, prudence, stewardship, urgency, or integrity.[50] Virtually all of them advocate for climate action in accordance with the scientific consensus

---

[47] UNESCO's 2017 Declaration of Ethical Principles in relation to Climate Change, https://en.unesco.org/themes/ethics-science-and-technology/ethical-principles.

[48] Details of SBTi can be found here: SBTi (2021, January). From ambition to impact: Science based targets initiative annual progress report. SBTi. https://sciencebasedtargets.org/resources/files/SBTiProgressReport2020.pdf; Another interesting case of enacting new business and trade standards based on moral environmental concerns was the effort to apply lessons learned from CITIES, the Convention on International Trade in Endangered Species of Wild Fauna and Flora, to the challenges of climate change: CITES (2022).

[49] Pope Francis, (2015). Laudato Si': On Care for Our Common Home [Encyclical]. http://www.vatican.va/content/francesco/en/encyclicals/documents/papa-francesco_20150524_enciclica-laudato-si.html.

[50] The tenets of some of these statements are highlighted in Hitzhusen's discussion of religion and climate change (Hitzhusen, 2019a, Sects. 9.1 and 9.2. https://ohiostate.pressbooks.pub/enr3470/), and many others are highlighted by the Forum on Religion and Ecology at Yale University, including a series of climate statements made by indigenous communities.

about climate change.[51] The sheer scope of religious statements is overwhelming, demonstrating that a wide range of moral values expressed across geographically diverse regions can serve as bases for climate action. Even with the great diversity in religious climate statements, one central theme of religious and non-religious statements alike is the issue of justice.

4.2.2 Climate Justice

Justice has long been a theme of moral concern, and particularly as concerns about racial equity have received greater attention in a COVID and post-COVID world, these remain as critical as ever. The disproportionate impact of climate change on vulnerable communities[52] is a quintessential example of environmental or climate injustice. Whether we think about which communities suffered the most disruption from Hurricane Katrina in New Orleans, or drought-stricken African farmers, or displaced Pacific Islanders or other climate refugees, or the disproportionate impact of climate change on mothers among black women,[53] these issues demand attention. One of the challenges in addressing these issues is revealed in the simple facts about who is causing climate change—the most powerful nations—and who is suffering most from climate change—the least powerful nations, as reflected in Figs. 5 and 6.

Note that China, which is the world's largest carbon emitter, is also the world's largest in population (China at 1.44 billion and India at 1.39 billion had the world's most people in 2021[54]). The U.S., large in carbon emissions, is quite small (around 4.2%) in world population, thus the per capita carbon footprint of Americans (14.86 tonnes CO_2 per capita) is nearly twice that of the Chinese (8.05 tonnes CO_2 per capita) and over 7 times more than people in India (1.93 tonnes CO_2 per capita).[55] So Americans are producing more than their fair share of carbon, while many wealthy nations whose standard of living exceeds that in the U.S. (like Germany and Japan)

---

[51] Official climate change policy statements from religious denominations overwhelmingly advocate for climate action. Only one denomination, within Christianity, the Southern Baptist Convention, supports a statement that expresses doubt about climate science, making it the exception that proves the rule that faith communities world wide support action on climate change (note also that a Southern Baptist *Environment and Climate Initiative* emerged to counter the denomination's skeptical climate statement). A few other "climate skeptic" faith-based climate statements can be found, but they do not represent any denominations, but rather are organized by free-market economics think tanks funded by fossil fuel money. Further discussion of these religious 'skeptic' patterns is found in: Hitzhusen (2019c).

[52] See, for example: Timperley (2021).

[53] See, for example: Bekkar et al. (2020).

[54] Population data are from in *Our World In Data*: https://ourworldindata.org/world-population-growth.

[55] 2021 Carbon dioxide emission data are from *Our World in Data*: https://ourworldindata.org/grapher/co-emissions-per-capita?tab=table&country=.

Table 3 Sources of climate change statements from religious communities

Dates	Source	Description
Multiple decades	Yale Forum on Religion and Ecology[56]	Includes links to climate statements of many world religions (e.g., Christianity, Hinduism, Islam, Jainism, Judaism) and indigenous communities
1971 to present	RESTORExchange Religion-Sustainability Database[57]	Database "climate change" search yields various religious statements and resources related to climate change
1993 to present	Interfaith Power and Light[58]	Website includes links to many faith-based climate statements, a handout with statement excerpts, and climate prayers and sermons
2011 to present	Christ the King Lutheran Church, Houston, TX[59]	Some local congregations have created their own creation care statements, such as this one, together with lists of creation and climate actions
2019, 2020	Iceland Conference on Faith for Earth[60]	A 2019 conference of Icelandic and other Nordic, US, and Canadian leaders generated a climate statement, and then helped spark a global conference of faith communities on all continents in 2020, with speeches from global faith leaders
2010	Lausanne Movement[61]	Similar to the Lausanne Covenant (one of the most influential evangelical Christian statements in the world), the Capetown Commitment led to the Lausanne/World Evangelical Alliance Creation Care Network (LWCCN)
1987, 1992	Au Sable Institute[62]	The AuSable Forum generated a 1987 statement of a Christian Land Ethic that led to a 1992 "Evangelical Christianity and the Environment" statement with the World Evangelical Fellowship Theological Commission, noting global warming
2008	Coalition on the Environment and Jewish Life[63]	This Jewish climate and energy statement was endorsed by a wide range of Jewish organizations

(continued)

---

[56] Yale Forum on Religion and Ecology, https://fore.yale.edu/.

[57] RESTORExchange Religion-Sustainability Database, https://restorexchange.org/s/ohio-state-senr/page/about.

[58] Interfaith Power and Light, https://www.interfaithpowerandlight.org/.

[59] Christ the King Lutheran Church, Houston, TX, https://ctkelc.org/caring-for-creation/.

[60] Iceland Conference on Faith for Earth, https://faithfornature.org/.

[61] Lausanne Movement, https://lausanne.org/networks/issues/creation-care.

[62] Au Sable Institute, https://www.webofcreation.org/DenominationalStatements/ausable.htm.

[63] Coalition on the Environment and Jewish Life, http://www.coejl.org/.

Table 3 (continued)

Dates	Source	Description
2015, 2001	Pope Francis, The Vatican[64]; US Conference of Catholic Bishops	*Laudato Si'*, Pope Francis' environmental encyclical letter, called for climate attention and action, echoing 2001 Bishops statement
2009, 2015	Global Buddhist Climate Change Collective, One Earth Sangha[65]	*The Time to Act is Now: A Buddhist Declaration on Climate Change* was adapted in 2015 on the eve of Paris climate talks from a 2009 statement, and included leading Buddhist signatories from around the world and as many as a million signatures of supporters

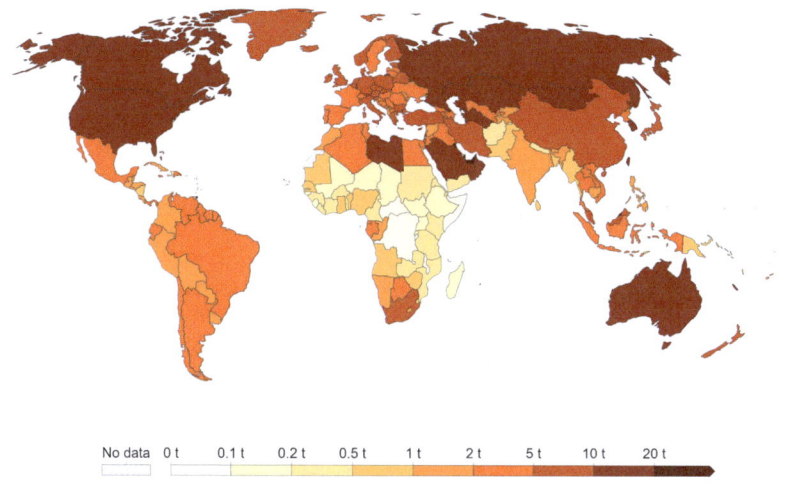

Fig. 5 Per capita CO_2 emissions from fossil fuels and industry in 2021 from *Our World in Data*[66]

---

[64] Pope Francis, (2015). Laudato Si': On Care for Our Common Home [Encyclical]. https://www.vatican.va/content/francesco/en/encyclicals/documents/papa-francesco_20150524_enciclica-laudato-si.html.

[65] Global Buddhist Climate Change Collective, One Earth Sangha, https://oneearthsangha.org/articles/buddhist-declaration-on-climate-change/.

[66] The World in Data site includes several interactive features, including historical data trends by country: https://ourworldindata.org/grapher/co-emissions-per-capita?time=latest&country=.

Share of population living in extreme poverty, 2021

Extreme poverty is defined as living below the International Poverty Line of $2.15 per day. This data is adjusted for inflation and for differences in the cost of living between countries.

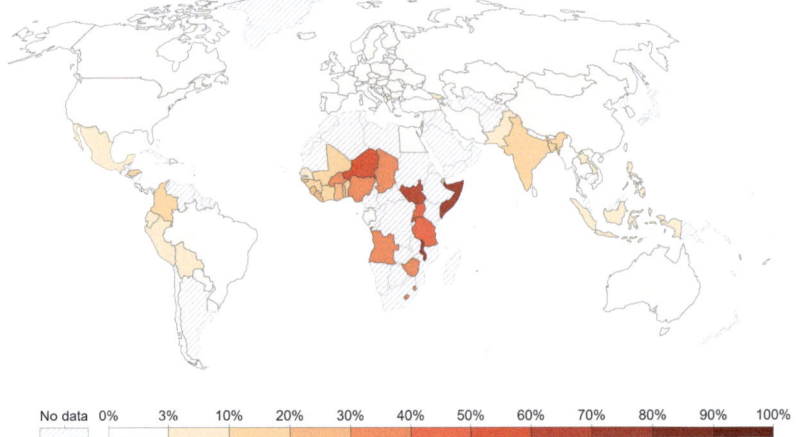

Source: World Bank Poverty and Inequality Platform
Note: This data is measured in international-$ at 2017 prices. Depending on the country and year, it relates to disposable income or consumption per capita.

CC BY

Fig. 6 Percent of 2021 population living in households with an income or consumption per person below $2.15 a day (a measure of extreme poverty) from *Our World in Data*[67]

have per capita CO_2 emissions rates just over half as much as the U.S.[68] In other words, Americans create much more carbon pollution than others without a corresponding increase in well-being. Meanwhile, African countries and India bear the largest global burden of poverty but have contributed the least (in total and per capita in Africa, per capita in India, and in total, historically) to global CO_2 emissions. South America, Central America, and South Pacific countries (e.g., Indonesia) are additional examples illustrating (see Figs. 5 and 6) this same point—there is a clear disparity between the 'Global South' and the 'Global North' in terms of CO_2 emissions and poverty.

Another important way to approach the moral implications of climate impacts is to simply look directly at the impacts climate change is having on people across many walks of life. Facts and figures sometimes obscure the more powerful stories that demonstrate the need for climate action. For instance, the stories of the plight of African farmers provide stark examples of injustice, as many poor, subsistence farmers lack any insurance to protect them from climate-induced drought and stand

---

[67] From *Our World in Data*, https://ourworldindata.org/poverty.

[68] Germany's per capita CO_2 emissions in 2021 were 8.09 tonnes. Japan's per capita CO_2 emissions in 2021 were 8.57 tonnes. Data from the same source as above, *Our World in Data*: https://ourworldindata.org/grapher/co-emissions-per-capita?tab=table&country=.

to lose their livelihoods.[69] The situation of Pacific Islanders has also been well documented, as rising sea levels cause the inundation of freshwater stores, and in some cases threaten the foundations of homes.[70] Extreme rainfall in an intensified 2022 summer monsoon season caused catastrophic flooding in Pakistan, covering nearly one-third of the country. As climate change makes such heavy rains more likely, calls for climate justice join the calls for disaster assistance.[71] And even in the American heartland, climate change impacts have been well documented, not as future projections, but as current impacts that affect people across a wide range of livelihoods, such as those seen in a series of climate change video documentaries from Wisconsin highlighting how farmers, beekeepers, foresters and others are being impacted.[72] In most of these cases, justice is one of the primary ethical concerns.

4.2.3 Harm and Risk Concerns for Future Generations

There are additional, farther-reaching ethical points to consider as well. For instance, as noted earlier in this chapter, climate ethicist Donald Brown has published a number of commentaries about climate denial, the phenomenon of industry and political money that has been directed at promoting disinformation campaigns about climate science.[73] Brown does not mince words in wondering, given what we know about the harms and risks caused by climate change, whether those industries and politicians supporting climate denial should be seen as committing a new kind of assault on humanity.[74] What Brown calls a moral travesty resonates with some of the stirring speeches made by Swedish youth activist Greta Thunberg, who told world leaders at a United Nations Climate Action Summit in New York in 2019 that she could not believe that current world leaders still are failing to take bolder action, "because if you really understood the situation and still kept on failing to act, then you would be evil, and that I refuse to believe."[75] Thunberg's voice has provided what might be

---

[69] Related stories about impact on African farmers have been highlighted here: One Acre Fund. (accessed 2023, May 24) *What Climate Change Means for Agriculture in Africa*. One Acre Fund. https://oneacrefund.org/blog/what-climate-change-means-agriculture-africa/; and here: Bounoua (2015); see also Pereira (2017).

[70] Some examples can be found in a video: Climate Reality. (2017, December 6) 24 Hours of Reality 2017: Weight of the World. Climate Reality, YouTube, https://www.youtube.com/watch?v=Swv1lOA5zDc&t=57s); and highlighted here: The Climate Reality Project (2019, March 14). *Trouble in paradise: how does climate change affect pacific island nations?* The Climate Reality Project. https://www.climaterealityproject.org/blog/trouble-paradise-how-does-climate-change-affect-pacific-island-nations.

[71] See: Morning Edition. (2022, September 13). Advocates call on U.S. to help flooded Pakistan in the name of climate justice. *NPR*. https://www.npr.org/2022/09/13/1122621405/advocates-call-on-u-s-to-help-flooded-pakistan-in-the-name-of-climate-justice.

[72] See: PBs-Wisconsin (2019).

[73] See: Brulle (2018); see also this New York Times article on the money behind climate denial: Root et al. (2019).

[74] See: Brown (2011).

[75] See: The Guardian (2019).

one of the most compelling moral statements among all the pronouncements about climate change: as a young person whose future is at stake, she has challenged and shamed current leaders for putting her future at risk, declaring repeatedly: "how dare you!" Indeed, this may be one of the most basic points, morally speaking, that current leaders should consider.

4.3 Moving Forward with Hope and Action

As this chapter has discussed, taking action on climate change always involves values and ethical motivations. Despite the obstructive efforts of those who oppose action on climate change, the determination of individuals and many groups to overcome this resistance through positive change is reason for hope. As Jane Goodall emphasized in her recent book, now is not a time for grief, but a time to get involved; action is needed at all levels, large and small, and taking action also fosters hope.[76]

What might *you* do to take action in response to climate change? Your agency to make changes is important at many levels: as an individual, as a community member, and as a citizen or leader. Perhaps you want to work on reducing your carbon footprint[77] by reducing your energy use and changing your eating habits,[78] or maybe you will do climate research, or you might use your power as a consumer to influence business and industry, or simply begin by starting conversations with your peers, your family, and your community to identify common-ground values around climate change issues that can motivate action. Recall that the primary question in ethics is "how should we live," but knowledge about how best to live is of little use if we don't do anything about it—if we don't move to action and actually live that way. *Phronesis* is the Greek term for practical wisdom or knowledge about *how* to live ethically, and perhaps that is one of our greatest needs now: more phronesis. So how do you think you should live given what we know about climate change, and what you believe about the values and moral and ethical issues related to climate change? What actions are you ready to take? After reflecting on your values and ethical concerns related to climate change, such as harm, justice, and concern for future generations, engage in respectful dialogue with others on these topics. One key

---

[76] Goodall and collaborators provide an empowering commentary on action in her book: Goodall et al. (2021); National Public Radio interviewed her recently about the book: Short Wave. (2021, December 3). Jane Goodall Says There's Hope For Our Planet. Act Now, Despair Later! *Short Wave, NPR.* https://one.npr.org/?sharedMediaId=1061020936:1061092609.

[77] There are several online carbon footprint calculators including: The International Student Carbon Footprint Challenge (for U.S. and international locations, based on a collaboration among universities in Sweden and the U.S.): https://depts.washington.edu/i2sea/iscfc/calculate.php; Carbon Footprint (a UK site but good for all locations): http://www.carbonfootprint.com/calculator.aspx; 'Cool-Climate' Network (A University of California-developed calculator): https://coolclimate.berkeley.edu/calculator; and the U.S. EPA-developed calculator: https://www3.epa.gov/carbon-footprint-calculator/. In addition see: van der Linden (2022).

[78] See: Vaidyanathan (2021); and see: Grunwald (2022).

way to better understand the moral/ethical dimensions of climate change is to discuss them with others, and considering at least three different ethical perspectives helps provide better understanding of any moral issue. Be willing to have unconformable, yet constructive, conversations with your peers, your family, and others in your larger social circles (e.g., your faith group, your neighbors). Some starting points for these conversations are in Box 3. Seek common ground—identify points of consensus about climate ethics. Use your gained knowledge from this book, and consider the relevant values premises to take action—to help others understand climate change, and to make informed decisions towards mitigation and adaptation.

Box 3 Exercises for further discussion

Exercise 1: Values premises and the Six Americas

Refer to the sections of this chapter that describe the Six Americas and K.D. Moore's insight that scientific premises need values premises to lead to action. Draft an imagined set of values premises and scientific premises that are likely operating for each of the Six America's categories, and consider adding any other specific perspectives (like a climate lobbyist; your university president; someone who has publicly stated their views on climate change, etc.) to the list. Once you have described the values premises for each group, make your own claim about which views *you* agree with. If you disagree with one or more perspectives in the list, be sure to note the *reasons* you disagree with those views. Which of your reasons are moral reasons (e.g., you think the view or those who hold it are dishonest, selfish, greedy)? Which of your reasons depends on how you understand what it means to *know* something? Some further questions to consider: What do you find to be most challenging about this exercise? Do you think your views on climate change are closer to absolutist, relativist, fallibilist, or ethical skepticist (or some other) views? Did your list include any values that are the same across all 6 categories? If not, do you think there **are** any values about climate change that can be universally accepted?

Exercise 2: Reflecting on ethical reasoning and climate change: best practices for potentially uncomfortable conversations

2A. Simple class discussion: Reflect on the various moral values stated in Table 2, and then write at least a paragraph describing which moral reasons are most compelling or important to you, and why. Once everyone has written down their thoughts, open up for discussion (or break up into small groups to share with a few others first before returning to your larger class group) to hear what others have said. Some further questions to consider: What additional questions does this discussion exercise raise? How did this discussion influence your thoughts on various moral reasons to care about climate change?

2B. Expanded discussion (in class, in the community, within other social groups): Consider the advice of K.D. Moore and Katharine Hayhoe in their encouragement of having discussions about climate change. Moore (see footnote 29) suggested bringing people together and asking whether we have a moral obligation to leave a world as rich in possibilities as our own; discuss this question in pairs or small groups first, and then with the whole group. Don't rank ideas, but just share what we each deeply believe. This provides a basis to move forward with respect for what one another deeply believes in (e.g., caring for our children/family, our health, something bigger than ourselves, spiritual beliefs about reality, etc.). People may disagree about these deeply held beliefs, but we can honor them in each other because they are honestly held beliefs. From here, discuss how to move forward in action to honor our obligations to the future and these deeply held beliefs. Hayhoe similarly suggests getting together to talk with others, and not starting with differences (esp. noting how culturally contentious climate science facts have been), but rather seeking to find common ground. Talk about what you value, and connect with your discussion partners over considering what we each value. Then discuss common actions you can take that are consistent with your values.

Exercise 3: Applying the eight key questions of ethical reasoning to climate change issues

3A. Responding to moral issues related to climate change: The following prompt can be used or revised to facilitate an in-class or online discussion that starts by letting students consider ideas individually, and then in small breakout groups before sharing one point from their group with the whole class. This exercise presumes that students have read one or more pre-assigned articles and responded in writing in a homework assignment in which they connect one or more of the JMU Eight Key Questions

on Ethical Reasoning (see Box 2) to what they read.[79]
In your breakout/small group:

- Discuss what you read and what you wrote about for your homework response. If you read different articles, compare and contrast what you learned from them about the ethical issues associated with climate change.
- With that as background, as well as your own perspectives on climate change, address the following question: What should we do about climate change and why?
    - For "we" consider this at a *personal* level (e.g., you and your family), at a *local* level (e.g., at your university, or your home town), and at a *national* level (e.g., your country).
    - When considering "do", think "taking specific actions." Include at least three specific actions total (one for each level).
    - Support your reasons why with ethical and other rationales.
    - Be prepared to share your ideas with the class.

3B. Climate Justice Scenarios: The previous exercise can be fit for any number of specific climate scenarios. Similar to the example in Box 2 about the Eight Key Questions of Ethical Reasoning, aim to apply the eight key questions (or at least 3 of them) to some particular dimension or issue (or local/regional situation) related to climate change. Examples include the issue of what to do about addressing salt water inundation in Pacific Island communities, which poses a threat to those communities being able to continue living on their home island. Consider the eight key questions in evaluating the desirability of different responses like evacuation, expensive sea wall infrastructure, and water purification systems, or other options. Alternatively, discuss the options for dealing with the problems faced by drought-stricken farm communities in Africa, where livelihoods are being lost as poor farmers who live on the margins are unable to make a living off the land. Or discuss what should be done about coastal communities in Alaska where sea level rise and the melting of permafrost is resulting in houses washing into the sea and threatening buildings with coastal erosion. Find articles to supply details about your scenario of choice.

---

[79] Various articles could serve to support this exercise. Some possible examples include:

- Paying for Climate Damage Is Not Charity by Ani Dasgupta, The New York Times, November 11, 2022.
- At Its Core, the Climate Crisis Is a Crisis of Politics and Justice by Jamie Males, *PLOS Blogs, Latitude*, October 13, 2021.
- The Disaster We Must Think About Every Day by Tressie McMillan Cottom, The New York Times, September 17, 2021.

References

Anderegg, W. R., Prall, J. W., Harold, J., & Schneider, S. H. (2010). Expert credibility in climate change. *Proceedings of the National Academy of Sciences, 107*(27), 12107–12109. https://doi.org/10.1073/pnas.1003187107

Autin, W. J., & Holbrook, J. M. (2012). Is the Anthropocene an issue of stratigraphy or pop culture? *GSA Today, 22*(7), 60–61. https://doi.org/10.1130/G153GW.1

Baer, R. A., Jr. (1976). Our need to control: Implications for environmental education. *The American Biology Teacher, 490*, 473–476.

Bekkar, B., Pacheco, S., Basu, R., & DeNicola, N. (2020). Association of air pollution and heat exposure with preterm birth, low birth weight, and stillbirth in the US—A systematic review. *JAMA Network Open, 3*(6), e208243. https://doi.org/10.1001/jamanetworkopen.2020.8243

Benestad, R. E., Nuccitell, D., Lewandowsky, S., Hayhoe, K., Hygen, H. O., van Dorland, R., & Cook, J. (2015). Learning from mistakes in climate research. *Theoretical and Applied Climatology, 126*, 699–703. https://doi.org/10.1007/s00704-015-1597-5

Bonneuil, C., & Fressoz, J. (2015). *The shock of the Anthropocene.* Verso, 306 p. ISBN-10: 1784785032.

Boran, I., & Katz, C. (2017). Climate change justice. *Routledge encyclopedia of philosophy.* https://doi.org/10.4324/0123456789-S113-1. https://www.rep.routledge.com/articles/thematic/climate-change-justice/v-1

Bounoua, L. (2015, May 12). Climate change is hitting African farmers the hardest of all. *The Conversation.* https://theconversation.com/climate-change-is-hitting-african-farmers-the-hardest-of-all-40845

Brown, D. (2010a, October 24). A new kind of crime against humanity?: The fossil fuel industry's disinformation campaign on climate change. *Ethics and Climate.* https://ethicsandclimate.org/2010/10/24/a_new_kind_of_vicious_crime_against_humanity_the_fossil_fuel_industrys_disinformation_campaign_on_cl/

- Call for Emergency Action to Limit Global Temperature Increases, Restore Biodiversity, and Protect Health by Lukoye Atwoli, Abdullah Baqui, Thomas Benfield, and others, The New England Journal of Medicine, 385:1134–1137, September 17, 2021.
- Climate Disaster Is the New Normal. Can We Save Ourselves? Spencer Bokat-Lindell, The New York Times, September 7, 2021.
- The Intergenerational Ethics of Climate Change by Steve Cohen, Columbia Climate School State of the Planet, Columbia Climate School, November 11, 2019.
- The Ethics of Climate Change: A Primer by Christine Emba, *The Washington Post*, January 5, 2016.
- Inaction on Global Warming Is as Reckless as Drunken Driving by Daniel Farber, *The Washington Post*, January 5, 2016.
- Pope Francis was Right on Climate Change by John Nagle, *The Washington Post*, January 5, 2016.
- America Has Been Duped on Climate Change by Robert Brulle, *The Washington Post*, January 5, 2016.
- If We're Going to Fix Climate Change, We'll Have to Get Creative by Thomas Kostigen, January 7, 2016.
- Why Climate Change Is an Ethical Problem by Stephen Gardiner, *The Washington Post*, January 9, 2016.
- When it Comes to Climate Change, Payback Isn't Enough by Eric Posner, *The Washington Post*, January 8, 2016.

Brown, D. (2011, December 2). An ethical analysis of the climate change disinformation campaign: Is this a new kind of assault on humanity? *Ethics and Climate*. https://ethicsandclimate.org/2011/12/02/an_ethical_analysis_of_the_climate_change_disinformation_campaign_is_this_a_new_kind_of_assault_on_h/

Brown, D. (2010b, November 1). Is climate science disinformation a crime against humanity? *The Guardian*. https://www.theguardian.com/environment/cif-green/2010/nov/01/climate-science-disinformation-crime

Brulle, R. J. (2013). Institutionalizing delay: Foundation funding and the creation of U.S. climate change counter-movement organizations. *Climate Change, 122*, 681–694. https://doi.org/10.1007/s10584-013-1018-7

Brulle, R. J. (2018). The climate lobby: A sectoral analysis of lobbying spending on climate change in the USA, 2000 to 2016. *Climatic Change, 149*, 289–303. https://doi.org/10.1007/s10584-018-2241-z

Caney, S. (2020). Climate Justice. In E. N. Zalta (Ed.), *The Stanford encyclopedia of philosophy*. https://plato.stanford.edu/archives/sum2020/entries/justice-climate/

CITES. (2022, October 28). CITES shares experiences with Climate Change Convention. CITES. https://cites.org/eng/news/sundry/2013/20131113_climate-change.php

Cook, J., Oreskes, N., Doran, P. T., Anderegg, W. R. L., Verheggen, B., Maibach, E. W., Carlton, J. S., Lewandowsky, S., Skuce, A. G., Green, S. A., Nuccitelli, D., Jacobs, P., Richardson, M., Winkler, B., Painting, R., & Rice, K. (2016). Consensus on consensus: A synthesis of consensus estimates on human-caused global warming. *Environmental Research Letters, 11*(4), 048002. https://doi.org/10.1088/1748-9326/11/4/048002

Crist, E. (2016). On the poverty of our nomenclature. In J. Moore (Ed.), *Anthropocene or Capitalocene?* (pp. 14–33). PM Press. ISBN: 9781629631486.

Decker, D. J., et al. (2019). An eye toward ethics: We lack tools to assess ethical issues in managing wildlife. *The Wildlife Professional*, 28–31.

Farrier, D., & Aeon. (2016, October 31). How the concept of deep time is changing. *The Atlantic*. https://www.theatlantic.com/science/archive/2016/10/aeon-deep-time/505922/

Feinberg, M., & Willer, R. (2013). The moral roots of environmental attitudes. *Psychological Science, 24*(1), 56–62. https://doi.org/10.1177/0956797612449177

Finney, S. C., & Edwards, L. E. (2016). "The Anthropocene" epoch: Scientific decision or political statement? *GSA Today, 26*(3–4), 4–10.

Fischer, D. (2013, December 23). "Dark money" funds climate change denial effort. *Scientific American*. https://www.scientificamerican.com/article/dark-money-funds-climate-change-denial-effort/

Goldberg, M. H., van der Linden, S., Maibach, E., & Leiserowitz, A. (2019). Discussing global warming leads to greater acceptance of climate science. *PNAS, 116*(30), 14804–14805. https://doi.org/10.1073/pnas.1906589116

Goodall, J., Abrams, D., & Hudson, G. (2021). *The book of hope: A survival guide for trying times* (272 p.). Celadon Books.

Graham, J., Haidt, J., & Nosek, B. A. (2009). Liberals and conservatives rely on different sets of moral foundations. *Journal of Personality and Social Psychology, 96*(5), 1029–1046.

Grunwald, M. (2022, December 15). No one wants to say 'put down that burger,' but we really should. *The New York Times*. https://www.nytimes.com/2022/12/15/opinion/food-diets-meat-biodiverstiy-cop15.html

Haraway, D. J. (2016). *Staying with the trouble: Making kin in the Chthulucene* (296 p.). Duke University Press. ISBN 10: 0822362244.

Hayhoe, K. (2018). The most important thing you can do to fight climate change: talk about it. *TED*. https://www.ted.com/talks/katharine_hayhoe_the_most_important_thing_you_can_do_to_fight_climate_change_talk_about_it?utm_campaign=tedspread&utm_medium=referral&utm_source=tedcomshare

Hayhoe, K. (2021). *Saving us—A climate scientist's case for hope and healing in a divided world* (307 p.). Simon and Schuster.

Hayhoe, K., & Farley, A. (2011). *A climate for change: Global warming facts for faith-based decisions* (224 p.). Faithworks.

Hitzhusen, G. E. (2019a). *Religion and environmental values in America.* OSU Pressbooks.

Hitzhusen, G. (2019b). Chapter 4: Epistemology and philosophy of science to help us think. *Religion and Environmental Values in America.* The Ohio State University Press. https://ohiostate.pressbooks.pub/enr3470/chapter/4-5-what-is-knowledge/

Hitzhusen, G. (2019c). Chapter 9.2: Faith and skepticism: A matter of morality. *Religion and environmental values in America.* The Ohio State University Press. https://ohiostate.pressbooks.pub/enr3470/chapter/9-2/

Hutchinson, I. (2011). *Monopolizing knowledge: A scientist refutes religion-denying, reason-destroying scientism* (p. 2). Fias Publishing.

Kahan, D., Peters, E., Wittlin, M., Slovic, P., Larrimore Ouellette, L., Braman, D., & Mandel, G. (2012). The polarizing impact of science literacy and numeracy on perceived climate change risks. *Nature Climate Change, 2*, 732–735. https://doi.org/10.1038/nclimate1547

Kipping, M., Kurosawa, T., & Westney, E. (2021). *The Oxford handbook of industry dynamics.* Oxford University Press. https://doi.org/10.1093/oxfordhb/9780190933463.013.26

Leiserowitz, A., Maibach, E., Rosenthal, S., Kotcher, J., Ballew, M., Goldberg, M., & Gustafson, A. (2019, January 22). *Climate change in the American mind: December 2018.* Yale Program on Climate Change Communication. https://climatecommunication.yale.edu/publications/climate-change-in-the-american-mind-december-2018/2/

Lynas, M., Houlton, B. Z., & Perry, S. (2021). Greater than 99% consensus on human caused climate change in the peer-reviewed scientific literature. *Environmental Research Letters, 16*(11), 114005. https://doi.org/10.1088/1748-9326/ac2966

Marlon, J., Neyens, L., Jefferson, M., Howe, P., Mildenberger, M., & Leiserowitz, A. (2022, February 23). *Yale climate opinion maps 2021.* Yale Program on Climate Change Communication. https://climatecommunication.yale.edu/visualizations-data/ycom-us/

Moore, J. (Ed.). (2016). *Anthropocene or Capitalocene?* (222 p.). PM Press. ISBN: 9781629631486.

Moore, K. D., & Nelson, M. (2010). *Moral ground: Ethical action for a planet in peril.* Trinity University Press. ISBN-10: 1595340858.

More, K. D. (2010) Climate change: A moral crisis [video]. Changing Climates @ Colorado State University. https://changingclimates.colostate.edu/movies/kathy_moore_796_kbits.mov

Oreskes, N., & Conway, E. M. (2011). *Merchants of doubt* (368 p.). Bloomsbury Publishing. ISBN-10:608193942.

PBs-Wisconsin. (2019). *Climate Wisconsin—Stories from a state of change.* PBS-Wisconsin. https://climatewisconsin.org/

Pereira, L. (2017). Climate change impacts on agriculture across Africa. *Environmental science Oxford research encyclopedia.* https://doi.org/10.1093/acrefore/9780199389414.013.292

Rainey, J. (2019, January 24). More Americans believe in global warming—But they won't pay much to fix it. *NBC News.* https://www.nbcnews.com/news/us-news/more-americans-believe-global-warming-they-won-t-pay-much-n962001

Root, T., Friedman, L., & Tabuchi, H. (2019, July 10). Following the money that undermines climate science. *The New York Times.* https://www.nytimes.com/2019/07/10/climate/nyt-climate-newsletter-cei.html.

Schendler, A. (2021, August 31). Worrying about your carbon footprint is exactly what big oil wants you to do. *The New York Times.* https://www.nytimes.com/2021/08/31/opinion/climate-change-carbon-neutral.html

Schneiderman, J. S. (2015). Naming the Anthropocene. *Philosophia. A Journal of Continental Feminism, 5*(2), 179–202.

Somerville, R. (2008, June 2). *The ethics of climate change.* Yale Environment 360. https://e360.yale.edu/features/the_ethics_of_climate_change

Stapczynski, S., Rathi, A., & Marawanyika, G. (2021, August 11). How to sell 'carbon neutral' fossil fuel that doesn't exist. *Bloomburg.* https://www.bloomberg.com/news/features/2021-08-11/the-fictitious-world-of-carbon-neutral-fossil-fuel

Strupp-Levitsky, M., Noorbaloochi, S., Shipley, A., & Jost, J. T. (2020). Moral 'foundations' as the product of motivated social cognition: Empathy and other psychological underpinnings of ideological divergence in 'individualizing' and 'binding' concerns. *PLoS ONE, 15*(11), e0241144. https://doi.org/10.1371/journal.pone.0241144

The Economist. (2016, September 3). Dawn of a new epoch? *The Economist*. http://www.economist.com/news/science-and-technology/21706227-people-may-have-propelled-earth-novel-episode-geological-time-dawn

The Guardian. (2019, September 23). Greta Thunberg to world leaders: 'How dare you – you have stolen my dreams and my childhood'—Video. *The Guardian*. https://www.theguardian.com/environment/video/2019/sep/23/greta-thunberg-to-world-leaders-how-dare-you-you-have-stolen-my-dreams-and-my-childhood-video

Timperley (2021, November 8). The world's fight for 'climate justice'. BBC. https://www.bbc.com/future/article/20211103-the-countries-calling-for-climate-justice

Tyson, A., Funk, C., & Kennedy, B. (2023, April 18). *What the data say about Americans' views of climate change*. Pew Research Center. https://www.pewresearch.org/fact-tank/2022/04/22/for-earth-day-key-facts-about-americans-views-of-climate-change-and-renewable-energy/

Vaidyanathan, G. (2021, December 1). What humanity should eat to stay healthy and save the planet. *Nature, 600*, 22–25. https://doi.org/10.1038/d41586-021-03565-5

van der Linden, S. (2022, December 15). Quiz: What's the best way to shrink your carbon footprint? *The New York Times*. https://www.nytimes.com/interactive/2022/12/15/opinion/how-reduce-carbon-footprint-climate-change.html

Voosen, P. (2022, May 5). Bogs, lakebeds, and sea floors compete to become Anthropocene's 'golden spike'. *Science*. https://www.science.org/content/article/bogs-lakebeds-and-sea-floors-compete-become-anthropocene-s-golden-spike

Voosen, P. (2024). The Anthropocene is dead. Long live the Anthropocene Panel rejects a proposed geologic time division reflecting human influence, but the concept is here to stay. *Science*. https://doi.org/10.1126/science.z3wcw7b

Waters, C. N., Zalasiewicz, J., Williams, M., Ellis, M. A., & Snelling, A. (Eds.). (2014). *A stratigraphical basis for the Anthropocene* (Vol. 395, 321 p.). The Geological Society Special Publication.

Williams, J. (2022, January 26). Why climate change is inherently racist. BBC. https://www.bbc.com/future/article/20220125-why-climate-change-is-inherently-racist

Witze, A. (2023). This quiet lake could mark the start of a new Anthropocene epoch. *Nature, 619*, 441–442. https://doi.org/10.1038/d41586-023-02234-z

Wolsko, C., Ariceaga, H., & Seiden, J. (2016). Red, white, and blue enough to be green: Effects of moral framing on climate change attitudes and conservation behaviors. *Journal of Experimental Social Psychology, 65*, 7–19. https://doi.org/10.1016/j.jesp.2016.02.005

Wood, A. (2002). Chapter 4: Relativism. In *Unsettling obligations: Essays on reason, reality, and the ethics of belief* (pp. 131–157). CSLI Publications.

Yang, W. (2017, February 14). Is the 'Anthropocene' epoch a condemnation of human interference— Or a call for more? *The New York Times*. https://www.nytimes.com/2017/02/14/magazine/is-the-anthropocene-era-a-condemnation-of-human-interference-or-a-call-for-more.html

Zalasiewicz, J., Waters, C. N., Williams, M., Barnosky, A. D., Cearreta, A., Crutzen, P., Ellis, E., Ellis, M. A., Fairchild, I. J., Grinevald, J., Haff, P. K., Hajdas, I., Leinfelder, R., McNeill, J., Odada, E. O., Poirier, C., Richter, D., Steffen, W., Summerhayes, C., … Oreskes, N. (2015). When did the Anthropocene begin? A mid-twentieth century boundary level is stratigraphically optimal. *Quaternary International, 385*(5), 196–203. https://doi.org/10.1016/j.quaint.2014.11.045

Zalasiewicz, J., Waters, C. N., Williams, M. M., & Summerayes, C. P. (Eds.). (2019). *The Anthropocene as a geological time unit*. Cambridge University Press. https://doi.org/10.1017/9781108621359

Zalasiewicz, J., Williams, M., Haywood, A., & Ellis, M. (2011). The Anthropocene: A new epoch of geologic time? *Philosophical Transactions of the Royal Society of London, 369*, 835–841. https://doi.org/10.1098/rsta.2010.0339

Zalasiewicz, J., Williams, M., Smith, A., Barry, T. L., Coe, A. L., Bown, P. R., Brenchley, P., Cantrill, D., Gale, A., Gibbard, P., Gregory, F. J., Hounslow, M. W., Kerr, A. C., Pearson, P., Knox, R., Powell, J., Waters, C., Marshall, J., Oates, M., … Stone, P. (2008). Are we now living in the Anthropocene? *GSA Today, 18*, 4–8. https://doi.org/10.1130/GSAT01802A.1

Greg Hitzhusen growing up in Ohio during the heyday of acid rain, young Greg suddenly realized that flicking on his light switch was causing sulfuric acid to fall from the skies in the Adirondacks. And that just seemed *wrong*. He studied ecology at Cornell to learn about how to help solve problems like acid rain, and realized it was the *ethical* dimensions of environmental problems (the part about something being "wrong," and why) that really drew his interest. He worked for the National Wildlife Federation's Outdoor Ethics division before pursuing joint theology-environment graduate studies at Yale, and then returned to Cornell for PhD research about environmental ethics in North American faith communities. After serving as Land Stewardship Specialist for the National Council of Churches Eco-Justice Programs in Washington, DC, and becoming the founding executive director of Ohio Interfaith Power and Light, he joined the faculty in the School of Environment and Natural Resources at The Ohio State University, where he teaches about religion and environmental values, environmental writing, and the philosophies of environmental and natural resource sciences. He serves on the steering team of OSUs Center for Ethics and Human Values, and coordinates a sustainability capstone program where student research teams collaborate with community partners to solve local and regional sustainability challenges; his students have now published over 130 reports and 30 book chapters pushing the boundaries of sustainability science and values. Greg is the primary author of Chap. 12.

Jill Schneiderman concerned about the welfare of humans and other living beings on Earth, Jill S. Schneiderman applies her geological knowledge in order to raise awareness about systemic injustices in the arena of the natural and built environment. Although she has conducted research on topics such as the formation of mountain belts, heavy minerals in deltas, and terrestrial microplastics, she considers herself primarily to be a teacher of the next generation who will need deep knowledge of the Earth System in order to secure a livable planet. She teaches courses such as The Solid Earth, Environmental Justice in the Anthropocene, and Feminist Approaches to Science and Technology. She earned her S.B at Yale University and her Ph.D. at Harvard University and is Professor of Earth Science in the Department of Earth Science and Geography at Vassar College in Poughkeepsie, New York. Jill is the author of Box 1 in Chap. 12.

Open Access This chapter is licensed under the terms of the Creative Commons Attribution 4.0 International License (http://creativecommons.org/licenses/by/4.0/), which permits use, sharing, adaptation, distribution and reproduction in any medium or format, as long as you give appropriate credit to the original author(s) and the source, provide a link to the Creative Commons license and indicate if changes were made.

The images or other third party material in this chapter are included in the chapter's Creative Commons license, unless indicated otherwise in a credit line to the material. If material is not included in the chapter's Creative Commons license and your intended use is not permitted by statutory regulation or exceeds the permitted use, you will need to obtain permission directly from the copyright holder.

MIX
Papier aus verantwortungsvollen Quellen
Paper from responsible sources
FSC® C105338

If you have any concerns about our products,
you can contact us on
ProductSafety@springernature.com

In case Publisher is established outside the EU,
the EU authorized representative is:
**Springer Nature Customer Service Center GmbH
Europaplatz 3, 69115 Heidelberg, Germany**

Printed by Libri Plureos GmbH
in Hamburg, Germany